Sensor Systems Simulations

Willem Dirk van Driel • Oliver Pyper
Cornelia Schumann
Editors

Sensor Systems Simulations

From Concept to Solution

Springer

Editors
Willem Dirk van Driel
Delft University of Technology
Delft, The Netherlands

Cornelia Schumann
Infineon Technologies Dresden GmbH & Co. KG
Dresden, Germany

Oliver Pyper
Infineon Technologies Dresden GmbH & Co. KG
Dresden, Germany

ISBN 978-3-030-16576-5 ISBN 978-3-030-16577-2 (eBook)
https://doi.org/10.1007/978-3-030-16577-2

This Springer imprint is published by the registered company Springer Nature Switzerland AG.
The registered company address is: Gewerbestrasse 11, 6330 Cham, Switzerland

Preface

In this book, we present the results of the European Union project IoSense (see www.iosense.eu). In this EU project, adequate and verified simulation environments are used to support the predevelopment, development and production ramp processes for heterogeneous sensor systems. Furthermore, it aims to verify the developed simulation strategies and conduct statistical sensitivity studies using the manufacturing-oriented simulations in order to support principles of design for manufacturability and design for testability (DfM and DfT). Simulations are means to improve the functional design and/or the processes creating them. The developed simulation environments will be used to provide application-oriented, multidomain, functional simulations in the area of:

- Device physics and related manufacturability
- Multi-level electrical functionality (top level, gate level, module level, device level, etc.)
- Thermal simulations
- Energy and power consumption aspects
- Other physical domains (pressure, stress, flow, sound, optical/light)
- Chemical domains (gas composition, liquid consistency)
- Runtime adaptivity and reconfiguration: design, representation and algorithm
- Design specifications for self-adaptivity and healing algorithms
- Implementation-oriented device simulations (mainly sensors)
- Interface between sensor model and device simulation
- Definition, modelling and evaluation of the interoperability security concept of the contactless secure coil-on-chip sensor configuration solution based on NFC (NFC-DIP) according to the defined interfaces of the system specification

This book provides the results of these simulation-based sensor system developments and may be used as a guideline for future sensor integrations concepts.

Delft, The Netherlands — Willem Dirk van Driel
Dresden, Germany — Oliver Pyper
Dresden, Germany — Cornelia Schumann

Acknowledgements

This work was supported by the European Union project "IoSense: Flexible FE/BE Sensor Pilot Line for the Internet of Everything". This project has received funding from the Electronic Component Systems for European Leadership Joint Undertaking under grant agreement No 692480. This Joint Undertaking receives support from the European Unions' Horizon 2020 research and innovation programme in Germany, the Netherlands, Spain, Austria, Belgium and Slovakia.

ECSEL Joint Undertaking

Personal Acknowledgements

Willem van Driel is grateful to his wife, Ruth Doomernik; their two sons, Juul and Mats; and their daughter, Lize, for their support on writing and editing this book. The coeditors, Oliver Pyper and Cornelia Schumann, would like to thank Willem van Driel for his outstanding commitment and energy in compiling this book. Furthermore, our thanks go to all the partners in the project for the great cooperation and, particularly, to the authors of the chapters, who present the achievements in the field of sensor systems simulations.

March 2019

W. D. van Driel
C. Schumann
O. Pyper

Contents

Chapter 1
From Si Towards SiC Technology for Harsh Environment Sensing

L. M. Middelburg, W. D. van Driel, and G. Q. Zhang

1.1 Introduction

Since it is obvious that Moore's Law in its classical way of scaling, which proved to be powerful over the last decades, is coming to an end, alternative routes towards technological progress are investigated [1]. One of the main fundamental reasons for this is that the smallest features size in newest technology nodes is approaching the level of only a few atom layers. As a result, the development and implementation of technology nodes based on a scaled-down version of the previous one, gets increasingly more expensive. An alternative approach to ensure technological progress of the microelectronics world and the semiconductor industry is described by a trend called "More than Moore" (MtM) [2], based on diversification and integration. In terms of diversification, materials beyond silicon can be considered for the development of sensors and electronics, while the integration aspects come to expression by combining different parts of a system in a smart and optimal way. Wide bandgap (WBG) materials, such as gallium nitride (GaN) or silicon carbide (SiC) are mature for power applications, but for other applications such as low-voltage (Bi)CMOS and/or VLSI they are still in the research phase. By integrating electronics monolithically on a sensor chip, improved system performance can be obtained by having signal amplifications close to the physical transducer. The integration aspects are strongly related to the packaging of microelectronic and

L. M. Middelburg · G. Q. Zhang
Delft University of Technology, EEMCS Faculty, Delft, The Netherlands
e-mail: G.Q.Zhang@tudelft.nl

W. D. van Driel (✉)
Delft University of Technology, EEMCS Faculty, Delft, The Netherlands

Signify, HTC48, Eindhoven, The Netherlands
e-mail: willem.van.driel@signify.com; willem.van.driel@philips.com

W. D. van Driel et al. (eds.), *Sensor Systems Simulations*,
https://doi.org/10.1007/978-3-030-16577-2_1

microfabricated devices, for example, when multi-physical sensors are considered. A recent example is the through polymer via enabling an optical channel through a package [3].

By investigating nonlinear effects in MEMS structures, the mechanical sensitivity can be boosted in a bulk micromachined and thus space limited chip. Furthermore, by the application of new materials such as SiC sensors and (low voltage) electronics can be yielded compatible with harsh environments and temperatures up to 500 °C. By investigating the monolithic integration of analog electronics with mature sensor technologies, the strengths of system integration can be implemented and exploited, resulting in more value from existing technologies. Especially, the latter two topics are examples of the "More than Moore" concept but still numerous challenges exist.

The development of SiC-based electronics to build up technologies for low-voltage CMOS, BJTs, or BiCMOS for both analog and digital circuits is still pre-mature. That this development is still in the research phase is illustrated by the given that numerous research works focus on device simulation and model extraction for SiC CMOS. Furthermore, when looking over the literature that is available on SiC technology, one could notice a trend in shifting interest from 6H-SiC to 4H-SiC [4–7].

Also, from junction formation by mesa etching on epitaxially grown layers to ion implantation techniques, which have been evolved during the last 10 years. Other challenges on the physical level are the chemical/physical effects in ohmic contacts and the long-term reliability and stability of metallization schemes [8]. So, even on the very basic physical level significant changes have taken place, illustrating that the SiC electronics development is in its pre-mature phase.

In addition to the development of SiC electronics, the compatibility of the fabrication processes is of utmost importance, when all SiC ASIC + MEMS monolithic system integration are considered. A cleanroom flowchart for the processing of MEMS can be very different from one for the processing of electronics in terms of thermal budget, contamination and topography. Silicon carbide being a harsh environment compatible material, and thus an inert material, involve more fabrication steps such as high temperature processing compared to standard silicon technology.

1.2 Silicon Technology and Its Limitations

Silicon technology has been mature for decades for the fabrication of a broad range of electronics, ranging from BJTs, analog CMOS, BiCMOS to digital integrated circuits such as VLSI. Based on this silicon technology, sensor technology is implemented. By reusing existing technologies and process steps such as oxidation, patterning, wet- and dry etching, and dopant formation, big steps are made and a powerful palette of fabrication methods are available for decades. The term "CMOS compatible" is very common in the field of sensors, which practically denotes that the sensors considered can be processed within a CMOS process flow, reducing costs dramatically. It should however be realized that this implicit choice for silicon and CMOS compatible technology can have some dramatic drawbacks in the sensor design for certain applications. For numerous applications, silicon is not the first-choice material, but is still chosen for its earlier mentioned widespread availability. Furthermore, a CMOS compatible process flow might put restrictions to the sensor design, negatively influencing overall design freedom and compromising sensor performance. Surface and bulk micromachining techniques, such as Deep Reactive Ion Etching (DRIE), have been developed based on etching technologies from silicon CMOS processing, for example, by extending etch times or increased etch power in case of plasma etching. This development has enabled the design and realization of Micro-Electro-Mechanical-Systems (MEMS) in silicon. Currently, the MEMS market covers application field such as Radar, Ultra Sonic, LiDAR, Chemical-, magnetic-, imaging-, and pressure sensors and has a value of around ten billion dollars [9].

A major field of application where silicon technology is not enough is the field of harsh environments. Harsh environments are considered environments with high temperatures, corrosive species, vibrations, or radiation. Silicon is in general not capable of being compatible with harsh environments. Firstly, because the electronic behavior is compromised at temperatures of 200 °C and beyond. This effect is caused by the intrinsic carrier density of silicon being several orders of magnitude larger than WBG counterpart materials, such as GaN or SiC. The intrinsic carrier concentration is increasing with increasing temperature and as soon as the intrinsic carrier density is exceeding the dopant concentration of the junctions defined by implantation, the electronic device fails in operation. It can thus be concluded that the electronic properties of silicon do not obey the high temperature requirements for harsh environments. The intrinsic carrier concentrations as function of temperature for silicon and the most common polytypes of silicon carbide are included in Fig. 1.1. It can be seen from this figure that the intrinsic carrier density for silicon is approaching typical dopant levels (1×10^{14} #/cm^3) for lowly doped regions, while the carrier density for 4H-SiC is around 13 orders of magnitude lower at the same temperature.

From a mechanical point of view, silicon is known to show plastic deformations under small loads from 500 °C and beyond, which limits the feasibility for the development of harsh environment microsystems.

To make silicon-based sensors as compatible as possible with harsh environments, Silicon On Insulator wafers are used and extensive packaging is typically required, resulting in higher cost and complexity. Interfacing a harsh environment physical transducer at high temperatures, typically involves fiber optics, because analog front-end electronics do not work at elevated temperatures for reasons described above.

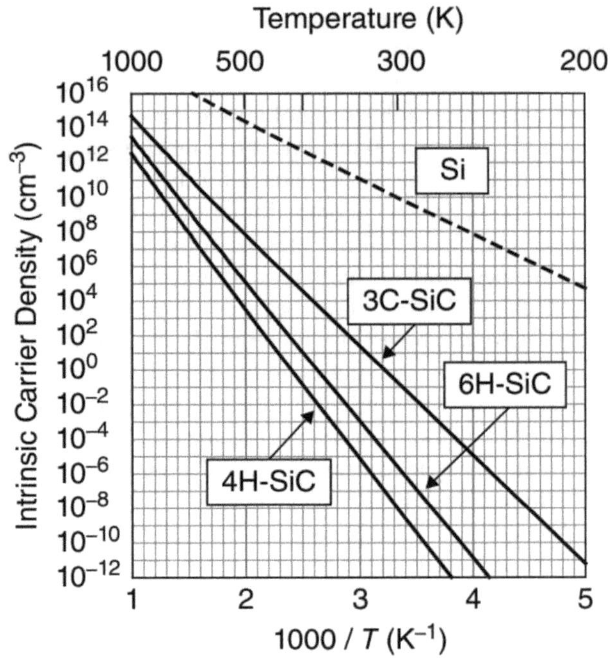

Fig. 1.1 An overview of the intrinsic carrier concentration of silicon, compared with the most common polytypes of SiC ([8], with permission)

1.3 Wide Bandgap Semiconductors

Apart from the well-known silicon, many more semiconductors exist. Wide bandgap (WBG) semiconductors (WBS) are semiconductors with an increased bandgap compared to silicon. Well-known examples are silicon carbide, gallium nitride (GaN), or gallium arsenide (GaAs). Both are the so-called compound semiconductors, SiC being a group IV–IV semiconductor, while GaN and GaAs is a group III–V semiconductor. These group number denotes the column in the periodic table. The wider bandgap manifests itself in the physical properties of the material. Here, the focus will be on silicon carbide. Thanks to the high critical electric field, the high thermal conductivity SiC technology is mature in the field of power electronics already. As a result of these two physical properties, higher switching speeds are feasible, resulting in lower losses and more compact form-factors of power electronics in general. Devices such as IGBTs and Power MOSFETs exploit largely from the high thermal conductivity and critical electric field. Larger voltages can be used, and less stringent cooling requirements simplifies the packaging of these devices.

1.3.1 Polytypes

One can distinguish the following different types of silicon carbide in terms of polytypes, indicated in Fig. 1.2. There are two main categories of SiC: α-silicon carbide and β-silicon carbide. When considering a mono-crystalline material, the structure is build up in unit cells, the smallest cell is called a primitive cell. In the case of α-silicon carbide, this privative cell is hexagonal, typically denoted by an H. The two main polytypes which are commercially available in wafer form are the 4H-SiC and 6H-SiC. In these polytypes, the stacking sequence of the hexagonal bi-layer consisting of silicon and carbon atom repeats itself after 4 and 6 layers, respectively [11]. Apart from α-SiC here is β-SiC. The main polytype in this category is 3C-SiC, showing a cubic primitive cell, similar to silicon. In addition to these mono-crystalline occurrences of silicon carbide, poly-crystalline and nano-crystalline silicon carbide exists, from which poly-crystalline 3C-SiC is most common, because it can be grown by CVD techniques [12]. Generally, it can be said that in existing silicon carbide technology, 4H-SiC and 6H-SiC are mainly occurring in wafer form and exploited for their excellent electronic properties, whereas poly-3C-SiC and amorphous SiC are mainly occurring in thin film deposition and strongly related to MEMS processing, thanks to the lower cost and larger flexibility.

1.3.2 Physical Properties

The unique physical properties are the result of the larger bandgap and the strong covalence bond between the silicon and carbon atom. A comparison is made between silicon, silicon carbide, and gallium nitride, while being the latter two common WBG semiconductors. The most important physical parameters are summarized in Table 1.1. It can be noted from this table that the bandgap of

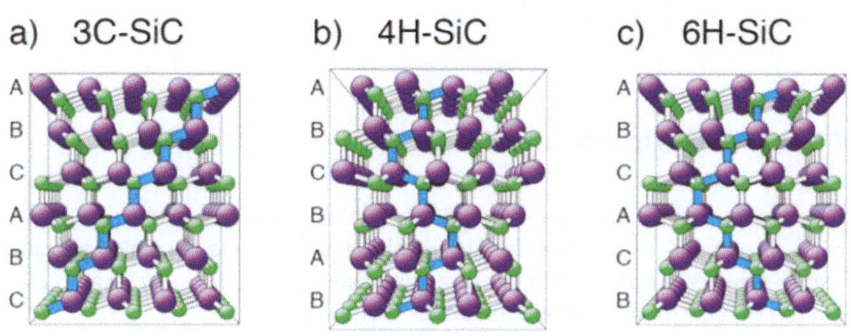

Fig. 1.2 An overview of the three common polytypes [10], with (**a**) 3C-SiC; (**b**) 4H-SiC and (**c**) 6H-SiC

Table 1.1 Comparison of physical properties of Si, SiC, and GaN

Figure	Unit	Si	SiC	GaN
Bandgap *Eg*	eV	1.12	3.0–3.6[a]	3.45
Maximum electric field	Vcm^{-1}	3×10^6	2.5×10^6	5×10^6
Relative dielectric constant ε_r	F/m	11.8	9.8	8.9
Thermal conductivity	$Wcm^{-1}\ K^{-1}$	1.5	4.9	1.5
Carrier mobility	$cm^2\ V^{-1}\ s^{-1}$	1350 and 480	980 and 200	1000 and 30
Density	kgm^{-3}	2330	3210	6150
Young's modulus	GPa	62–202	410–700	398
Acoustic velocity	m/s	8415	13,100	8044

[a]Depends on polytype

silicon carbide is with 2.3 eV for 3C-SiC, 3.26 eV for 4H-SiC, and 3.03 eV for 6H-SiC significantly larger than the 1.12 eV of silicon. The larger bandgap dramatically reduces the electron-hole pairs caused by external thermal energy, which reduces leakage currents at elevated temperatures which is one of the major issues with silicon technology in the application of electronics. The wider bandgap also influences the spectral band where the semiconductor can be used to detect the light, which will be discussed later onwards. As mentioned above in the section introduction, silicon carbide shows a high electric breakdown field, 1.8–3.8 MV/cm (depending on the polytype) compared to 0.3 MV/cm for silicon [13], which is in combination with its high thermal conductivity advantageous for power applications. Further, silicon carbide has a high chemical inertness making it an excellent material for transducers in corrosive environments, while it relaxes stringent packaging requirements. The high Young's modulus opens new design opportunities for the development of silicon carbide MEMS, looking into miniaturization of structures that have mechanically not been feasible in silicon MEMS. This aspect is especially exploitable since it is known that silicon carbide exhibits a piezo-resistive effect [14].

Silicon carbide is moreover known to have a high radiation tolerance, ideal for space applications [11] and a high acoustic velocity: $11.9 \times 10^3\ ms^{-1}$ for SiC vs. $9.1 \times 10^3\ ms^{-1}$ for Si [15].

1.4 Harsh Environment Sensing

As sensors fabricated in silicon carbide technology have advantages in terms of harsh environment compatibility, in this section the fields and possible applications for such sensors are discussed. Before discussing specific sensor examples, the definition of a harsh environment is further specified.

1.4.1 Harsh Environments

Because the term harsh environments can be interpreted in different ways, it will be quantified in this section. Harsh environments are seen by this work as environments with elevated temperatures. A temperature of 200 °C and beyond can already be seen as harsh, silicon-based electronics start to fail namely, but the temperature range can even go up to 800 °C. Other aspects which make environments harsh in this context are the presence of corrosive species (gasses, liquids), vibrations, radiation, and/or a high pressure. Examples of harsh environments include the geographical poles, very arid deserts, volcanoes, deep ocean trenches, upper atmosphere, Mt. Everest, outer space, and the environments of every planet in the Solar System except the Earth. In applications, examples are boreholes, automobiles under the hood (motor area), and/or power applications like seen in energy grids.

1.4.2 Overview of Applications

From an application perspective, there are many challenges in our technological world, for example, food security, pollution, global warming, growing demand for energy, health, and well-being. This results in applications such as environmental sensing, air quality monitoring, gas sensors for cleaner combustion, sensors for the Smart Grid and Structural and Mechanical Health Monitoring. A more structured overview of fields, specific applications, and corresponding sensors is given in Table 1.2.

1.5 Harsh Environment Pressure Sensing

1.5.1 Applications

Numerous applications can be thought of like measurements of absolute pressures and pressure changes in combustion engines, gas turbines, and jet- and rocket engines. Furthermore, reaction containers and vessels in the industry can be applications where pressure sensors can have added value when they are harsh environment compatible. The harsh environments for pressure sensors come most to expression in applications where both high temperature and corrosive environments are included. This is the case in applications where combustion is involved such as aviation jet engines or space applications.

Most pressure sensors consist of a membrane which is basically a transducer of a difference in air pressure to a stress on the membrane. The stress is transferred to a strain by the Young's modulus and needs to be read out. From this reasoning it could be stated that pressure sensors and strain/strain sensors are closely related to each

Table 1.2 An overview of field, applications, and sensor classifications

Field	Application	Sensor classification
Automotive	Exhaust gas monitoring Engine instrumentation Combustion monitoring Electrical drive monitoring Particulate filter monitoring	Temperature Humidity Gas species Vibrations, resonance Particulates
Aviation	Jet engine monitoring Structural health monitoring Mechanical health monitoring	Temperature, pressure, flow Stress, strain, pressure Vibrations, flow
Space	Navigation/position sensing	UV blind photo detectors
Earth monitoring	Gravimetry Ocean behavior Oil/gas well monitoring	Deviations in gravity Flow, pressure, temperature Sensing systems
Environment sensing	IoT distributed monitoring System	Particulates 10 nm–10^1 m VOCs Humidity CO, NOx, SOx UV index Occupation/traffic density Radiation level, etc.
Health/sport/well-being	Monitoring in vivo Excitation	Blood pressure, heartbeat pH, salinity Pacemaker, neurostimulation
Food security	Measuring toxic substances	Heavy metals Pesticide residues
Industry	Process monitoring	Temperature, gas species pH value, ion concentration Pressure Flow, radiation
Energy	Power grid sensors Offshore wind parks Nuclear energy monitoring	Current, hall sensor EM fields Temperature Stress Strain
Communication	High frequency/RF compatible Sensors Antenna/base station Monitoring	Latency Response time
Consumer electronics	RF cooking	Temperature

other. Health monitoring of electrical and electronic devices such as power devices, smart grid components, or electrical drives are possible application areas. But health monitoring of mechanical systems like bearings could be a possible application for harsh environment pressure sensors.

Regarding the multi-sensor platform, the monolithic integration of a temperature sensor and/or a photo sensor can be especially relevant in processes like combustion

Table 1.3 Specifications based on applications

Application	Pressure	Temperature
Medical	69 mbar [18]	50 °C
Oil wells	344 bar [18]	
Combustion engine	0–100 bar [19]	574 °C [20]
Geothermal wells	14 bar [16]	350 °C [21]
Oil and gas exploration		275 °C
Aircraft/turbine engines	1–50 bar [22]	−50 up to 650 °C [18]
Industrial gas turbines	345 bar [18]	450–600 °C

monitoring. The integration of multiple sensors on wafer scale, and thus all-SiC has the huge advantage of a system which is harsh environment compatible. Regarding the integration of electronics, this is especially powerful in the application of harsh environments, since on-chip electronics can modulate and amplify the measured signals and can simplify read out of the sensor, which is currently commonly done with optical fibers. Examples are:

- Combustion monitoring for automotive
- Jet engines for aviation
- Health monitoring by stress measurement in SiC electronic components, for example, in electric power domain
- Smart grid health monitoring, for example, transformer oil pressure
- Pressure measurement on drill-heads for the oil/gas industry
- Geothermal wells [16]
- Improve jet engine testing (NASA)
- Space applications, such as the VENUS project KTH [17]

Specification based on applications are listed in Table 1.3 and an overview is depicted in Fig. 1.3.

1.5.2 State-of-the-Art Harsh Environment Pressure Sensors

For overview and clarity reasons, the found literature is tabulated in Table 1.4. To identify each pressure sensor, the substrate material was listed, along with the main membrane dimensions and the sensor performance and the transduction type.

Beker et al. [16] write on a surface micromachined circular concentrically matched capacitive pressure intended for measurements on geothermal wells. The substrate material used was silicon, but the structural layer is poly-crystalline SiC. Chen et al. [19] uses a poly-SiC substrate enabling an all-SiC device. Hung et al. [24] did a comparative study on both mono-crystalline and poly-crystalline SiC and concluded that the gauge factor in the case of a piezo-resistive pressure sensor

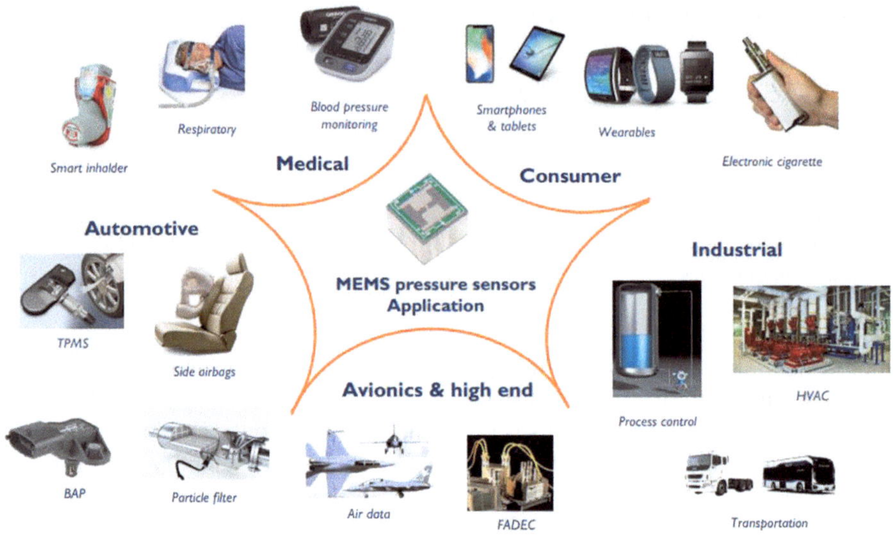

Fig. 1.3 An overview of pressure sensors by Yole Development ([23], with permission)

implementation is significantly larger for 3C-SiC. Eickhoff et al. [25] used an SOI substrates to isolate the different poly-SiC piezo-resistive elements on the (silicon) membrane. Fonseca et al. [26] uses a rather classical approach to make their devices compatible with elevated temperatures, in this case 400 °C, namely the usage of ceramic materials. Fricke et al. [27] claim pressure sensor operation at 800 °C by making use of platinum strain gauges in a sapphire substrate. Jin et al. [18] write on the capacitive surface micromachined capacitive structures on a standard silicon substrate, resulting in a pressure sensor for in-cylinder pressure measurements on 574 °C. Lalinksy et al. [28] show a pressure sensor based on an active device, a high-electron mobility transistor in AlGaN/GaN technology on a silicon substrate. Okojie et al. [5, 7] showed a fully mono-crystalline SiC implementation based on the piezo-resistive effect. Thanks to full exploitation of the electronic and mechanical properties, the devices in these two works show operation up to 600 °C, resp. 800 °C. Jiang et al. [34] described an optical pressure sensor based on an Fabry-Perot cavity. The readout is done with an optical fiber to interface the harsh environment.

When analyzing all the literature found, it can safely be said that the majority of the harsh environment pressure sensors work with non-SiC substrate materials such as Si, SOI, sapphire, or another ceramic. The high temperature compatibility of the silicon substrate-based implementation is compromised in general, except for the work of Jin et al., which uses a capacitive surface micromachined implementation and claims operation up to 574 °C. The three implementations based on 4H-SiC

Table 1.4 An overview of harsh environment pressure sensors in literature

Substrate	Transduction	Membrane thickness	Sensitivity	Pressure range	Temp.	Ref
Si	Capacitive SiC-poly	2 μm	Circular R = 120 μm	1.03 fF/kPa:0–1.4 MPa	180° C	[16]
Poly-SiC	Capacitive SiC-poly	2.8 μm	272 μV/psi	5 MPa	574° C	[19]
Si	Piezo-resistive poly-SiC	15 μm	177 mV/V.psi	0.5 MPa	25–450 °C	[24]
SOI	Piezo-resistive 3C SiC (poly)	100 μm	3.5 mV/V.bar	0.35 MPa	200 °C	[25]
Ceramic	LC tank w. Antenna	100 μm	−141 kHz/bar	10 MPa	400 °C	[26]
Sapphire	Pt strain gauges	200 μm	10 μV/V.bar	3 MPa	800 °C	[27]
Si	Poly-SiC capacitive	2.7 μm	7.2 fF/psi		574 °C	[18]
Si	AlGaN/GaN HEMT	1.9 μm		1 MPa		[28]
6H-SiC	Piezo resistive	50 μm, circular $R = 600$ μm	32.5 μV/V.Psi	1.4 MPa	600 °C	[5]
4H-SiC	Piezo resistive	50 μm		1.38 MPa	800 °C	[7]
6H-SiC	Piezo resistive		200 μV/bar	12 MPa	400 °C	[29]
Si	Capacitive, mono-3CSiC	0.5 μm, circular $R = 400$ μm	7.7 fF/torr	146 kPa–235 kPa	400 °C	[30]
SiC wafer	Capacitive cavity	18 μm	10.6 kHz/kPa	200 kPa	600 °C	[31]
GaAs	Res.Tunnel diode	1 μm	6 kHz/kPa	1–50 kPa	Around RT	[32]
Si	Piezo-resistive 3C-SiC poly	30 μm	3.9 mV/psi	83 kPa	Around RT	[33]
SiC	SiC Fabry-Perot cavity	50 μm circular $R = 1.5$ mm		0.1–0.9 MPa	RT	[34]

and 6H-SiC, respectively, do not describe the etching process or the way the membrane was fabricated or formed. It is known from literature that etching of mono-crystalline SiC, and silicon carbide in general, is very challenging. Dry etching methods require in general metal hard-masks, which can in turn result in micromasking issues. Furthermore, the etch rate relatively low, in conventional ICP etchers only up to 500 nm/min [35].

1.6 SiC System Integration: Advantages and Challenges

To fully exploit the advantages of WBG semiconductors in harsh environments, the monolithic integration of readout and communication electronics in SiC is a major advantage. In this way, both the sensor itself and electronics for readout and the communication are harsh environment compatible. In order to make the entire sensor system harsh environment compatible, an optical readout can be used to interface the transducer in its hostile environment [34]. In such a case, conventional silicon-based electronics for amplification, processing and further communication are then placed in less hostile environments.

When electronics can be integrated with the physical transducer, being the sensor, signal amplification can be done directly in the physical location of the transducer by analog front-end electronics, thereby boosting signal power. These electronics do not necessarily have to be complex circuitry, already an output-buffer or relatively simple differential amplifier can be of great value in terms of increasing signal power. In this way, noise contributions caused by interference on the interconnect to the sensor is compromising the analog signal to a smaller extend. This would result in a significantly increased Signal-to-Noise Ratio (SNR).

When more extensive and complex SiC circuitry is considered, also circuits like data converters can be considered and an even larger part of the sensor system might be integrated in a single chip, including both analog and digital signal processing as well as communication.

The advantages of monolithic integration lie in the nature of dealing with "one-piece-of-substrate." Integration on package level typically requires the combination of multiple dies, yielding the so-called System-in-Package (SiP) solution. Such an approach requires interconnects between different dies, by for example 3D packaging or wire-bonding techniques. Such solutions are undesirable from a reliability perspective in case of the applications in harsh environments. Different CTEs of the used materials in such a SiP in combination with extremely large temperature variations and vibrations will influence the durability and reliability of such a solution dramatically. When monolithic integration of the ASIC part with the sensor, i.e., MEMS, part is considered, one dies has to be packaged.

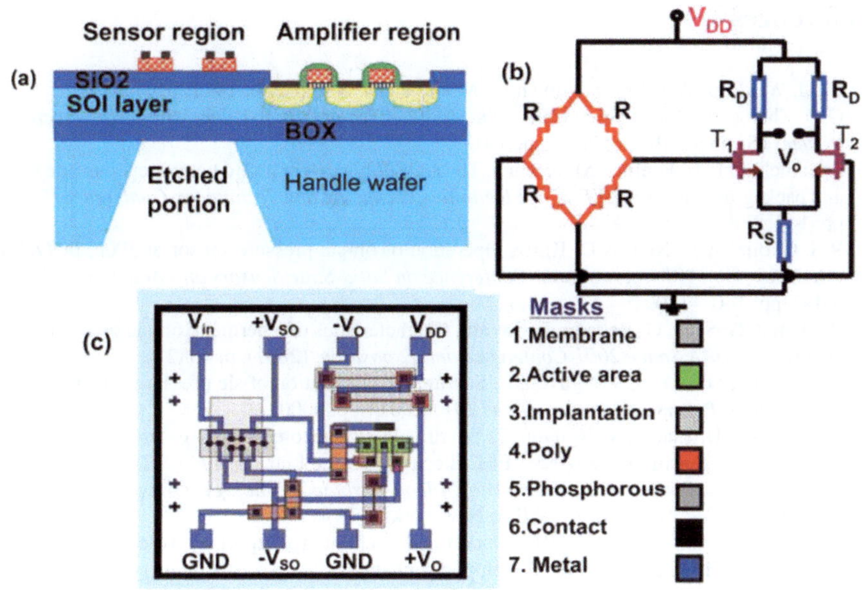

Fig. 1.4 An example from literature for the concept of monolithic integration of a pressure transducer and readout electronics in silicon [37], with (**a**) cross-section view, (**b**) schematic and (**c**) top view on the IC

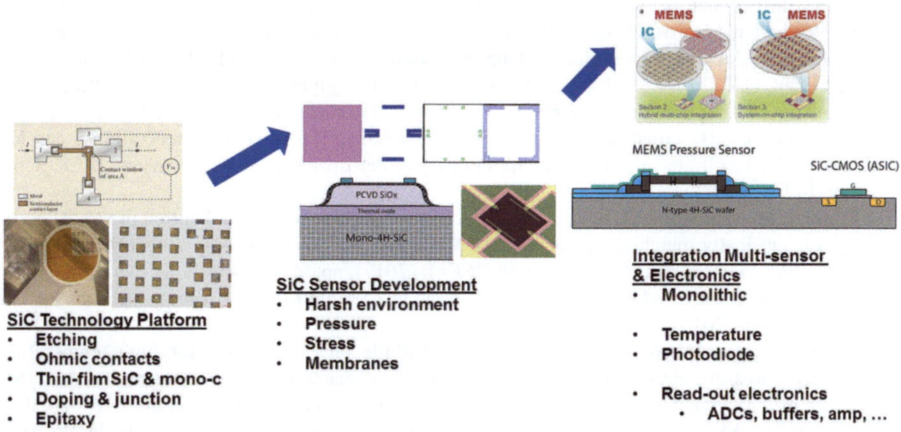

Fig. 1.5 The technology roadmap for the SiC pressure sensor platform

Some research is conducted on a CMOS SiC process [36] and the monolithic integration of front-end electronics with a pressure sensor for silicon technology is demonstrated in literature, see Fig. 1.4.

The technology roadmap for the SiC pressure sensor platform is indicated in Fig. 1.5.

References

1. M.M. Waldrop, The chips are down for Moore's law. Nature **530**, 144 (2016)
2. G.Q. Zhang, A. Roosmalen, *More than Moore: Creating High Value Micro/Nanoelectronics Systems* (Springer, Berlin, 2009), p. 330
3. J. Hamelink, R.H. Poelma, M. Kengen, Through-polymer-via for 3d heterogeneous integration and packaging, in *2015 IEEE 17th Electronics Packaging and Technology Conference (EPTC)*, pp. 1–7
4. R.S. Okojie, A.A. Ned, A.D. Kurtz, Operation of 6h-sic pressure sensor at 500c, in *TRANSDUCERS '97. 1997 International Conference on Solid-State Sensors and Actuators, Chicago*, vol. 2, pp. 1407–1409
5. R. Okojie, G. Saad, G. Beheim, E. Savrun, Characteristics of a hermetic 6h-sic pressure sensor at 600 c, in *AIAA Space 2001 Conference and Exposition*, (2001), p. 4652
6. R.S. Okojie, D. Lukco, V. Nguyen, E. Savrun, Demonstration of sic pressure sensors at 750c, in *Additional Papers and Presentations 2014*, (2014), p. 000028
7. R.S. Okojie, D. Lukco, V. Nguyen, E. Savrun, 4h-sic piezoresistive pressure sensors at 800c with observed sensitivity recovery. IEEE Electron. Device Lett. **36**, 174 (2015)
8. T. Kimoto, J.A. Cooper, *Fundamentals of Silicon Carbide Technology: Growth, Characterization, Devices and Applications* (Wiley, New York, 2014)
9. Status of the Mems Industry 2018 Market and Technology Report by Yole Développement. https://www.slideshare.net/Yole_Developpement/status-of-the-mems-industry-2018-market-and-technology-report-by-yole-dveloppement
10. M. Wijesundara, R. Azevedo, *Silicon Carbide Microsystems for Harsh Environments*, vol 22 (Springer Science & BusinessMedia, Berlin, 2011)
11. N.G. Wright, A.B. Horsfall, Sic sensors: A review. J. Phys. D. Appl. Phys. **40**, 6345 (2007)
12. M. Eickhoff, M. Möller, G. Kroetz, M. Stutzmann, Piezoresistive properties of single crystalline, polycrystalline, and nanocrystalline n-type 3 c-sic. J. Appl. Phys. **96**, 2872 (2004)
13. M. Willander, M. Friesel, Q.-u. Wahab, B. Straumal, Silicon carbide and diamond for high temperature device applications. J. Mater. Sci. Mater. Electron. J. **17**(1), 1–25 (2006)
14. H.P. Phan, D.V. Dao, K. Nakamura, S. Dimitrijev, N.T. Nguyen, The piezoresistive effect of sic for mems sensors at high temperatures: A review. J. Microelectromech. Syst. **24**, 1663 (2015)
15. D.G. Senesky, B. Jamshidi, K.B. Cheng, A.P. Pisano, Harsh environment silicon carbide sensors for health and performance monitoring of aerospace systems: A review. IEEE Sens. J. **9**, 1472 (2009)
16. L. Beker, A. Maralani, L. Lin, A.P. Pisano, A Silicon Carbide differential output pressure sensor by concentrically matched capacitance, in *Micro Electro Mechanical Systems (MEMS), 2017 IEEE 30th International Conference on (IEEE)*, (2017), pp. 981–984
17. M. Ericson, J. Silverudd, *Design of Measurement Circuits for Sic Experiment: Kth Student Satellite Mist* (2016)
18. S. Jin, S. Rajgopal, M. Mehregany, Silicon carbide pressure sensor for high temperature and high-pressure applications: Influence of substrate material on performance, in *2011 16th International Solid-State Sensors, Actuators and Microsystems Conference*, pp. 2026–2029
19. L. Chen, M. Mehregany, A silicon carbide capacitive pressure sensor for in-cylinder pressure measurement. Sens. Actuat. A Phys. **145**, 2–8 (2008)
20. C. Li, M. Mehregany, A silicon carbide capacitive pressure sensor for high temperature and harsh environment applications, in *TRANSDUCERS 2007–2007 International Solid-State Sensors, Actuators and Microsystems Conference*, pp. 2597–2600
21. S. Shao, *4h-Silicon Carbide Pn Diode for Harsh Environment Sensing Applications* (2016)
22. Flowmeters & pressure sensors. http://www.flowmeters.com/differential-pressure-technology
23. Y. Dévellopment, *Mems Pressure Sensor 2018—Market & Technologies Report* (2018)
24. W. Chien-Hung, C.A. Zorman, M. Mehregany, Fabrication and testing of bulk micromachined silicon carbide piezoresistive pressure sensors for high temperature applications. IEEE Sens. J. **6**, 316 (2006)

25. M. Eickhoff, H. Möller, G. Kroetz, J. v. Berg, R. Ziermann, A high temperature pressure sensor prepared by selective deposition of cubic silicon carbide on soi substrates. Sens. Actuat. A Phys. **74**, 56 (1999)
26. M.A. Fonseca, J.M. English, M.v. Arx, M.G. Allen, Wireless micromachined ceramic pressure sensor for high-temperature applications. J. Microelectromech. Syst. **11**, 337 (2002)
27. S. Fricke, A. Friedberger, H. Seidel, U. Schmid, A robust pressure sensor for harsh environmental applications. Sens. Actuat. A Phys. **184**, 16 (2012)
28. T. Lalinský, P. Hudek, G. Vanko, J. Dzuba, V. Kutiš, R. Srnánek, P. Choleva, M. Vallo, M. Držík, L. Matay, I. Kosti˘c, Micromachined membrane structures for pressure sensors based on algan/Gan circular hemt sensing device. Microelectron. Eng. **98**, 578 (2012)
29. G. Wieczorek, B. Schellin, E. Obermeier, G. Fagnani, L. Drera, Sic based pressure sensor for high-temperature environments. IEEE Sens. J., 748–751 (2007)
30. D.J. Young, D. Jiangang, C.A. Zorman, W.H. Ko, High-temperature single-crystal 3c-sic capacitive pressure sensor. IEEE Sens. J. **4**, 464 (2004)
31. R. Zhang, T. Liang, Y. Li, J. Xiong, A novel mems sic pressure sensor for high-temperature application, in *2015 12th IEEE International Conference on Electronic Measurement & Instruments (ICEMI)*, vol. 3, pp. 1572–1576
32. K. Fobelets, R. Vounckx, G. Borghs, A gaas pressure sensor based on resonant tunneling diodes. J. Micromech. Microeng. **4**, 123 (1994)
33. M.A. Fragaa, H. Furlan, M. Massia, I.C. Oliveiraa, L.L. Koberstein, Fabrication and characterization of a sic/sio2/si piqaezoresistive pressure sensor. Proc. Eng. **5**, 609 (2010)
34. Y. Jiang, J. Li, Z. Zhou, X. Jiang, D. Zhang, Fabrication of all-sic fiber-optic pressure sensors for high-temperature applications. Sensors **16**, 1660 (2016)
35. K.M. Dowling, E.H. Ransom, D.G. Senesky, Profile evolution of high aspect ratio silicon carbide trenches by inductive coupled plasmaetching. J. Microelectromech. Syst. **26**, 135 (2017)
36. A. Rahman, A.M. Francis, S. Ahmed, S.K. Akula, J. Holmes, A. Mantooth, High temperature voltage and current references in silicon carbide cmos. IEEE Trans./Electron Devices **63**, 2455 (2016)
37. K. Bhat, M. Nayak, MEMS pressure sensors-an overview of challenges in technology and packaging. J. Smart Struct. Syst. **2**, 1–10 (2013)

Chapter 2
Electro-Thermal-Mechanical Modeling of Gas Sensor Hotplates

Raffaele Coppeta, Ayoub Lahlalia, Darjan Kozic, René Hammer, Johann Riedler, Gregor Toschkoff, Anderson Singulani, Zeeshan Ali, Martin Sagmeister, Sara Carniello, Siegfried Selberherr, and Lado Filipovic

2.1 Introduction

2.1.1 Historical Overview

Before the application of semiconducting materials and the discovery of gas sensors, canaries were taken into mines as an alarm for the presence of harmful gases, such as methane, carbon dioxide, and carbon monoxide. A canary is considered to be a songful bird, but it stops singing when exposed to these types of gases, signaling to the miners to exit the mine immediately.

By the middle of the previous century, it was demonstrated for the first time that certain semiconducting materials show changing conductivity when exposed to some gas molecules, especially when heated to an elevated temperature [1]. Electrical properties of these materials change when the chemical composition of its ambient gas changes. In the early 1960s, Seyama proposed a gas-sensing device based on a thin ZnO film [2]. With a simple electronic circuit, along with a thin film-sensitive layer operating at 485°C, it was demonstrated that the detection of a variety of gases such as propane, benzene, and hydrogen was possible. In 1967, Shaver described a new method to improve the sensing properties of some semiconducting metal oxide (SMO) materials towards reducing gases by an addition of small amounts of noble metals, namely, platinum, rhodium, iridium, gold, and

R. Coppeta (✉) · G. Toschkoff · A. Singulani · Z. Ali · M. Sagmeister · S. Carniello
ams AG, Premstaetten, Austria
e-mail: raffaele.coppeta@ams.com

A. Lahlalia · S. Selberherr · L. Filipovic
Institute for Microelectronics, TU Wien, Vienna, Austria

D. Kozic · R. Hammer · J. Riedler
Materials Center Leoben Forschung GmbH, Leoben, Austria

W. D. van Driel et al. (eds.), *Sensor Systems Simulations*,
https://doi.org/10.1007/978-3-030-16577-2_2

palladium [3]. Since then, research has intensified for the development of new sensitive materials and micro-hotplates have been designed and optimized with the aim to commercialize the new generation of the SMO gas sensors.

In July 1970, Taguchi filled a patent application in the United States for the first SMO gas sensor device dedicated to safety monitoring [4]. A porous SnO_2-sensitive thick film was used for this first-generation due to its promising sensing performance. To further enhance its sensitivity, palladium was added to the sensitive layer as a metal catalyst. Afterwards, the sensor was commercialized by Figaro Inc. in alarms for the detection of flammable gases to prevent fires in domestic residences.

Over the last five decades, due to the small footprint, low cost, high sensitivity, and fast response time of the SMO gas sensor, the device has been applied in a variety of applications and in different fields, including food and air quality monitoring, healthcare, electronic nose, agriculture, and so on [5, 6]. The SMO sensor is able to be integrated into a simple electronic circuit, making the potential application of this technology so widespread that specific needs have arisen, which must be satisfied at an industrial level.

Recently, the desire for SMO gas sensors suitable for portable devices such as smartphones and smartwatches has notably increased. New scaling challenges must be overcome in order to enable the practical integration into wearable devices. Low power consumption, high selectivity, and high device reliability are the most common issues considered during gas sensor development. A massive research and development effort is under way to fulfil all the requirements for a good gas sensor performance. The research activities are divided into two main topics: the electro-thermal-mechanical performance of the micro-hotplates and the sensing capability of the sensitive SMO films. This chapter deals with the electro-thermal-mechanical performance and modeling of SMO sensors.

2.1.2 MEMS Gas Sensor

2.1.2.1 Definitions

Micro-Electro-Mechanical Systems (MEMS) refers to technologies used to fabricate miniaturized integrated devices, which combine mechanical and electro-mechanical elements. They are fabricated using micro-fabrication techniques, such as thermal oxidation, photolithography, and chemical vapor deposition (CVD). The physical size of MEMS devices can range from the nanometer to the millimeter scale. These types of devices are used as actuators, controllers, and even sensors in the micrometer range, thereby generating effects on the macroscale. It should however be noted that MEMS devices do not always include mechanical elements; for instance, the SMO gas sensors are fabricated using bulk micromachining, which is a process used to produce micromachinery or MEMS, but have no moving parts. The SMO gas sensor is included in the MEMS fabrication family with the aim to reduce the power consumption without using mechanical elements. By forming a static membrane as a last step during sensor fabrication, the heat losses from the heated area to the substrate are dramatically reduced.

MEMS gas sensors are broadly based on metal oxides such as $ZnSnO_4$, Nb_2O_5, In_2O_3, ITO, and CdO. Among these materials SnO_2, WO_3, and ZnO are the most commonly used in the commercial market since they fulfil all the requirements for a good gas-sensing performance at reasonable fabrication costs [7, 8]. The films are deposited on top of suspended micro-hotplates using a variety of techniques and in different forms, namely: thick film, nanobelt, nanotubes, nanowires, thin film, and nanocompound. The operating principle of the MEMS gas sensor relies on heating the sensitive material to high temperatures between 250°C and 550°C using Joule heating of an integrated microheater. The working temperature required depends on the sensitive material used and the target gas species. To enable the adsorption and electron exchange between the chemical composition of the ambient gas and the sensitive material, the device must operate at elevated temperatures in the presence of oxygen [9].

2.1.2.2 Significance

The market size of gas sensors for consumer applications is expected to reach USD 1297 million by 2023, with a 6.83% compound annual growth rate (CAGR) between 2017 and 2023 [10]. This sector is about to experience the highest growth rate of the sensor market. The main factors responsible for the growth of this business are increasing pollution regulations laid down by governments in developed countries, which mandate the use of gas sensors in potentially hazardous environments, increasing the use of MEMS-based sensor worldwide, and raising awareness of air quality control among users. In May 2018, the World Health Organization (WHO) reported that around seven million people die each year, one in eight of total global deaths, as a result of exposure to air pollution [11]. New data reveal that 90% of the world's population is exposed to fine particles in polluted air, leading to cardiovascular diseases and lung diseases, including heart disease, stroke, lung cancer, respiratory infections, and chronic obstructive pulmonary diseases. Note that, ambient air pollution has caused around 4.2 million deaths, whereas household air pollution has caused about 3.8 million deaths in 2017 alone [11].

Today, wearable devices contain a variety of micro-sensors, such as a light sensor, a pressure sensor, a proximity sensor, an inertial sensor, a hall sensor, and many more. It is very likely that gas sensors will be the next sensor to be integrated in portable devices [12]. Consumer applications are forcing the new generation of gas sensors to minimize size, power consumption, and cost, especially with the use of MEMS technologies. Making gas sensors available to everyone through integration with handheld devices, such as smartphones and wrist watches, allows to monitor air quality easily at any time and from anywhere, thus leading to further increasing awareness about the impacts of climate change. Monitoring indoor and outdoor air quality in real time helps improve the health and quality of life of all human beings.

2.1.2.3 Applications

The detection of gases at an affordable price, low power consumption, and with a fast response time, is essential in numerous high-technology fields. This is why the MEMS gas sensor is generating phenomenal interest due to its broad application potential in healthcare, military, industry, agriculture, space exploration, cosmetics, and environmental monitoring. Among other requirements for practical gas-sensing devices, high reliability, low operating temperature, and high selectivity and sensitivity are desired.

One of the major problems faced by gas sensors dedicated to practical applications is to estimate the concentration of a target gas in a realistic ambient, meaning improved selectivity towards a target gas. Unfortunately, MEMS gas sensors are characterized by high sensitivity but have a poor selectivity. To overcome this limitation, an array of gas sensors is used to form an artificial olfactory system. The so-called electronic-nose (E-nose) gathers multiple gas sensors in the same device simultaneously. Each sensitive material is heated to a specific and uniform temperature, as the sensitivity of metal oxide to gases relies on the operating temperature. Measured responses of all sensors are treated using non-parametric analyses in order to distinguish between gases, thus enhancing the sensor selectivity.

Nowadays, the MEMS gas sensor can be found in different applications across the market. Some of the most significant application fields of this sensor are mentioned below.

- Automotive applications: SMO gas sensors can be used to control motor functioning and to help reduce the emissions of harmful gases coming from combustion engines [13]. Indeed, a special packaging must be conceived for theses sensors in order to not be influenced by high temperatures in the exhausts.
- Environmental applications: Due to their outstanding features compared to other sensors available in the market, the MEMS gas sensor can also be used to measure and monitor trace amounts of volatile organic compounds (VOCs) in the air [14]. In this area, it is necessary to develop a simple and low-priced device able to monitor indoor and outdoor air quality.
- Medical applications: MEMS gas sensors can be used for clinical diagnostics. The detection of target gases coming from biochemical processes, taking place in the human body, leads to the rapid diagnosis of several diseases [15]. The analyses can be carried out either directly from the patient's skin or from their breath.
- Agricultural applications: To detect rotting fruits and vegetables during storage, MEMS gas sensor can be employed [16].

2.1.3 FEM Simulations of MEMS Gas Sensors

The Finite Element Method (FEM) is a numerical tool which allows solving a continuum physics problem by discretizing the space into a set of subdomains. For example, the geometrical structure of the MEMS sensor is discretized by finite elements in the shape of tetrahedra or hexahedra elements in the 3D case. In this procedure, the field variables like electrical, thermal, or displacement field are approximated by a set of basic functions, for which frequently Lagrange polynomials are used. The mostly used order of polynomial or synonymously order of element is linear or quadratic, which allow linear or quadratic behavior of the field variable within the element. The set of resulting element equations is assembled into a global system of equations and is solved together with the given initial and boundary conditions. From the results of the field variables, relevant parameters like the thermal response time, temperature uniformity, heat losses, and mechanical stresses can be obtained.

The most well-known commercial FEM software tools in the market are Comsol Multiphysics, ANSYS, CoventorWare, MEMS+, and IntelliSense. These tools can be used to apply models which predict how the sensors react to real-world forces, heat, fluid flow, and other physical effects. Before fabrication, MEMS devices are often designed, simulated, and optimized using these Technology Computer Aided Design (TCAD) tools, leading to a reduction in the manufacturing costs and a reduction of the prototype development cycle. TCAD tools contribute significantly in the development of novel and optimized MEMS devices with higher yields. Regarding MEMS gas sensors, these software tools are primarily used to study the mechanical stability of the membrane, the temperature uniformity over the active area, and the power consumption of the sensor.

2.1.3.1 Temperature Distribution

The appropriate choice of the heater and membrane design are essential to achieve a uniform temperature over the active area, where the sensitive material is deposited. Materials with high thermal conductivities, together with an optimized heater geometry, are usually adopted to achieve the desired temperature distribution. However, using high thermal conducting films increases thermal leakage from the heated area to the Si substrate, thus leading to an increase in the overall power consumption of the device, which is a crucial requirement if the sensors should be integrated with embedded and portable systems. In addition, improving the heater geometry layout with the help of FEM simulations may be difficult in some cases due to the stringent mesh requirements for complex geometrical designs. One practical solution is presented in a recent publication from Lahlalia et al. describing how to efficiently enhance the temperature distribution [17].

The authors in [17] managed to improve the temperature uniformity over the active area without increasing the power consumption of the device. This was achieved by using a novel design, the so-called dual-hotplate, which is based on a single circular microheater along with two passive micro-hotplates. The operating principle of this novel structure depends on the high thermal conductivity of the microheater material compared to the membrane materials. It should be noted that a uniform temperature over the active region is a crucial part for baseline stability since a small change in the temperature over the sensitive material leads to baseline drift, which impacts the accuracy of the gas sensor measurement [18]. To further decrease the heat losses to the substrate, and thereby reduce the power consumption down to a few mW, a new membrane shape is implemented in the dual-hotplate sensor. Curved micro-bridges are used instead of simple beams to enlarge the distance between the active region and the substrate, while preserving the same membrane size.

2.1.3.2 Thermal Transient Response

The thermal transient response refers to the time consumed to heat up the micro-heater from room temperature T_c to the target temperature T_h. It is calculated from a simple expression (2.1), where the actual temperature distribution inside the sensor is neglected, while the overall thermal resistance R_{th} and the overall thermal capacitance C_{th} of the micro-hotplate are taken into account. The heat balance between the input power P_{in}, which corresponds to a steady-state heater temperature T_h and heat losses, can be described by

$$C_{th}\frac{\partial T(t)}{\partial t} = \frac{T_h - T_c}{R_{th}} + P_{in}. \tag{2.1}$$

This equation is solved using Fourier and Laplace analysis with boundary conditions $T_{(t=0)} = T_c$ and $T_{(t=\infty)} = T_h$. The temperature of the microheater is observed to follow exponential behavior with

$$T(t) = (T_h - T_c)\, e^{-\frac{t}{\tau}} + T_c, \tag{2.2}$$

where $\tau = R_{th}C_{th}$ is the thermal time constant. As can be seen from the previous equations the thermal time constant depends linearly on the thermal resistance and heat capacity. However, a micro-hotplate with materials having a small thermal resistance and low thermal mass will give a faster response. A small heater exhibits a faster thermal response thanks to the smaller heat capacity, allowing the microheater to operate at very short pulse times, which reduces the power consumption drastically. Note that reducing the sensor's active area and membrane thickness also improves the thermal response time of the micro-hotplate, as the total heated mass is reduced.

2.1.3.3 Thermal Simulation

After the design and meshing of the MEMS gas sensor geometry within a TCAD tool, verification of the thermal performance, including the temperature distribution, thermal response time, temperature gradient, heat losses, and heat exchange between the sensor and its environment, are obtained with the help of a design validation software. Indeed, measuring these parameters without the help of FEM tools may be quite challenging, especially if the temperatures are changing quickly, or need to be measured inside the sensor. This means that TCAD software with FEM analysis is an indispensable tool to engineers interested in the detailed thermal performance of their devices.

To model the entire sensor, each part of the structure is represented by a corresponding mesh. The proper choice of the mesh is essential to obtaining accurate approximations. As mentioned earlier, the mesh is a set of elements for which the temperature versus time is calculated. Within each element the temperature is approximated by an ansatz function. One idea is to derive the equation for the temperature at the nodes, which are the centers of the elements. For this approach, temperature and flow variations within the elements are neglected and the node temperature is regarded as representative of the whole element.

This lower order approximation is of linear convergence order. If the heat flow is balanced by the continuity equation of heat energy, we arrive at the finite volume approach. Another concept is to replace the differential equation within each element using finite differences, which is known as the finite-differences method.

All these approaches require equations to be solved at each node at every temperature, generating a large set of equations, which must be solved. One alternative for the thermal problem is to describe the thermal parameters by their electrical equivalent as shown in Table 2.1. For this lumped thermal network, the equations can be solved analytically. For effortless equation solving, standard circuit analysis tools such as LTSpice can be used.

To simulate the heat transfer in a MEMS gas sensor, three mechanisms must be simultaneously taken into account, namely, conduction, convection, and radiation. Generally, radiation is considered to be negligible for temperatures below 600°C

Table 2.1 Thermal to electric parameter equivalence

Thermal parameter	Electrical equivalent
Temperature T (K)	Voltage V (V)
Specific heat C_p (J/kg K)	Permittivity ε (F/m)
Thermal resistivity ρ_{th} (K m/W)	Electric resistivity ρ_{el} (Ω m)
Resistance R_{th} (K/W)	Resistance R (Ω = V/A)
Heat flow P (W)	Current I (A)
Heat Q (J = W s)	Charge W (C = A s)
Thermal conductivity k (W/K m)	Electric conductivity σ (S/m)
Capacitance C_{th} (J/K)	Capacitance C (F)

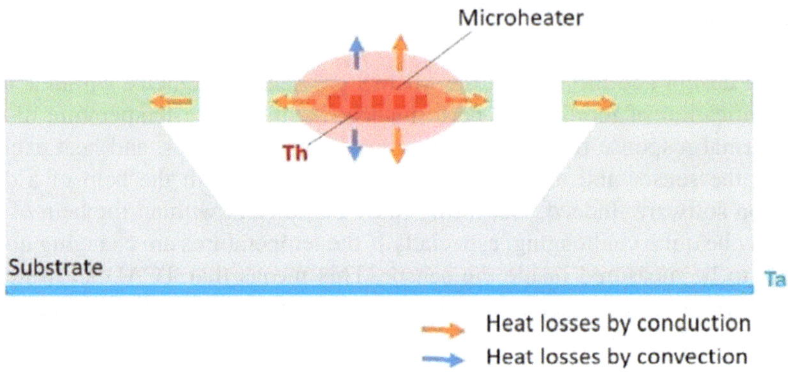

Fig. 2.1 Heat loss mechanisms through the MEMS gas sensor. **Th** is the temperature of the microheater; **Ta** is the ambient temperature

compared to the heat losses by conduction and convection; heat losses in the MEMS gas sensor are caused mainly by heat conduction through the micro-hotplate and the air, and by heat convection, through heat exchange between the external face of the heated membrane and the surrounding air (Fig. 2.1). It must be noted that the amount of heat lost by convection is proportional to the temperature difference between the sensor surface and the surrounding fluid, and to the area of the face exchanging the heat. In addition, natural convection can only occur in the presence of gravity since air movement is dependent on the difference between the specific gravity of cold and hot air. Through this entire discussion, one can deduce that the choice of the membrane and microheater materials and the chosen structure play integral roles in defining the sensor's power consumption.

2.1.3.4 Mechanical Behavior

The design of an effective and reliable MEMS gas sensor is not only a challenge of having a good thermal performance and high sensing capability but also of having an excellent thermo-mechanical stability. To consider mechanical issues during the fabrication stage of the MEMS sensor, one has to analyze the internal stress accumulated in the sensor micro-hotplate. This is one of the major concerns impacting the performance and long-term mechanical reliability of the device. In order to minimize the internal stresses, an appropriate set of process parameters must be found and the fabrication process must be well controlled. Mechanical properties such as density, stoichiometry, orientation, and the average grain size of each layer of the sensor are defined by the specific deposition conditions. In this context, it should be noted that the mechanical characteristics of the sensor layers can be shifted by annealing for one or more cycles. Fortunately, it is possible to adjust these properties by a further annealing step at a specific temperature.

Another problem to be considered during sensor design is the thermal stress. It is introduced on top of the residual stress during operation at high temperature, produced by the difference in the thermal expansion coefficients between membrane materials and by the non-uniform temperature distribution. Thermal stress may lead to a significant increase in membrane deformation and undesirable bimetallic warping effects, which reduces the lifetime of the sensor. Indeed, the operating temperature impacts the mechanical behavior of the sensor, but other thermal effects also play a role. The ultra-short heat pulses influence the mechanical properties since a fast temperature ramp-up may lead to adherence problems or to membrane instability, which may even collapse due to excessive stress changes [19].

2.2 Gas Sensor Micro-Hotplate

2.2.1 Introduction

The SMO sensor, one of the most widely used sensors for gas detection, requires being heated to an elevated temperature in order to enable a reaction between the sensitive material and a target gas. Therefore, a micro-hotplate, which is a common structure in a MEMS-based gas sensor system, is an essential component for these devices. Additionally, it is required to thermally insulate the active area and the electrical components in order to integrate the sensor with the appropriate analog and digital circuitry.

A micro-hotplate is a miniaturized suspended thin membrane which is thermally insulated from the silicon substrate, usually containing a microheater to heat up the sensitive material, a resistive temperature detector (RTD) to estimate the changes in the temperature over the active area, and interdigitated electrodes to measure the electrical resistivity of the sensitive material. Gas sensors based on this type of MEMS structure are very useful for the purpose of minimizing the overall power consumption, enabling the MEMS gas sensor to be applied in the field of chemical micro-sensing. The MEMS-based heating structure can be used for gas-sensing applications after coating its surface with a sensing metal oxide film, which can be deposited, either by liquid phase or by vapor phase deposition, as shown in Fig. 2.2.

2.2.2 Microheater

2.2.2.1 Heater Materials and Geometries

The microheater is the key component of the SMO gas sensor, as its primary function is to raise the temperature and maintain a uniform temperature profile over the sensitive material. The area where the sensing layer is deposited is known as the "active region" or "active area." The level of the operating temperature

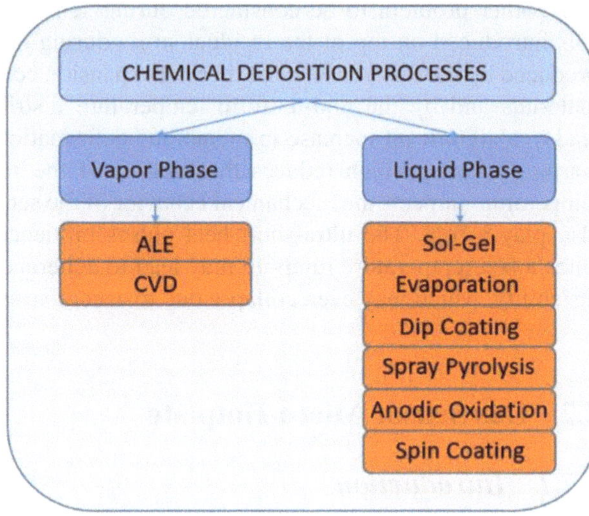

Fig. 2.2 Metal oxide thin film deposition techniques

is dependent on the target gases and the sensitive material used. Therefore, the appropriate choice of the microheater material plays a crucial role in determining the gas sensor's performance and its reliability [20]. The desirable characteristics for a good microheater material are low thermal conductivity, high melting point, high electrical resistivity, low fabrication cost, low thermal expansion coefficient, low Poisson's ratio, and most importantly, high compatibility with MEMS and complementary metal oxide semiconductor (CMOS) fabrication technologies [21].

At the onset of the micro-hotplate development, the commonly used metals for electrical connection in standard integrated circuit (IC) technology such as aluminum and gold were used as a microheater element [22, 23]. Gradually, it was observed that these materials have several drawbacks like oxide formation, low resistivity, poor contact properties, and electromigration effects at high temperatures [21]. In this respect, platinum, which is a very popular heating element for temperature below 500°C, offers better performance such as the ability to deal with high current density, chemical inertness, and high stability of the temperature coefficient of resistance (TCR) up to 650°C [24]. Nevertheless, this material is quite expensive, and its electrical contact formation with other materials is also critical. In addition, platinum has a positive TCR, which magnifies the effect of hotspots, leading to an adverse impact on the long-term reliability of the microheater [21].

Research is currently ongoing to find new materials to overcome the limitations and drawbacks mentioned previously. More recently, nickel and iron-nickel have been used as a microheater, thanks to their low TCR and thermal conductivity [25, 26]. Materials like tungsten [27, 28], nickel-chromium alloys [29], Dilver P1 [30], molybdenum [31], hafnium diboride [32], titanium nitride [33], silicon carbide [34], and Sb-doped SnO_2 [35] have also found to be promising as a heating element owing to the several positive features of these materials, namely, low thermal expansion, resistance to humidity, high Young's modulus, and their non-magnetic nature. Tungsten was reported by Ali et al. [36] as a good high temperature material for a heater element. Lahlalia et al. [37] presented a Tantalum-Aluminum (TaAl) layer as a resistive microheater on a perforated membrane in silicon nitride. TaAl is characterized by its ability to retain its mechanical strength at high temperature and by its negative TCR of about -100 ppm/°C, leading to minimal hotspot formation and a stable temperature versus input power curve. The bottom line for choosing a particular heater material is to fulfil the desired requirements; therefore, there are no simple design rules. However, the heater geometry plays a critical and active role to define sensor performance.

Sensitivity, selectivity, and response time are partially dependent on the thermal behavior of the micro-hotplate. Therefore, the proper choice of the microheater design is a crucial factor in determining the sensing performance of the SMO gas sensor. Low power consumption, temperature stability, and temperature uniformity over the sensitive material are three parameters desired while designing the micro-heater element. To achieve the optimal aforementioned requirements, one simple solution is to alter the microheater geometry. Note that, it is also important to consider the stress induced in the microheater while testing different geometries.

A high stress in the heater element leads to a reduced lifetime of the device. Moreover, current crowding in the corners of the microheater lines is another factor which should be taken into account when choosing the geometry of the heater element. Localized electron accumulation may lead to the generation of microcracks and localized deformations. To overcome this issue, circular type heater structures are reported to be a good alternative to conventional microheater geometries such as the meander shape [38]. Figure 2.3 shows different microheater geometries investigated so far in previous research [39–41].

A new generation of integrated solid-state gas sensors embedded in Silicon on Insulator (SOI) micro-hotplates offer ultra-low power consumption (under 100 mW), high sensitivity, low noise, low unit cost, reproducibility, and reliability through the use of the on-chip integration. The micro-hotplate lies on a SOI membrane and consist of Metal Oxide Semiconductor Field Effect Transistor (MOSFET) heaters which elevate the operating temperature, through self-heating, of a gas-sensitive material. The sensors are fully compatible with SOI CMOS or biCMOS technologies, In addition, the new integrated sensors offer a nearly uniform temperature distribution over the active area at its operating temperature at up to about 300–350°C. This makes SOI-based gas-sensing devices particularly attractive for use in hand-held battery-operated gas monitors [42].

2.2.2.2 Heat Losses

MEMS gas sensor-based micro-hotplate dissipates power through three different mechanisms as already mentioned in Sect. 2.1.3.3.

Free or natural convection is the heat transfer occurring between the heated surface of the membrane and the surrounding fluid, including air and other gases. This mechanism is partly described by fluid motion and partly by heat conduction

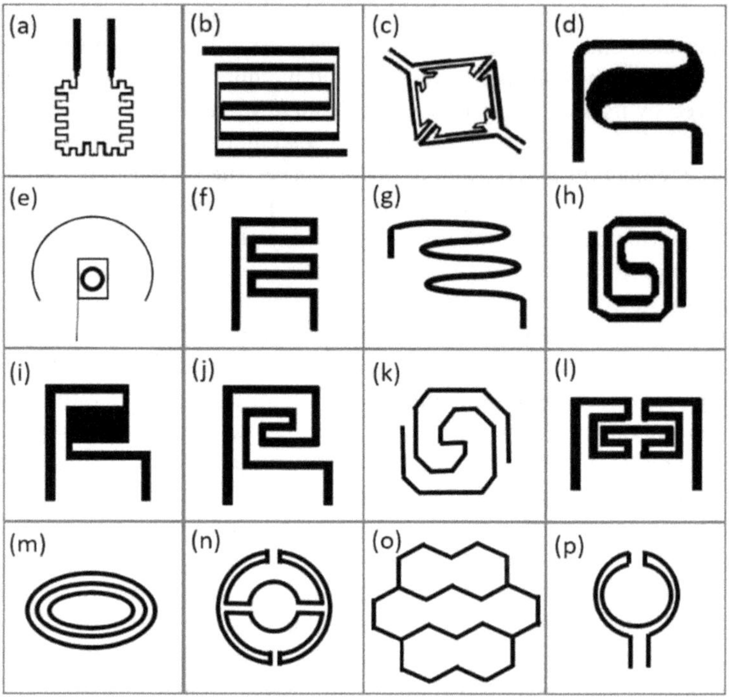

Fig. 2.3 Different microheater geometries used in MEMS gas sensors. (**a**) Loop shape. (**b**) Irregular shape 1. (**c**) Irregular shape 2. (**d**) S shape with rounded corner. (**e**) Plane plate with center square hole. (**f**) Meander. (**g**) Curved meander. (**h**) Curved double spiral. (**i**) S shape. (**j**) Double spiral. (**k**) Double spiral with rounded corner. (**l**) Fan shape. (**m**) Elliptical. (**n**) Drive wheel. (**o**) Honeycomb. (**p**) Circular

through the air [43]. Fluid motion can be caused by a gravitational force, as is the case in natural convection, which is the result of temperature differences, or by external energy in the case of forced convection. For convection simulations of the MEMS gas sensor, forced convection is always excluded. Nevertheless, the calculation of the heat losses by natural convection is still very difficult due to the complexity in coupling between the density field, the temperature field, and the fluid field. The total heat flow in natural convection from a heated membrane to the surrounding air can be expressed by Newton's law of cooling as

$$Q_{\mathrm{conv}} = h \cdot A\,(T_{\mathrm{h}} - T_{\mathrm{c}})\,, \tag{2.3}$$

where h is the mean heat transfer coefficient and A is the exposed area from which the heat flows.

Heat conduction must be considered in MEMS gas sensors. This is the heat transfer which occurs between the heated area of the membrane and the substrate. To simplify the model which describes this mechanism, the heat conduction perpendicular to the membrane is neglected due to the small thickness of the layers which compose the membrane stack. This leads to a one-dimensional heat conduction problem in cuboid coordinates. If the entire suspended membrane is heated to a uniform temperature, the heat conduction occurs only in the suspension beams. For suspended membranes with three suspension beams, heat losses by conduction can be expressed as

$$Q_{\text{cond}} = \frac{3 \cdot \lambda_T \cdot A_{\text{beam}} \left(T_{\text{h}} - T_{\text{c}}\right)}{l}. \tag{2.4}$$

Here, A_{beam} and l are the sectional area and length of the beam, respectively, and λ_T is the thermal conductivity of the membrane stack with an n-multilayer system, which can be calculated by

$$\lambda_T = \sum_{k=1}^{n} \lambda_k \times t_k / \sum_{k=1}^{n} t_k, \tag{2.5}$$

where t_k is the thickness of the layer k.

Radiation is the heat transfer which takes place in the form of electromagnetic waves primarily in the infrared region. Radiation is emitted by a body as a consequence of thermal agitation of its composing molecules. In the MEMS gas sensor, radiation is considered only on the surface of the heated membrane area as the radiation emitted from the interior regions can never reach the surface. Under the assumption that the heated membrane area behaves like a grey body, the heat losses by radiation can be expressed as

$$Q_{\text{rad}} = \in \sigma \left(T_{\text{h}}^4 - T_{\text{c}}^4\right), \tag{2.6}$$

where σ is the Stefan–Boltzmann constant, which equals to 5.67×10^{-8} W/m^{2}K^4. For this type of theoretical model, where the frequency-dependent emissivity is lower than that of a perfect black body, the emissivity $\in$ must be included. It should be noted that the heat losses through radiation are often neglected since they represent only a few percent of the total heat losses. Nevertheless, due to the T^4 dependency, radiation must be taken into account if the sensor operates at very high temperatures.

2.2.3 Membrane Types and Materials

In order to achieve a high temperature with low power consumption, different types of the membranes have been adopted instead of using only Si bulk [21]. A cavity below the membrane of the gas sensor is essential to minimize the vertical heat losses, as the thermal conductivity of the air is much lower than

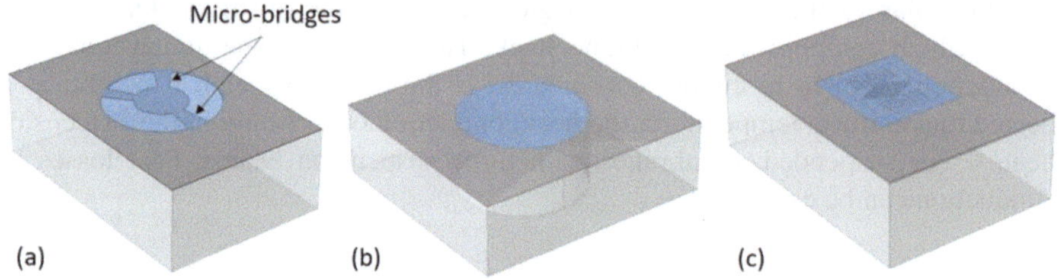

Fig. 2.4 Schematic of different membrane-type gas sensors. (**a**) Suspended membrane-type. (**b**) Closed membrane-type. (**c**) Perforated membrane-type. The blue color shows where the membrane is formed

the materials used in MEMS and CMOS technology. Up to now, three different structures for the membrane based on SMO gas sensors have been researched, namely, suspended membrane-type gas sensor, closed membrane-type gas sensor, and perforated membrane-type gas sensor. Three-dimensional (3D) views of those structures are depicted in Fig. 2.4.

2.2.3.1 Suspended Membrane-Type

The suspended membrane-type gas sensor can be formed by etching the substrate from the front-side using either standard wet etchant with EDP or KOH, or by sacrificial etching [44]. For standard etchant, the sensor surface is first coated with a dielectric layer such as silicon nitride and/or silicon dioxide, then patterned to form the etch windows. The membrane is subsequently released by etching the silicon. In some cases, this step is performed after packaging the sensors with the aim to avoid harsh environments when dicing the wafers, leading to an increase in the fabrication yield. Sacrificial etching is another way to release a suspended membrane, which depends on using the selective formation of porous silicon in a p-type silicon wafer to form a cavity below an implemented n-type silicon layer [44].

One of the primary benefits of using the suspended membrane-type is that it provides low power consumption compared to the closed membrane-type, which is achieved by reducing the thermal mass of the membrane. In addition, this type of membrane is more suitable for CMOS fabrication since there is no need to align the top and bottom sides during the lithography process. Despite these advantages, the suspended membrane gas sensor is less mechanically stable, as the membrane is supported only by micro-bridges. In fact, the residual and thermal stress in the micro-bridge was found to be high and the vertical displacement of the entire membrane was also considerable during operation at high temperature [39].

2.2.3.2 Closed Membrane-Type

The closed membrane-type gas sensor can be formed by an anisotropic etching of the silicon from the backside. The etching stops when a solution like KOH or EDP2 contacts the membrane materials. Alternatively, deep reactive ion etching (DRIE), also known as the Bosch process, can be used to etch the wafer backside. DRIE is a multi-cycle process, where each cycle consists of an isotropic deposition followed by an anisotropic etching, usually performed in an ion-enhanced plasma.

This membrane structure is chosen as an alternative to the suspended membrane-type, as the high temperature reached in the beams results in high thermal stress, which may cause the generation of microcracks, leading to a shorter lifetime of the sensor. Recently, novel nanomaterials have been investigated for gas-sensing applications, enabling the detection of gases at low operating temperatures leading to a further exploitation of the closed membrane-type instead of suspended membrane-type to improve the sensor reliability [45].

Usually, the membrane stack of the closed membrane is composed of thin silicon dioxide and silicon nitride layers in order to minimize the intrinsic stress. The silicon dioxide layer has a tensile stress, while the silicon nitride has a compressive stress. For stress compensation, the silicon nitride layer is generally stacked between two silicon dioxide layers. Other materials which have been attempted as a membrane include aluminum oxide [46], polyimide [47], porous silicon [48], silicon oxynitride [49], and silicon carbide [50].

2.2.3.3 Perforated Membrane-Type

In order to combine the advantages of both of the aforementioned membrane-types in one structure, the perforated membrane-type was suggested by Lahlalia et al. [37]. This structure is formed by etching the sacrificial layer from the front-side while getting better mechanical performance than a suspended membrane. Moreover, membrane holes which are used for etching the layer below also minimize the lateral heat losses, which reduces the total power consumption compared to a closed membrane-type. The perforated membrane may be classified as a sub-type of the closed membrane since there are no beams supporting the micro-hotplate.

Lahlalia et al. presented for the first time a novel micro-hotplate structure for the SMO gas sensor which includes a pierced TaAl plate as a microheater and a perforated Si_3N_4 membrane [37]. To form this membrane structure, polyimide HD8820 is first deposited in a cavity after etching the SiO_2, then cured for 1 h at 350°C to get the desired thickness of 3 μm. Micro-hotplate materials are deposited thereafter, namely, Si_3N_4, TaAl, AlCu, TiW, and Pt. Finally, the polyimide is etched to release the membrane after opening the holes in the micro-hotplate by highly selective dry etching using a CH_4, CHF_3, and Ar plasma.

2.2.4 Electrodes and Thermometer

2.2.4.1 Electrodes

The so-called interdigitated electrodes (IDEs), implemented in the MEMS gas sensor, are used to make the measurements of the gas concentration possible. These electrodes have two functions. On one hand, they measure the resistance change in the sensitive layer, which represents the gas concentration. On the other hand, they act as a catalyst when the electrodes are made by noble materials or their alloys. The catalytic materials promote the interaction between the gas and the sensitive material, leading to an improvement in the sensitivity and response time of the sensor. Usually, when a thick porous-sensing film is used the electrodes are deposited below the sensitive material; however, with a thin film-sensitive layer, they are deposited on top in order to measure the resistance changes on the surface of the sensing layer and to ease the interaction when catalytic electrodes are used. Among the most popular materials used for IDEs are gold [51], platinum [52], AlCu [37], and Al [53]. These materials are often deposited with an adhesion layer, such as a 20 nm of titanium-tungsten, in order to act as a dielectric, to achieve good adhesion, and to prevent delamination [54].

Another aspect which should be considered when using IDEs for MEMS gas sensors is the output signal strength. This is controlled mainly by the design of the active area, and the pitch of the electrode fingers. The proper design and positioning of the electrodes is essential to maximize the gas sensor response towards desired gases. It should be noted that it is very important to maintain a uniform temperature over the IDEs, which must be the same as the microheater temperature for the sensing gases. This can be achieved by forming the microheaters, thermometer, and IDEs in different layers, as the temperature uniformity is better for this combination. Heater and thermometer structures can also be integrated in the same layer with IDEs to ease the fabrication process, but unfortunately, the temperature is less uniform for this configuration compared with the vertical approach [17, 38].

2.2.4.2 Thermometers

Since the operating temperature of the microheater depends on the target gases and the sensitive material, an accurate measurement of the micro-hotplate temperature as a function of the applied power is essential for an optimum performance of the MEMS gas sensor. The extraction of the temperature in these tiny devices is highly challenging due to the extremely small surfaces of the active region. Moreover, this surface is often heated to a non-uniform temperature, which impacts the accuracy of the measurements. Over the past years, different approaches have been adopted for this purpose, including simulations, light glowing, RTD, analytical modeling, and thermoreflectance (TR) [55].

Platinum, a commonly used material for the microheater, has also been used as an RTD since its resistance linearly increases with the temperature [24]. Lahlalia et al. used three types of Pt RTDs instead of one [37], deposited on three different locations over the active region, to additionally characterize the temperature distribution. To further support the obtained results by Pt RTD, the temperature was characterized using CrSi as a temperature sensor. This material has the ability to detect temperatures above 450°C by acting as a phase-change material. Hannef et al. experimentally determined the temperature of the micro-hotplate using TR thermography [27] and proved the accuracy of this method by comparing the obtained results with simulations and silicon RTD.

2.3 Material Properties of Micro-Hotplates

2.3.1 Introduction

In order to achieve the real-world physical performance of the micro-hotplate in a simulation environment, whether relating to power consumption or mechanical behavior of the membrane, it is essential to use the correct values for the material properties as a function of the operating temperature for all micro-hotplate materials. These material properties can vary due to many factors, such as material thickness, deposition process, and environment conditions. In this section, the main properties, which define the electro-thermal-mechanical performance of the micro-hotplate, are discussed, namely, thermal conductivity, specific heat capacity, Young's modulus, and tensile strength.

2.3.2 Thermal Properties

The thermal conductivity and specific heat capacity are the two essential material properties which define the capability of a material to store and transfer heat. Understanding these properties is fundamental for devices which experiences rapid changes in the operating temperature, for any deposition process, or for material which are exposed to large temperature gradients. Precise values for these properties are indispensable for micro-hotplate modeling, for heat management to decrease the power consumption of the device, and for the capability of the device to withstand high temperatures and fast temperature changes in the membrane.

2.3.2.1 Thermal Characterization

Thermal characterization of the materials involved in the MEMS gas sensors is crucial for understanding the device behavior and for developing physical models, essential for obtaining predictive simulation results.

For thin layers, mainly electrical resistance thermometry (ERT) or laser-based methods are used [56, 57]. For ERT, an electrically conductive material, acting as a heater, must be placed on top of the layer of interest. If an electrically conductive layer has to be characterized, a thin insulating layer must be placed in between. Alternatively, a p-n junction is used as heater and thermometer. For the named methods, the general principle is the same. Heat is introduced locally into the sample by switching a heat source on and off. The transient temperature curve is determined by the thermal conductivity and the heat capacity.

In the case of ERT, the 3ω method and the thermal impedance are the dominating methods. The principle of the 3ω method is that alternating current with frequency ω is used for heating, which implies that the power, and therefore also the temperature, shows a 2ω modulation. Since electrical resistance is a function of temperature, 3ω frequency components show up in the measured voltage. From the latter, the thermal conductivity can be computed using an analytical model [58]. Recently, this principle was also exploited for scanning probe microscopy, where the scanning thermal microscope (SThM) approaches a nanometer resolution [59]. The thermal impedance method works with the transient temperature response after a step like switching off of the power of a heater [60]. It also allows to characterize a component's thermal response and to calibrate the thermal properties of the materials when an accompanying thermal simulation is performed [61].

In the time-domain thermoreflectance (TDTR) method, a pump laser heats a thin metallic layer (~50 nm) on top of the layer of interest. The reflectance of a probe laser is dependent on the temperature. By using fast lasers, e.g., in the picosecond regime, a very fine resolution in time can be obtained. This allows the measurement of very thin layers down to a thickness of a few tens of nanometers. For example, by varying the pump and probe laser spot diameter, the in-plane, as well as out of plane, thermal conductivity can be measured [62]. This method potentially can also identify the interface conductance and contributions of ballistic transport. This is of upmost importance since as layers become thinner and thinner, the system's behavior can no longer be described using a bulk thermal conductivity alone.

2.3.2.2 Thermal Modeling

To calculate the heat flux in the micro-hotplate, the thermal conductivity of materials used must be known, as it provides an indication of the rate at which energy is transferred from the heated area to the substrate. The thermal conductivity depends on the physical structure of matter, which is associated with the state of the matter. From Fourier's law, the thermal conductivity is expressed as

$$\vec{q} = -k\nabla T, \tag{2.7}$$

where $\vec{q}$ is the local heat flux density in $W \cdot m^{-2}$, k is the thermal conductivity in $W \cdot m^{-1} \cdot K^{-1}$, and ∇T is the temperature gradient in $K \cdot m^{-1}$. From the above Eq. (2.7), it is clear that the heat flux increases with increasing thermal conductivity.

The transport of the thermal energy in the materials which compose the micro-hotplate can be attributed to two mechanisms, namely, migration of free electrons and vibrational movements of the atoms/ions. The latter are called phonons in the case of periodic crystalline structures. Even if strictly speaking there are no phonons in solids without periodicity in structure and composition, e.g., in amorphous materials, with the propagons and diffusons, there are the so-called non-localized modes which behave similar to phonons [63]. From kinetic theory, thermal conductivity is defined as [64]

$$k = \frac{1}{3} C \cdot \overline{c} \cdot \lambda_l. \tag{2.8}$$

For conducting materials, C is equivalent to the electron-specific heat (C_e), $\overline{c}$ is the mean electron velocity, and λ_l is equivalent to the electron mean-free path (λ_e), whereas in non-conducting materials, C is equivalent to phonon-specific heat (C_{ph}), $\overline{c}$ is the average speed of sound, and λ_l is the phonon mean-free path (λ_{ph}). Regarding materials where the electrons and phonons carry thermal energy at the same time, thermal conductivity may be expressed as

$$k = k_e + k_{ph}, \tag{2.9}$$

where k_e represents the heat transfer in the solid caused by electrons and k_{ph} represents the heat transfer in the solid caused by phonons. In alloys, with a high electrical resistance, the interaction between the atoms and the lattices decreases, leading to increased k_{ph} in such a way that it becomes no longer negligible compared to k_e. It should however be pointed out that the regularity of the lattice arrangement has a significant effect on k_{ph}. For instance, quartz, which is a crystalline material with well-ordered lattices, exhibits a higher thermal conductivity than amorphous materials like glass.

The calculation of the bulk thermal conductivity is described above for material with relatively large dimensions. However, in many fields of technology like MEMS and CMOS gas sensors, the materials' dimensions can be in the order of few micrometers down to several nanometers. In that case, care must be taken to account for the possible alteration of the thermal conductivity as the dimensions of the sensor layers are smaller.

Figure 2.5 shows the cross section of silicon dioxide with thickness t along with phonons which contribute to the transport of the thermal energy. When t is significantly larger than λ_l, as is the case for a bulk material, the effect of the boundaries on thermal conductivity is negligible. However, when the ratio t/λ_l is close to one, as is the case for thin films, boundaries of the film act to scatter the

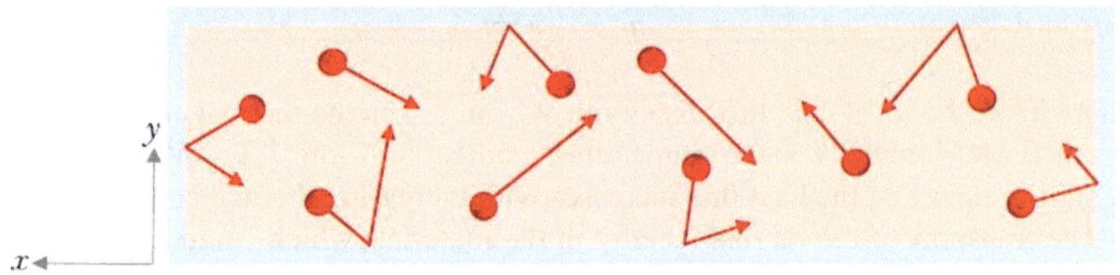

Fig. 2.5 Phonon trajectories in a silicon dioxide thin film with boundary effects

phonons, which eventually redirect their propagation. As a result, the average net distance traveled by the phonons decreases, which reduces the thermal conductivity of the thin film silicon dioxide. It should be noted that the effect of boundaries on phonons moving in the y-direction is more significant than phonons moving in the x-direction. Therefore, for thin films with small t/λ_l, k_y is less than k_x, where k_x and k_y are the effective conductivity in parallel and perpendicular to the average flow in the film, respectively.

According to the analysis of Flik and Tien [64], k_x and k_y can be obtained as follows:

$$\lambda_x = k\,(1 - 2\lambda_l/\,(3\pi t)) \tag{2.10}$$

$$\lambda_y = k\,(1 - \lambda_l/(3t)) \tag{2.11}$$

The use of the bulk thermal conductivity rather than the effective conductivity results in an error greater than 5% if $t < 7\lambda_l$ (for k_y) and $t < 4.5\lambda_l$ (for k_x). When $\lambda_l < t < t_c$, the microscale effect must be taken into account, as k_x and k_y are reduced from the bulk value k. For instance, the critical film thickness t_c of silicon dioxide in the y-direction ($t_{c,y}$) and x-direction ($t_{c,x}$) are 4 nm and 3 nm, respectively, and λ_l is 0.6 nm at 300 K [65]. If 0.6 nm $< t <$ 4 nm, the effective conductivity should be calculated using (2.10) and (2.11). One must keep in mind that, in a solid, λ_l decreases as the temperature increases, which results in a reduction of the thermal conductivity. Typical values for the thermal conductivity and capacitance of W, SiO_2, and Si_3N_4 for different thicknesses and under a variety of temperatures, obtained through a literature study, are provided in the Appendix Tables 2.3 and 2.4 respectively.

2.3.3 *Mechanical Properties*

2.3.3.1 Mechanical Testing

The mechanical properties of semiconductors and metals have become the subject of an extensive research field to understand the mechanical behavior of the gas sensor during operation. The harmful effects that stresses cause in the membrane, electrodes, thin film-sensitive layer, and the microheater have prompted a plethora of research to determine the origin and the magnitude of stress as well as the means of minimizing stresses. Stress is almost always present in the thin films which constitute the MEMS gas sensor. Note that stress exists even when thin films are not heated or loaded due to the presence of residual stress. The residual stress directly impacts a variety of phenomena, including the generation of crystalline defects, adhesion, and the formation of film surface growths. Another issue which must be addressed is the stress effect in very small regions, which must be measured, understood, and controlled as the gas sensors continue to shrink in size. The build-up of stress and its high concentration in small regions can lead to cracking and delamination, ultimately resulting in failure. The methods to model fracture mechanics and crack evolution are given in Sect. 2.5.2.

A variety of methods have been used in the past with the aim to determine the mechanical properties of thin films, especially their strength [66]. The most common techniques are depicted in Fig. 2.6. Each technique gives values of the Young's modulus (E), strength, and internal stress in films. In this subsection, the testing techniques are separately described to evaluate the information gained and compare their advantages.

Bulge testing [67, 68] is commonly used to determine the mechanical properties of membrane and thin films in general. This testing method is based on fixing the membrane ends to the ends of a hollow cylindrical tube. To extract the mechanical performance, the membrane is pressurized with gas, then the maximum deflection corresponding to the height of the hemispherical bulge in the membrane is measured

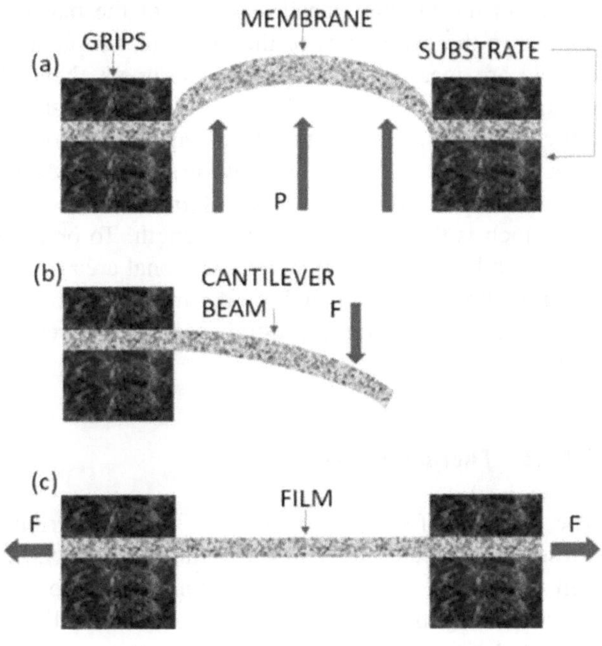

Fig. 2.6 Methods for mechanical properties testing for thin films. (**a**) Bulge testing. (**b**) Microbeam bending. (**c**) Tensile testing

with an interferometer or a microscope and converted to the strain. The equation which governs the relationship between the height (h) and the applied differential pressure (P) is given by

$$P = \frac{4th}{r^2}\left[\sigma_i + \frac{2Eh^2}{3(1-v)r^2}\right], \tag{2.12}$$

where t and r are the thickness of the membrane and specimen radius, respectively; σ_i is the residual stress in the membrane under a zero pressure differential, and v is the Poisson's ratio.

Microbeam bending or **focused ion beam** [69–74] is used to extract the mechanical properties of the cantilever microbeams, fabricated using micromachining techniques involving photolithography and etching processes. Typical dimensions of the beams are smaller than 1 μm thick, 20 μm wide, and 100 μm long, allowing the cantilever microbeams to be deposited at different locations on the wafer. This allows for a local monitoring of stress variations in subsequently deposited thin films. To determine the Young's modulus and the yield strengths of the thin film materials which comprise the cantilever beam, load–deflection data along with beam theory are used. This data are obtained with the aid of a submicron indentation instrument which continuously monitors the load and deflection caused by a nanoindenter [69].

Tensile testing [75, 76] is used to determine how materials will behave under a tension load in order to evaluate both the elastic and plastic response of thin films. In a simple tensile test, a thin film is typically pulled to its breaking point to extract the ultimate tensile strength of the material. The amount of force (F) applied to the extremities of the thin film sample and the elongation (ΔL) of the sample are measured during the experiment. Note that the force is applied by electromagnetic force transducers, and the elongations are generally measured by optical methods. In micro-tensile testing, an electron microscope is used to enable a direct observation of the defects during straining. Mechanical properties of materials are often expressed in terms of stress (σ), which is the force per unit area, and strain (ε), which is the percent change in length. To obtain the stress, the applied force is divided by the thin film's cross-sectional area ($\sigma = F/A$). Strain is measured by dividing the change in length by the initial length of the sample ($\varepsilon = \Delta L/L$). The numerical values of stress and strain can then be presented as a graph called a stress–strain curve.

2.3.3.2 Thermal Stress

The thermal effect is another issue which contributes dramatically to the film stress. Films heated to high temperatures and then cooled to room temperature will experience a thermal stress. Films which are cooled from room temperature to cryogenic temperatures or thermally cycled will also be thermally stressed. The magnitude of the thermal stress can be expressed as

$$\sigma = E\alpha\,(T - T_0)\,, \tag{2.13}$$

where α is the coefficient of thermal expansion, and $T - T_0$ is the temperature difference.

For bilayer combinations such as film/substrate, the strain in the film and the substrate are, respectively, expressed by

$$\varepsilon_{\text{film}} = \alpha_{\text{film}}\Delta T + F_{\text{film}}\,(1 - v_{\text{film}})\,/\,(E_{\text{film}}t_{\text{film}}w)\,, \tag{2.14}$$

$$\varepsilon_{\text{sub}} = \alpha_{\text{sub}}\Delta T + F_{\text{film}}\,(1 - v_{\text{sub}})\,/\,(E_{\text{sub}}t_{\text{sub}}w)\,, \tag{2.15}$$

where F_{film} is the thermal mismatch force, w is the width, and t is the thickness. When the strain of film and substrate are identical $\varepsilon_{\text{film}} = \varepsilon_{\text{sub}}$, F_{film} can be obtained by

$$F_{\text{film}} = w\,(\alpha_{\text{sub}}\alpha_{\text{film}})\,\Delta T/\left\{\frac{1 - v_{\text{film}}}{t_{\text{film}}E_{\text{film}}} + \frac{1 - v_{\text{sub}}}{t_{\text{sub}}E_{\text{sub}}}\right\}. \tag{2.16}$$

In the case $\frac{t_{\text{sub}}E_{\text{sub}}}{1} - v_{\text{sub}} \gg \frac{t_{\text{film}}E_{\text{film}}}{1} - v_{\text{film}}$, the thermal stress in the film can be expressed as

$$\sigma_{\text{film}}(T) = \frac{F_{\text{film}}}{t_{\text{film}}w} = \frac{(\alpha_{\text{sub}} - \alpha_{\text{film}})\,\Delta T\;E_{\text{film}}}{1 - v_{\text{film}}}. \tag{2.17}$$

It should however be noted that the signs are consistent with dimensional changes in the film and substrate. In the case of $\alpha_{\text{film}} < \alpha_{\text{sub}}$, thin films prepared at high temperature will be residually compressed when measured at ambient temperature, as the film shrinks less than the substrate.

2.3.3.3 Intrinsic Stress

Thin film deposition techniques are widely used for the fabrication of the components of the SMO gas sensors, including microheater, membrane, sensitive layer, insulation layer, and electrodes. During deposition of these thin components, an intrinsic stress σ_i is generated, which, in combination with the unavoidable thermal stress, $\sigma_{\text{film}}(T)$, forms the internal or the so-called residual stress (σ_{film}). Thus to extract σ_i, the thermal stress given in (2.17) must be subtracted from the measured value of σ_{film}.

Over the past 40 years, the intrinsic stress has prompted scientists to search for a universal explanation of their origins. According to Dorner and Nix [77], the intrinsic stress observed in thin films comes from the following main sources:

Grain Growth Intrinsic stress (compressive stress) in a thin granular film is built up by grain growth, which occurs by normal boundary migration when the initial grain is below the critical value in terms of size [78]. Small islands of depositing film form on the layer below and as they grow, stress builds up at their interfaces.

Grain Coalescence During the thin film growth, cohesion starts to occur between the islands as they grow and the gaps between them are small enough, making the participating islands elastically strained, which causes a tensile stress in the thin film [79]. This process creates grain boundaries in the film.

Annihilation of Excess Vacancies Intrinsic stress (tensile stress) also builds up due to the gaps caused by vacancy annihilation at grain boundaries and inside grains. A gap in matter causes atoms in the crystals to move towards each other in order to fill this void, leading to a planar contraction in the thin film, if it is not deposited on the substrate. Otherwise, the substrate prevents contraction, which leads to the build-up of tensile stress [77].

Annealing of the Film To produce a better crystalline arrangement and an increase the material density, an annealing step is required after the deposition of metals on the substrate [80]. This step results in a shrinkage in the thin film, which eventually leads to build up a tensile stress as the substrate works to prevent the film from shrinking.

Insertion of Excess Atoms One of the ways to add atoms to the film during a film growth process is to incorporate excess atoms into the grain boundaries [81]. This mechanism leads to the development of a compressive stress in thin films [82].

Misfit Stress Intrinsic stress can be built up during the deposition process as the lattice constant of the thin film is different for the lattice constant of the substrate. The crystal lattice of the thin film and the substrate are forced to line up perfectly at the interface, especially during the initial stage of thin film deposition [83]. Note that the misfit stress can also arise between the grain boundaries since the neighboring grains have different crystal orientations.

2.4 Electro-Thermal and Mechanical Simulations

Simulations are performed using FEM to model the thermo-electrical-mechanical behavior of the SMO device during use. The geometry of the calibration device was designed using a commercial Computer Aided Design (CAD) software SolidWorks and imported into the FEM simulation software (COMSOL). One example of a micro-hotplate is shown in Figs. 2.7 and 2.8.

In general, a membrane made of silicon oxide is suspended over a cavity in the silicon substrate. The membrane contains the heater, which is made of tungsten and is connected with two arms to the electrical circuit. In addition to the heater, the membrane contains one heat spreading plate made of tungsten and one layer of silicon nitride, forming the passivation layer of the membrane. The electrodes for gas sensing are deposited above the membrane. Further connections between the heater and the power source can be embedded in the arms. The heater, which is embedded in the membrane, can have various shapes, as discussed in Sect. 2.2.2.1. In general, the heater and the other embedded layers are very thin compared to

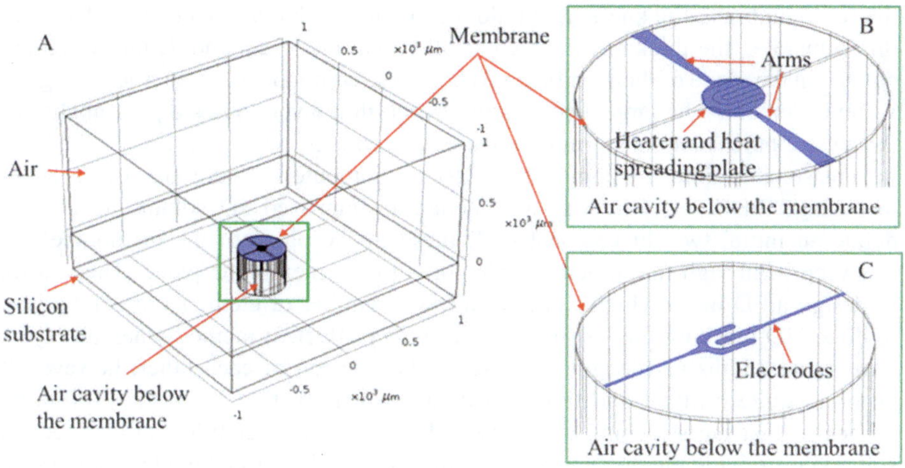

Fig. 2.7 Simplified geometry of the device used in the FEM model. In inset **A**, the hotplate is surrounded by the silicon substrate and the air. In inset **B**, the heater and the heat spreading plate inside the membrane are shown together with the air cavity below it. In inset **C**, the electrodes above the membrane are shown

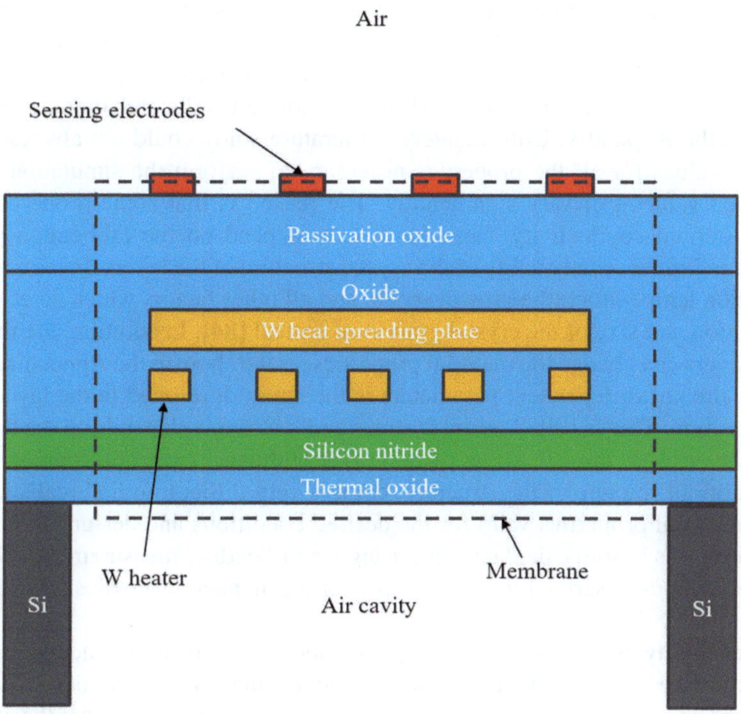

Fig. 2.8 Schematic of the layers composing the membrane of the hotplate

the full membrane thickness. It should also be noted that all layers are effectively thin films, i.e., the thickness is much smaller than the width and depth, which are defined in the plane of the membrane. When these components are meshed using 3D volume elements, the large aspect ratio of each thin layer causes a poor quality of the mesh elements or an enormous number of elements, which cannot be calculated in a reasonable time, on commercial desktop computers. There are several ways to manage this issue. One way to generate an improved 3D volume mesh is to define an initial two-dimensional (2D) mesh in the plane of the membrane and to sweep it through the direction orthogonal to the membrane plane. This allows to design a 3D mesh which can have a lower number of elements in comparison to direct 3D meshing the volume. This is a very effective method when all layers have equal widths and lengths and are stacked on top of each other; however, it becomes quite difficult and time consuming when the thin layers have different shapes so that the sweeping of the mesh becomes not straightforward. A second solution is drawing the thin layers as bi-dimensional objects and models them using the shell elements feature, as described later in this section. This solution simplifies sensibly the geometry of the structure and saves calculation time and effort, but the conditions to use the shell elements have to be carefully reviewed to avoid non-physical results.

Once the device geometry is finalized, the FEM software requires the definition of the material properties. It is paramount to choose the appropriate material parameters that the FEM model needs. In the case of the micro-hotplate simulation, a literature study of the characteristics of the materials composing the micro-hotplate is necessary, which was performed, and the results are summarized in the tables in the Appendix. Unfortunately, a literature study could not always provide realistic values for all the properties necessary to perform the simulations, as the published values can vary significantly. The reason is that some properties, such as the thermal conductivity, can significantly depend on the fabrication process, which can influence the quality of the crystal structure of the layers. In particular, the deposition temperature, the growth speed, and all other factors which determine the distribution and size of the crystal grains are critical [84]. In addition, the thickness of the layers can change the material properties, as we show in the Appendix. One of the key thickness-dependent parameters is the stress developed in the layer during its deposition. This is called intrinsic stress and it strongly depends on the thickness of the layer. The intrinsic stress of thin layers influences the deformation and the eventuality of fracture of the structure, but it is very difficult to find realistic values of the material properties valid for the desired conditions and designs in literature. Therefore, it is best to calculate them using wafer bending measurements [84] or to use the methods described in Sect. 2.3.3.1 for mechanical properties testing of thin films.

A sensitivity analysis of the input parameters is suggested to understand which of them have the highest influence on the results. Based on our experience, important parameters for the electro-thermal behavior of the device are the electrical conductivity of the heater and the thermal conductivity of the materials composing the membrane. In addition to those, eventual metallic paths connecting the heater, or the heater area, to the chip can remove more heat from the active area, thereby increasing the power consumption; therefore, the properties of the metallic paths become of increased importance.

When the device is biased, its temperature is increased by a value dependent on the applied bias. Many material properties, such as the electrical and thermal conductivity, but also the Young's modulus, can vary with the temperature, as discussed in Sect. 2.3 and shown in the Appendix. Most FEM tools allow for the implementation of the dependence of the material properties on the temperature. This allows for more realistic results, at the cost of increased simulation time.

Once the geometry and the material properties of the device are defined, it is necessary to model the physics of the problem. When the device is biased, an electrical current runs through the circuit, reaching the heater. This is a resistive element and some power is dissipated due to Joule heating. The material and the shape of the heater are chosen in order to provide the desired amount of power dissipation while improving the heat distribution at the desired location, as discussed in more detail in Sect. 2.2.2.1. The Joule heating causes an increase in the temperature, heating the sensitive material, which can interact with the gas molecules. The increasing temperature also has a collateral effect. The material forming the membrane and the sensing layer has different coefficients of thermal expansion (CTEs). This causes mechanical deformation when the device is biased and could lead to fracture if the design is not optimized to take this into consideration.

All these considerations lead to the need to model both the electro-thermal and the thermo-mechanical behavior of the device. The procedure is described in the following. At first, the electrical component is modeled as an AC/DC circuit, depending on the bias applied. Then Joule heating is calculated and transferred to the thermal model, as defined by the heat equation. The Joule heat is the energy source term which has to be included in the heat equation. Subsequently, the heat equation is solved and the temperature distribution is transferred to the mechanical model of the simulation. In particular, the temperature increase is used to calculate the deformation caused by the difference of the CTEs. This forces the membrane to bend with respect to the original position. Finally, Hooke's law relates this deformation to the stress distribution inside the device. The possibility of fracture caused by the stress and the methods to limit it are discussed in Sect. 2.5.2.

The first step to model the electrical behavior is to set the terminals which define the difference in potential across the circuit. Based on our experience, it is not necessary to draw and model the entire electric circuit but only the part embedded into the membrane. In relation to Fig. 2.7, this is the heater with the two arms, depicted in inset B. Subsequently, the extremities are set as the terminals of the circuit, where the bias is applied. The circuit can be modeled using simple three dimensional elements if the thickness of the heater has comparable dimensions with the width and the depth. If not, the meshing operations could be difficult and produce elements with a poor quality. Consequently, it is necessary to refine the mesh by increasing the number of elements, which increases the simulation time. To overcome this issue, it is faster to draw each thin layer as a shell element defining its thickness in the shell interface. This may reduce the accuracy of the simulation, but only minimally, considering that the normal component plays almost no role in a thin film's behavior. When a DC bias is applied, the electric behavior is described by

$$\nabla \cdot \vec{j} = Q_{j,V}, \tag{2.18}$$

$$\vec{j} = \sigma \vec{E}, \tag{2.19}$$

and

$$\vec{E} = -\nabla V, \tag{2.20}$$

where $\vec{j}$ is the current density, $Q_{j,V}$ is the current source, σ is the electrical conductivity, $\vec{E}$ is the electric field, and V is the applied potential. When the shell element is used, the equations are modified to use the tangential gradient operator, which removes the normal component from the gradient operation, so that only tangential components remain. This is mathematically expressed for the potential field V as

$$\nabla_{\vec{T}} V = \nabla V - \left(\nabla V \cdot \vec{n}\right) \cdot \vec{n}. \tag{2.21}$$

Thus, the normal gradient operator is

$$\nabla_{\vec{N}} V = \left(\nabla V \cdot \vec{n}\right) \cdot \vec{n}. \tag{2.22}$$

In case a shell element is used, the equations are modified as

$$\nabla_{\vec{T}} \cdot \left(d_s \vec{j}\right) = d_s Q_{j,V} \tag{2.23}$$

and

$$\vec{E} = -\nabla_{\vec{T}} V, \tag{2.24}$$

where the subscript s denotes the layer and d_s is the thickness of the layer.

After the electrical behavior is modeled, it is necessary to describe the thermal problem using the heat equation, which is defined over the entire structure in the stationary case as

$$\nabla \cdot (k \nabla T) = -Q_j, \tag{2.25}$$

where

$$Q_j = \vec{j} \cdot \vec{E}. \tag{2.26}$$

Here, T is the temperature, k is the thermal conductivity, Q_j is the Joule heat, and the heat flux is represented by

$$\vec{q} = -k\nabla T. \tag{2.27}$$

The layers, which are much thinner compared to the total thickness of the membrane, can be drawn and modeled as bi-dimensional elements. The boundary conditions of these thin layers require the definition of their real thickness. If a thin layer is a bad thermal conductor compared to adjacent elements, the tangential heat flux can be neglected and only the heat flux across the layer's thickness is considered. When the thin layer is instead a good thermal conductor, the temperature difference and the heat flux across the layer's thickness are neglected and only the tangential flux is considered. In our simulations, the general case, where both the tangential and normal heat fluxes are considered in the calculation, is assumed:

$$\nabla_{\vec{T}} \cdot \vec{q_s} = Q_s \tag{2.28}$$

$$\vec{q_s} = -k_s \left(\nabla_{\vec{T}} T_s + \nabla_{\vec{N}} T_s \right) \tag{2.29}$$

Here, Q_s is the Joule heat and $\vec{q_s}$ is the heat flux in the layer.

The heat equation as presented before includes the conduction, but not the convection of the heat, which could take place in the air. In the simulated cases, the convection has a negligible impact on the results; therefore, it is not included in the FEM model. A deeper discussion on the sources of heat loss is given in Sect. 2.2.2.2.

Once the mathematical equations describing the electro-thermal behavior of the device are defined, it is necessary to choose appropriate boundary conditions (BCs) to ensure realistic results. Appropriate BCs have to be assigned also to the physical boundaries of the FEM model. In fact, it is not possible to include the entire electronic circuit around the sensor in the FEM model as it would require an unmanageable computational cost. Therefore, it is necessary to exclude it and assign appropriate BCs to the extents of the model.

In our simulations, a good match between the FEM results and the experimental data is obtained when the hotplate is surrounded by a thick layer of silicon. The reason is that, in reality, the hotplate is connected to the heat source through the arms and is isolated by the silicon from other metallic paths. Therefore, there are no other metal interconnections which could cause heat losses to the system.

The external surfaces of the silicon domain are assumed to be thermal insulators, i.e., the heat flux through them is zero. Our simulations show that this condition ensures realistic results of the FEM simulation if the distance between the sensor and the external surfaces of the silicon domain is at least double the membrane diameter of the gas sensor. This implies that the system can dissipate the heat only through the air. This is a reasonable assumption because the heater is in the central area of the membrane and the membrane is thin and wide. Thus, most of the heat is dissipated through the interface between the membrane and the air. A small amount of heat is transferred to the silicon surrounding the membrane; however, this is insignificant compared to the dissipation through the air. If the external surfaces of the silicon domain are far enough from the membrane, the heat is again dissipated through the interface between the silicon and the air. In addition, the FEM simulations show that at such conditions, no relevant heat flow is present in the silicon substrate far from the membrane; therefore, it is assumed that no thermal leak happens through the silicon substrate. The system with such boundary conditions could represent in reality a membrane realized on a silicon die, which is embedded inside an insulating material. However, if the membrane is close to a metal line connected to an external circuit, the metal line would be an additional path for the heat to dissipate. In such situations, the metal line has to be considered in the simulation, together with different boundary conditions for the external surfaces of the system.

Above the sensor, air is represented by a one-millimeter-thick domain (Figs. 2.7 and 2.8). The temperature of the external boundary of the air domain is constant at room temperature. With such boundary conditions, our FEM simulations show that a thinner air domain influences the temperature and power dissipation of the hotplate, making the results unrealistic. Hence, it is necessary to keep the external border of the air domain far enough from the hotplate. In addition, in the FEM model the cavity below the membrane is filled with air, and the temperature of the bottom surface of the air cavity is constant at room temperature as well.

The meshed FEM model of the structure in Fig. 2.7 is composed of 290,250 elements and it was solved in about 30 min using the previously described mathematical equations and relevant BCs. The hotplate is subjected to a different bias and the resistance and the dissipated power of the conductive layers, i.e., the heater and the arms, are calculated and compared with the measured data. Figures 2.9 and 2.10 show a good match between the FEM results and the experimental data of both the resistance and the dissipated power, respectively. The resistance is measured at every applied bias. The experimental value of the dissipated power is calculated by multiplying the measured electrical current with the measured resistance. Figure 2.11 shows the comparison between the temperature of the hotplate and the one obtained from the experimental data. In particular, the latter is calculated based on the following formula

$$R = R_{25}\left(1 + \alpha\,(T - 25) + \beta(T - 25)^2\right), \tag{2.30}$$

where R is the measured resistance for a specific bias, R_{25} is the measured resistance at 25°C, T is the temperature expressed in °C. α and β are empirical parameters, which are determined through an additional experiment. A probe, made of the same metal as the hotplate heater, is located on a chuck. The chuck is heated to different temperatures. Subsequently, the measured values of the resistance and the

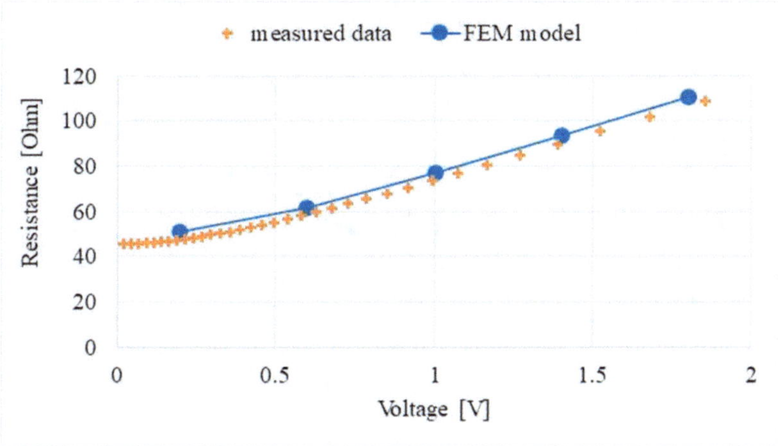

Fig. 2.9 The resistance of the conductive layer of the sensor is measured and simulated for different applied biases

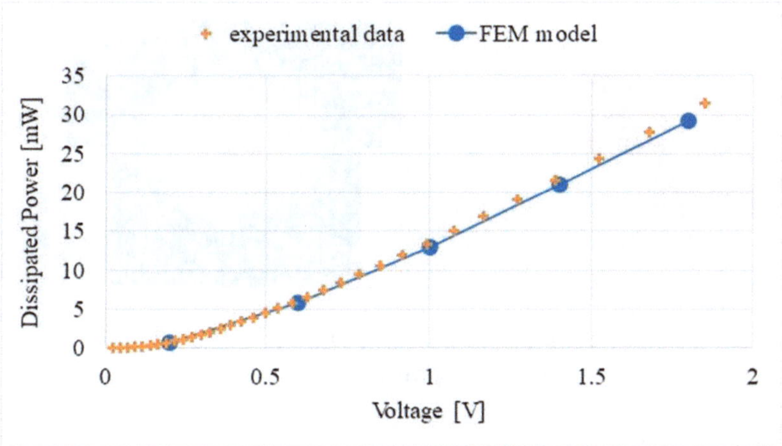

Fig. 2.10 The power dissipated by the sensor is measured and simulated for different applied biases

temperature define a curve, which is fitted with the previous curve, from which the parameter values for α and β are extrapolated.

The difference between the temperature profiles obtained from the FEM model and from the experiments at high bias, shown in Fig. 2.11, is most likely caused by the approximated method to define the parameters α and β.

The FEM model can also accurately reproduce the temperature distribution around the sensor, as shown in the cross section in Fig. 2.12. The clear gradient of the temperature in the air around the sensor suggests that the air takes an active

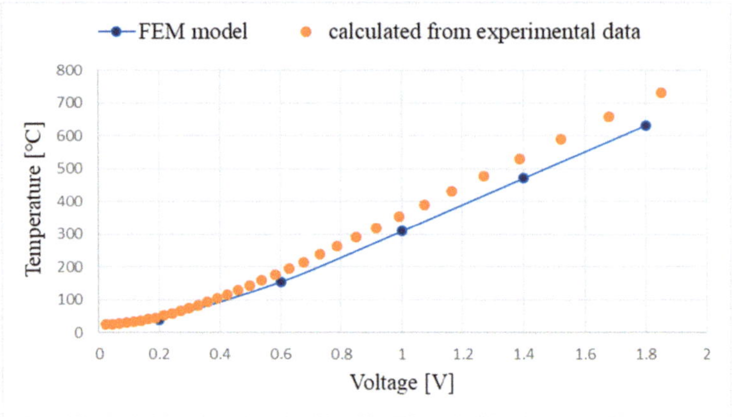

Fig. 2.11 The temperature obtained from the FEM model is compared with the one calculated using the measured resistance

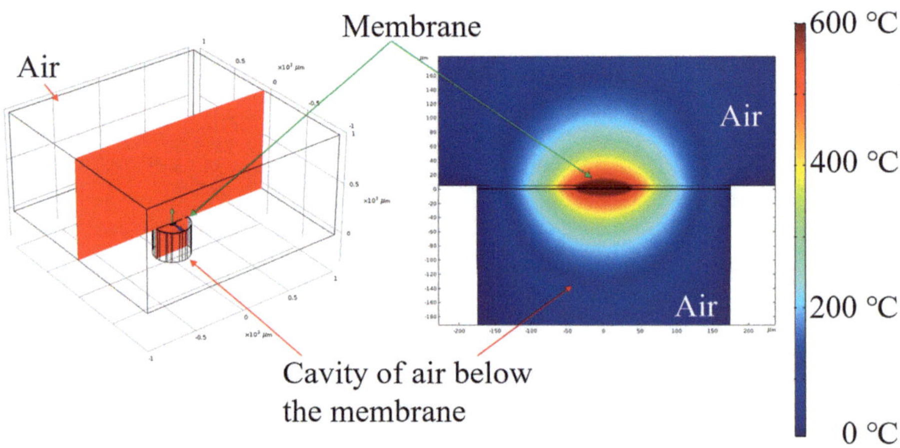

Fig. 2.12 Cross section temperature distribution

role in heat conduction. It is known, from literature, that the thermal conductivity of the dry air increases with temperature [85]. However, if the humidity is taken into consideration, the thermal conductivity decreases at high temperatures. In reality, the physical dynamic of the air around the hotplate is more complicated than pure dry air or humid air. In reality, there would be a dynamic of vapor being created and annihilated by convection and conduction while colder humid air replaces it. Close to the hotplate of the sensor, the dry air is a good approximation of the real conditions because the temperature gradient is really small. Instead, for regions far from the hotplate, this dynamic would not be negligible because the heat conduction worsens as humid air comes close to the sensor. Therefore, the influence of the humidity on the temperature around the hotplate was simulated.

In our simulations, the relative humidity of the air surrounding the hotplate was varied from 0.25 to 0.75 and the variation of the thermal conductivity of the air was calculated. Figures 2.13 and 2.14 show the thermal conductivity and the temperature, respectively, at a distance of 75 μm away from the heater. In particular, their variation with the applied bias and the relative humidity of the air is evaluated. The results show that a variation of about 15% in the thermal conductivity of air causes a variation in temperature by approximately 5°C, leading us to consider the influence of the humidity on the temperature distribution of the air as insignificant and almost negligible. In reality, it could be the case that the velocity of the air and vapor around the hotplate would increase the impact of the humidity.

The mechanical behavior of the device under operating conditions is critical because the long use, coupled with the high temperatures, can potentially lead to high levels of deformation, such as a bending of the membrane and, in the worst case, to cracking and eventual device failure. The membrane is the most critical area of the device because of its bending during device operation. The bending is related to the stress accumulated inside it. In particular, the membrane is a multilayered structure, as depicted in Fig. 2.8, whose layers deform differently during heating because they have different coefficients of thermal expansion. This type of stress is often called thermal stress, but it is not the only component causing the deformation. As previously described in section 2.3.3.3, the intrinsic stress is an additional type of stress, which develops during the deposition of one layer over a substrate.

The estimation of the intrinsic stress in a layer is based on the measurement of the bending of a substrate, in general a silicon wafer, above which the layer is deposited [84]. However, the intrinsic stress, obtained with such a method, can

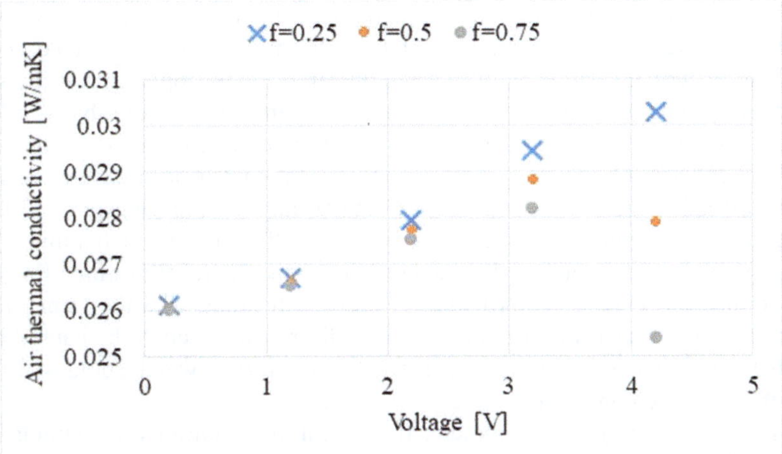

Fig. 2.13 The thermal conductivity of the air as a function of the applied bias for different values of the relative humidity **f** of the air surrounding the device

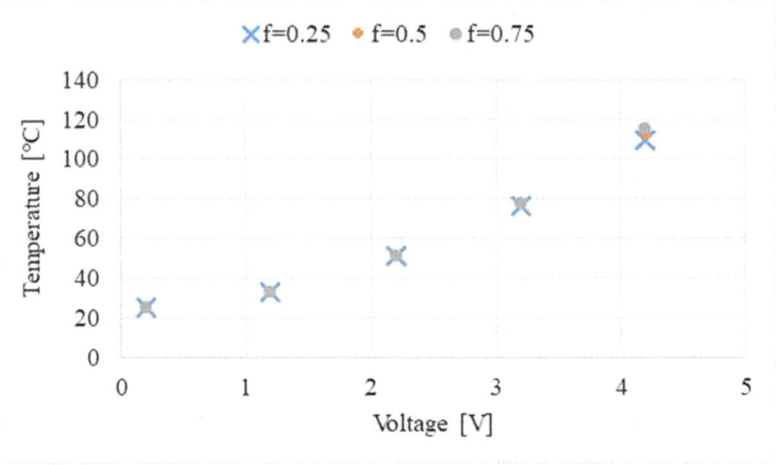

Fig. 2.14 The temperature of the air as a function of the applied bias for different values of the relative humidity **f** of the air surrounding the device

differ significantly from the one in a layer with a different and complex shape and embedded inside an insulating material; this is the case for the heater embedded in silicon oxide (Figs. 2.7 and 2.8). In such conditions, the measurement of the intrinsic stress is difficult and, in order to ensure accurate simulations, it is necessary to calibrate its value in the FEM model with the measured out-of-plane displacement of the unbiased hotplate membrane. The calibration procedure is quite straightforward when the geometry of the structure is simple, but it could become cumbersome and near-impossible when the interfaces among the layers are many and have complex shapes. The intrinsic stresses are added to the FEM model as a boundary condition.

The out-of-plane displacement of a hotplate membrane, similar to the one depicted in Fig. 2.7, along the diameter of the membrane top surface at room temperature was measured with a profilometer prior to applying any bias. The measured profile showed that the membrane bends downward by ~100 nm. Simulations of the mechanical behavior of the membrane prior to any bias being applied were performed using measured values of the intrinsic stress for the silicon oxide 280 MPa, the silicon nitride −260 MPa, and for the tungsten 1330 MPa. These values are obtained by measuring wafer bending above which a thin layer of tungsten, oxide, or silicon nitride was deposited [85]. The results of the FEM model prior to any bias (Fig. 2.15) provide a membrane bending with the same order of magnitude of the measured profile. Eventual discrepancies could be caused by the fact that there is some variation in the real intrinsic stresses, which could be different from the calibrated ones used in the simulation.

Once the calibration is complete, the operating conditions are simulated by applying a bias to the device. The stress and the deformation of the membrane are subsequently calculated with the following hypothesis: All materials composing the structure are assumed to behave linearly elastic. Eventual plastic deformation or

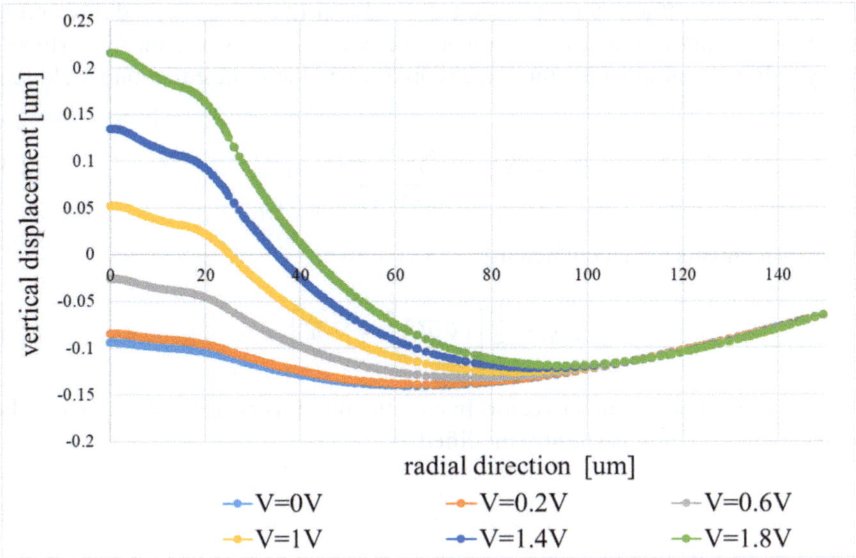

Fig. 2.15 FEM model results of the out-of-plane displacement of the top surface of the membrane along its radius of the membrane. The hotplate is biased and the deformation is caused by the intrinsic and the thermal stress caused by the temperature increase due to the Joule effect. Zero displacement indicates the un-deformed state of the hotplate

creep phenomena of the heater and the other metallic layers inside the membrane are negligible when they are made of tungsten. In fact, we performed some experiments of temperature-driven stress relaxation of polycrystalline tungsten films, deposited on a silicon wafer, close to the tungsten deposition temperature. The results show that the creep and the consequent stress relaxation are negligible in the first 5 h of experimental observation. Afterwards, these two phenomena are no longer negligible. When the heater of the micro-hotplate is made of polycrystalline silicon, no plastic deformation or creep behavior are expected with normal operating conditions. Therefore, in our devices, both creep and plasticity are neglected on the basis of the experimental observations and the linear elasticity theory is used in the FEM model, as described below.

The Joule heating causes the deformation of the structure because it is composed of materials having different coefficients of thermal expansions. In particular the elastic strain of a layer, with the CTE α, caused by heating ΔT is

$$\varepsilon_{th} = \alpha \cdot \Delta T \tag{2.31}$$

Assuming that the materials behave linearly elastic, they are described by Hooke's law, which relates the elastic strain $\varepsilon_{\mathrm{el}}$ and the stress S as

$$S = S_0 + C : \varepsilon_{th}, \tag{2.32}$$

where C is the stiffness tensor, which depends on the Young's modulus and the Poisson ratio only in the case of an isotropic body and S_0 is the intrinsic stress of the layer. In a notation where the indices are shown, the same equation would read

$$\sigma_{ij} = S_{0,ij} + \sum_{k=1}^{3} \sum_{l=1}^{3} c_{ijkl} \varepsilon_{kl}. \tag{2.33}$$

The displacements are calculated as

$$\varepsilon_{th} = \frac{1}{2} \left[\left(\nabla \vec{u} \right)^T + \nabla \vec{u} \right], \tag{2.34}$$

where $\vec{u}$ is the displacement vector. In case the thin layers are modeled using shell elements, the equation is slightly modified as

$$\varepsilon_{th} = \frac{1}{2} \left(\nabla_t \vec{u} \right). \tag{2.35}$$

One boundary condition of the mechanical model is that the bottom surface of the silicon domain (Fig. 2.7) is physically fixed, i.e., the displacements in the three orthogonal directions are zero. In addition, the intrinsic stress of each layer could be neglected, as a result of the calibration. If the device is not directly subjected to any externally applied loads, the intrinsic and thermal stresses are the only stress types considered in the FEM model.

Considering the hotplate membrane shown in Figs. 2.7 and 2.8, thin layers (i.e., the heater, the heat spreading plate, and silicon nitride) are simulated using shell elements in our FEM model. Using the equations of the mechanical behavior, coupled with the electro-thermal model described previously, the simulation provides the profile of the membrane displacement on its out-of-plane direction for different applied biases (see Fig. 2.15).

In general, the membrane bends downward at 0 V as a result of the relaxation of the intrinsic stress. The center of the membrane, which contains the W heater and heat spreading plate, has an upward displacement of about 100 nm.

Increasing the applied bias up to 1.8 V, the displacement of the central area increases from −100 nm to 200 nm as a result of the Joule effect. Since the heating concerns more the central area of the membrane, as it contains the heater, the sections of the membrane far from the center are less displaced.

The results show another interesting effect of Joule heating. At room temperature, i.e., when no bias is applied, the maximum principal stress along the heater and the arms is approximately equal to the intrinsic one, inserted as a boundary condition. Increasing the bias, i.e., the temperature, the stress in the W heater decreases. This happens because the W heater has a higher CTE with respect to the oxide, thus the heater expansion is limited by the oxide and a compressive stress is created and it compensates the initial high tensile stress. The consequence is that the total stress is gradually reduced by increasing the bias, i.e., increasing the temperature.

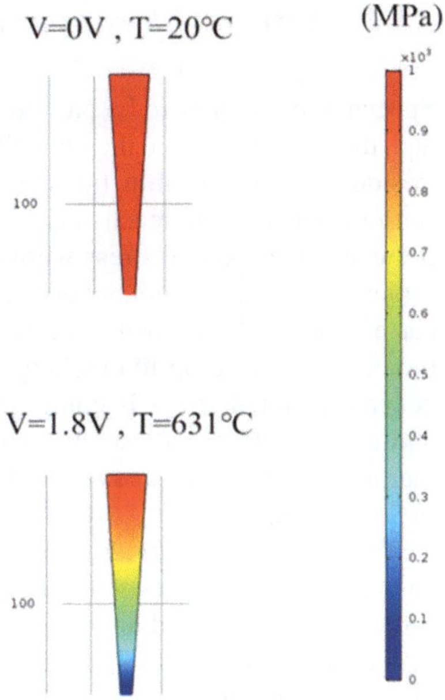

Fig. 2.16 FEM model results of the maximum principal stress along the arms of the heater

Despite this, Fig. 2.16 shows that the stress in the arms remains high at high bias, instead of decreasing. This happens because the Joule effect is weaker in the arms than in the heater, thus no compressive stress is created during the heating. During the operating lifetime of the hotplate, the device is continously switched on and off, but this does not reduce the stress in the arms; therefore, the arms could be subjected to damage deriving from cycling fatigue. These phenomena deserve more experimental investigations in order to be proven and a more sophisticated mechanical model should be developed.

2.5 Fracture Modeling for Mechanical Robustness of Sensor Components

Thermo-electrical characterization and modeling of components for microelectronic applications, such as gas sensor hotplates, are very well established in the research activities of the microelectronic industry. With the progress towards new materials and complex designs, the need for mechanical stability of the components is increasing. The reliability of thin film stacks can be improved if the material properties and internal mechanical loading conditions are properly utilized. Initial defects, due to manufacturing, cannot be excluded and for that reason a device should be designed in order to be failure tolerant. The two major questions which have to be addressed are: Is there a fracture parameter that can properly describe crack propagation in such systems and how can we determine the crack driving force in thin film stacks.

2.5.1 Residual Stresses in Thin Films for Gas Sensor Hotplates

Special attention has to be paid to the role of residual stresses on failure processes and their influence on the reliability and functionality of components [84, 86]. Residual stresses in thin films used in microelectronic components, such as gas sensor hotplates, can reach very high values, i.e., in the order of GPa [87–90]. Even if the average residual stress within the layer or in layer stacks does not seem to be critical, strong stress gradients can appear [89, 90]. The local residual stress values can even exceed the tensile, compressive or shear strength and interface strengths of the materials, leading to cracking and delamination within the component [91]. In general, residual stress is composed of intrinsic stress, formed during deposition, and extrinsic (thermal) stress stemming from the mismatch in the coefficient of thermal expansion between the individual layers and the substrate [84], as discussed in Sect. 2.3.3.

Two groups of methods, which enable local and depth resolved residual stress characterization have evolved over the last years. Within the first group we find X-ray methods performed in grazing incidence or in transmission. Recent development of the latter uses focused synchrotron X-ray beams [92–95]. The second group includes methods where a specific region with characteristic features is imaged before and after local removal of a stressed material using, for example, ion beam milling. The form of the processed area can have numerous geometrical shapes [70, 71]. This results in a local relief of residual stress and a consequential deformation of the sample, from which the strains and stresses can be reconstructed analytically or using finite element modeling, when the stiffness is known [72, 73, 96, 97]. In contrast to the X-ray methods, these methods work without the need for the presence of crystalline phases and can also be applied to amorphous materials.

In many applications, small amounts of residual stresses are tolerable and do not have a significant influence on the performance or reliability of a device. On the other hand, residual stresses have been shown to have decisive impact on the crack driving force in a material when high enough. While tensile residual stresses, in general, increase the crack driving force, compressive residual stresses are desired because of their crack stopping capabilities. In the case of compressive residual stress, however, the risk for delamination increases [84]. Thus, in order to increase the reliability of microelectronic components in general, and gas sensor hotplates in particular, it is of uttermost importance to work on increasing the failure resistance of the components on the one hand (e.g., layer design for fracture toughness, interface strength) and the reduction of failure driving forces (e.g., CTE mismatches and intrinsic residual stress engineering) on the other.

2.5.2 *Fracture Mechanics for Thin Films in Gas Sensor Hotplates*

The topic of fracture mechanics is one of the most active fields of research and was already extensively investigated in the past. The basics of fracture mechanics are collected in different books and publications [98, 99]. The book by Kolednik [100] gives a very good overview of fracture mechanics. In the course of this section, the most relevant concepts and terms of fracture mechanics are introduced, along with how fracture mechanics applies to thin film structures for sensing applications.

2.5.2.1 Definition of the Crack Driving Force

Imagine a homogeneous body with an initial crack under loading, as shown in Fig. 2.17. Generally, this crack with an initial length a_0 will extend if the crack driving force (CDF) D equals or exceeds the crack growth resistance R [100, 101]. The CDF D describes a loading parameter deduced from the strain energy and the work from external loading. The crack growth resistance R will prevent crack propagation. It depends on the fracture toughness of the material, on the crack extension Δa (because of the increasing size of the plastic zone) and on the geometry (because of the constraint effect), shown in Fig. 2.17. If $D < R$, the crack will remain stationary. However, if $D = R$, the crack will exhibit stable growth, and if $D > R$, crack growth becomes unstable.

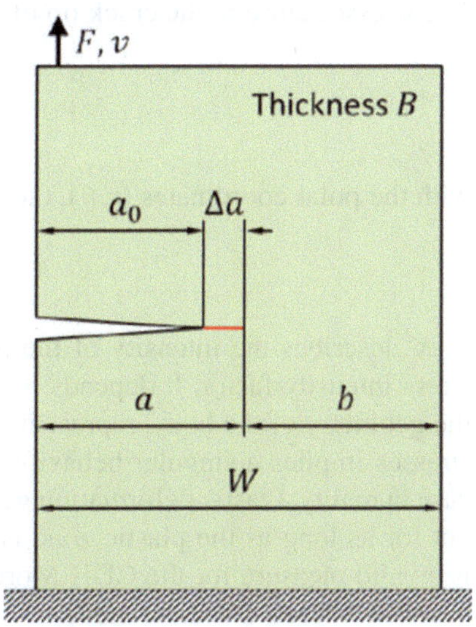

Fig. 2.17 A homogeneous body under Mode I loading. The initial crack length is a_0 and the current crack length is denoted by α and together with the ligament length b it gives the specimen thickness W

2.5.2.2 Regimes of Fracture Mechanics

In a simple distinction, fracture mechanics can be divided into linear elastic fracture mechanics (LEFM) and elastic–plastic or nonlinear fracture mechanics (E–PFM, NLFM) [100]. The concept of LEFM applies if the plastic deformation during crack growth is zero or limited. In this case, the size of the plastic zone r_{pl} is very small compared to the crack length a and the ligament length b, $r_{\mathrm{pl}} \ll a, b$, in which case small-scale yielding (ssy) conditions apply, shown in Fig. 2.18a.

E–PFM, on the other hand, is used if the body experiences significant plastic deformation. In a homogeneous body with a long crack, the prevailing conditions are either large-scale yielding (lsy) or general yielding (gy), as shown in Fig. 2.18b. Under the lsy conditions, the onset of plastic deformation is at the back face of the body, while gy applies when the crack tip plastic zone and back face plasticity merge into one plastic zone. In a special case where the crack is very short, as shown in Fig. 2.18c, E–PFM has to be applied, since also a small plastic zone does not comply with the conditions from Fig. 2.18a. In layered composites, such as thin metallic film stacks shown in Fig. 2.18c, the materials deform in a different way and often the lsy condition cannot be defined by the onset of back face plasticity. In this case, lsy applies if the size of the plastic zone r_{pl} compares to or is higher than the crack length a.

2.5.2.3 Linear Elastic Fracture Mechanics

The stresses close to the crack tip of a linear elastic body are calculated by [102]

$$\sigma_{ij} = \frac{K}{\sqrt{2\pi r}} f_{ij}(\theta), \tag{2.36}$$

with the polar coordinates (r, θ), the angular stress functions $f_{ij}(\theta)$ and

$$K = \sigma_{\mathrm{ap}} \sqrt{\pi a}\, f_K. \tag{2.37}$$

K describes the intensity of the near-tip stress field and is therefore called the stress intensity factor. It depends on the applied stress σ_{ap}, the crack length a, and the geometry of the body, represented by f_K. However, the equation for the near-tip stresses implies a singular behavior of the stress field, as $r \to 0$, which is not the case in reality. Plastic deformation generated at the crack tip keeps the stresses finite, but for as long as the plastic zone is reasonably small the stress intensity factor K is a valid measure for the CDF. Moreover, as the stresses or strains at the crack tip equal or exceed a critical value $K = K_{\mathrm{C}}$, fracture occurs. The material parameter K_{C} is called fracture toughness and describes the resistance of the material against fracture. Connecting this stress and strain deformation concept to the energy-based perspective of a CDF D, one finds the relation

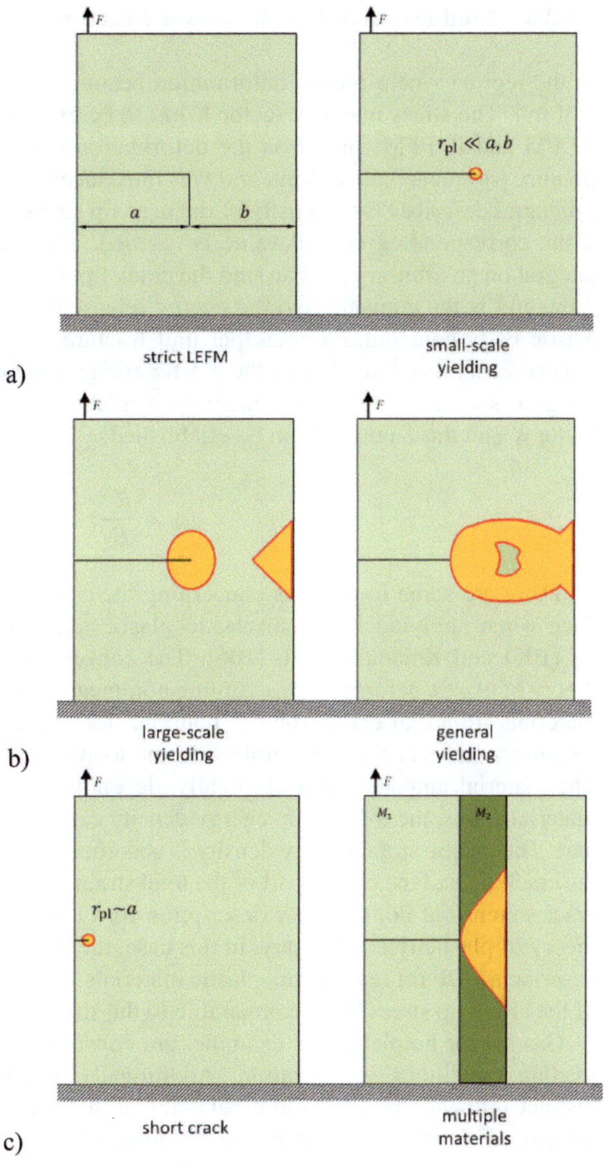

Fig. 2.18 Different regimes of fracture mechanics. (**a**) Linear elastic fracture mechanics. (**b**) Elastic–plastic fracture mechanics. (**c**) Special cases of E–PFM and fracture mechanics in composites

$$G = \frac{K^2}{E_b}, \tag{2.38}$$

where G is the LEFM equivalent to the generalized definition of the CDF D, called elastic energy release rate [103]. $E_b = E/(1 - \nu^2)$ is the biaxial Young's modulus for plane strain conditions, where ν is the Poisson's ratio.

2.5.2.4 Nonlinear and Elastic–Plastic Fracture Mechanics

In the regime where plastic deformation becomes relevant, the concept of LEFM will fail. The stress intensity factor K has to be extended into a parameter valid for NLFM and E–PFM. Based on the deformation theory of plasticity from 1968, a fracture parameter called J-integral was introduced by Rice [104]. Similar to K, the J-integral describes the intensity of the near-tip stress field and crack growth occurs if the corresponding critical value is reached. The J-integral describes a contour integral on an arbitrary path around the crack tip. It was shown by Rice [99] that the J-integral is the equivalent to the energy release rate of work done on a nonlinear elastic body containing a crack per unit fracture area. This can be understood as a more generalized version of the elastic energy release rate G. For LEFM, the J-integral is identical to G; consequently, a relationship between the stress intensity factor K and the J-integral can be established:

$$J_{\mathrm{el}} = \frac{K^2}{E_{\mathrm{b}}} \tag{2.39}$$

There are some limitations concerning the conventional J-integral proposed by Rice when applying it to real elastic–plastic materials, pointed out by Simha et al. [105] and Kolednik et al. [106]. The conventional J-integral is based on the theory of plastic deformation, assuming nonlinear elastic behavior for elastic–plastic materials. In a real elastic–plastic material, for example, the unloading portion of the stress–strain curve is not following the loading curve, as is the case in NLFM. The material unloads linear elastically, depending on the elastic properties of the material. Thus, the total strain energy density can be split into an elastic and plastic part. The plastic strain energy density is absorbed for plastic deformation in front of the crack tip and the elastic part of the total strain energy density is available to drive crack extension. For a proper description for elastic–plastic materials incremental theory of plasticity is necessary. In this case, the NLFM J-integral does not directly describe a CDF for real elastic–plastic materials. It acts as a measure of the intensity of the crack tip stress field, comparable to the stress intensity factor K in LEFM.

Gas sensor hotplates, for example, are constantly subjected to nonproportional loading conditions in operation. Additionally, several inhomogeneities have a distinct influence on the failure behavior in the thin film systems, and render the determination of critical fracture parameters and critical loading conditions difficult.

2.5.2.5 Configurational Force Concept and the J-Integral

Material imperfections, smooth or discontinuous material variations, as well as residual stresses, are known to have an influence on the CDF in a body [106–114]. A very convenient way to describe the impact of material inhomogeneities on the fracture behavior is provided by the configurational force concept, which is based on the ideas of Eshelby [115], and adopted by Gurtin [116] and Maugin [117].

This concept makes it theoretically possible to describe any kind of inhomogeneous behavior in a body. Examples for inhomogeneities or imperfections in a material are cracks, voids, dislocations, and sharp interfaces.

In the following, the basic idea of the configurational force concept is given. For comprehensive derivations of the following equations and definitions, the reader is referred to [106, 108, 115–117].

From a thermodynamical point of view, a configurational force (CF) tries to push a defect into a configuration where the total potential energy of the system has its minimum. In general, a CF vector **f** can be calculated at each material point in a body. It is defined as the divergence of the configurational stress tensor **C** and becomes non-zero only at positions of a defect in the body

$$\mathbf{f} = -\nabla \cdot \mathbf{C} = -\nabla \cdot \left(\phi \mathbf{I} - \mathbf{F}^{\mathrm{T}} \mathbf{S}\right), \tag{2.40}$$

where ϕ is the Helmholtz-free energy or strain energy density, **I** represents the identity tensor, and $\mathbf{F}^{\mathrm{T}}$ and **S** are the transposed deformation gradient and the first Piola–Kirchhoff stress tensor, respectively [106, 108].

If, for example, a two-dimensional homogeneous elastic body with a sharp crack is considered, the crack tip exhibits a CF vector $\mathbf{f}_{\mathrm{tip}}$, determined by

$$\mathbf{f}_{\mathrm{tip}} = -\lim_{r \to 0} \int_{\Gamma_{\mathrm{r}}} \left(\phi \mathbf{I} - \mathbf{F}^{\mathrm{T}} \mathbf{S}\right) \mathbf{m} dl. \tag{2.41}$$

Here, Γ_r is the contour around the crack tip at the distance r away from the crack tip and **m** denotes the unit normal vector to the contour. The corresponding energy dissipated per unit crack extension is a projection of $\mathbf{f}_{\mathrm{tip}}$ in the direction of crack extension **e** and gives

$$J_{\mathrm{tip}} = -\mathbf{e} \cdot \mathbf{f}_{\mathrm{tip}}, \tag{2.42}$$

where J_{tip} is the scalar near-tip J-integral and it represents the CDF. If the body is externally loaded the CDF is equal to $J_{\mathrm{tip}} = J_{\mathrm{far}}$, where J_{far} is the far-field J-integral, which can be understood as the driving force induced by the external load in the body.

2.5.2.6 Material Inhomogeneity Term

Suppose we have a layer stack where two materials M1 and M2, as shown in Fig. 2.19, are separated by a sharp interface (IF) Σ. The material properties, such as the Young's modulus, experience a jump and, therefore, CFs $\mathbf{f}_{\Sigma}$ are induced at the sharp IF, shown in Fig. 2.19, given by [118]:

$$\mathbf{f}_{\Sigma} = -\left([\![\phi]\!] \, \mathbf{I} - [\![\mathbf{F}^{\mathrm{T}}]\!] \, \langle \mathbf{S} \rangle \right) \mathbf{n} \tag{2.43}$$

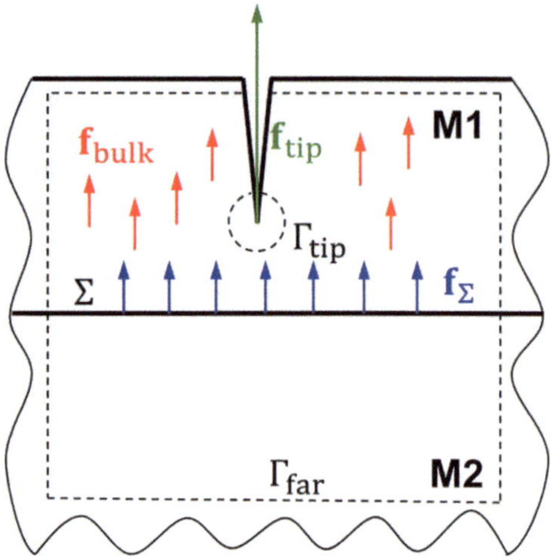

Fig. 2.19 An example of a two-material body with a crack. Configurational forces (CFs) $\mathbf{f}_{\Sigma}$ at the sharp interface $\boldsymbol{\Sigma}$ and bulk CFs $\mathbf{f}_{\text{bulk}}$ inside the material layer **M1** are induced, respectively, because of material property variations. The material layer **M2** does not exhibit bulk CFs, as the material properties are constant

In (2.43), $\mathbf{n}$ denotes the unit normal vector to the IF. A jump of a quantity at the IF is designated by $[\![q]\!] = (q^+ - q^-)$ and $\langle q \rangle = (q^+ + q^-)/2$ represents the average of q across the IF, where q^+ and q^- are the limiting values of q on each side of the IF.

A continuous variation of material properties can also occur in a body. This circumstance induces additional CFs $\mathbf{f}_{\text{bulk}}$ inside of the material, as is the case in material M1 in Fig. 2.19. The bulk CFs are given by the relation [118]

$$\mathbf{f}_{\text{bulk}} = -\nabla_{\mathbf{x}}\phi\left(\mathbf{F}, \mathbf{x}\right). \tag{2.44}$$

The strain energy density ϕ in (2.44) depends on the reference coordinate $\mathbf{x}$ where $\mathbf{x} = \mathbf{x}(x, y, z)$ and $\nabla_{\mathbf{x}}$ denotes the explicit gradient in the reference frame. If the material properties exhibit only a variation in the y-direction, the sole contribution from $\mathbf{f}_{\text{bulk}}$ is $\mathbf{f}_{\text{bulk},y}$.

The CFs induced at the IF $\mathbf{f}_{\Sigma}$ and in the bulk $\mathbf{f}_{\text{bulk}}$ strongly affect the CF at the crack tip and, therefore, the magnitude of the CDF. Two terms can be introduced to quantify those effects [118]:

- **Interface inhomogeneity term** C^{IF}, being the sum of all CFs $\mathbf{f}_{\Sigma}$ at the IF

$$C^{\text{IF}} = -\mathbf{e} \cdot \int_{\Sigma} \mathbf{f}_{\Sigma} dl. \tag{2.45}$$

- **Bulk term** C^{B}, which is the sum of all CFs $\mathbf{f}_{\text{bulk}}$ inside of a material layer

$$C^{\text{B}} = -\mathbf{e} \cdot \int_{\text{M}} \mathbf{f}_{\text{bulk}} dA. \tag{2.46}$$

The sum of $C^{\text{IF}}$and C^{B} results in the material inhomogeneity term C_{inh}:

$$C_{\mathrm{inh}} = C^{\mathrm{IF}} + C^{\mathrm{B}} \tag{2.47}$$

The balance of CFs is therefore fulfilled if the material inhomogeneity term C_{inh} is introduced, which renders the CDF to be [108, 118]

$$J_{\mathrm{tip}} = J_{\mathrm{far}} + C_{\mathrm{inh}}, \tag{2.48}$$

where J_{far} is calculated around the far-field contour Γ_{far} as shown in Fig. 2.19. In this context, C_{inh} corresponds to the CDF caused by the material inhomogeneities at a given loading J_{far} in the body. An anti-shielding effect is described by a positive shielding by a negative value of C_{inh}.

In the same way, a material inhomogeneity effect is induced if, instead of the material properties, the eigenstrains or the resulting eigenstresses that exhibit a jump or smooth variation [114, 118].

2.5.3 *Crack Driving Force in Thin Film Stacks for Micro-Hotplates*

In the following, some results for an application of the introduced fracture mechanics concept are presented. The CDF is determined numerically for a tri-layer stack, where a 0.3 μm thin tungsten (W) heater-layer is sandwiched between two 1.5 μm thick silicon dioxide (SiO_2) layers, and positioned on a 5.0 μm thick silicon (Si) substrate, as shown in Fig. 2.20. After performing fracture simulations on a micro-cantilever beam [119], the CDF can be calculated by applying Eqs. (2.43)–(2.48).

Linear elastic and elastic–plastic simulations are performed where tensile residual stresses for SiO_2 and W are considered as internal loading, denoted by LERS and EPRS, respectively, depicted in Fig. 2.21a.

The material properties used for modeling are given in Table 2.2. Note that after a crack is introduced into the system, the residual stresses have to redistribute in order to reach an equilibrium state, causing a finite CDF value. The influence of the Young's modulus and yield strength inhomogeneity on the CDF is demonstrated for an increasing crack length. For a better understanding, it has to be mentioned that in a homogeneous linear elastic material the CDF, in form of the J-integral, would increase linearly for an increasing crack length while under the same load. Additionally, the upper and lower dashed lines represent critical J-integral values J_{C} for W and SiO_2, respectively, calculated from literature fracture toughness values [120, 121].

As the crack transitions from the compliant SiO_2 layer to the stiffer W layer, see Table 2.2, the CDF is reduced by the shielding effect from the Young's modulus inhomogeneity at the first IF. When elastic–plastic properties are considered the yield strength inhomogeneity causes a slight increase of J_{tip} very close to the first IF compared to the LERS calculation, which was expected as SiO_2 behaves

Fig. 2.20 Example for a thin film stack for numerical determination of the crack driving force

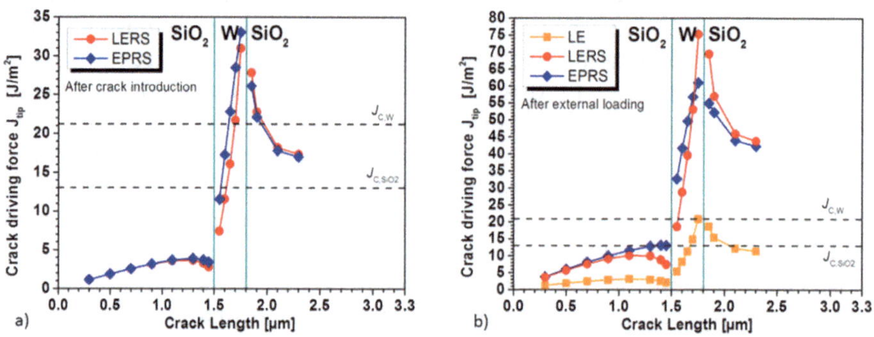

Fig. 2.21 Crack driving force J_{tip} variation in the SiO_2/W/SiO_2 thin film stack. The critical J-integral values for both materials are represented by dashed lines. (**a**) Crack driving force caused by residual stress redistribution after introducing a crack into the system. (**b**) Crack driving force for the different model considerations after external loading is applied

Table 2.2 Material properties for the SiO_2, W, and Si layers in the thin film stack

Material	E [GPa]	ν	σ_y [GPa]	n [−]	α [%]	σ_{RS} [GPa]
SiO_2	73	0.25	–	–	–	0.28
W	411	0.28	1.91	13.3	0.2	1.6
Si	170	0.28	–	–	–	–

Young's modulus E, Poisson's ration ν, yield strength σ_y, hardening exponent n, and the yield offset α. The constant value of the residual stresses σ_{RS} in the SiO_2 and W layers infinitely hard. SiO_2 is considered to behave linear elastically and W is the material for which a stress–strain behavior following a Ramberg-Osgood material model is introduced [122]. The mechanical parameters used in the fracture simulations are given in Table 2.2, where it can be noted that the Young's modulus values are within the range discussed in Sect. 2.3.3 and given in Appendix Table 2.5. However, the internal loading does not induce component failure, as the CDF never exceeds the critical J-integral J_{C,SiO_2} in the SiO_2 layer.

The CDF is strongly increased when the crack is elongated into the W layer and towards the second IF. The anti-shielding effect from the yield strength inhomogeneity effect at the first IF is much more evident here, as the CDF from the EPRS calculation is significantly higher. While the crack is extended towards the second IF, the slope of the LERS/EPRS curve increases/decreases. From theory we know that $J_{tip} \to \infty$ if the crack tip is situated directly at an IF with an anti-shielding effect. In the present case, the second IF has an anti-shielding effect from the Young's modulus and a shielding effect from the yield strength inhomogeneity; J_{tip} increases faster towards the second IF when the elastic–plastic material properties are omitted. Most importantly, the CDF reaches and surpasses the $J_{C,W}$ value, which would result in unstable crack propagation and critical failure of the entire structure.

For a clearer picture regarding the shielding/anti-shielding effect at the second IF, the system was externally loaded in addition to the internal loading. Displacement controlled loading was used in this test case, as it resembles the loading conditions from micro-cantilever beam fracture experiments [119]. The CDF close to the second IF is notably lower when calculated from the EPRS model, as shown in Fig. 2.21b. Interestingly, in a linear elastic calculation where the residual stresses are disregarded (LE in Fig. 2.21b) the CDF is significantly lower compared to the results from the simulations with internal loading. It is clear that the tensile residual stresses and the stress field in front of the crack tip from external loading add up to a higher value of the CDF. But especially in the W layer, the CDF calculated from the LE case does not even reach $J_{C,W}$ and would therefore resist failure if the W layer would not suffer from such high tensile residual stresses (Table 2.2).

It seems that the high tensile residual stress state has the largest impact on the CDF in the material system. Even in the case without external loading, as shown in Fig. 2.21a, the resulting CDF is too high for the W layer to facilitate failure resistance.

We can conclude that it is crucial to consider the residual stress state as well as the proper material behavior in such thin film stacks. Although the flow behavior is not the deciding factor for critical failure in this particular system, in a different scenario it will, together with the residual stresses, definitely play an important role when it comes to the optimal design of material stacks.

2.6 Conclusions

This chapter offers an overview of the most important aspects for the modeling of semiconductor metal oxide gas sensor hotplates. After a historical overview, describing the need and development of the gas sensor, Section 2.1 describes the importance and the application of gas sensors in different markets. Subsequently, the importance of the Finite Element Method to model the mechanical stability of the hotplate, the temperature distribution over the active area, and the power consumption of the sensor is described.

Section 2.2 describes how specific geometries of the hotplate can maximize its performance by limiting the heat losses. Different concepts for the hotplate designs are analyzed, along with a list of most frequently used materials to realize the components of the sensor. Section 2.3 puts in evidence the material properties, which influence mostly the heat conduction and convection, and therefore strongly influence the electro-thermal performance and, consequently, play a significant role in the ability to perform accurate simulations of the device. A description of the main techniques used to characterize these properties concludes this section.

Section 2.4 describes our simulations of the hotplate during operation. The electro-thermal model matches very well the measured resistance variation and power dissipation of the hotplate, showing the validity of the assumptions made and the hypothesis of our FEM model. Deviations of the simulation from the characterized temperature profile are observed at high temperatures. A possible reason is the application of several approximations used in order to extract the temperature from the experimental resistance data.

The end of Section 2.4 and Section 2.5 deal with the mechanical behavior of the hotplate membrane. The limited knowledge for the values of the residual stress of the layers forming the membrane did not prevent the FEM model to predict a realistic membrane deformation, in agreement with the experimental observations. The tendency of the crack failure of the multilayer forming the membrane has been analyzed using the configurational force concept. This increases the general understanding of how the interfaces between silicon dioxide and tungsten could shield from crack propagation. The calculations also show that the high level of tensile stress in the tungsten layer has a strong influence on the crack driving force. This means that knowing the exact value of the residual stress in the layers forming the hotplate are of critical importance to understanding the device's mechanical reliability.

Appendix: Thermo-Mechanical Properties of W, SiO_2, and Si_3N_4

Table 2.3 Thermal conductivities of selected materials

	Property (W/m·K) at various temperatures (K)						
Composition	300	400	500	600	800	1000	1200
W (bulk)	174	159	–	137	125	118	113
W (240 nm)	50–60	–	–	–	–	–	–
W (193 nm)	50	–	–	–	–	–	–
SiO_2 (bulk)	1.39	1.51	–	1.75	2.17	2.87	4
SiO_2 (1.74 μm)	0.68	0.57	0.48	–	–	–	–
SiO_2 (3.04 μm)	0.98	0.8	0.68	–	–	–	–
Si_3N_4 (bulk)	16	13.9	–	11.3	9.88	8.76	8
Si_3N_4 (0.6 μm)	12	12.7	–	–	–	–	–
Si_3N_4 (1.4 μm)	8.1	8.5	–	–	–	–	–

Numerical values of the thermal conductivity of selected materials are taken from [123–128]

Table 2.4 Thermal capacitance of selected materials

	Property (J/kg·K) at various temperatures (K)						
Composition	300	400	500	600	800	1000	1300
W (bulk)	–	140	139	141	144	–	145
W (bulk)	160	245	255	–	–	–	–
W (bulk)	132	–	–	–	–	–	–
SiO_2 (bulk)	725	–	–	–	–	–	–
SiO_2 (bulk)	1000	–	–	–	–	–	–
SiO_2 (glass)	–	–	~960	~1010	~1040	–	–
Si_3N_4 (bulk)	700	–	–	–	–	–	–
Si_3N_4 (1.5 μm)	500	720	~900	–	–	–	–
Si_3N_4 (bulk)	800	–	–	–	–	–	–

Numerical values of the thermal capacitance of selected materials are taken from [126, 128–135]

Table 2.5 Mechanical properties of selected materials

	Young's modulus (GPa) at various temperatures (K)				Tensile strength (GPa)
Composition	300	600	800	1100	Property at 300 K
W (bulk)	395	–	378	349	1.51
W (bulk)	391	$E = 391 - 1.3e^{-2}T - 1.4e^{-5}T^2$			0.88 (pure)
W (sheet)	397	–	–	–	1.39 (1.57 mm)
SiO_2 (bulk)	78	71	78	–	0.2
SiO_2 (1 μm)	60.1	–	–	–	0.6 in air
SiO_2 (bulk)	59	–	–	–	1.2 in vacuum
Si_3N_4 (bulk)	300	280	280	270	0.345–0.483
Si_3N_4 (bulk)	320	$E = 320.4 - 0.0151Te^{-445/T}$			2.4 (thin film)
Si_3N_4 (800 nm)	280	–	–	–	0.39 (1 μm)

Numerical values of the Young's modulus and the tensile strength of selected materials are taken from [136–152]

Acknowledgements Financial support by the Austrian Federal Government (in particular from Bundesministerium für Verkehr, Innovation und Technologie and Bundesministerium für Wissenschaft, Forschung und Wirtschaft) represented by Österreichische Forschungsförderungsgesellschaft mbH and the Styrian and the Tyrolean Provincial Government, represented by Steirische Wirtschaftsförderungsgesellschaft mbH and Standortagentur Tirol, within the framework of the COMET Funding Programme is gratefully acknowledged.

References

1. W.H. Brattain, J. Bardeen, Surface properties of germanium. Bell Syst. Tech. J. **32**(1), 1–41 (1953)
2. T. Seiyama et al., A new detector for gaseous components using semiconductive thin films. Anal. Chem. **34**(11), 1502–1503 (1962)
3. P.J. Shaver, Activated tungsten oxide gas detectors. Appl. Phys. Lett. **11**(8), 255–257 (1967)
4. N. Taguchi, Gas-detecting device. U.S Patent 3,631,436, 28 Dec 1971
5. K. Kalantar-Zadeh et al., Intestinal gas capsules: A proof-of-concept demonstration. Gastroenterology **150**(1), 37–39 (2016)
6. E. Abad et al., Flexible tag microlab development: Gas sensors integration in RFID flexible tags for food logistic. Sensors Actuators B Chem. **127**(1), 2–7 (2007)
7. M. Ortel et al., Spray pyrolysis of ZnO–TFTs utilizing a perfume atomizer. Solid State Electron. **86**, 22–26 (2013)
8. M. Prasad et al., Design and fabrication of Sidiaphragm, ZnO piezoelectric film-based MEMS acoustic sensor using SOI wafers. IEEE Trans. Semicond. Manuf. **26**(2), 233–241 (2013)
9. D.D. Lee et al., Environmental gas sensors. IEEE Sensors J. **1**(3), 214–224 (2001)
10. MarketsandMarkets, Gas Sensors Market worth 1,297.6 Million USD by 2023, 2018. [Online]. https://www.marketsandmarkets.com/PressReleases/gas-sensor.asp. Accessed Jul 2018
11. World Health Organization, 9 out of 10 people worldwide breathe polluted air, but more countries are taking action, 2018. [Online]. http://www.who.int/news-room/detail/02-05-2018-9-out-of-10-people-worldwide-breathe-polluted-air-but-more-countries-are-taking-action. Accessed Jul 2018

12. Hemming Fire, Looking to the future of gas sensing—a new galaxy of possibilities, Hemming Group Ltd, 08 April 2010. [Online]. http://www.hemmingfire.com/news/fullstory.php/aid/844/Looking_to_the_future_of_gas_sensing__96_a_new_galaxy_of_possibilities__.html. Accessed May 2018
13. J. Riegel et al., Exhaust gas sensors for automotive emission control. Solid State Ionics **152**, 783–800 (2002)
14. G.F. Fine et al., Metal oxide semi-conductor gas sensors in environmental monitoring. Sensors **10**(6), 5469–5502 (2010)
15. E. Kanazawa et al., Metal oxide semiconductor N2O sensor for medical use. Sensors Actuators B Chem. **77**(1–2), 72–77 (2001)
16. T. Konduru et al., A customized metal oxide semiconductor-based gas sensor array for onion quality evaluation: System development and characterization. Sensors **15**(1), 1252–1273 (2015)
17. A. Lahlalia et al., Modeling and simulation of novel semiconducting metal oxide gas sensors for wearable devices. IEEE Sensors J. **18**(5), 1960–1970 (2018)
18. S.Z. Ali et al., Nanowire hydrogen gas sensor employing CMOS micro-hotplate, in *Proceedings of IEEE Sensors 2009 Conference*, (2009)
19. H.M. Low et al., Thermal induced stress on the membrane in integrated gas sensor with micro-heater, in *Proceedings of the 1998 IEEE Electron Devices Meeting, Hong Kong*, (1998)
20. D.-D. Lee et al., Low power micro gas sensor, in *Solid-State Sensors and Actuators and Eurosensors IX.. Transducers' 95, IEEE*, (1995)
21. I. Simon et al., Micromachined metal oxide gas sensors: Opportunities to improve sensor performance. Sensors Actuators B Chem. **73**(1), 1–26 (2001)
22. R. Phatthanakun et al., Fabrication and control of thin-film aluminum microheater and nickel temperature Sensor, in *Electrical Engineering/Electronics, Computer, Telecommunications and Information Technology (ECTI-CON), IEEE*, (2011)
23. K. Zhang et al., Fabrication, modeling and testing of a thin film Au/Ti microheater. Int. J. Therm. Sci. **46**(6), 580–588 (2007)
24. L. Xu et al., Development of a reliable micro-hotplate with low power consumption. IEEE Sensors J. **11**(4), 913–919 (2011)
25. P. Bhattacharyya et al., A low power MEMS gas sensor based on nanocrystalline ZnO thin films for sensing methane. Microelectron. Reliab. **48**(11), 1772–1779 (2008)
26. U. Dibbern, A substrate for thin-film gas sensors in microelectronic technology. Sensors Actuators B Chem. **2**(1), 63–70 (1990)
27. I. Haneef et al., Thermal characterization of SOI CMOS micro hot-plate gas sensors, in *Thermal Investigations of ICs and Systems (THERMINIC), IEEE*, (2010)
28. S.Z. Ali et al., Tungsten-based SOI microhotplates for smart gas sensors. IEEE J. Microelectromech. Syst. **17**(6), 1408–1417 (2008)
29. W. Yan et al., Nickel membrane temperature sensor in micro-flow measurement. J. Alloys Compd. **449**(1–2), 210–213 (2008)
30. D. Monika et al., Design and simulation of MEMS based microhotplate as gas sensor. Int. J. Adv. Eng. Res. Technol. **2**, 2487–2492 (2013)
31. L. Mele et al., A molybdenum MEMS microhotplate for high-temperature operation. Sensors Actuators A Phys. **188**, 173–180 (2012)
32. V. Balakrishnan et al., Steady-state analytical model of suspended p-type 3C–SiC bridges under consideration of Joule heating. J. Micromech. Microeng. **27**(7), 075008 (2017)
33. J.F. Creemer et al., Microhotplates with TiN heaters. Sensors Actuators A Phys. **148**(2), 416–421 (2008)
34. G. Benn, Design of a Silicon Carbide Micro-Hotplate Geometry for High Temperature Chemical Sensing, M.S. thesis (MIT, Cambridge, 2001)
35. J. Spannhake et al., High-temperature MEMS heater platforms: Long-term performance of metal and semiconductor heater materials. Sensors **6**(4), 405–419 (2006)
36. S.Z. Ali et al., A low-power, low-cost infra-red emitter in CMOS technology. IEEE Sensors J. **15**(12), 6775–6782 (2015)

37. A. Lahlalia et al., Electro-thermal simulation & characterization of a microheater for SMO gas sensors. J. Microelectromech. Syst. **27**(3), 529–537 (2018)
38. I. Elmi et al., Development of ultra-low-power consumption MOX sensors with ppb-level VOC detection capabilities for emerging applications. Sensors Actuators B Chem. **135**(1), 342–351 (2008)
39. J.C. Belmonte et al., High-temperature low-power performing micromachined suspended micro-hotplate for gas sensing applications. Sensors Actuators B Chem. **114**(2), 826–835 (2006)
40. J. Li et al., Dynamic characteristics of transient boiling on a square platinum microheater under millisecond pulsed heating. Int. J. Heat Mass Transf. **51**(1/2), 273–282 (2008)
41. S.M. Lee et al., Design and optimisation of a high-temperature silicon micro-hotplate for nanoporous palladium pellistors. Microelectron. J. **34**(2), 115–126 (2003)
42. F. Udrea et al., Design and simulations of SOI CMOS micro-hotplate gas sensors. Sensors Actuators B Chem. **78**(1–3), 180–190 (2001)
43. Y. Çengel et al., *Fundamentals of Thermal-Fluid Sciences* (McGraw-Hill, New York, 2001)
44. C. Dücsö et al., Porous silicon bulk micromachining for thermally isolated membrane formation. Sensors Actuators A Phys. **60**(1–3), 235–239 (1997)
45. A.I. Uddin et al., Low temperature acetylene gas sensor based on Ag nanoparticles-loaded ZnO-reduced graphene oxide hybrid. Sensors Actuators B Chem. **207**, 362–369 (2015)
46. R. Artzi-Gerlitz et al., Fabrication and gas sensing performance of parallel assemblies of metal oxide nanotubes supported by porous aluminum oxide membranes. Sensors Actuators B Chem. **136**(1), 257–264 (2009)
47. M. Aslam et al., Polyimide membrane for micro-heated gas sensor array. Sensors Actuators B Chem. **103**(1–2), 153–157 (2004)
48. T. Taliercio et al., Realization of porous silicon membranes for gas sensor applications. Thin Solid Films **255**(1–2), 310–312 (1995)
49. S. Astié et al., Design of a low power SnO2 gas sensor integrated on silicon oxynitride membrane. Sensors Actuators B Chem. **67**(1–2), 84–88 (2000)
50. G. Wiche et al., Thermal analysis of silicon carbide based micro hotplates for metal oxide gas sensors. Sensors Actuators A Phys. **123**, 12–17 (2005)
51. T. Zhang et al., Electrochemically functionalized single-walled carbon nanotube gas sensor. Electroanalysis **18**(12), 1153–1158 (2006)
52. J. Li et al., A gas sensor array using carbon nanotubes and microfabrication technology. Electrochem. Solid-State Lett. **8**(11), H100–H102 (2005)
53. K.D. Mitzner et al., Development of a micromachined hazardous gas sensor array. Sensors Actuators B Chem. **93**(1–3), 92–99 (2003)
54. V. Guarnieri et al., Platinum metallization for MEMS application: Focus on coating adhesion for biomedical applications. Biomatter **4**(1), e28822 (2014)
55. Q. Zhou et al., Fast response integrated MEMS microheaters for ultra low power gas detection. Sensors Actuators A **223**, 67–75 (2015)
56. D.G. Cahill et al., Thermometry and thermal transport in micro/nanoscale solid-state devices and structures. J. Heat Transf. **124**(2), 223–241 (2002)
57. D.G. Cahill, Analysis of heat flow in layered structures for time-domain thermoreflectance. Rev. Sci. Instrum. **75**(12), 5119 (2004)
58. F. Claro, Theory of resonant modes in particulate matter. Phys. Rev. B **30**(9), 4989–4999 (1984)
59. S. Gomès et al., Scanning thermal microscopy: A review. Phys. Status Solidi A **212**(3), 477–494 (2015)
60. V. Szekely, Identification of RC networks by deconvolution: Chances and limits. IEEE Trans. Circ. Syst. Fund. Theor. Appl. **45**(3), 244–258 (1998)
61. L. Mitterhuber et al., Validation methodology to analyze the temperature-dependent heat path of a 4-chip LED module using a finite volume simulation. Microelectron. Reliab. **79**, 462–472 (2017)

62. A.J. Schmidt et al., Pulse accumulation, radial heat conduction, and anisotropic thermal conductivity in pump-probe transient thermoreflectance. Rev. Sci. Instrum. **79**, 114902(9) (2008)
63. P.B. Allen et al., Diffusons, locons and propagons: Character of atomie yibrations in amorphous si. Philos. Mag. B **79**(11–12), 1715–1731 (1999)
64. M. Flik et al., Heat transfer regimes in microstructures. J. Heat Transf. **114**(3), 666–674 (1992)
65. G. Chen, Nonlocal and nonequilibrium heat conduction in the vicinity of nanoparticles. J. Heat Transf. **118**(3), 539–545 (1996)
66. J.Å. Schweitz, Mechanical characterization of thin films by micromechanical techniques. MRS Bull. **17**(7), 34–45 (1992)
67. V.M. Paviot et al., Measuring the mechanical properties of thin metal films by means of bulge testing of micromachined windows. MRS Online Proc. Libr. Arch. **356**, 579–584 (1994)
68. S. Mahabunphachai et al., Investigation of size effects on material behavior of thin sheet metals using hydraulic bulge testing at micro/meso-scales. Int. J. Mach. Tools Manuf. **48**(9), 1014–1029 (2008)
69. T.P. Weihs et al., Mechanical deflection of cantilever microbeams: A new technique for testing the mechanical properties of thin films. J. Mater. Res. **3**(5), 931–942 (1988)
70. X. Song et al., Residual stress measurement in thin films at sub-micron scale using focused ion beam milling and imaging. Thin Solid Films **520**(6), 2073–2076 (2012)
71. M. Krottenthaler et al., A simple method for residual stress measurements in thin films by means of focused ion beam milling and digital image correlation. Surf. Coat. Technol. **215**, 247–252 (2013)
72. N. Sabaté et al., FIB-based technique for stress characterization on thin films for reliability purposes. Microelectron. Eng. **84**, 1783–1787 (2007)
73. S. Massl et al., A direct method of determining complex depth profiles of residual stresses in thin films on a nanoscale. Acta Mater. **55**, 4835–4844 (2007)
74. G. Moser et al., Sample preparation by metallography and focused ion beam for nanomechanical testing. Pract. Metallogr. **49**(6), 343–355 (2012)
75. D. Kiener et al., Source truncation and exhaustion: Insights from quantitative in situ TEM tensile testing. Nano Lett. **11**(9), 3816–3820 (2011)
76. D. Kiener et al., Strength, hardening, and failure observed by in situ tem tensile testing. Adv. Eng. Mater. **14**(11), 960–967 (2012)
77. M.F. Dorner et al., Stresses and deformation processes in thin films on substrates. CRC Crit. Rev. Solid State Mater. Sci. **14**(3), 225–267 (1988)
78. P. Chaudhari, Grain growth and stress relief in thin films. J. Vac. Sci. Technol. **9**(1), 520–522 (1972)
79. R.W. Hoffman, Stresses in thin films: The relevance of grain boundaries and impurities. Thin Solid Films **34**, 185–190 (1976)
80. E. Klokholm et al., Intinsic stress in evaporated metal films. J. Electrochem. Soc. **115**(8), 823–826 (1968)
81. B.W. Sheldon et al., Intinsic compressive stress in polycrystalline films with negligible grain boundary diffusion. J. Appl. Phys. **94**(2), 948–957 (2003)
82. E. Chason et al., Origin of compressive residual stress in polycrystalline thin films. Phys. Rev. Lett. **88**(15), 156103 (2002)
83. K. Cholevas, Misfit dislocation patterning in thin films. Phys. Status Solidi B **209**(10), 295–304 (1998)
84. L.B. Freund et al., *Thin Film Materials: Stress, Defect Formation and Surface Evolution* (Cambridge University Press, Cambridge, 2003)
85. P. Tsilingiris, Thermal conductivity of air under different humidity conditions. Energy Convers. Manag. **49**, 1098–1110 (2008)
86. A. Moridi et al., Residual stresses in thin film systems:Effects of lattice mismatch, thermal mismatch and interface dislocations. Int. J. Solids Struct. **50**(22–23), 3562–3569 (2013)

87. H. Köstenbauer et al., Annealing of intrinsic stresses in sputtered TiN films: The role of thickness-dependent gradients of point defect density. Surf. Coat. Technol. **201**, 4777–4780 (2007)
88. R. Machunze et al., Stress and strain in titanium nitride thin films. Thin Solid Films **517**, 5888–5893 (2009)
89. R. Treml et al., High resolution determination of local residual stress gradients in single- and multilayer thin film systems. Acta Mater. **103**, 616–623 (2016)
90. R. Hammer et al., High resolution residual stress gradient characterization in W/TiN-stack on Si(100): Correlating in-plane stress and grain size distributions in W sublayer. Mater. Des. **132**, 72–78 (2017)
91. R. Konetschnik et al., Micro-mechanical in situ measurements in thin film systems regarding the determination of residual stress, fracture properties and Interface toughness. Microsc. Microanal. **23**, 750–751 (2017)
92. J. Keckes et al., X-ray nanodiffraction reveals strain and microstructure evolution in nanocrystalline thin films. Scr. Mater. **67**, 748–751 (2012)
93. C. Genzel, X-ray residual stress analysis in thin films under grazing incidence–basic aspects and applications. Mater. Sci. Technol. **21**, 10–18 (2005)
94. J. Todt et al., X-ray nanodiffraction analysis of stress oscillations in a W thin film on through-silicon via. J. Appl. Crystallogr. **49**, 182–187 (2016)
95. M. Stefenelli et al., X-ray nanodiffraction reveals stress distribution across an indented multilayered CrN–Cr thin film. Acta Mater. **85**, 24–31 (2015)
96. R. Schöngrundner et al., Critical assessment of the determination of residual stress profiles in thin films by means of the ion beamlayer removal method. Thin Solid Films **564**, 321–330 (2014)
97. M. Sebastiani et al., Depth-resolved residual stress analysis of thin coatings by a new FIB–DIC method. Mater. Sci. Eng. A **528**, 7901–7908 (2011)
98. T.L. Anderson, *Fracture Mechanics: Fundamentals and Applications* (CRC, Boca Raton, 2017)
99. M. Kuna, *Finite Elements in Fracture Mechanics: Theory—Numerics—Applications. Solid Mechanics and Its Applications* (Springer, Dordrecht, 2015)
100. O. Kolednik, *Fracture Mechanics, Wiley Encyclopedia of Composites* (Wiley, New York, 2011)
101. X.K. Zhu et al., Review of fracture toughness (G, K, J, CTOD, CTOA) testing and standardization. Eng. Fract. Mech. **85**, 1–46 (2012)
102. G. Irwin, Analysis of stresses and strains near the end of a crack traversing a plate. J. Appl. Mech. **24**(3), 361–364 (1957)
103. A.A. Griffith, The phenomena of rupture and flow in solids. Philos. Trans. R. Soc. A Math. Phys. Eng. Sci. **221**(582–593), 163–198 (1921)
104. J.R. Rice, A path independent integral and the approximate analysis of strain concentration by notches and cracks. J. Appl. Mech. **35**(2), 379–386 (1968)
105. N.K. Simha et al., J-integral and crack driving force in elastic-plastic materials. J. Mech. Phys. Solids **56**(9), 2876–2895 (2008)
106. O. Kolednik et al., A new view on J-integrals inelastic–plastic materials. Int. J. Fract. **187**(1), 77–107 (2014)
107. R.O. Ritchie, Mechanisms of fatigue crack propagation in metals, ceramics and composites: Role of crack tip shielding. Mater. Sci. Eng. **103**(1), 15–28 (1988)
108. N.K. Simha et al., Inhomogeneity effects on the crack driving force in elastic and elastic-plastic materials. J. Mech. Phys. Solids **51**(1), 209–240 (2003)
109. R.O. Ritchie et al., Fatigue crack propagation in ARALL® LAMINATES: Measurement of the effect of crack-tip shielding from crack bridging. Eng. Fract. Mech. **32**(3), 361–377 (1989)
110. O. Kolednik et al., Improvement of fatigue life by compliant and soft interlayers. Scr. Mater. **113**, 1–5 (2016)
111. Y. Sugimura et al., Fracture normal to a biomaterial interface: Effects of plasticity on crack-tip shielding and amplification. Acta Metall. Mater. **43**(3), 1157–1169 (1995)

112. J. Predan et al., On the local variation of the crack driving force in a double mismatched weld. Eng. Fract. Mech. **74**(11), 1739–1757 (2007)
113. O. Kolednik et al., Modeling fatigue crack growth in a bimaterial specimen with the configurational forces concept. Mater. Sci. Eng. A **519**(1–2), 172–183 (2009)
114. N.K. Simha et al., Material force models for cracks—influences of eigenstrains, thermal strains & residual stresses, in *11th International Conference on Fracture*, (2005)
115. J.D. Eshelby, *Energy Relations and the Energy-Momentum Tensor in Continuum Mechanics BT* (Springer, Berlin, 1999)
116. M.E. Gurtin, *Configurational Forces as Basic Concepts of Continuum Physics* (Springer, New York, 2000)
117. G.A. Maugin, *Configurational Forces: Thermodynamics, Physics, Mathematics, and Numerics* (CRC, Boca Raton, 2010)
118. N.K. Simha et al., Crack tip shielding or anti-shielding due to smooth and discontinuous material inhomogeneities. Int. J. Fract. **135**(1), 73–93 (2005)
119. R. Treml et al., Miniaturized fracture experiments to determine the toughness of individual films in a multilayer system. Extreme Mech. Lett. **8**, 235–244 (2016)
120. B. Merle et al., Fracture toughness of silicon nitride thin films of different thicknesses as measured by bulge tests. Acta Mater. **59**, 1772–1779 (2011)
121. E. Harry et al., Mechanical properties of W and W(C) thin films: Young's modulus, fracture toughness and adhesion. Thin Solid Films **332**, 195–201 (1998)
122. D. Kozic et al., Extracting flow curves from nano-sized metal layers in thin film systems. Scr. Mater. **130**, 143–417 (2017)
123. G. Klemes, *Thermal Conductivity: Metallic Elements and Alloys* (Plenum, New York, 1970)
124. J. Hostetler et al., Thin-film thermal conductivity and thickness measurements using picosecond ultrasonics. Microsc. Thermophys. Eng. **1**(3), 237–244 (1997)
125. L. Xiang, Thermal conductivity modeling of copper and tungsten damascene structures. J. Appl. Phys. **105**(9), 094301 (2009)
126. T.L. Bergman et al., *Fundamentals of Heat and Mass Transfer* (Wiley, New York, 2011)
127. H.A. Schafft et al., Thermal conductivity measurements of thin-film silicon dioxide in microelectronic test structures, in *Microelectronic Test Structures (ICMTS), IEEE*, (1989)
128. X. Zhang et al., Thermal conductivity and diffusivity of free-standing silicon nitride thin films. Rev. Sci. Instrum. **66**(2), 1115–1120 (1995)
129. Texas Instruments, Thermal conductivity and thermal diffusivity, Report (2014)
130. P.I. Dorogokupets et al., Optimization of experimental data on the heat capacity, volume, and bulk moduli of minerals. Petrology **7**(6), 574–591 (1999)
131. S. Andersson, Thermal conductivity and heat capacity of amorphous SiO2: pressure and volume dependence. J. Phys. Condens. Matter **4**(29), 6209 (1992)
132. A.S. Grove, *Physics and Technology of Semiconductor Devices* (Wiley, New York, 1967)
133. T. Ohmura et al., Specific heat measurement of high temperature thermal insulations by drop calorimeter method. Int. J. Thermophys. **24**(2), 559–575 (2003)
134. C.H. Mastrangelo et al., Thermophysical properties of low-residual stress, silicon-rich, LPCVD silicon nitride films. Sensors Actuators A Phys. **23**(1–3), 856–860 (1990)
135. A. Jain et al., Measurement of the thermal conductivity and heat capacity of freestanding shape memory thin films using the 3ω method. J. Heat Transf. **130**(10), 102402 (2008)
136. J. Harrigill et al., Method for Measuring Static Young's Modulus of Tungsten to 1900 K (1972)
137. J.W. Davis et al., ITER material properties handbook. J. Nucl. Mater. **233**, 1593–1596 (1996)
138. G.P. Škoro et al., Dynamic Young's moduli of tungsten and tantalum at high temperature and stress. J. Nucl. Mater. **409**(1), 40–46 (2011)
139. D. Makwana et al., Review of miniature specimen tensile test method of tungsten at elevated temperature. Int. J. Eng. Dev. Res. **4**(4), 132–139 (2016)
140. S. Krimpalis et al., Comparative study of the mechanical properties of different tungsten materials for fusion applications. Phys. Scripta **2017**(T170), 014068 (2017)

141. F.F. Schmidt et al., *The Engineering Properties of Tungsten and Tungsten Alloys, No. DMIC191* (Battelle Memorial Institute, Defense Metals Information Center, Columbus, 1963)
142. T. Shinoda et al., Young's modulus of RF-sputtered amorphous thin films in the SiO2-Y2O3 system at high temperature. Thin Solid Films **293**(1–2), 144–148 (1997)
143. O. Morozov et al., Mechanical strength study of SiO2 isolation blocks merged in silicon substrate. J. Micromech. Microeng. **25**(1), 015014 (2014)
144. W.N. Sharpe et al., Strain measurements of silicon dioxide microspecimens by digital imaging processing. Exp. Mech. **47**(5), 649–658 (2007)
145. T. Tsuchiya et al., Tensile testing of insulating thin films; humidity effect on tensile strength of SiO2 films. Sensors Actuators A Phys. **82**(1–3), 286–290 (2000)
146. J.-H. Zhao et al., Measurement of elastic modulus, Poisson ratio, and coefficient of thermal expansion of on-wafer submicron films. J. Appl. Phys. **85**(9), 6421–6424 (1999)
147. E. Sánchez-González et al., Effect of temperature on the pre-creep mechanical properties of silicon nitride. J. Eur. Ceram. Soc. **29**(12), 2635–2641 (2009)
148. aZo Materials, Sintered Silicon Nitride (Si_3N_4), [Online]. https://www.azom.com/properties.aspx?ArticleID=260
149. R.J. Bruls et al., The temperature dependence of the Young's modulus of MgSiN2, AlN and Si3N4. J. Eur. Ceram. Soc. **21**(3), 263–268 (2001)
150. A.E. Kaloyeros et al., Silicon nitride and silicon nitride-rich thin film technologies: Trends in deposition techniques and related applications. ECS J. Solid State Sci. Technol. **6**(10), 691–714 (2017)
151. A. Khan et al., Young's modulus of silicon nitride used in scanning force microscope cantilevers. J. Appl. Phys. **95**(4), 1667–1672 (2004)
152. G.F. Cardinale et al., Fracture strength and biaxial modulus measurement of plasma silicon nitride films. Thin Solid Films **207**(1–2), 126–130 (1992)

Chapter 3
Miniaturized Photoacoustic Gas Sensor for CO_2

Horst Theuss, Stefan Kolb, Matthias Eberl, and Rainer Schaller

3.1 Introduction and Motivation

Consciousness for environmental topics increases worldwide. It drives the need for probing of respective data, such as atmospheric CO_2-concentration. For thousands of years, the atmospheric CO_2-concentration remained stable at a level of approximately 280 ppm. Driven by industrialization, it showed a drastic increase to 400 ppm in the nineteenth and twentieth century—while today growth rates are still increasing.

Besides environmental topics, air quality monitoring applications in buildings, homes, etc. will profit from accurate measurements of CO_2-concentrations. Air conditioning systems based upon liquid CO_2 as cooling agent will definitely need a CO_2 leakage detection sensor for safety reasons. Such systems are currently under discussion in the automotive industry. The door for highest volume production will be opened once such sensors will be integrated into smartphones and other mobile devices. The next big step will be gas sensors, which are able to detect multiple substances in parallel.

This chapter describes a gas sensor specifically designed for CO_2, but its underlying physical principle—the photoacoustic effect—can as well be applied to a variety of other gases.

The chapter is organized as follows: An overview over gas sensing principles is followed by explaining the basics of the photoacoustic effect. This leads to the concept of the sensor and the requirements, which physics expose to the system and the package. Mathematical models help to translate these requirements into a sensor

H. Theuss (✉) · S. Kolb · M. Eberl · R. Schaller
Infineon Technologies AG, Neubiberg, Germany
e-mail: horst.theuss@infineon.com

W. D. van Driel et al. (eds.), *Sensor Systems Simulations*,
https://doi.org/10.1007/978-3-030-16577-2_3

system. Assembly and characterization of a respective demonstrator represent the focus of this chapter, which is completed by a summary and an outlook.

3.2 Gas Sensing Principles

Gas sensing concepts are manifold and cannot be treated comprehensively in this chapter. However, in terms of CO_2-measurements, steady discussion focuses on two basic sensor principles with potential for cost-efficient manufacturability.

3.2.1 Chemo-Resistive Sensors

Gas molecules adhere to a gas-sensitive sensor layer. Interaction of the gas molecules with the sensor material directly impacts the electrical resistivity of the layer. Most mature sensor materials are metal oxides (MOX-sensors). Reversibility of the effect is achieved by thermal desorption of the gas from the sensor material.

3.2.2 Sensors Based Upon IR Absorption

These sensors measure the absorption of infrared (IR) light by CO_2-molecules along a defined absorption path. All systems need an IR-emitter, which—depending on the specific sensor type—can be either a monochromatic or a broadband source. The emitted spectrum has to contain one or more wavelengths, which interact with the CO_2-molecules, resulting in an observable absorption effect. The actual measurement in such Non-Dispersive IR sensors (NDIR) is then performed by a detector, e.g., a bolometer, thermopile, or photo-sensitive semiconductor. The photoacoustic effect, however, takes advantage of the following phenomenon: IR absorption in a defined volume increases the total energy within this volume and thus creates a pressure pulse. In resonant systems, the IR beam is modulated with a frequency equal to the resonance defined by the atmosphere in the detector volume and its length. Since these frequencies typically are in the audio range, a sensitive microphone can be used as detector. The resonance frequency represents the actual sensor signal since it varies with the gas composition in the resonator volume. The sensor described in this chapter refers to the so-called non-resonant system: The sensor measures individual pressure pulses, which are created by IR signals chopped with very low (quasi static) frequencies—typically below 30 Hz. The microphone is just used as a highly sensitive pressure sensor. Details are described in the following sections.

While MOX-sensors are supposed to show high potential for miniaturization and integration, they are considered to be limited with respect to reliability and long-term stability. For applications requiring higher robustness, the market tends to prefer optical systems rather than chemo-resistive principles.

3.3 Basics: The Photoacoustic Effect

Depending on their complexity, gas molecules can show a variety of vibrational modes. Only a selection of modes which go along with a dipole moment can interact with electromagnetic radiation. The relevant IR-active oscillation mode for our CO_2 sensor is an antisymmetric stretching mode as illustrated in Fig. 3.1. Quantum mechanics and overlap of the vibration with rotational modes lead to complex absorption bands (Fig. 3.1).

These absorption bands are very unique for different molecules. Thus, IR absorption can be used for selective detection of specific substances. If this absorption happens within a closed volume by a pulsed incoming IR beam, the respective absorbed energy creates a measurable pressure pulse inside this volume. A very high CO_2-concentration (close to 100%) in this reference cell increases this signal and at the same time cares for the gas selectivity of the sensor: In case a light source with a continuous spectrum (black body emitter) is used as IR source, this reference cell concept selects the relevant absorption lines in an elegant way. Only the absorption lines associated with the spectrum of the relevant gas (Fig. 3.1) contribute to the height of the created pressure pulse. This concept makes the implementation of a high-quality filter obsolete, which would be necessary in case of a standard NDIR sensor.

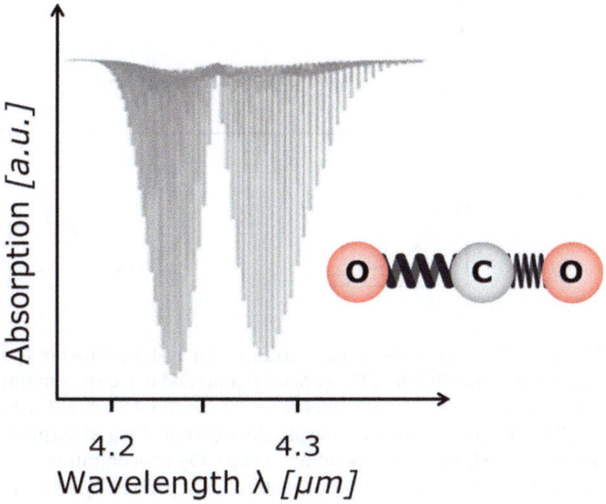

Fig. 3.1 Absorption band of the antisymmetric stretching oscillation in a CO_2-molecule

3.3.1 Concept of the Sensor

Figure 3.2 illustrates the overall sensor concept: An emitter (Fig. 3.2a), left) emits a pulsed IR beam. The intensity spectrum $S_0(\lambda)$ can be a broadband spectrum, but it must contain the relevant gas-specific wavelengths. The beam travels through the absorption path where environmental CO_2 causes absorption. Figure 3.2b schematically illustrates the absorption and mentions absorption coefficients α_1 and α_2 corresponding to CO_2-concentrations $c_1 < c_2$. As we are dealing with a whole absorption band (Fig. 3.1) rather than a monochromatic beam, a simple Lambert-Beer type absorption law $I(x)/I(0) = e^{-\alpha\, c\, x}$ for the intensity $I(x)$ is not applicable: The coefficients α strongly depend on the wavelength λ. In other words: Individual lines in Fig. 3.1 experience different absorption strengths. This fine structure is not shown in Fig. 3.2c, but rather summarized in one absorption peak. The IR-pulse corresponding to spectrum S_a eventually hits the reference volume, where it

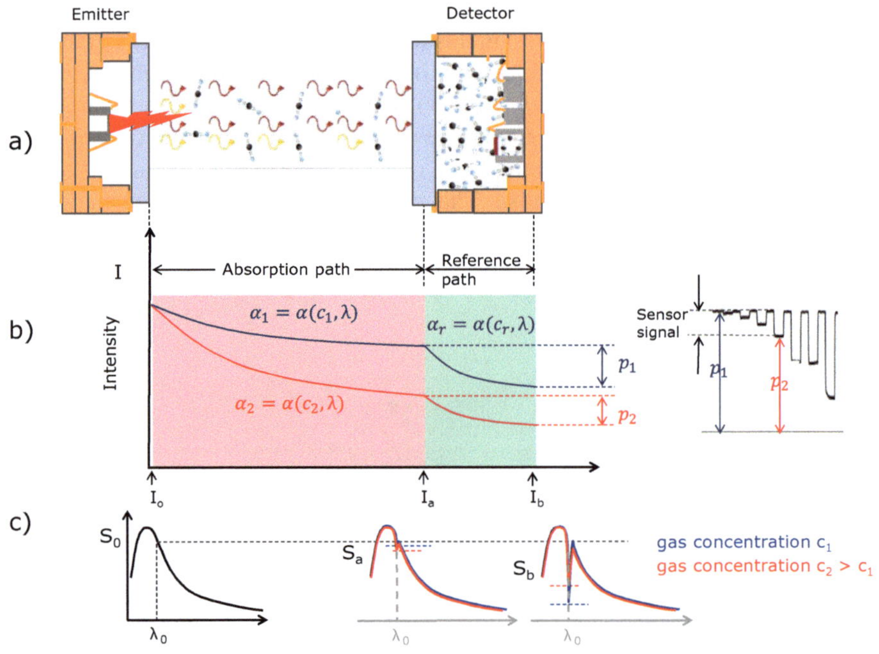

Fig. 3.2 Concept of the sensor. (**a**) Physical system showing broadband emitter (left), absorption path (with typically low CO_2-concentration) and detector containing MEMS microphone and other components. Detector components are enclosed into a cavity which contains a high concentration of CO_2, (**b**) Illustration of intensity distribution along absorption and reference path. Also shown are pressure signals corresponding to two CO_2-concentrations $c_1 < c_2$. The steps in the chart on the right refer to increasing CO_2-concentrations separated by evacuation steps. (**c**) Schematic spectrum as provided by the emitter (left), after penetration through the absorption path (middle) and after passing through the reference cell (right)

experiences a much higher absorption due to the high CO_2-concentration $c_r >> c_1$, c_2. In the reference cell, molecules in the excited vibration state very quickly relax by transferring their oscillation energy into translation energy via hits with other molecules (typical relaxation time $\sim 10^{-4}$ s). In consequence, the temperature T increases. Part of the absorbed energy then translates into a pressure pulse via $pV_r \sim T$ (where V_r is the volume of the reference cell). This is the photoacoustic pressure as detected by the MEMS microphone (Micro Electro Mechanical System). Another part of the absorbed energy escapes by heat dissipation into and through the walls of the reference cell. The spectra S_a and S_b in Fig. 3.2c show the condition of the IR-pulse prior and after passing through the reference cell. These spectra correspond to the respective intensities I_x ($x = 0$, a, b) mentioned in Fig. 3.2b. Figure 3.2c nicely explains the gas selectivity of the sensor: Only the gas-specific absorption lines contribute to the photoacoustic effect. The spectrum outside the absorption band of CO_2 does not.

An increase of the concentration in the absorption path from c_1 to $c_2 > c_1$ leads to increased absorption (lower I_a). However, this translates into a lower photoacoustic pressure pulse p_2 since the absorption in the reference cell starts at this lower level of I_a. Thus, higher CO_2-concentrations in the absorption path (environment) create lower photoacoustic pressure pulses. In order to actually probe the concentration, increase from c_1 to c_2, the sensor needs to resolve the photoacoustic pressure difference $p_1 - p_2 \sim c_2 - c_1$.

3.4 Translation into Package and System Requirements

The necessary emitter of such a system needs to emit a spectrum containing the relevant absorption bands in a relatively high intensity. Luckily, this is the case for a variety of MEMS emitters. These provide a spectrum close to a black body radiation with a maximum around 4 μm.

A suitable MEMS microphone must be tuned with respect to its cut-off frequency, which needs to fit to the modulation frequency of the emitter (typically 10–30 Hz).

In order to ensure the functionality of the sensor and to avoid drift effects, the CO_2-concentration in the reference cell needs to be constant through the lifetime of the sensor. In consequence, the reference cell must be sealed hermetically. At the same time, it must be filled with a high concentration of CO_2.

Lid materials for emitter and detector package must be highly transparent for wavelengths in the 4.3 μm range in order not to lose too much intensity by reflection and absorption.

Finally, the dimensions of the sensor, especially the lengths of absorption and reference paths must be compromised with the requirement to make a miniaturized system.

3.5 Ceramic Package Concept

As polymer materials still show a certain permeability for gases, plastic encapsulation is not an option to fulfill the required hermeticity level. The remaining options are metal can packages and ceramic packaging. Through hole metal can technology—frequently called TO-packaging (Transistor Outline)—is rather old fashioned but provides a reliable sealing process by resistance welding. The concept in Fig. 3.1 has already been demonstrated on the basis of TO-packaging [1]. Here the target is to further miniaturize the system by making use of ceramic cavity packaging, which offers flexible routing capability along with SMT mount ability (Surface Mount Technology).

Both emitter and detector components were die–/wire-bonded into standard HTCC cavity packages (High Temperature Cofired Ceramics)—size approx. 9×9 mm^2. Optical sapphire windows were chosen due to their good IR-transmission properties. Hermetic sealing was achieved by a soldering process using respective metal stacks both on the ceramic cavity and the sapphire lid together with a frame-shaped AuSn preform. The solder process for the lid attach of the detector was performed in CO_2-atmosphere in order to enclose the gas in the reference cell.

Now emitter and detector need to be aligned face to face to each other in order to reproduce the situation schematically illustrated in Fig. 3.2. If the emitter is chosen as top component, a simple but efficient approach is shown in Fig. 3.3: The two electrical contacts are routed to the bottom using appropriate metal pieces. This enables later SMD-assembly (Surface Mount Device) onto a printed circuit board (PCB).

3.6 Hermeticity

Hermeticity is the effectiveness of the seal of microelectronic devices with designed internal cavities [2] and by design, a hermetic seal prevents gases and liquids from entering the package cavity where the die is mounted [3]. Hermeticity can be described by a supportive quality property called leak rate.

There are different backgrounds for the emitter and detector package why a hermetic seal is required. While the emitter needs to enclose a protective gas (e.g., N_2) inside the cavity for reliability reason the detector needs to provide the reference gas filling inside the sealed cavity for functionality of the sensor device.

There are three major reliability aspects which potentially can harm the inert gas atmosphere within the hermetic sealed cavity with respect to its initial state. First, there are capillary leaks leading to viscous or molecular gas flow. Second, there is permeation which is a function of adsorption on a package material surface, diffusion through this permeate, and desorption on the opposite surface to gas interface. And last, there are outgassing effects of polymers or trapped molecules in bulk materials. The initial state of the enclosed gas atmosphere is represented

Fig. 3.3 Assembly of detector and emitter into ceramic cavity packages, sealed with IR-permeable optical sapphire windows. Top left: Detector containing MEMS microphone, ASIC for signal amplification and conditioning, temperature sensor, and a photodiode to observe fluctuations in the emitted IR-intensity. Bottom left: Emitter containing a micromachined MEMS-chip. The emitter basically is a resistor designed to emit a maximum amount of input energy as IR-radiation. Thus, it has only two electrical contacts. Right: Example for a subassembly with emitter and detector facing each other. The distance in between the optical lids provides the absorption path

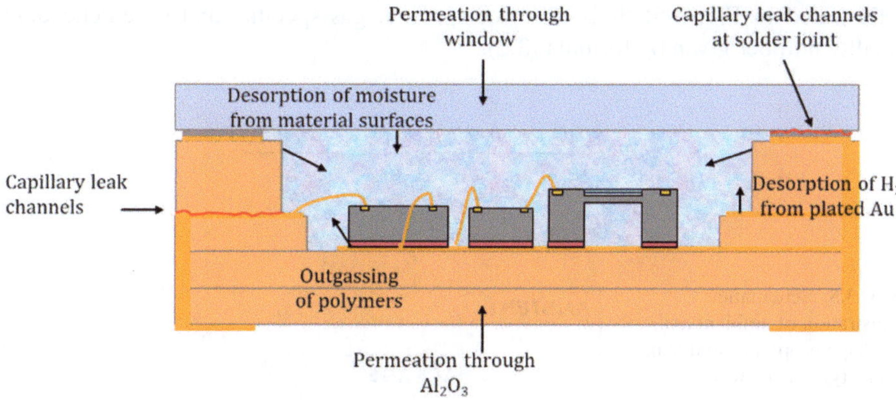

Fig. 3.4 Sealing relevant failure mechanisms

by the condition directly after cooling down to room temperature after the solder seal process. Critical zones of the ceramic package which are endangered to cause gas impurity or pressure changes are shown in Fig. 3.4. Besides other failure mechanisms like optical haze or particle contamination, loss of hermeticity is most critical for the ceramic PAS concept. Leakage fails can be distinguished between construction limits in principle like solder quality or material permeability on the one hand and on the other hand degradation fails like thermo-mechanic stress induced loss of interface integrity. Both are limiting the components lifetime.

The sealing was achieved by a soldering process. To avoid remelting at board assembly, the eutectic solder alloy Au80/Sn20 with a melting temperature of T_eut = 280 °C has been used for soldering. The liquidation shifts to even higher temperature by consuming Au from the ceramic seal ring and the lid metallization during solder process. Another advantage by using AuSn is that there is no need for formic acid or solder flux to protect the solder surfaces from oxidation. AuSn solder alloy in preformed shape as small frame structure has been used to clamp between ceramic cavity and lid. Wetable metallization on both the ceramic cavity seal ring as well as on the lid is a prerequisite for soldering. In this case, both were supplying Au finish. The soldering process itself was done in a gas tight process chamber of a solder oven. The oven is providing vacuum capability till p_min = 3 mbar as well as gas filling like N_2 or CO_2 up to p_max = 1 atm. During the process, a heat profile is applied which ramps the temperature up to T_max = 310 °C to achieve homogenous melting of the solder alloy. Simultaneously, the preferred gas atmosphere is filled into the process chamber. To avoid tilting or rotating during soldering the devices to be soldered consisting of ceramic cavity, preform and lid are mechanically fixed and pushed together by force for good solder quality. After cooling down, the devices are hermetically sealed featuring a defined gas filling inside the cavity.

The requirements of the hermetic sealing of the package are derived from a leak rate model for the PAS sensor. The leak rate q_l is characterizing the mass flow of gases caused by a partial pressure difference between two volumes per time (illustrated in Fig. 3.5). It is a device-specific, gas-specific, and time-dependent quality number given by formula (3.2).

$$q_{\mathrm{l}} = \frac{|p_{\mathrm{c}}(t) - p_{\mathrm{c}}(t_0)| \cdot V_{\mathrm{c}}}{t - t_0} = \frac{\Delta p_{\mathrm{c}} \cdot V_{\mathrm{c}}}{\Delta t} \tag{3.1}$$

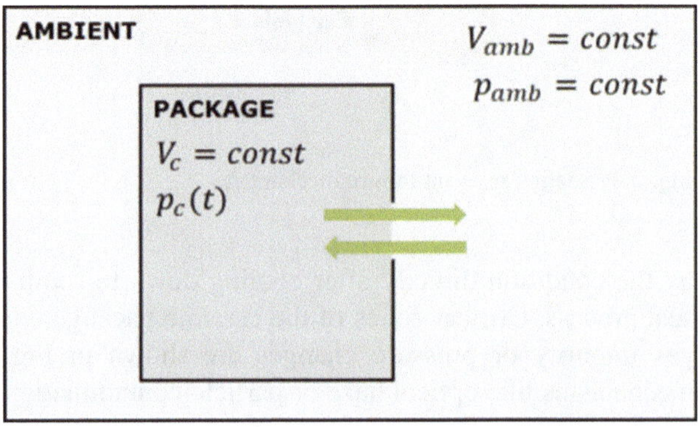

Fig. 3.5 Schematic illustration of ambient and package properties and leak paths (green arrows)

$$\lim_{\Delta t \to 0} q_l = \frac{dp_c}{dt} \cdot V_c \tag{3.2}$$

q_l = leak rate
p_c = partial pressure in the cavity
V_c = internal free-volume of cavity
t_0 = startpoint of measurement
t = endpoint of measurement
Δt = time of measurement

Gas transport mechanism is divided into three different regimes: the viscous, the molecular, and the transition flow. Which flow mechanism is dominating is depending on geometric aspects of the flow channel, e.g., radius r for an idealized cylindrical leak channel and the mean-free path of the gas molecules λ. The relation between both is represented by the Knudsen number K_n as described by formula (3.3). If $K_n < 1$, a molecular flow is dominating [5].

$$K_n = \frac{\lambda}{d} \tag{3.3}$$

λ = mean-free path of gas molecules
d = diameter of capillary

Already a rough estimation shows that $q_m = 10^8$ mbar*L/s is too high for gas sensor hermeticity to ensure proper functionality over product lifetime. The dominating gas flow mechanism for capillary leaks therefore is the molecular gas flow which can be derived from Fig. 3.6 because the Knudsen number is $K_n > 1$ for all leak rate $q_m < 10^8$ mbar*L/s. The gas flow model accordingly is based on the molecular gas flow which is given by:

$$q_m = \frac{2}{3}\pi \frac{r^3}{l} v_m \left|(p_{amb} - p_c)\right| \tag{3.4}$$

q_m = molecular leak rate
r = radius of leak channel
l = length of the leak channel
p_{amb} = ambient pressure
p_c = pressure in cavity
v_m = mean molecular speed of the gas

Gas exchange or moisture increase due to permeation like it is discussed in [4] has not yet been included into the leak model so far. On top also outgassing of certain species cannot be quantified in a specific manner yet. The moisture content of filling gas is negligible. Desorption of moisture or H_2 from package materials can be mitigated by dry bake before sealing.

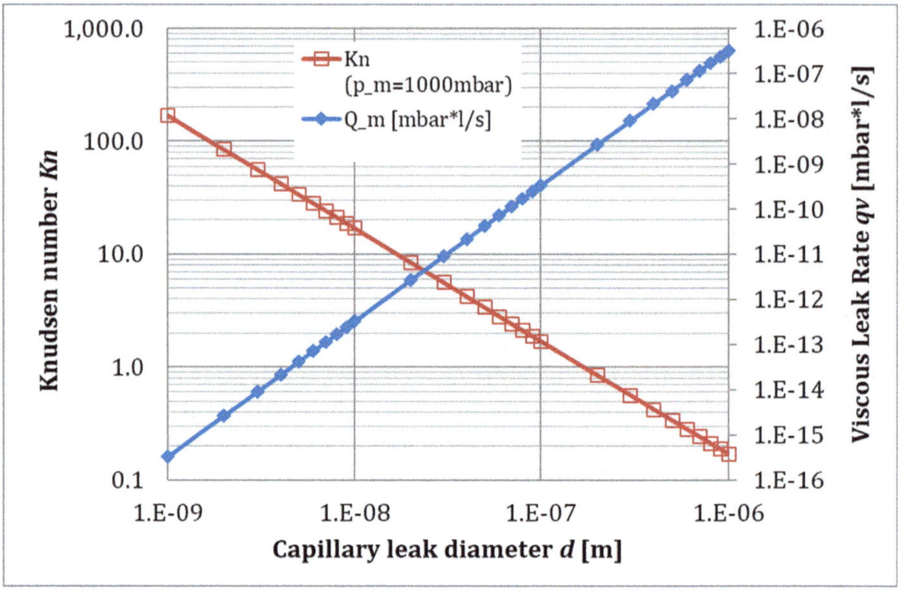

Fig. 3.6 Knudsen number and viscous leak rate vs. capillary leak diameter

For molecular gas flow the pressure behavior within in the sealed cavity can be modeled as follows from [7]:

$$p = p_{\text{amb}} + (p_0 - p_{\text{amb}}) \cdot e^{-\frac{(t-t_0) \cdot q_1 \cdot \sqrt{\frac{M_{\text{air}}}{M_{\text{gas}}}}}{V}} \quad (3.5)$$

p_{amb} = ambient partial pressure of specific gas species
p_0 = partial pressure of specific gas species at time t_0
p = partial pressure of specific gas species at time t

The behavior of the partial pressures of different species inside the cavity over time is shown in Fig. 3.7. The underlying corresponding, He leak rate of the calculation is q_l(He) $= 10^{-9}$ mbar*L/s. Within this simulation the package is assumed to be exhausted to air at p_amb = 1013 mbar. The absolute pressure jumps at $t = 0$d is caused by the sealing process. Sealing is taking place at $T = 310$ °C with p_ CO_2 = 1 bar. After cooling down to room temperature, the absolute pressure inside the cavity is nearly cut into halves.

The correlation of concentration change in detector cell and measurement sensitivity is estimated as follows: A change of CO_2-concentration Δc within the detector cell causes the same change in energy absorption E_D of the IR-radiation as the sum $\sum_{i=1}^{n} E_i$ caused by change of CO_2-concentration $\Delta c/n$ within the absorption path. Meaning that it is not possible to distinguish between a change

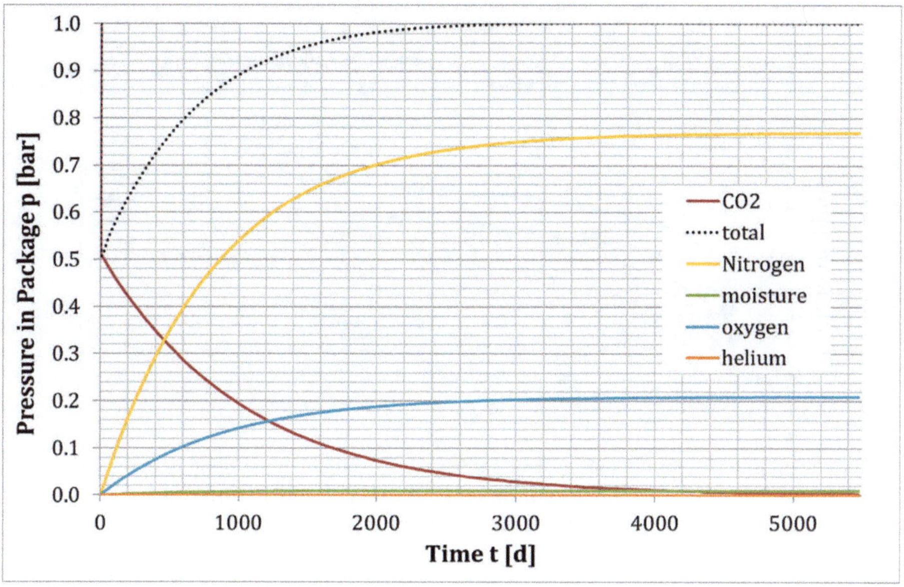

Fig. 3.7 Simulation of partial pressures of different gas species in soldered package

of detector concentration and concentration in absorption path. A gas sensor-specific correlation coefficient a which is defined per formula (3.6) and a linear correlation approach is used like shown in formula (3.7) to estimate the impact of a certain leak rate on sensor performance drift.

$$a = \frac{\Delta c_c\,(CO_2)}{\Delta c_{abs\ path}\,(CO_2)} \tag{3.6}$$

$$q_{CO_2}\left(c_{abs\ path}\right) = \frac{\sqrt{M_{CO_2}}}{\sqrt{M_{He}}} \cdot a \cdot \left[c_{abs\ path}(t) \cdot p_c(t) - c_{abs\ path}\,(t_0) \cdot p_c\,(t_0)\right] \cdot \frac{V_c}{\Delta t} \tag{3.7}$$

Figure 3.8 is illustrating the CO_2 leak rate q_ CO_2 as well as the analog He leak rate q_He over the deviation of measured CO_2-concentration. The calculation is based on the assumption of $a = 0.25$.

The requirement for the detector is a minimum leak rate of q_ CO_2_min = $7.4*10^{12}$ mbar*L/s to achieve less than 200 ppm sensor output signal drift over 2 years.

Due to the fact that the target for the detector seal quality is the more stringent one, the seal requirement for the emitter is automatically achieved when using equivalent sealing technology.

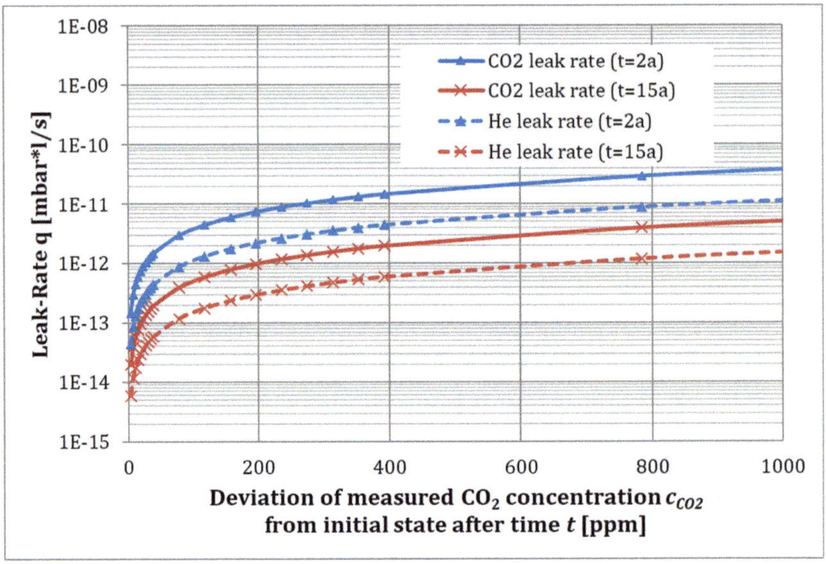

Fig. 3.8 Leak rate impact on sensor output signal drift

a) 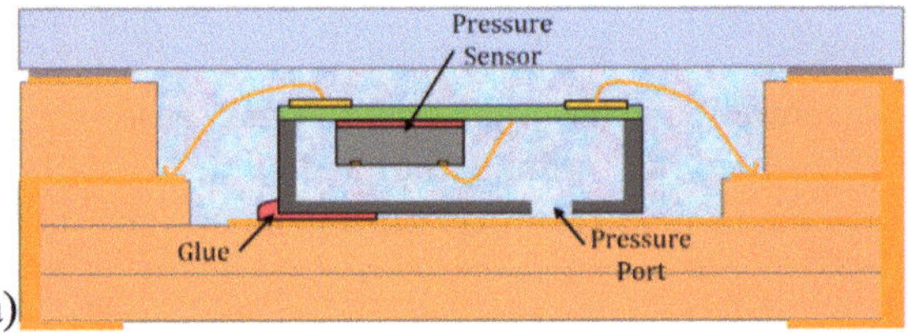

b) 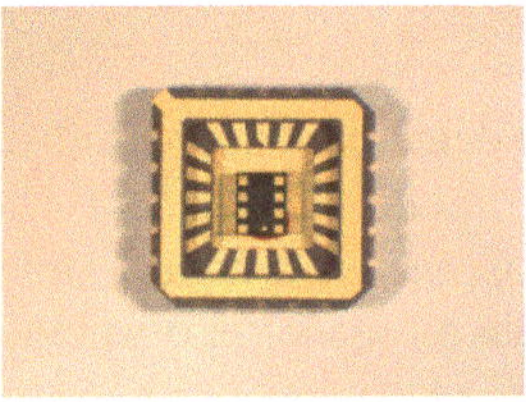

Fig. 3.9 (**a**) Schematic of the sealed ceramic cavity with pressure sensor included (**b**) Picture of an assembled sample: Glued and wire bonded pressure package in ceramic cavity package before sealing

To verify the high requirement for the hermetic seal, standard He-leak testing is not enough because of the limited minimum leak rate detection [6, 7]. Therefore, an advanced hermeticity test method was adopted for our needs: By integrating a pressure sensor into the cavity, the absolute pressure condition during the storage of the device can be monitored in situ. Figure 3.9 is showing an example of a device with soldered lid which was measured in different time steps and stored in a low humid office environment at room temperature. The pressure development inside the cavity is listed in Table 3.1. After a storage of 219 days, the pressure is still in the range of the inaccuracy of the pressure sensor itself which is +/−1 mbar.

By comparing this pressure increase of 1 mbar in 219 days, a lower limit estimation for q_air can be given: q_air$\approx$2*10^{-12} mbar*L/s. Subsequently, the seal has the potential to achieve a quality needed for the gas tight detector cell.

Table 3.1 Pressure development in cavity after sealing

Time after sealing t [d]	Temperature T [K]	Pressure p_c [mbar]	$\frac{p_c}{T}$ [mbar/K]
0	583.2	815	Na
15	299.7	487	1,63
24	297.2	484	1,63
35	299.4	487	1,63
219	297.6	481	1,62

Start pressure at $t = 0$ is the pressure during sealing process

3.7 The Demonstrator

The sensor and detector cell are held in place by a metal bracket (Fig. 3.10) and separated by an optical shielding which suppresses ambient light. As this light might influence the measurement signal and disturb the sensor, an additional shielding is required. The shielding frame in different heights of 5 and 10 mm and their cross-section is shown in Fig. 3.11a.

The 3D-printed parts were made out of an aluminum reinforced polymer which is able to resist temperatures up to 130 °C and absorbs most of the radiation energy (Fig. 3.11b). Long-term stability is insured by avoiding any reflective parts and mirrors in the absorption path. IR-light, which does not enter the detector on a direct path, is absorbed by the walls of the frame. Because the distance between the windows of the emitter and detector is set by the frame, any variation in height by the solder process is omitted.

A PCB setup is necessary to support the emitter and detector assembly with the frame in between which is connected by the metal clamp. Besides the mechanical mounting, the platform also provides voltage supply and conditioning, a microcontroller for communication, evaluation of the measurement signal and controlling of the emitter as well as secondary sensors to compensate cross influences. The system is not only used for development but also as a demonstrator to show the new sensor technology. The requirements for both kinds of systems are to be considered in the design. A demonstrator asks for a small form factor whereas a development system needs larger components to allow an easy exchange of them. The resulting system concept is shown in Fig. 3.12.

The central component is the XMC4400 Cortex M4 microcontroller with 120 MHz clock frequency from Infineon AG. The processor is responsible for the communication over the USB interface, controls the emitter as well as reads and evaluates all sensors on the board. A microphone with a digital output is used inside of the detector and connected in a direct way with the controller. This replaces an analog equivalent and therefore a large analog amplification and filter circuit with a considerable need for components and PCB-space. Analog filtering was replaced by a digital filter function located partly in the microphone itself as well as in the microcontroller. The analog temperature sensor in the detector is measured by the ADC without any additional amplification. The current generated by the photodiode

Fig. 3.10 Photoacoustic Gas Sensor demonstrator

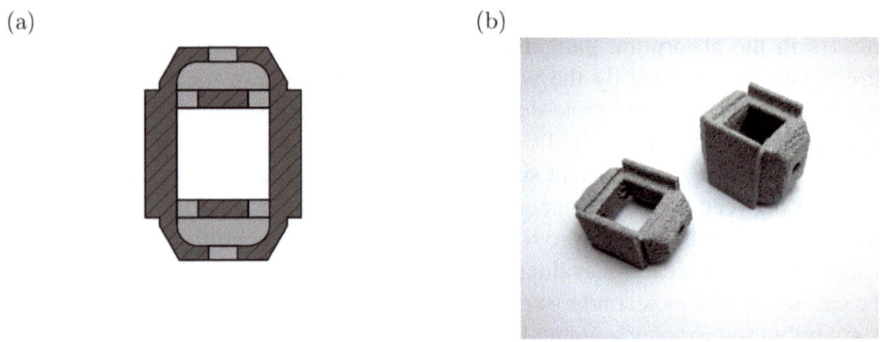

Fig. 3.11 Shielding frame, (**a**) schematic cross section and (**b**) parts with different heights (5 and 10 mm) to suppress ambient light

is transformed by a transimpedance amplifier into voltage and measured by the integrated ADC of the XMC controller like the emitter current and the emitter voltage. The PCB is supplied by 5 V of the USB port. From this voltage 3.3 V is generated by a line regulator to supply the controller and 12 V by a step-up converter to drive the emitter. These voltages are also monitored by the controller. The emitter voltage waveform is set by a H-bridge with 500 kHz switching frequency. With this component it is feasible to drive any voltage and current waveform in dc or alternating direction and thus to reduce electromigration effects on the emitter-membrane. The current of the infrared emitter is measured with a shunt resistor, amplified and digitalized by the ADC of the controller.

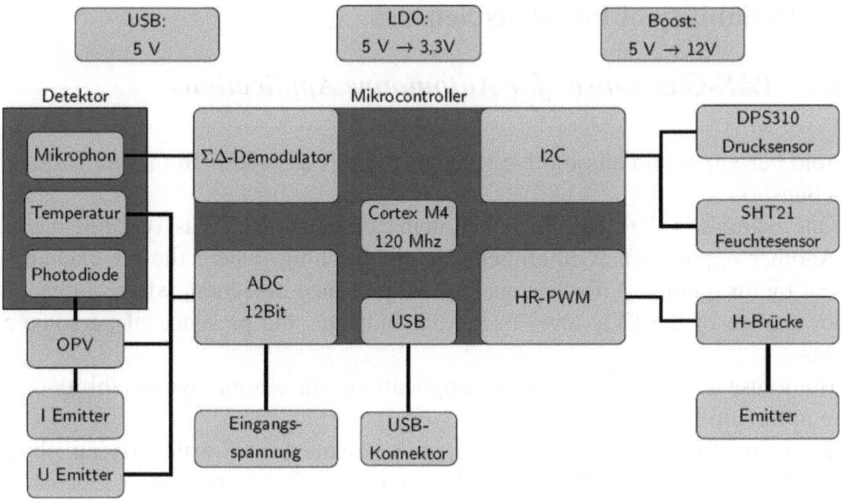

Fig. 3.12 Overview of the demonstrator system and its main components and sensors

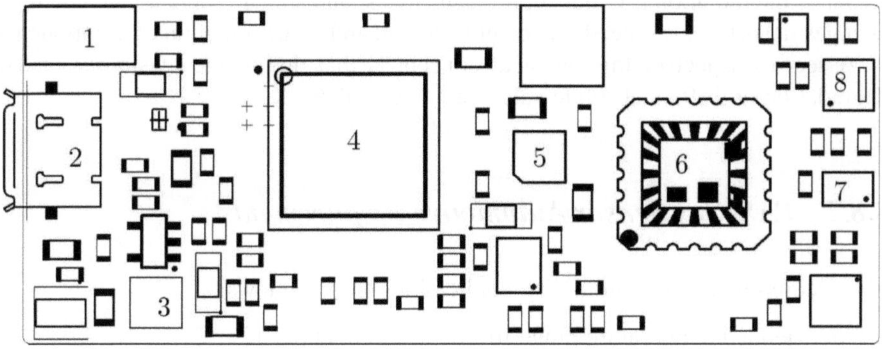

Fig. 3.13 (*1*) Debug connector (*2*) USB connector (*3*) Voltage regulator (*4*) Microcontroller (*5*) H-Bridge (*6*): CePas sensor (*7*) Pressure sensor (*8*) Humidity sensor

Secondary sensors are also added on the board to acquire additional ambient air data such as pressure, humidity, and temperature. Both sensors are connected with the system controller via I2C. The Infineon DPS310 is used because of its small form factor and high accuracy to measure the air pressure and temperature. A SHT21 allows the system to determine the humidity in the atmosphere.

Figure 3.13 shows the demonstrator PCB and the location of the core components. Due to the high component density, a small form factor of the design was accomplished with dimensions of 50 × 20 mm.

3.8 Reliability of PAS-Gas Sensors

3.8.1 PAS-Gas Sensor for Automotive Applications

As told before, within automotive vehicles there is a broad field of applications for Gas Sensors.

One example is CO_2 leakage detection in case of using CO_2 as a cooling medium.

Another application is intelligent air conditioning, where the air exchange is driven by air quality. A further use case is presence detection, where a change or a not change of the CO_2 level in the car indicates the presence of persons, e.g., babies.

But to use a PAS sensor for these applications, the automotive reliability criteria have to be fulfilled.

To address thus topic, first of all an assessment of the underlying physical principles has to be done. The physical principle determines the failure mechanisms.

In the case of PAS-Gas sensor, the principle is a pure physical volume concept. Radiation is absorbed and converted into heat and heat is converted into a pressure signal. If the radiation is chopped, it results in a sound signal. There are no chemical changes in surfaces, no need for reconditioning, and no degradation of components.

A second aspect of this physical concept is that due to the feasibility of low degradation or drift an absolute value can be calculated.

3.8.2 PAS-Gas Sensor Automotive Requirements

Main requirements for automotive applications are as follows:

- An extended temperature range from −40 °C to about 125 °C.
- Temperature robustness for standard MEMS devices and integrated circuits in the detector is given. No degradation expected here.
- Extended humidity range up to 95% rH.
- Closed cavity emitter and closed cavity detector sensor components give environmental robustness.
- Limited supply voltage below 12 V.
- 12 V is a supply voltage which can be handled can be provided in standard integrated circuits.
- Limited power consumption in standby mode < 1 mA.
- The PAS sensor using high sensitive microphones allows a broad variation of adaption to different applications. High emitter temperatures in combination with a high measurement frequency enable high resolution in concentration and time. Low emitter temperatures in combination with a low measurement frequency enable high robustness and long lifetime.
- Strong requirements about chemical robustness of the system in an automotive environment.

- Closed cavity emitter and closed cavity detector components enable a high environmental robustness.
- Failure rate is according to automotive standards.
- Usage of standard integrated components like the control ASICs and standard MEMS components like the microphone and ASIC and well-known packages enable sensor with known failure rates over lifetime.
- Feasible performance of the sensor over a lifetime of 15 years.
- Usage of standard integrated components (like microphone and ASIC, standard packages, etc.) with known failure rate.

3.9 Reliability Providing Building Blocks

A main advantage of the PAS-Gas Sensor is its underlying concept.

The physics on which the concept is based is radiation absorption in a gas volume. Therefore, there is no surface which changes, which can degrade, which can be contaminated, or which has to be refreshed.

A second building block, also based on the concept, is that the development of a physical model.

A model, which can explain the performance and the functionality of the sensor, can also be used to explain deviation. In addition, such a model can help to easily adapt to the setup of the sensor, to avoid failure mechanisms and to make the sensor more robust.

A third building block of reliability is the qualification of the sensor according to automotive rules.

This must be topic of further PAS sensor development. In addition, in this case the model will help to find appropriate tests. A main task here will be the development of test, which can accelerate the aging of the sensor. The development of tests with known activation energies regarding aging will be used to show the feasibility of a robust sensor. In addition, the availability of those tests accelerates the development of optimized sensors and sensor modules.

3.10 Characterization of the Demonstrator

The system was placed in an aluminum housing with two gas fittings. This allows to measure the response time due to its small volume and to pressurize the system due to its rigid design (Fig. 3.14).

Systematic measurements to characterize the system behavior were conducted. Therefore, different gas concentrations of CO_2 at different environmental conditions like temperature, pressure, and humidity were flown into the sample cell.

The measurement of the sensitivity shows a standard deviation of 25 ppm at 100 ppm CO_2-concentration and up to 60 ppm at 40,000 ppm. A significant but

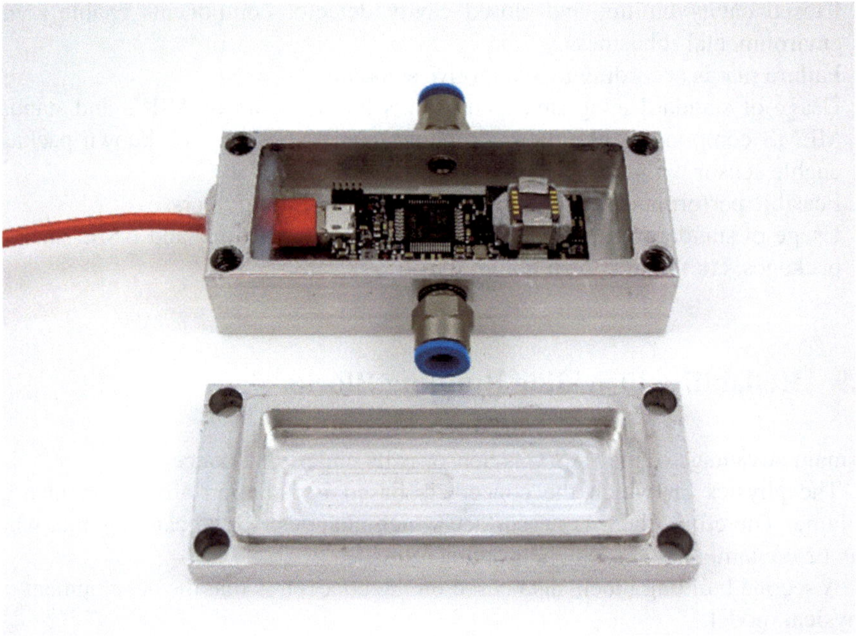

Fig. 3.14 For the measurements a system in an aluminum housing with two gas fittings and the USB interface is used

linear temperature influence was determined in the range from −40 to 90 °C. Due to its linear behavior, it can be compensated to achieve an accuracy of 10% of the readout value.

The pressure impact as well as the temperature impact was measured with the same system in a range from 700 to 1100 hPa. The detected deviation was relative to the measured concentration in a linear manner. The influence of the pressure can be compensated with the pressure sensor on the board. In an automotive application, it is feasible to use the pressure data from the engine or the interior control bus.

The impact of humidity is negligible in a wide range and smaller as the system noise. Only high H_2O concentrations at high temperature and high relative humidity could alter low CO_2-concentrations up to 10%, which is still acceptable. Further investigations about how very high humidity concentration affects the thermal behavior needs to be done.

Additional environmental interference factors like noise and vibration have also been analyzed because of their appearance in the automotive setting. Sound shows no impact at all on the system, whereas an influence of vibration is observable. If the vibration is outside of the range of the IR-emitter frequency, it is suppressed by digital and analog filters. But it leads to a beat of the two frequencies if it is right next to it. First results show only small distortions, but further investigations need to be done to identify the impact of this behavior.

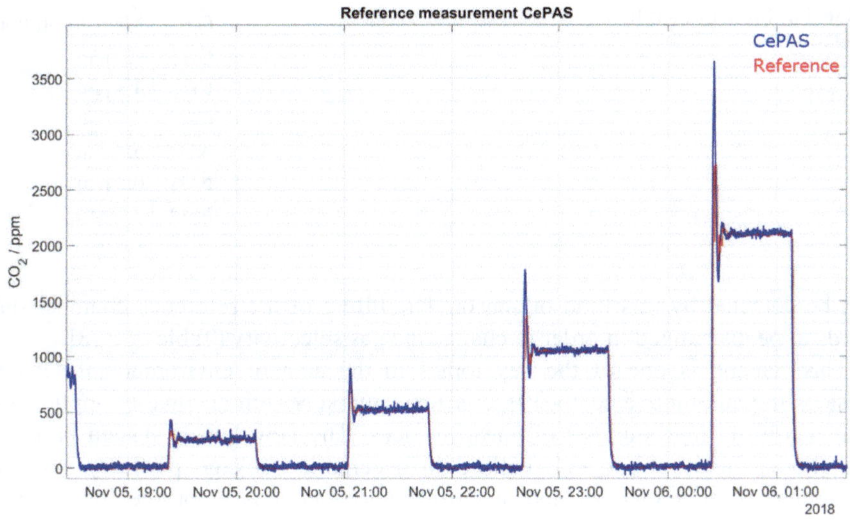

Fig. 3.15 Reference measurement of the CePas Sensor

The reaction time of the system is defined as the T90 time, the time until the output changes up to 90% of the settled concentration. Because of the very short absorption range of 4 mm, only a small amount of gas needs to be exchanged and therefore a very fast system reaction can be achieved. Measurements at different concentrations show a reaction time of 20 s which is mostly limited by the exchange of the gas inside of the whole aluminum measurement setup. Figure 3.15 shows a reference measurement of the CePas sensor and the result of a reference sensor. The spikes in the graphs are caused by the measurement setup which uses a PID algorithm to set the concentration with the mass flow controller.

3.11 Summary and Outlook

A PAS sensor is an IR sensor which uses a highly sensitive pressure sensor—in this case a MEMS microphone—as detector. The concept provides advantages in size, resolution, manufacturability, and big potential regarding adaption to different use cases. The concept is an approach for a small sensor device with the potential of further miniaturization and applicability in various markets including automotive. Several parts of the sensor have been optimized in a certain way to increase the performance. The sensor was assembled in a measurement system and systematic measurements to characterize the system behavior have been conducted. With the integration of the components and the circuit into a dedicated ASIC, a further reduction in cost and form factor can be achieved. CO_2 is not the only gas that

Table 3.2 Sensor absorption levels

Gas	Main absorption
CH_4	3.3 μm
CO_2	4.3 μm
CO	4.6 μm
NO	5.3 μm
NO_2	6.2 μm
N_2O	7.7 μm

can be detected by this kind of sensor. The filling of the reference chamber only needs to be exchanged in order to change the gas selectivity (Table 3.2) [8].

Gas sensing is one of the key topics in the area of environmental sensing. Besides the automotive sector [9], consumer markets with air quality applications also require a huge number of cheap devices. The new sensor system based on the photoacoustic effect and realized in a ceramic package perfectly suits the requirements of a modern and state-of-the-art gas sensor for the high-volume market.

References

1. T.H. Tille, *Automobil-Sensorik: Ausgewählte Sensorprinzipien und deren automobile Anwendung. 1. Auflage* (Springer, Berlin, 2016). ISBN 978-3-6624-8943-7
2. Department of Defense, *MIL-STD-883K: Test Method Standard Microcircuits* (Department of Defense, Columbo, OH, 2016)
3. National Semiconductor Corporation, *Semiconductor Packaging Assembly Technology* (National Semiconductor, Santa Clara, CA, 1999). www.national.com
4. D. Stroehle, On the penetration of gases and water vapour into packages with cavities and on maximum allowable leak rates, in *15th International Reliability Physics Symposium, Las Vegas (Nevada)*, (1977), pp. 101–106. https://doi.org/10.1109/IRPS.1977.362778
5. S. Costello, M.P.Y. Desmulliez, *Hermeticity testing of MEMS microelectronic packages*, 1st edn. (Artech House, Norwood, 2013). ISBN 978-1-6080-7527-0
6. S. Costello, M.P.Y. Desmulliez, S. Mccracken, Review of test methods used for the measurement of hermeticity in packages containing small cavities, in *IEEE Transactions on Components, Packaging and Manufacturing Technology*, ed. by S. Costello et al., vol. 2, (2012), pp. 430–438. https://doi.org/10.1109/TCPMT.2011.2176122
7. H. Greenhouse, *Hermeticity of Electronic Packages*, 2nd edn. (Elsevier, Amsterdam, 2011). ISBN 978-1-4377-7877-9
8. "HITRAN spectral database," 2016 [online] https://www.cfa.harvard.edu/hitran/
9. S.A. Abdul-Wahab, S.C.F. En, A. Elkamel, L. Ahmadi, K. Yetilmezsoy, A review of standards and guidelines set by international bodies for the parameters of indoor air quality. Atmos. Pollut. Res. **5**, 751–767 (2015)

Chapter 4
Multipath Ray-Tracing-Based Modelling of Time-of-Flight Cameras

Method and Application to Semi-Transparent and Scattering Media

Marcus Baumgart, Norbert Druml, and Cristina Consani

4.1 Introduction and Related Work

The intent of this section is to give a brief overview on optical Time-of-Flight (ToF) sensors and technologies, with a special focus on implementations for ranging. After a brief introduction on current ranging technologies, the working principle of ToF sensors is discussed in more detail. In particular, we will briefly review existing simulation frameworks that allow modelling a ToF sensor response, and the challenges imposed by multipath interference effects not only on the sensor accuracy, but also on the realization of an accurate and predictive sensor simulation environment.

4.1.1 Optical Range Detection

Fast and reliable classification of the environment is considered an evolutionary advantage. Range detection based on light started evolving millions of years ago and since then continuously specialized and improved in nature. A very familiar example of light-based range sensing and 3D perception is given by the human eyes. The human eye can be classified as a passive optical range detector because it only relies on illumination from environmental light sources, like the sun, to detect objects and estimate their positions. The light is converted in the eye's retina

M. Baumgart · C. Consani (✉)
Photonic Systems, CTR Carinthian Tech Research AG, Villach, Austria
e-mail: cristina.consani@ctr.at

N. Druml
Infineon Technologies Austria AG, Graz, Austria

W. D. van Driel et al. (eds.), *Sensor Systems Simulations*,
https://doi.org/10.1007/978-3-030-16577-2_4

into electrical neural impulses which are decoded by the brain to perform visual perception. A single eye uses a variety of strategies known as *monocular visual cues* to estimate the distance of an object and the relative position of different objects in space. Among such strategies are the size of a familiar object on the retina, the relative sizes of familiar objects, visual overlap of objects, linear perspective, motion parallax, relative position, and/or interposition of shadows, as well as defocusing and refocusing techniques which allow inferring the distance of an object from the contraction of the muscles of the eye. Although most of these strategies only provide cues on the relative distances between different objects present in the scene, continuous training and the resulting experience enables good absolute distance estimations even for monocular vision [1–3].

With the combination of the visual information from both eyes (*binocular vision*, or *stereovision*) the human vision system becomes a very good range sensor. When we observe an object first with one eye and then with the other, we realize that the images from the two eyes differ. This phenomenon is called "retinal disparity": Since the eyes are horizontally separated, the projection of the same object on the retina differs between the two eyes, with a difference that decreases with longer distances. Our brain "decodes" the information contained in this retinal disparity and returns a single image with a good estimation of the location of the objects in the 3D space. This principle is of course known and very well understood, and it is at the basis of triangulation techniques. Most "passive" optical ranging techniques were initially inspired by human vision, and later on further optimized.

Range detection systems can be divided into "passive illumination" and "active illumination" systems. Passive illumination systems are the direct analog to human eyes: they use external illumination sources (e.g., sunlight) and adopt specific strategies to evaluate the distance of surrounding objects. Just like people can infer distance information from monocular cues, several techniques using a single camera aided by specially designed smoothing and image-recognition algorithms have been developed for ranging. Examples are depth from focus and defocus [4], or monocular Simultaneous Localization and Mapping (SLAM) [5]. However, the most common strategies are inspired by binocular vision. The main idea behind *stereovision* is to observe the same scene with two cameras that are separated by some distance in space. Completely analogous to what happens in human eyes, the projection of the same scene on the two cameras differs due to the distance between them, and a disparity map can be constructed (which, in the simplest case, is computed from the difference between the pixel locations of the same object position on the two cameras). As the disparity is inversely proportional to the distance of the objects, the depth of the scene can be reconstructed. Such *triangulation* schemes allow reconstructing the positions of object points within a three-dimensional scene, provided that several unique corresponding points in both images (within the common field of view of the two cameras) can be identified. This requirement is also known as the "correspondence problem" and it can be difficult to satisfy in practice, especially for multi-camera systems in front of arbitrary scenes. Much software development in the direction of image processing (e.g., to correct for lens distortions), image recognition, and modelling is active in this directionbreak

[6–8]. Nowadays stereo cameras combined with advanced algorithms can be used for detection, recognition, and distance estimation of static scenes, as well as for motion tracking for a large variety of scenes. An in-depth introduction to this topic is given in [9–11]. Extending this scheme to include more cameras at different angles will result in a more precise description of the three-dimensional scene. This is the idea behind *lightfield imaging*. In most implementations, a lightfield camera includes a micro-lens array in front of a conventional 2D pixelated detector so that a large number of images from different angular perspectives can be acquired in a single shot.

All these techniques rely on external light sources. If the available environmental light sources are not sufficient, a dedicated light source can be used (e.g., using a torch on a night walk enables us to see and detect objects in the absence of sufficient environmental light). Since the source characteristics do not affect the detection scheme, no other requirement is imposed on the source other than being intense enough to illuminate the scene at a level that is sufficient for detection.

Active illumination detection systems by contrast require a dedicated illumination source and specific knowledge about the projected light. The term "active" implies that not only the source is part of the sensor itself, but that different illumination patterns (*structured* or *coded light*) are projected (in space and/or in time) onto the scene. The 3D coordinates of the objects in the scene are then reconstructed using the response of the sensor to the multiple illumination patterns. In general, three main principles can be employed for active illumination systems: triangulation-based schemes, round-trip time-dependent schemes and wave-mixing schemes.

In the torch example above, further information about the torch, like the illumination characteristic and the position relative to the eyes, may enable a more precise 3D reconstruction (*active triangulation*) compared to purely passive detection methods. For example, the appearance and shape of shades in combination with the light source parameters will improve the robustness of the estimation of the object positions (*image pattern recognition*). Alternatively, the projection of an artificial pattern placed in front of the light source can introduce further structure into the scene (*pattern projection*). The knowledge of the pattern's structure adds an overlay "mask" to the objects being imaged. Upon proper design of the pattern, a unique correspondence between emitted and observed features can be established, which allows for a precise reconstruction of distances. These special coding/decoding strategies are known as *coded structured light* (CSL) and are at the basis of several well-known ranging systems [12], including the first (2010) Microsoft Kinect™ camera [13]. When a second pattern is inserted in front of the detector in addition to the projection pattern, object geometry reconstruction based on *shadow Moiré* interferences can be performed [14].

An additional opportunity to measure distances with active illumination becomes evident if we realize that there is more information encoded in a reflected light beam than just its intensity and angular distribution. As light travels through air at a constant finite speed, the time it takes for a short light pulse to travel to a distant object and back is a direct measure of the object distance. For pulsed sources, this time delay can be directly measured with a very fast opto-electrical detector synchronized with the light source although in most common applications indirect methods (also called correlation methods) using slower detectors are implemented. The round-trip time measurement with a *pulsed* or *modulated* (intensity modulated or frequency modulated) light source is known as *optical Time-of-Flight* (ToF) and will be the focus of the rest of this chapter.

While the ToF schemes mentioned above use external modulation of the light source, the intrinsic oscillation of the light electric field can also be employed to measure distances. Since the oscillation frequencies are much higher than the response times of available detectors, interferometry has to be used. In the simplest interferometric scheme, a beam is split into a reference and a measurement beam travelling through different optical paths. The difference in optical path length between the two beams is obtained by measuring the interference obtained when recombining the two beams on the detector. For a monochromatic source with wavelength λ, measuring the intensity at the detector only allows to retrieve a relative phase difference between the two beams, corresponding to a maximum unambiguous depth range of $\lambda/2$. Continuously counting the number of interference fringes allows tracking larger distance changes, provided the direction of the distance change is known. The unambiguity in the direction can be removed with little effort using quadrature detection. The distance measurement remains, however, a relative measurement.

Performing interferometry at different light frequencies simultaneously (*multi-wavelength interferometry*) allows extending the unambiguity range. If sufficiently many different light frequencies are used, the unambiguous range can be extended to span all expected target distances (thus being equivalent to an absolute distance measurement) while maintaining interferometric accuracy as determined by the shortest used wavelength. This is the basic principle of optical coherent tomography (OCT) in the frequency domain. The approach requires a highly complex sensor system and may become unfeasible if a large measurement range is required. *Heterodyne interferometry* often provides a good alternative. Here two different light frequencies from the measurement and reference paths, respectively, are brought to interfere. In practice, this is realized by generating a sideband of a laser source using an optical modulator. After recombination, this generates a modulated signal with the beat frequency $\nu = |\nu_1 - \nu_2|$. The unambiguous range can be made arbitrarily large if the frequency difference is made sufficiently small. The disadvantage is the lower accuracy now determined by the heavily enlarged synthetic wavelength. Implementations where more than two frequencies are used or where the distance is measured at several beat frequencies by changing the frequency in the reference or measurement arm allow for getting a large unambiguous range while still maintaining good accuracy.

Another approach employs frequency sweeping of a narrowband light source and uses interferometry to determine the frequency shift of the returning light, which is proportional to the round-trip time. The frequency shift between the reference arm and the measurement arm is measured as a beat frequency and depends on the absolute path length difference between light travelling through the two interferometer arms. The technique is known under the name of frequency modulated continuous wave (FMCW) round-trip time measurement or optical frequency domain reflectometry (OFDR) and is heavily used in fiber sensing and radar ranging.

Most of these interferometric concepts can also be extended to map a 3D scene either by scanning or through optical expansion of the beam (*holographic techniques*). Finally, several of the principles described above can be combined within a single sensor to achieve a more accurate 3D perception. Of course, for most applications *ranging* implies the measurement of an extended area in front of the sensor. Not all the mentioned principles cover this out-of-the-box, and in some cases some extra "trick" needs to be included. For example, the combination of a collimated light source with a scanning device is common practice and yields very precise results in Time-of-Flight imaging or interferometry.

4.1.2 Optical Round-Trip Time Measurement Methods and Devices: Lidar and Time-of-Flight

The idea of determining a distance by measuring the round-trip time of an emitted wave is at the basics of widespread ranging techniques, including sonar (SOund Navigation and Ranging) and radar ("RAdio waves Detection And Ranging"). Historically, the idea to use the round-trip time of a light pulse at the place of radiofrequency waves was introduced in the 1950s [15] under the name of lidar[1] (LIght Detection And Ranging) although the concept became only attractive after the invention of the laser in 1960 [16–19], which granted a source of (short) narrowband and collimated light pulses.

The lidar principle can be considered the optical equivalent of radar: a conventional radar emits electromagnetic pulses in the radiofrequency range and measures the return time of the radiation reflected from surrounding objects (echoes), whereas a conventional lidar emits light pulses and measures the return time of the reflected light (round-trip time or Time-of-Flight) in a stop-watch scheme. In practice however, as it is described below, both methods are often operated with amplitude and/or frequency modulated continuous wave (CW) schemes [20].

Typical wavelengths used by lidar span from 250 to 2000 nm, depending on the application. As the frequency of the light waves is between four and eight magnitudes higher than the radiofrequencies used by radars, range detection with a lidar provides a higher spatial resolution with a much smaller receiver. The drawback is that light at optical wavelengths interacts much stronger with particles and gases in the atmosphere than radiofrequency waves. While this can be an advantage for applications related to atmospheric analysis, the same effect can be

[1]Other names of lidar sometimes found in literature are optical radar, laser radar, and ladar (laser detection and ranging).

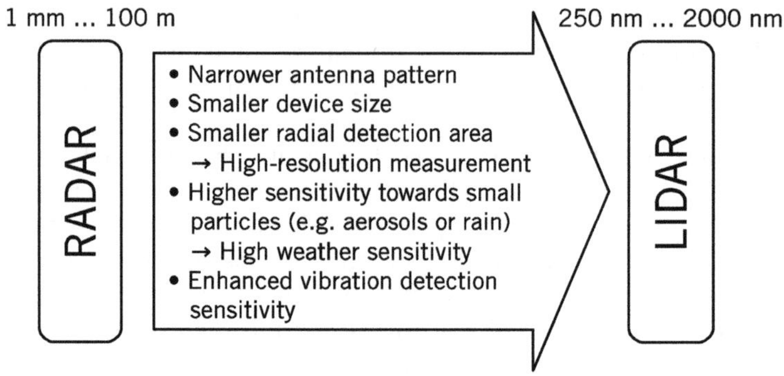

Fig. 4.1 The decrease in wavelength from radar to lidar is accompanied by significant changes regarding the device properties and the interaction with the environment

detrimental for applications where accurate ranging in any circumstance is expected. A non-exhaustive comparison between radar and lidar is given in Fig. 4.1. Besides its use in metrology, lidar finds application in topography and landscape mapping [21–24], bathymetry, and remote monitoring and analysis of the atmosphere [25]. In particular, topics like cloud research [26], analyzing the composition of aerosols, including the presence of ozone and pollutants [27, 28], and monitoring wind speeds and the presence of turbulence [29, 30] triggered the development of a variety of lidar implementations. Examples include Raman Lidar, Fluorescence Lidar, Differential Absorption Lidar, Backscattering Lidar, and Doppler Lidar. As these applications and methods are beyond the scope of this chapter, we refer the interested reader to the excellent review book of Takashi Fujii and Tetsuo Fukuchi [31]. In the following, we will focus on ranging applications.

A simple lidar device consists of a pulsed laser light source, a receiver optics and a detector with their respective electronics and data-treatment system. Depending on the application, some optics might be included after the laser source (e.g., a beam expander); schemes with beam-steering devices or multiple lasers and/or detectors also exist. Emitted light pulses encountering an obstacle are partially backscattered towards the receiver optics, with an efficiency depending on the material and surface properties of the object. The time delay τ between the emitted and received light pulse is a measure of the obstacle distance D as given by

$$D = \frac{c \cdot \tau}{2} \tag{4.1}$$

where c is the speed of light in vacuum (which is a reasonable approximation for a homogenous air medium). Since the round-trip time is directly measured by the detector, the method is known as *direct Time-of-Flight* (d-ToF). The use of short light pulses has several advantages, among them the reduction of sensitivity to background light and the increase in signal-to-noise ratio enabled by the possibility of using high pulse-energies while still keeping the total emitted power low enough to allow for eye-safe operation. A direct measurement of the time delay, however, requires picosecond time resolution for millimeter accuracy (the round-trip time for 1 mm path length is 6.7 ps) and, thus, puts high demands on the detector and its electronics in terms of bandwidth, dynamical range, and noise requirements.

An often-preferable alternative is directly inspired by interferometry (see Sect. 4.1.1). The idea is to use a non-coherent continuous-wave light source and add a modulation of the light intensity in the radio-frequency (RF) range. The beam returning to the detector contains the same RF modulation as the emitted light, with an additional phase shift that is proportional to the round-trip time. The phase shift is extracted in a demodulation step by mixing the modulation of the returning light with the original modulation. Initially, the signal mixing and correlation of the received signal with the reference was performed using separate electronic amplifiers and mixers implemented after the detector. However, their temporal delays and drifts could cause significant errors on the extracted phase delay. While the use of electro-optical modulators provided a good solution for the demodulation step (an exemplary scheme is shown in Fig. 4.2a), the introduction of demodulation pixels (in particular photonic mixer devices or PMDs [33]) allowed to perform

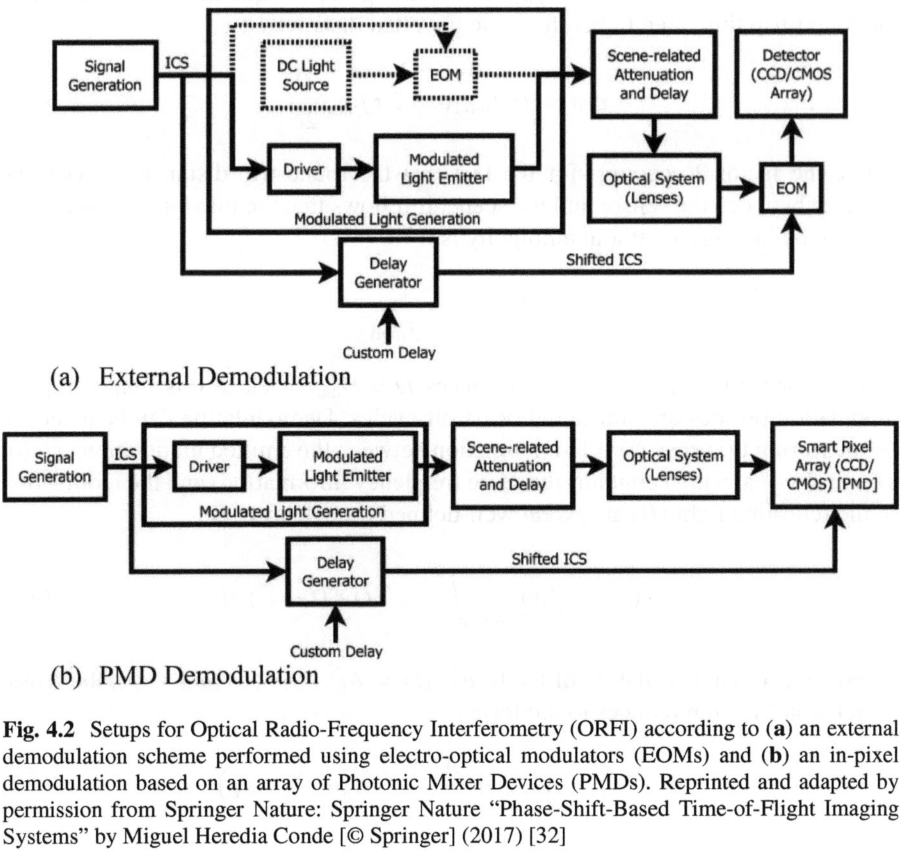

Fig. 4.2 Setups for Optical Radio-Frequency Interferometry (ORFI) according to (**a**) an external demodulation scheme performed using electro-optical modulators (EOMs) and (**b**) an in-pixel demodulation based on an array of Photonic Mixer Devices (PMDs). Reprinted and adapted by permission from Springer Nature: Springer Nature "Phase-Shift-Based Time-of-Flight Imaging Systems" by Miguel Heredia Conde [© Springer] (2017) [32]

the mixing directly at the detector, with a significant increase in detection speed and decrease in detector size, and without the need of an additional demodulation element in the beam path (Fig. 4.2b). This technique is known as *correlation-based ToF* (c-ToF) or *optical RF interferometry* (ORFI) [34], and is at the basis of the modern detection schemes of most lidar and Time-of-Flight devices. As the second name suggests, it can be considered an interferometric method, where the frequency information is shifted from the frequency of the light to the RF modulation applied to it. As for interferometry, the detection of the round-trip time reduces to a correlation measurement between an electrical signal containing the original modulation (control signal) and the modulation of the returning light. Many different implementations exist, for both homodyne and heterodyne detection schemes, using continuous or discontinuous modulation waveforms.

Here, we will focus on the basics of homodyne-detected amplitude-modulated (AM) c-ToF, which is similar to phase-measuring interferometry [35]. In such a scheme, the brightness of the light source is modulated as a function of time with a modulation frequency f_{mod} in the RF range. Since the frequency of the returning light coincides with the frequency of the control signal, the information on the distance difference is coded into the phase-difference $\Delta\varphi$ between emitted and returning wave. In fact, known f_{mod}, determining $\Delta\varphi$ is equivalent to determining the round-trip time τ and, therefore, the travelled distance D:

$$\Delta\varphi = 2\pi f_{\text{mod}}\tau = 2\pi f_{\text{mod}}\frac{D}{2c} \tag{4.2}$$

where the factor 2 accounts for the fact that the round-trip distance is twice the distance between the sensor and the scene. It follows that the maximum range D_{max} that can be measured without ambiguity is

$$D_{\text{max}} = \frac{c}{2f_{\text{mod}}} \tag{4.3}$$

and that phase-wrapping projects distances $D > D_{\text{max}}$ back to this range. Typical modulation profiles are sinusoidal or on/off cycles. Demodulation can be achieved by measuring the cross-correlation function between the emitted modulation $q(t)$ (or a control signal which contains the same frequency information) and the modulation of the returning light $r(t)$ at several well-defined time delays t':

$$c_{q,r}\left(t'\right) = \lim_{T\to\infty}\frac{1}{2T}\int_{-T}^{T} q^*(t)\, r\left(t - t'\right)\, dt \tag{4.4}$$

For sinusoidal modulation of the form $q(t) = A_0^q + A^q\cos(\omega\, t - \varphi)$, the cross-correlation function reduces to the form

$$c_{q,r}\left(t'\right) = A_0^q A_0^r + \frac{A^q A^r}{2}\cos\left(\omega\, t' - \Delta\varphi\right) \tag{4.5}$$

Unique determination of $\Delta\varphi$ requires measuring $c_{q,r}(t')$ for at least two-time delays. Note that measuring the cross-correlation function at several delays t'_i is equivalent to repeating the measurement by applying a phase offset $\varphi_i = \omega t'_i$ to the control signal. If the modulation waveform is not sinusoidal, the cross-correlation function contains also harmonics of the fundamental frequency.[2]

For a sinusoidal RF modulation, a minimum of 2 phase offsets between the reference and returning signal are sufficient to determine both the phase difference $\Delta\varphi$ and the returning amplitude. In practice, however, the typical acquisition scheme uses four-step phase cycling for the control signal with 90° phase steps ($\varphi = 0°$, 90°, 180°, 270°), corresponding to the four-step algorithm used in phase-shifting interferometry.

The phase difference can be retrieved from the cross-correlation signals by

$$\Delta\varphi = \tan^{-1}\left(\frac{C_{270^\circ} - C_{90^\circ}}{C_{180^\circ} - C_{0^\circ}}\right) \tag{4.6}$$

Compared to the two-step scheme, the four-step scheme reduces the noise contributions, allows retrieving the magnitude of an eventual DC offset and, for non-sinusoidal modulations, also the contribution of the higher harmonics [32].

Most of the current c-ToF systems use PMD pixels or arrays for detection. While a variety of technologies exist for PMDs, both in terms of materials and physical implementations, these devices are all based on the generation of electron-hole pairs in a photoelectric-active layer. Electrode pairs located in proximity of the photoactive layer are used to separate and transport the charges to a specific "storage area," where readout takes place. Recent PMDs feature more than one collection area to improve measurement speed (simultaneous measurement of two or more phase offsets). Demodulation occurs because the charges, which are generated according to the modulation characteristics (amplitude, frequency, and phase) of the incident light, are shifted to the storage areas using the modulation of the control signal. In the simplest and most intuitive scheme, a photo-gate PMD (Fig. 4.3a–c) pushes the charges to the storage areas A or B depending on the voltage applied to the corresponding electrode. If the voltage signal is periodically modulated, the process can be seen as a "dynamic seesaw" [37] moving the charges alternately to the two storage areas. Any DC light component will generate an identical amount of charges in the two storage areas. By defining the quantity K_φ as the difference between the PMD signals $I_\varphi^{A,B}$ from channels A and B, the contribution of background light can be eliminated. If, instead, the modulation frequency of the light matches that of the control signal, the amount of charges collected in the two storage areas depends only on the amplitude and the phase of the returning optical

[2]It can be shown, however, that it is sufficient that only one of the two signals in Eq. (4.4) is sinusoidal to eliminate the contributions of the harmonics from the correlation [32, 36].

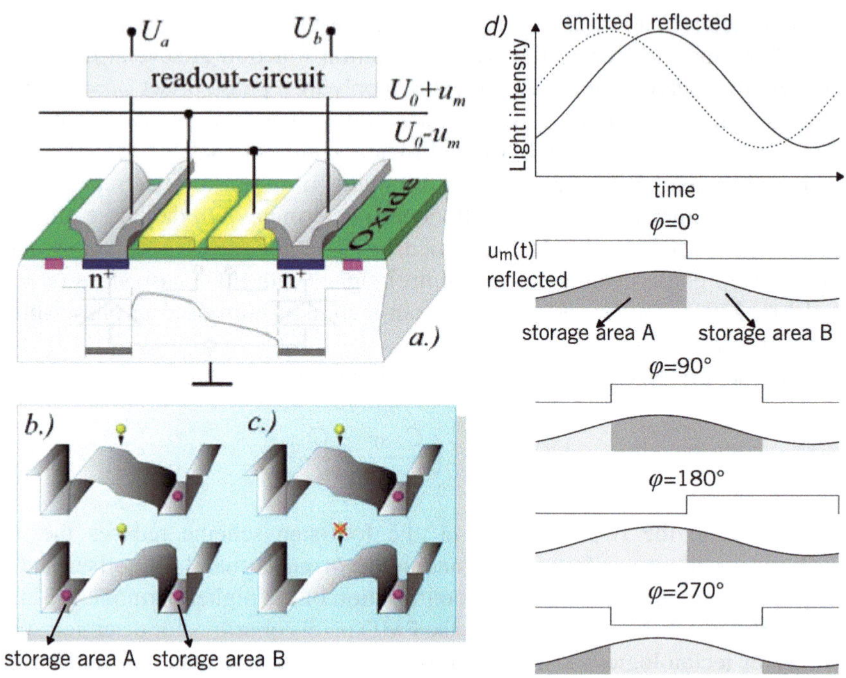

Fig. 4.3 Functionality of a photo-gate Photonic Mixer Device (PMD). (**a–c**) The charges generated in the PMD pixel by illumination are shifted to the storage areas A or B by the electric field generated by the difference of potential between the photo-gates. DC light (**b**) builds up an identical charge in the two storage areas, while (**c**) modulated light having the same modulation frequency as the control signal $u_m(t)$ results in a different population of the two storage areas. The population difference depends on the phase difference between the two signals, as exemplified in (**d**) for the four-step phase scheme. Panels **a–c** are reproduced from [37] with kind permission from pmdtechnologies ag on behalf of the authors

signal. Figure 4.3d shows an example where the applied voltage follows the four-step phase modulation scheme[3].

Having defined the differential signals K_φ, Eq. (4.6) can be rewritten as

$$\tan(\Delta\varphi) = \frac{K_{270^\circ} - K_{90^\circ}}{K_{180^\circ} - K_{0^\circ}} \tag{4.7}$$

[3]A photo-gate PMD pixel with two storage areas can acquire the information for two steps (e.g., 0° and 180°) in parallel, as one storage area per phase step is needed. The remaining two steps can be acquired in a second measurement (serial process) or by a second pixel in close vicinity to the first one.

It can be shown that the amplitude A of the returning signal and any DC offset A_0 can also easily be determined with this modulation scheme as:

$$A = \frac{\sqrt{(K_{270^\circ} - K_{90^\circ})^2 + (K_{180^\circ} - K_{0^\circ})^2}}{4} \tag{4.8}$$

$$A_0 = \frac{I_{0^\circ} + I_{90^\circ} + I_{180^\circ} + I_{270^\circ}}{4} \tag{4.9}$$

State-of-the-art PMD designs achieve a demodulation contrast decrease of 50% at 100 MHz [38]. Improvement in performance typically has to deal with a trade-off between high frequency operation and large light detection region (large pixel size and/or high array fill factor).

More advanced amplitude modulation schemes also exist. In the simplest case, adding a second (or more) RF modulation frequencies allows extending the unambiguous range. A somewhat more sophisticated version of the same concept is realized by chirped AMCW lidar [39], where the amplitude of the light source is modulated not at a fixed RF, but using a chirped RF modulation. Similar to single-frequency AMCW lidar, the returning light is mixed electrically with the control signal at the detector.

An alternative to AMCW lidar schemes is presented by frequency-modulated (FM) CW ToF. The technique is also referred to as coherent ToF since the returning light is mixed optically with a reference beam (local oscillator or LO) in a heterodyned scheme. The FMCW scheme is completely analogous to the optical frequency domain reflectometry described in Sect. 4.1.1. The wavelength of a CW light source is modulated with a chirped or periodically varying RF modulation, i.e., the frequency of the light changes (typically linearly or sinusoidally) with time. The instantaneous frequency of the returning light is therefore a direct measurement of the round-trip time, and it is extracted from the beat frequency resulting from the mixing of the returning light with the local oscillator. FMCW ToF requires a coherent light source and a suitable optoelectronic system to achieve a repeatable frequency modulation, yet it provides several advantages compared to incoherent schemes, including higher sensitivity and lower noise as amplification is performed optically by mixing the signal with the LO and, therefore, thermal noise is not amplified in the process.

An excellent and comprehensive review of ToF principles, including the mathematical derivation of the Eqs. (4.4)–(4.9) and an in-depth description of PMD technologies, measurement schemes and limitations can be found in [32]. For a discussion of more advanced schemes employing a combination of AM and FM modulation waveforms (e.g., AM with selective frequency switching or FM with selective amplitude switching) or even pseudo-noise modulation we refer the reader to [34, 40, 41].

Until now the functionality of d-ToF and c-ToF sensors was described under the assumption of measuring the return time of a collimated source illuminating a single point in front of the sensor. As mentioned above, different strategies exist to turn a point distance measurement into a 3D imaging system. Historically, the *scanning approach* formed the basis of the very first optical ranging system [42]. Such an approach requires the same hardware as for the single-point measurement with the addition of a scanning device that scans the direction of the emitted light through the scene. Clearly, the spatial resolution of a scanning device is limited by the size of the illuminated spot, thus a well-collimated light source, e.g., a laser, should be used for such applications. Using a collimated source has the additional benefit that all the light intensity is directed to the small area to be measured, thus the amount of light returning to the detector is maximized. Therefore, scanning devices feature typically good signal-to-noise ratio even at distances of several tens of meters. However, the serial point-by-point scanning results in a slow framerate and even in motion artifacts between different points within one frame if the object is moving. The fundamental limit for the maximum achievable framerate is determined by the depth range, due to the finite propagation speed of light. In practice, the framerate may be further reduced by the adjustment speed of the scanner and the electronics of the detector and the processing unit. Consider, for example, a d-ToF device with 50 m depth range. This corresponds to a maximum round-trip time of 0.33 μs which results in a maximum repetition rate of 3 MHz. Typical object tracking applications demand a framerate faster than 30 Hz and a wide field-of-view. In the example at hand, a maximum of 333×333 data points can be measured at 30 Hz framerate, neglecting additional delays arising from the electronics and the mechanical parts. The operation of an analogous c-ToF scanning device is even slower since at least two measurements per point are needed to recover the phase. Recently, the possibility to use integrated optical phased arrays [43] as a beam-steering device for lidars has been proposed [44, 45] and realized [46]. The approach is very promising for the realization of a robust, cheap, and fast scanning device, which would additionally provide large flexibility in the design of customized scanning patterns. The fundamental limit on the maximum framerate, however, remains. Analogous to multi-wavelength interferometry, the best strategy to overcome this limit is to implement multiple-frequency measurements although a trade-off exists between the number of frequencies used and the increase in the number of required measurements for each scene point.

Framerate and spatial resolution can be improved by arranging multiple scanning units together and operating them in parallel. This is often not realistic due to the cost and size of the emitting units. A better alternative is the *direct imaging* approach, where an extended area of the scene is illuminated and the returning light is imaged on a pixelated detector, where each pixel has to be able to measure the depth independently. In d-ToF, this has been realized with streak-cameras [47], photodiode arrays [12], or single-photon avalanche diode (SPAD) arrays, which provide up to single-photon sensitivity and have been successfully implemented with good performance in CMOS technology [48, 49]. In c-ToF devices, mostly PMD arrays are used although also SPAD-based sensors have been reported [50, 51]. The direct imaging approach requires very few optical elements and allows for high integration and massively parallel data acquisition and processing. The approach requires, however, a light source that can be modulated at several MHz and

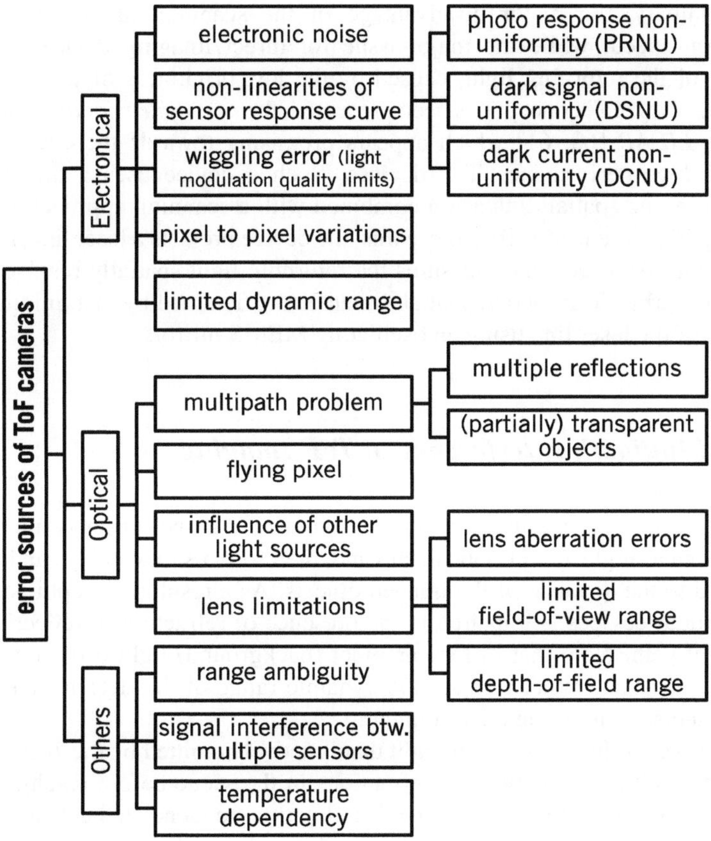

Fig. 4.4 Important error sources in camera-based Time-of-Flight sensors (based on error investigations given in [53–57]. Reproduced from [58]

can illuminate the whole scene with a reasonable brightness. Additionally, ranging errors caused by multipath interference (see Sect. 4.1.4) and mixed pixels are more pronounced in ranging sensors featuring direct imaging [52], and lens aberrations might have a significant impact on the accuracy and need to be compensated for in a post-processing step. Another issue is the cross-talk between independent devices. When parts of the scene are illuminated by two devices with identical modulation frequency, the reflected intensity modulation is biased and may lead to wrong distance measurements for a large area, or even the whole scene. The issue is of course present also for scanning devices, but the extent of perturbation is typically restricted to smaller areas of the scene by the simultaneous scanning of emitter and received direction. A more detailed list of potential errors present in a ToF camera, which either stem from the sensor's measurement principle or are due to technical limitations, are given in Fig. 4.4. A recent overview comparing existing devices can be found in [59].

In summary, the major disadvantage of the scanning approach is the low acquisition speed, while the major issue for direct imaging devices is the low intensity of the returning light, caused by the low brightness of the light source. Recently, a "mixed" concept has been presented, the *one-dimensional MEMS micro-scanning LIDAR* [60, 61] which appears promising towards the development of low-cost, high-performance ToF devices. Such a device uses a direct imaging approach in one spatial dimension combined with a scanning approach to map the second spatial dimension. To achieve this, the device projects a laser line, rather than a point, onto the scene and measures the returning light spatially resolved using a photodiode array. The second spatial dimension is obtained by fast unidimensional scanning of the laser line using an oscillating MEMS mirror.

4.1.3 Multipath Interference in ToF Imaging

In a Time-of-Flight measurement, the light returning at the sensor is interpreted as arising from a single, direct optical path and, therefore, its time delay can be directly correlated to the distance of the imaged objects. As a result, the depth accuracy of ToF and lidar devices is impaired in the presence of refraction or reflections along the optical path. Note that diffuse ambient (background) light does not affect the depth accuracy but rather reduces the dynamic range of the detector and thus the signal-to-noise ratio of the measurement.

More severe effects arise from light that is actively emitted by the ToF device and undergoes multiple reflections or refractions in the scene before reaching the ToF detector. This light carries the "correct" modulation frequency to be demodulated by the ToF detection scheme, but the distance information does not correlate trivially with the distance of the object imaged on the camera. As such light paths overlap with light mapping the "direct" emitter-object-detector path onto the detector, they deteriorate the ranging accuracy. The situation, which is also known as *global illumination* or *multipath interference* (MPI) problem, is illustrated in Fig. 4.5. The MPI problem is of course very well-known and has been deeply investigated by many groups. An overview of the major works in the field and their approaches is presented in Table 4.1. Three major strategies were shown to be able to either mitigate or to separate out global illumination effects.

(a) *Transient time-of-flight imaging* is a direct-ToF approach using typically short laser pulses and fast detectors. The basic idea behind transient imaging is to collect several "slices" of the 3D scene, each corresponding to a different distance from the sensor. If "slicing" is fast enough and ultrafast pulses are used, this corresponds to measuring the light transport matrix. The simplest implementation of transient imaging is time-gating. Two-dimensional slices of the scene are recorded sequentially using an ultrafast detector, which only collects light returning to the sensor within a very short time window (typically few hundreds ps or less). The method was first demonstrated by Busck and

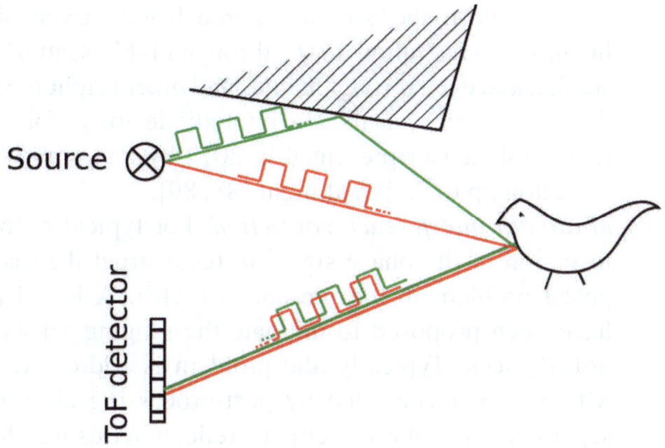

Fig. 4.5 Schematics of the MPI problem. Light following different optical paths (and therefore carrying different phase information) is incident on the same pixel on the ToF sensor, causing an erroneous estimation of the object distance

Table 4.1 Overview of the different approaches that address the problem of MPI in ToF imaging

Work	Multipath type	Solution type	Hardware modifications
Fuchs [62]	Continuous	Iterative	None
Dorrington et al. [63]	2-sparse	Iterative	Frequency sweep
Godbaz et al. [64]	2-sparse	Closed-form	Frequency sweep
Kadambi et al. [65]	K-sparse	Iterative	Custom code
Kirmani et al. [66]	K-sparse	Iterative	Frequency sweep
Heide et al. [67]	K-sparse	Sparse regularization	Frequency sweep
Freedman et al. [68]	K-sparse	Iterative	None
Jiménez et al. [69]	K-sparse	Iterative	None
O'Toole et al. [70]	Continuous	None	Extensive
Lin et al. [71]	Continuous	Closed-form	Frequency sweep
Gupta et al. [72]	Continuous	Closed-form	Extensive
Naik et al. [73]	Continuous	Closed-form	External projector
Peters et al. [74]	K-sparse	Closed-form	Frequency sweep
Qiao et al. [75]	K-sparse	Sparse regularization	Frequency sweep
Kadambi et al. [76]	Continuous	Closed-form	Frequency sweep
Whyte et al. [77]	Continuous	Closed-form	Custom code

Reproduced from the excellent review paper [78] with permission from the authors

Heiselberg in 2004 [79], and further extended in the following years to image hidden objects [80–82], even in motion [83]. Less than 10 years later, the use of a streak camera for detection, combined with scanning mirrors or special imaging optics, allowed to reduce the time resolution to a couple of ps and gave rise to the field of femto-photography [84, 85]. The use of single-photon avalanche diode (SPAD) arrays at the place of a streak camera was also demonstrated [86]. The transient imaging approach was further extended to incoherent amplitude-modulated (AM) CW light sources and cheap PMD detectors at the price of using a very large number of modulation frequencies to reconstruct a single scene and limiting the time resolution to the ns timescale

[67]. Thanks to the "slicing" approach and the very short capture times, transient imaging devices allow to see through highly scattering media, including fog and smoke, and even through flames and other bright media [87]. Recent advances in the technology of avalanche photodiode arrays form the basis of the novel long-range ToF sensor presented in 2018 by Panasonic Corporation, which showed detection up to 250 m at night [88, 89].

(b) *Multipath interference correction*: For typical c-ToF acquisition schemes, the inversion of the phase signal to reconstruct the distance information is an ill-posed problem in the presence of MPI. A lot of approaches and algorithms have been proposed to mitigate the ranging errors caused by MPI effects on ToF devices. Typically, the problem is addressed from two perspectives: the MPI is either corrected by post-processing algorithms, or its contribution is separated from the direct-path reflections using clever measurement schemes and/or specific hardware implementations. In general, pure post-processing strategies [62, 90, 91] show limited applicability because of the large variability of MPI contributions in real-world scenarios. Nevertheless, very recently a post-processing correction algorithm based on a convolutional neural network demonstrated a reliable correction for MPI effects on several scenes, with a correction time of about 10 ms per frame [92]. Mixed methods, employing specific mathematical description of the light transport to identify strategies capable of separating MPI contributions, have shown more success. While initial works imposed strict conditions for the scene to be modelled [62, 91], recent approaches focused on more generalized formulations, which are mostly based on either sparsity or compressibility of light transport in the time domain and typically exploit measurements with several modulation frequencies [63, 93]. Sparse models are, however, not accurate in the presence of scattering media. Nevertheless, a recent novel approach based on convolutional sparse coding showed a very good performance through scattering media, albeit at the price of acquiring a large set of data for reconstructing the image [94]. An alternative approach focuses on the fact that the time-domain light transport can be described and analyzed in the frequency domain without loss of information. This is the basic principle behind the *Phasor Imaging Approach* proposed by Gupta and coworkers [72], which relies on the observation that, for most scenes and global illumination circumstances, there is a threshold modulation frequency such that at higher frequencies the global illumination component does not act on the phase of the returning light, but only appears as a constant offset. This allows reconstruction of a scene by measuring the full range at low frequencies and then removing MPI contributions using high modulation frequencies. A more detailed review of such approaches is given in the recent paper of Jarabo and coworkers [78].

(c) *Coded light*: By operating a ToF device with the addition of clever temporal or spatio-temporal modulation sequences, multipath interference effects can be separated or "filtered out." Among the most relevant examples, adding a spatial modulation to the temporal modulation makes it possible to analyze and retrieve 3D scenes more accurately in the presence of global illumination effects [70, 73]. For 1D-scanning devices, epipolar imaging projecting laser lines on the scene also show some improvement in terms of reduction of global illumination artifacts [95]. Another possibility is to use specific temporal pulse sequences to selectively reconstruct the 3D scene. For example, Kadambi and coworkers showed that, by using customized temporal coding, a functionality analogous to that of time-gating d-ToF cameras can be achieved, i.e., the scene can be "sliced" in 2D images corresponding to different distances [65]. In general, coded-light approaches are very promising due to their typical low-cost applicability, but they usually require a few seconds to fully reconstruct a scene.

4.1.4 Simulation Frameworks for ToF Devices

The typical computer graphics rendering pipeline for 3D scenes already brings parts of the functionality needed for a ToF camera simulator. It takes 3D data (camera, objects, surfaces, textures, and light sources) as input and generates a 2D image with brightness per pixel and/or color values. In recent years, a lot of progress in terms of algorithm development and processing speed was achieved in the field of physical based rendering (PBS). A comprehensive review with accompanying source code for many example scenes is available on this topic [96]. In this context "physical"—sometimes also referred to as "predictive"—means that rendering of a scene is performed using light interactions on surfaces and in the bulk material that are based on optical models rather than using algorithms which aim to give a visually convincing effect ("plausible" or "visually rich" rendering). Such an approach naturally allows including the effects of global illumination. There are several ray-tracing-based strategies to model global illumination. According to [97], the following methods are most widely used:

- *Path tracing*: Monte Carlo rays are sent from the eye to scene objects.
- *Photon tracing*: Monte Carlo rays are sent from light source(s) to scene objects.
- *Bidirectional path tracing*: Path and photon tracing is performed and match-making between partial paths from each direction is applied to get full light-to-eye paths.
- *Metropolis*: The initial set of seeds is taken from any of the above-mentioned methods (bidirectional is most popular). Then some slight modifications to the existing paths are applied. Thus, nearby paths, which are difficult to find, can be detected with a lower number of rays. This method uses the Metropolis algorithm, which is a special implementation of the Monte-Carlo concept, to perform ray seeding and weighting for the overall brightness calculation.
- *Radiosity*: A full finite element solution is performed, and the results are projected to the screen.

Such renderers are very suited to simulate realistic brightness (intensity) distributions, but—to our knowledge—lack the ability to calculate the average path length per pixel. In contrast to rasterization rendering methods[4], ray-tracing-based rendering methods calculate ray–object intersections making them the ideal base for ToF camera simulations. Several strategies of different complexity exist to determine the path length between the ToF camera and the objects in the scene. The simplest approach consists in using a set of test rays launched from the sensor position. This approach uses the implicit assumption that a light source is located at the position of each detector pixel to justify reverse ray-tracing from the detector towards the scene. Collision detection algorithms are used to check if and which surfaces are located in the direction of a test ray. The intersection coordinates are used to calculate the length of the corresponding test ray. The scene might be defined by several surfaces, where one is placed behind another. Thus, the length calculation has to be repeated for every surface which intersects with the test ray. The shortest distance, which belongs to the closest and thus non-occluded surface, is taken as the valid value. This procedure is repeated for every test ray. Pre-filtering strategies of possible occluded surfaces allow (further) reduction of the calculation time. This procedure only accounts for direct "line-of-sight" distances and ignores reflection and scattering of light between surfaces. Various ToF simulators, like ROS/Gazebo [98, 99] or PMDSim [100, 101], implement this nearest objects distances detection.

The Blensor [102] algorithm, which is based on the Blender engine, enhances this method for rudimentary support of optical properties of surfaces and objects. If a test ray intersects with a surface exhibiting specular reflective (mirror-like) properties, the ray is rotated at the intersection according to the law of reflection and collision detection is performed again for the new direction. The total path length is then the sum of the individual segments. This is repeated until the ray intersects with a diffusively reflecting surface or reaches a predefined maximum length. The second case leads to an invalid ray. In the first case, a valid ray is returned with a total length as the sum of all individual segments as described above while keeping the original direction of the ray (i.e., the direction from the pixel coordinate on the sensor through the focal point of the ToF camera). Artifacts in depth arising from mirror-like reflective surfaces can thus be simulated. Blensor currently does not support transmission with refraction and true multipath analysis for realistic surfaces and objects (with partial reflectance and transmittance). Another approach has been developed, based on the MatLab platform, which can distinguish between the direct and global component of scenes with reflective surfaces [72].

The most advanced ToF camera simulator—to our knowledge—is the CamSim toolbox [103]. It is a framework for evaluating ToF camera performance using OpenGL as backend for 3D scene creation and rendering and supports amplitude-modulated continuous-wave Time-of-Flight cameras with 4 phase images. It features an arbitrary number of light sources, even with measured intensity distribution

[4]In rasterization rendering methods, scene objects are created from a mesh of virtual triangles or polygons. Then, this mesh is projected onto a 2D screen and pixelated. The result is the 2D image of the 3D scene. This approach is typically computational less demanding than ray-tracing approaches.

and structured light, and a large number of surface properties for the objects in the scene, includes lens distortions and vignetting, and allows modelling simple animation effects. The illumination model even enables rudimentary global illumination effects with "single-bounce indirect illumination via reflective shadow maps" [104]. Unfortunately, CamSim only supports surface interaction effects and, thus, ignores refraction.

4.2 Description of the Simulation Procedure

Applications for ranging cameras can be found in the fields of enhanced reality [105], robotics [106, 107], healthcare [108, 109], and autonomous mobility, both to monitor the alertness of the driver [110, 111] and the vehicle surrounding.

Despite their great potential, active ranging cameras face several challenges with respect to safe and reliable operation in outdoor scenarios. In particular:

- the necessity to monitor a large area on very short time scales, which requires either fast scanning techniques or very sensitive detectors for direct imaging implementations
- the presence of background illumination, which reduces the dynamic range of the sensor
- the capability to detect objects with very different surface properties (e.g., semi-transparent, highly absorbing, highly reflecting)
- the need to maintain good performance in challenging environmental conditions such as rain, snow, or fog.

For an efficient development of technology best equipped to deal with these challenges, it is necessary to develop simulation approaches capable of predicting the performance of different technologies. Ideally, these simulation approaches can reliably model objects with very different optical properties as well as different environmental conditions and can simultaneously return not only the extent of the ranging perturbation as measured by the sensor, but also a precise evaluation of the origin of such error. Finally, the approach must also allow for a quantitative comparison between different hardware and technologies.

This chapter presents a simulation approach for ToF cameras based on Zemax OpticStudio® and Python that attempts to fulfil such goals. The current implementation is limited to direct imaging ToF cameras but can easily be extended to include LIDAR scanning schemes. The application of our approach to different objects and environmental conditions will be presented in Sects. 4.3 and 4.4, while Sect. 4.5 will focus on a detailed comparison of our approach with existing ToF simulation algorithms including a discussion of the advantages and drawbacks of our approach.

4.2.1 The Simulation Approach and Its Implementation

The simulation approach we present here is based on a two-step strategy. In the first step (*simulation phase*), a steady-state ray-tracing simulation is run for the model ToF sensor in front of an arbitrary scene. Within the simulation phase, the intensity and the full optical path of each ray reaching the ToF detector are recorded including the details of all ray segments and all the interactions that took place within the ToF optics or within the scene. The master parameter for all the following evaluations of the ToF signal is the optical path length (OPL), defined as

$$\mathrm{OPL} = \sum_i n_i d_i \tag{4.10}$$

where d_i is the geometrical length of the ith ray segment, n_i is the refractive index of the medium in which the ith ray segment propagates and the sum runs over all the ray segments from the source to the detector. Given the OPL, using the speed of light c, the travel time t of each ray can be reconstructed as $t = \mathrm{OPL}/c$.

The simulation phase is performed with Zemax OpticStudio® (ZOS). ZOS is a development software for optical system design supporting Monte Carlo methods for ray-tracing. Multiple light sources, 3D objects and detectors can be placed within the 3D scene. The software allows defining the optical properties independently for each surface (transmission, absorption, reflectivity, and scattering) and volume (refractive index, volumetric scattering). Additionally, the ray-tracing kernel keeps track of the absolute energy values of each ray at the ray-segment level and can account for polarization (each ray is split into its electric and magnetic field vectors), enabling, for example, the calculation of the surface reflectance from the object's index of refraction without any further assumption. Rays hitting a certain object or detector can be stored and saved with their intensity and their complete ray path, including information on the occurrence and position of reflection and/or scattering events. The possibility to simulate all optical effects (at least within the geometric optics model) and the built-in functionality of ray-segment fine data storage make ZOS an ideal choice for ToF ray-tracing simulations. Nevertheless, the complete procedure, as shown in Fig. 4.6 and described in the following chapters, is valid for any ray-tracing-based simulation software as long as the OPL of each ray can be stored or extracted.

It's important to note that the method presented here does not support real-time simulation of the ToF scene since the simulation is based on steady-state ray-tracing. Effects arising from moving objects and time-dependent scenes can be modelled by performing a simulation sequence, where the position of certain objects in the scene is systematically varied; however, the procedure can be slow due to the computational time required by the individual simulations.

The second step (*processing phase*) deals with post-processing of the ray-tracing results and reconstruction of the full 3D point-cloud. The steps described here, which are presented in Fig. 4.6, only deal with the reconstruction of the 3D point-

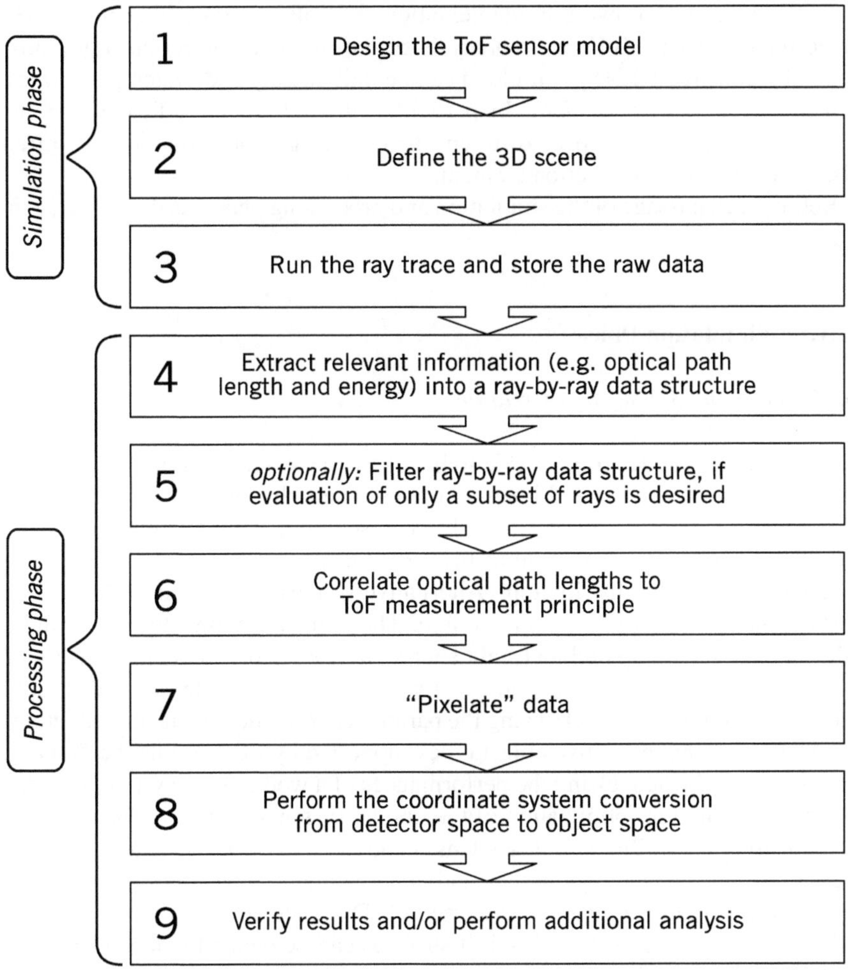

Fig. 4.6 Flowchart of the procedure adopted to simulate the Time-of-Flight signals. Figure is adapted from [58]

cloud and the range image as seen by the sensor, but further algorithms for advanced data analysis or to model ToF software features can be easily implemented. In the procedure described herein, the processing phase is implemented in Python and requires the numpy, struct, and mmap plugins. The choice of Python is mostly motivated by its straightforward high-level syntax and the free of charge Spyder scientific development environment plugin, but other programming languages for scientific computing and image processing can be used instead.

Among the major advantages of this approach is the clear separation between the ray-tracing simulation and the processing phase. By defining the OPL as a master parameter and by collecting all the information on the interactions experienced by each ray, different hardware implementations of ToF cameras, frequency analysis for correlation-based ToF, or an analysis of the influence of certain ray paths on the total reconstructed 3D point-cloud can be individually and systematically tested *with the same simulation data.* In particular, this allows for efficient systematic testing of the performance of different ToF hardware (or software) for different "risk" scenarios or perturbation elements.

Below, each passage of the simulation and processing phases is described in more details.

4.2.1.1 Simulation Phase

ToF Camera and 3D Scenery Creation

The ToF camera model comprises a light source, a two-dimensional pixelated (and absorbing) detector array, and imaging optics. The light source type can be chosen from a list of supported sources, which are based on mathematical models and are defined by parameters determining their spatial extension and the characteristics of the emitted light distribution. Alternatively, measured or user-defined light distributions can be used as a ray source. The software allows to define multiple light sources (including ambient light source models) and, of course, to distinguish rays originating from different sources. Optical elements like lenses and objectives can be either imported in ZOS using the parameters provided by the manufacturer or modelled using the respective object types supported by ZOS. While the first option is recommended when testing the performances of a well-known system, the second option is suitable for optimization of the optical system of interest, due to the wide customization possibilities. Besides lenses and objectives, ZOS supports numerous internal 3D object types like cubes and spheres that can be used to model a ToF scene, but it can also operate on imported CAD scenes or on a mixture of both. In both cases, the properties of all object surfaces can be defined independently. As a small drawback, the light source(s) always have to be placed separately and cannot be imported as part of a CAD file.

Ray-Tracing and List of Rays Containing Detector Coordinates, Intensity, and OPL

The ray parameters are typically created with a rigorous random or a Sobol-based pseudo-random algorithm. Alternatively, a regular grid or a user-defined distribution of rays can be used.

Upon ray-tracing, a ray-trace raw data file is generated, including energy and optical path length information for each ray and on a ray-segment to ray-segment level. In particular, not only the length of each ray segment but also the coordinates of each interaction are stored, including the coordinates where the ray hits the detector. Such information allows estimating the signal as a function of the pixel size (number) in the post-processing step without need for repeating the simulations.

4.2.1.2 Processing Phase

Ray Information Extraction, OPL Reconstruction, and Ray Filtering

The raw-data file containing all the ray information is processed using a parser script. For each ray, the wavelength, light energy, total OPL, and the full list of interactions with the objects of the scene are stored for further processing. The total OPL is calculated using the coordinates of each ray segment.

Additionally, parameters like wavelength and the list of interactions can be used to filter or categorize the ray data. This helps examining how certain ray paths, wavelengths, or object interaction types (refraction, reflection, scattering, etc.) contribute to the overall intensity and/or depth signal.

If multiple light-source types were included within the 3D scenery, the separation of their contributions should be done in this filtering step, as they have to be processed differently in the following steps.

Correlating OPL to ToF Measurement Principle

This step deals with the reconstruction of the ToF signal depending on the arrival time of each ray, which is mapped by the OPL. Depending on the simulated ToF hardware, the complexity of the reconstruction varies.

The d-ToF principle is the simplest to model, since the distance is automatically defined by the OPL parameter. Only objects whose distance exceeds $c/(2f_{\mathrm{mod}})$, where f_{mod} is the frequency of the illumination pulses, need to be projected back to a detected distance $d = \mathrm{OPL} - m\ c/(2f_{\mathrm{mod}})$, where $m \in \mathbf{N}$ such that $d \in [0, c/(2f_{\mathrm{mod}})]$.

C-ToF cameras operating at a single frequency can be modelled using a ray separation technique, which takes the ray's OPL and the sensor's modulation frequency into account. As described in Sect. 4.1.2, also for c-ToF cameras the modulation frequency f_{mod} defines the maximum detectable distance, as phase-wrapping occurs for longer distances. Additionally, a phase delay discretization has to be defined. This discretization influences the accuracy of the depth reconstruction: at least two-phase delay ranges are necessary, but a higher number of ranges lead to more-precise depth reconstruction, albeit at the cost of having longer measurement times per frame. In practice, for each ray the phase delay is calculated from the OPL and the modulation frequency using Eq. (4.2). The intensity of the ray is then distributed between the phase measurements according to its phase delay. The contributions of different rays hitting the same pixel are added, and finally, for the widespread four-phase scheme, the distance signal is calculated according to Eq. (4.7). Note that this procedure inherently assumes a rectangular modulation waveform for both control and returning signals.

Multi-frequency c-ToF sensors are interesting as they can partially overcome the limitation of the modulation-frequency bound measurement range and, to some extent, can also limit the influence of global illumination on the ToF ranging accuracy [52]. Such sensors can be modelled via repeating the procedure of the single-frequency ToF for every frequency.

Pixelate Data

This step mimics the functionality of a pixelated detector. Accordingly, the detector area is divided in a pixel matrix, and the ray information for all rays arriving on the detector within a certain pixel area is combined. Optionally, the space between pixels can be also accounted for by eliminating the rays arriving on such gaps from the following analysis.

As a result, the ray-based path length and intensity information is converted into a matrix-like pattern for a given number of horizontal and vertical sensor pixels in this discretization process. In the simplest case (no multi-frequency sensor, no direct vs. global illumination differentiation), this procedure returns two matrices. The first matrix contains the integrated intensity at each pixel, which is the sum of the intensities of all rays arriving at each pixel. The second matrix contains the average path length. For each pixel, the latter is obtained as the average of the OPLs of each ray impinging on the pixel, weighted by the ray intensity. Different smoothing strategies, where the ray data also contributes to neighboring pixels, can be introduced within this step.

Coordinate System Conversion

Finally, a coordinate system conversion from the image space (defined in pixel units) to the object space is necessary. The behavior of the modelled imaging objective has to be taken into account to properly perform this conversion. The simplest approach to define such a camera function is to use a global magnification factor for all x- and y-coordinates (pinhole camera). This gives a first approximation of the 3D scene. This approach does not take into account magnification changes due to defocusing and lens aberration effects for objects at different lateral and axial positions. To that end, more complex coordinate mapping and calibration approaches can also be implemented. For a detailed description of such mapping functions and their implementation, we refer the reader to [112–114].

4.3 Simulation Results of a ToF Camera Observing a Test Scene

In this section, we present the simulation of a ToF camera in front of a simple scene following our approach. In particular, we will focus on some key advantages of this method, such as the capability of filtering and analyzing single ray paths and the possibility of modelling different ToF hardware without having to re-trace the scene. This allows for a quick and direct comparison between different ToF technologies.

The same ToF model and the same test scene described in this section will be used in the next chapters to demonstrate the capabilities of our approach in the presence of refracting and scattering media.

4.3.1 Simulation Setup

The simulation setup, shown in Fig. 4.7, consists of a model of the ToF camera and a scene located in front of it. Please note that all 3D plots shown in this publication use parallel projection for the 3D effect.

The ToF camera model comprises a monochromatic point source at 850 nm with a Gaussian angular intensity distribution (18° full angle divergence @ $1/e^2$). The total power of the light source is set to 1 W. Alternatively, a source array can be used. A Petzval lens design (see inset in Fig. 4.7) was chosen for the receiver optics because this lens type offers good light collection efficiency. An antireflective

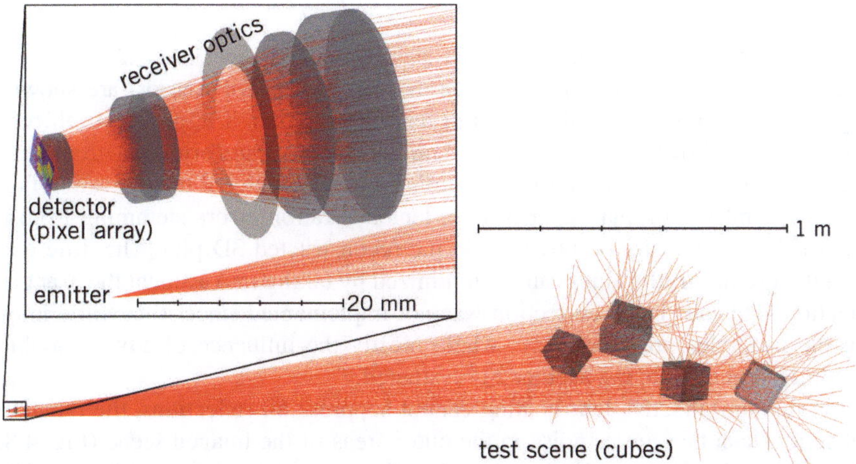

Fig. 4.7 Model for the ToF sensor and a simple scene. The sensor consists of a point source and a pixelated detector array preceded by a Petzval lens optics. The scene consists of four cubes with diffuse-scattering surfaces. Figure is adapted from [3]

coating optimized for 850 nm was applied on the lens surfaces. The ToF sensor is described by a pixelated absorptive detector. Since the outcome of the simulation is given by the coordinates of each ray at the detector, the number of detector pixels in the simulation can be kept low.

The scene itself consists of four cubes of different sizes, located at distances between 1.8 and 2.5 m from the ToF camera and forming different angles with respect to it. For the examples shown here, the cubes have identical surface properties, as this allows investigating the effect of different orientations of the scene on the ToF reconstruction. The surface of the cubes is chosen to be partly absorbing (90%) and partly scattering (10%). The scattering distribution is modelled as Gaussian, with a Bi-Directional Scatter Distribution Function (BSDF) defined as:

$$\mathrm{BSDF}\left(\vec{x}\right) = A \exp\left(-\frac{\left|\vec{x}\right|^2}{\sigma^2}\right) \tag{4.11}$$

where the vector $\vec{x}$ is the difference between the projection of the specular ray vector and the scattered ray vector onto the surface plane, σ is the standard deviation of the Gaussian distribution on the projected plane ($\sigma = 1$ in our case), and A is a normalization factor. While changes in the BSDF function can only be investigated by running a new simulation, the percentage of absorption and scattering from each surface can be addressed in the processing phase and its effect can be investigated without repeating the ray-tracing simulation phase.

4.3.2 Scene Reconstruction Using Ray-Tracing Data

The scene reconstruction is performed as described in Sect. 4.2.1.2. The reconstruction results (intensity and distance as seen by the ToF sensor) are shown in Figs. 4.8a, b. Figure 4.8c plots the full data point-cloud. The ground-truth object positions are shown as grey shades. The coordinate transformation from image space to object space is performed assuming a constant global magnification factor for all x- and y-coordinates. Therefore, lens aberration errors stemming from the Petzval lens can easily be observed in the reconstructed 3D plot. The differences with the ground-truth values can be minimized by taking into account the exact lens function. This advanced correction was not implemented since it has little impact for the examples presented here, where mostly the influence of adverse ambient conditions on the ranging accuracy is evaluated.

As the antireflective lens coating cannot suppress all reflections, the remaining reflections can be seen as noise in the outer areas of the imaged scene (Fig. 4.8c). The magnitude of such reflections is so low that it has no significance for a real ToF

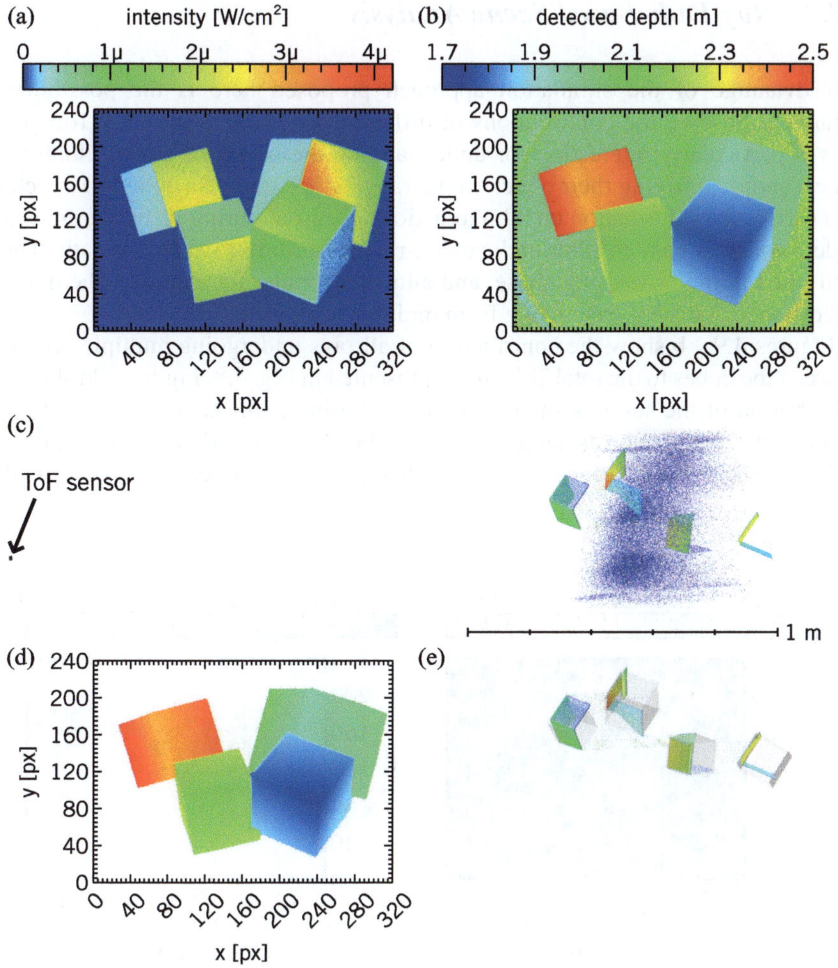

Fig. 4.8 Simulated intensity and depth for the ToF-sensor model and scene shown in Fig. 4.7. (**a**) Intensity, (**b**) depth and (**c**) 3D point-cloud combining the information of (**a**) and (**b**). Colors follow the same intensity scale of panel (**a**). Grey cubes represent the ground-truth position of the objects. (**d**) Depth and (**e**) 3D point-cloud plot for the same test scene when assuming a 12-bit analog-to-digital conversion of the ToF signal. The contribution from internal lens reflections is suppressed. Color scale is the same of panels (**a**) and (**c**). Figure is adapted and extended from [58]

camera. This is demonstrated in Figs. 4.8d, e, where the effect of a 12-bit analog-to-digital conversion of the ToF signal is modelled, which results in filtering out low-intensity signals. In this example, the camera is assumed to automatically scale the intensity range to the maximum detected incident intensity.

4.3.3 Ray Path-Based Scene Analysis

An advantage of the simulation approach proposed here is the possibility to separately observe the contributions of different ray paths to the total ToF signal. This is particularly interesting for understanding the origin of global illumination effects and quantifying their effect on the ranging accuracy. Since in ZOS each ray data carries the information on all interactions occurred during its path from source to detector, it is easy to filter and catalog rays according to their ray path. This is performed in the processing phase, and allows mapping, for example, the intensity and distance contributions arising from multiple scattering.

Figures 4.9a, b show the contribution of all rays undergoing multiple scattering between the cubes to the total ToF signal presented in Fig. 4.8. Figure 4.9d shows the contribution of the sub-ensemble of multi-scattering paths having their first hit on the orange-colored cube as depicted in Fig. 4.9c. As expected, rays interacting with multiple objects travel a longer distance; therefore, their effect is an overestimation

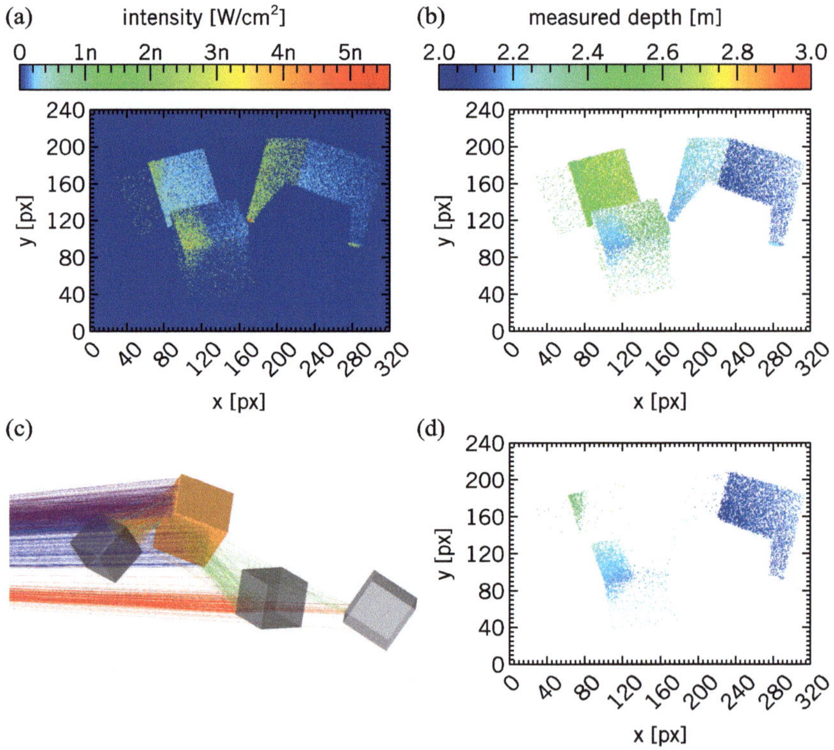

Fig. 4.9 Example of a multipath analysis. (**a**) Intensity and (**b**) depth contribution from all rays hitting at least two cubes. (**c**) Selection of the sub-ensemble of rays undergoing multiple interactions with the cubes and having the first interaction with the orange cube, and (**d**) depth contribution from this sub-ensemble of rays. Figure is adapted and extended from [58]

of the camera-scene distance. However, as can be seen by comparing Fig. 4.9a with Fig. 4.8a, multipath contributions remain two to three orders of magnitude smaller than the contribution of direct paths in the absence of highly reflecting objects.

4.3.4 Influence of the ToF Sensor Model on the Depth Calculation

The majority of ToF cameras available today use the Amplitude Modulated Continuous Wave (AMCW) principle, which is typically realized with a photon mixer device (PMD) pixel array. To illustrate the capability of our approach to model different ToF technologies, a c-ToF scheme was implemented, where a rectangular continuous waveform was used for the amplitude modulation of the illumination and, upon proper delay, as a control signal for the pixels of the detector. The modelled detector structure mimics the function of PMD pixels with a rectangular integration and a four-step phase scheme. The ray-tracing data used to model the c-ToF scheme is the same as previously used for the d-ToF model (Sect. 4.3.2). The AMCW principle was implemented according to the description in Sect. 4.2.1.2 for a single modulation frequency of 50 MHz, corresponding to a maximum unambiguous range of 3 m.

The phase shift of the control signal φ_s was varied in steps of 90° (0°, 90°, 180°, 270°) according to the four-step phase cycling scheme and the signal in each range K_{φ_s} was reconstructed as explained in Sect. 4.2.1.2. Finally, for each pixel the depth signal D is calculated as:

$$D = \frac{c}{2f} \cdot \frac{1}{2\pi} \cdot \arctan 2\,(K_{270^\circ} - K_{90^\circ}, K_{180^\circ} - K_{0^\circ}) \tag{4.12}$$

The calculated depths from the d-ToF and the c-ToF models are compared in Fig. 4.10. Figure 4.10a shows the reconstructed depth using the four-phase c-ToF algorithm, while Fig. 4.10b shows the depth difference with respect to the d-ToF model (Fig. 4.8d). Again, a 12-bit analog-to-digital conversion of the ToF signal is assumed. The intensity distribution is independent from the ToF hardware and is therefore not shown again. A better visualization of the difference between the range reconstructions of the four-phase c-ToF and the d-ToF can be obtained by plotting the depth obtained from the c-ToF model as a function of the depth obtained from the d-ToF model for each pixel (Fig. 4.12c). The depth obtained from the d-ToF model is used here as a reference (x-axis) instead of the ground-truth values because no correction for the lens aberrations is implemented in the reconstruction procedure.

Most of the data points (dark curve in Fig. 4.12c) show a deviation with respect to the d-ToF distance. The difference between the depth recovered by the d-ToF and c-ToF algorithms stems from the rectangular waveform used by the c-ToF illumination and control signals. In this case, the argument of the arctangent function contains the higher order harmonics of the fundamental frequency, which in turn result in

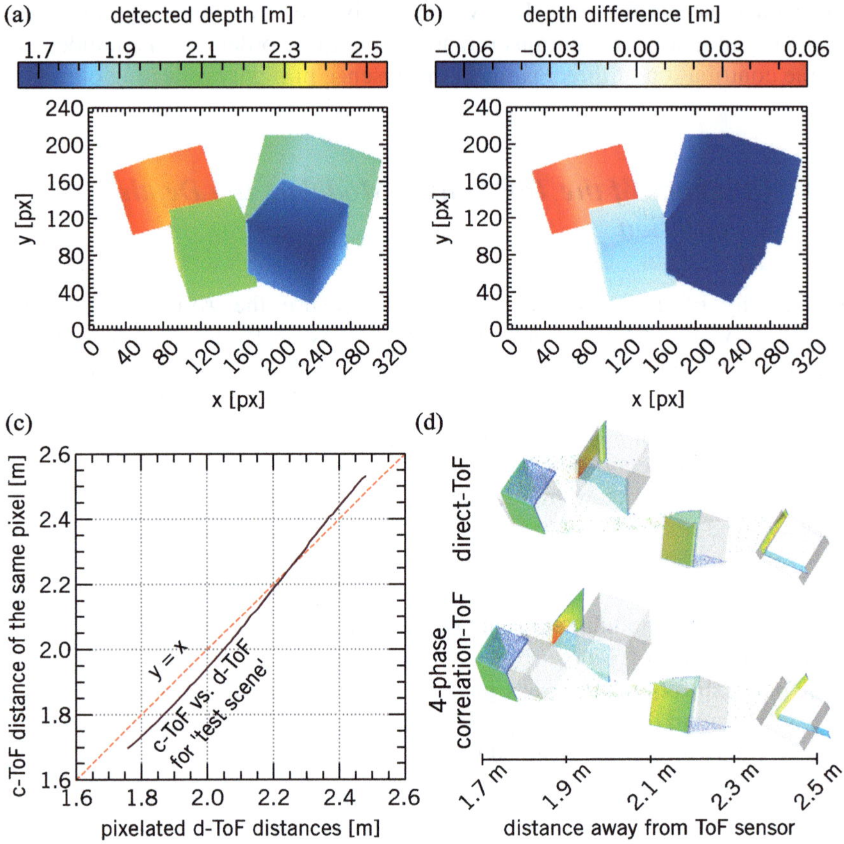

Fig. 4.10 Comparison between range reconstruction using d-ToF and rectangular waveform (single-frequency) four-phase c-ToF for the test model shown in Fig. 4.8. (**a**) Calculated depth using the four-phase c-ToF sensor model. (**b**) Difference between the depth mapping of the scene obtained from c-ToF and d-ToF. (**c**) C-ToF depths as a function of the d-ToF depth for all pixels imaging an object of the scene, emphasizing the nonlinearity of the c-ToF depth reconstruction. (**d**) Full data point-cloud highlighting the additional distance error introduced by the rectangular waveform c-ToF sensor model in comparison to the d-ToF model for the test scene. The intensity scale is identical to that of Fig. 4.8a. Figure is adapted and extended from [58]

additional phase terms (this follows from substituting a rectangular function at the place of $q(t)$ and $r(t)$ in Eq. 4.4). The full derivation can be found in [35], while [41, 115, 116] contain a detailed examination of these aliasing effects for selected higher order contributions. The resulting nonlinearities are typically corrected in an additional calibration step in commercial c-ToF cameras. For clarity, the difference is displayed in Fig. 4.12d as a 3D point-cloud, including the ground-truth positions of the objects (grey cubes).

4.3.5 Time-Gated ToF

Other hardware implementations besides d-ToF and c-ToF can be simulated as well. Time-gated ToF is another example where our approach allows reconstructing the ToF signal in a very straightforward way. As discussed in Sect. 4.1.3, time-gated ToF (TG-ToF) technologies can be implemented as direct ToF or as correlation ToF by using specially designed coded illumination schemes. The main idea is to reconstruct the 3D scene "slice-by-slice," using a collection of 2D images mapping different distances. In our approach, time-gated d-ToF can be implemented by using the information on the distance travelled by each ray reaching the detector. The measurable range is divided in time intervals, which describe the time resolution of the ToF sensor (e.g., 70 ps). The rays are thus assigned to the different time intervals depending on the time at which they reach the detector. Since each time interval corresponds to a certain range of camera-scene distances, the intensity image from the different time intervals can be used to reconstruct the full 3D scene. Time-gated c-ToF can also be realized by first mapping the distance into phase information, applying the decoding algorithm used from the ToF sensor, and finally mapping back the phase information into a distance. This whole process allows to account for systematic effects introduced by the ToF signal readout and its processing, similar as it was shown in Sect. 4.3.4 for the nonlinearity of the depth mapping in four-phase ToF sensors.

4.4 Application Examples: Influence of Transparent Objects and Environmental Conditions on ToF Accuracy

As discussed in the previous sections, multipath interference effects of light actively emitted by the ToF device are among the major causes of ranging errors for ToF cameras. In contrast to scanning Lidar systems, direct imaging ToF cameras are heavily affected by such global illumination effects, due to the wide illumination angle [52].

Sections 4.2 and 4.3 showed that the ray-tracing approach presented here is particularly suitable to investigate global illumination effects since (1) multiple reflections and scattering ray paths are retained and (2) these ray paths can be analyzed independently. In this section, we will investigate ranging performance of ToF devices for three different scenarios, which are particularly relevant for outdoor operation and autonomous mobility: highly transparent objects, rain and fog. The examples will illustrate the capability of our proposed method to analyze ToF signals in the presence of refractive and scattering media. For simplicity, the d-ToF implementation is used everywhere for the distance reconstruction. The ranging errors are computed as the difference between the scene with and without the "perturbing" element in order to minimize additional aberration effects from the lens system.

4.4.1 Influence of a Glass Window in Front of the Scene

The antireflective coating of the ToF camera lens reduces the influence of lens reflections on the depth calculation to a minimum. In a real-world scenario, however, it is plausible to encounter semi-transparent or highly transparent objects without antireflective coating. To examine their influence on the ToF ranging accuracy, the test scene was modified by adding an (uncoated) glass window in front of part of the scene.

The window, modelled as a 1 cm thick N-BK7 quartz glass plate, is positioned 1 m away from the ToF camera and covers only the left half of the scene (Fig. 4.11a). Around 4% of the incoming light perpendicular to the glass surface is reflected as given by the Fresnel equations. Additionally, a 1% Gaussian scattering (with parameters analogous to the surfaces of the cubes, see also Eq. 4.11) was added to mimic surface roughness and glass impurities. The glass window is oriented perpendicular to the ToF sensor and is positioned in such a way that light directly reflected from its surface does not hit the receiving optics. In fact, if the direct reflection would hit the sensor, the intensity of the directly reflected light would completely exceed the scattered light from the scene, and the scene behind the glass would not be detected. The modelled scenario, instead, deals with a more subtle situation, where the direct signal from the glass window is expected to be very small.

Three main effects are expected to influence the depth estimation. The first is caused by light backscattered from the window prior to reaching the scene (direct backscattering). These rays image the presence of the window on the sensor, but when the window image overlaps with the image of the cubes, they cause an underestimation of the distances of the cubes. The second effect arises from rays which, after reaching the scene, hit the glass window, are either reflected or scattered back towards the cubes and finally are scattered back to the ToF detector. The last effect is caused by rays undergoing multiple reflections within the glass plate. The last two effects will both contribute to an overestimation of the scene distance.

Figure 4.11b shows the point-cloud data. Figures 4.11c, d show the intensity detected by the sensor and the reconstructed distance, respectively. Figure 4.11e shows the ranging difference induced by the presence of the glass window. The presence of the window is evident where its image does not overlap with the image of other cubes, as it can be seen in both the intensity map and in the distance map. The size of the window is overestimated because it is not in the focus of the ToF camera optics. In the distance map, the window appears as an obstacle located about 1.2 m away from the scene, while only the outer borders of the glass are correctly estimated at a distance of 1 m. The overestimation of the distance of the window can be explained by observing that the window is imaged by rays that are directly backscattered towards the sensor before any interaction with the rest of the scene. Since the window is highly transparent and weakly reflecting, these backscattered rays overlap with rays undergoing multiple beam reflections between the cubes and the window, and therefore contribute with a significantly longer optical path. Only at the glass edges, where the direct backscattering contribution is dominant and the

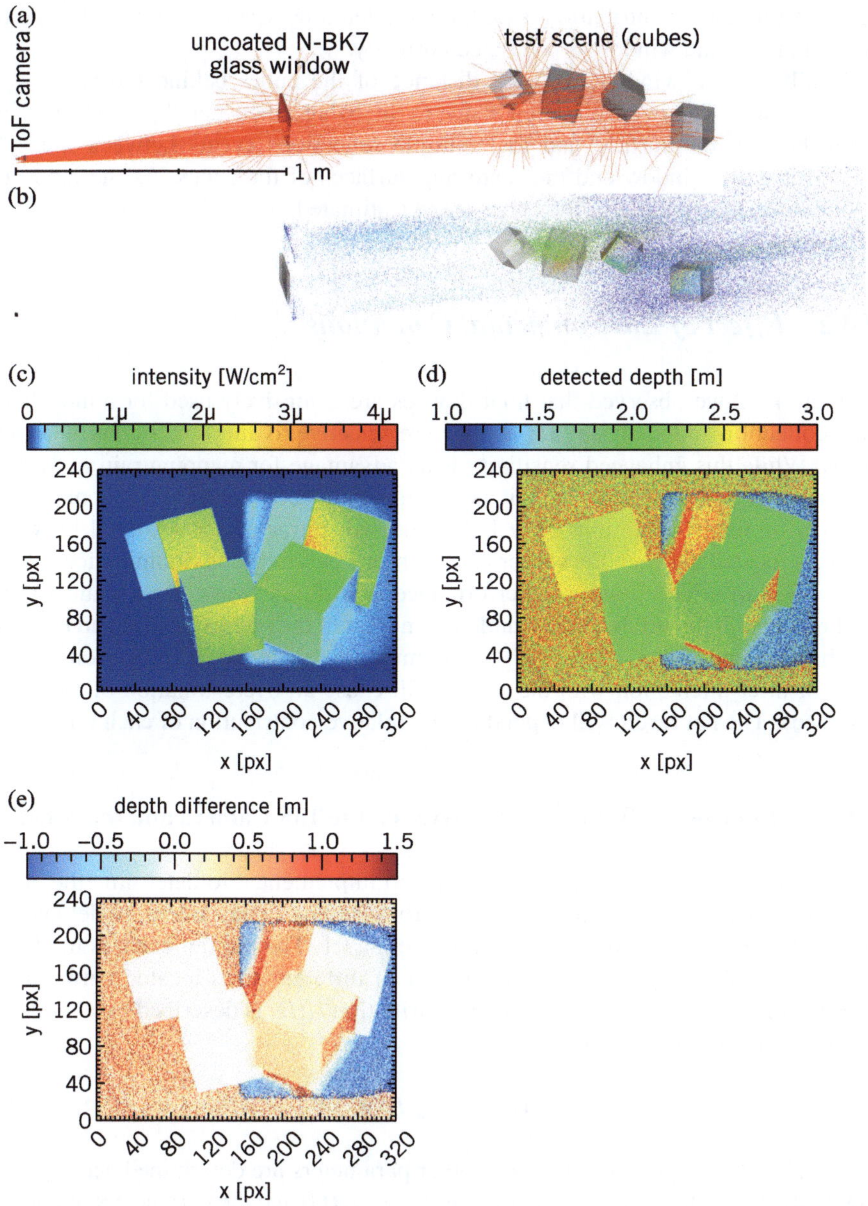

Fig. 4.11 Glass window influence. (**a**) Test scene from Fig. 4.7 after adding an uncoated N-BK7 window in front of part of the scene. (**b**) 3D point-cloud data. (**c**) Intensity and (**d**) calculated depth images using the global magnification factor correspondence. (**e**) Depth difference between the depth retrieved with (panel d) and without (Fig. 4.8b) glass window

multiple reflection contribution is partially hindered by scattering at the glass edge, the distance of the window is estimated correctly.

Similar to the window, also the distance of the cubes behind the window is overestimated. The overestimation is due to the balance between the backscattering from the window surface and the multiple reflections events between the 4%-reflective glass window and the scattering surfaces of the cubes. As the latter are dominating, the distance of the cubes is overestimated.

4.4.2 *Effect of Environmental Conditions*

In Sect. 4.1.2 we observed that lidar devices are extensively used for atmospheric data collection and analysis [31] as they are sensitive to atmospheric particles and gases. While this enhanced sensitivity is an advantage for mapping rain or aerosol distributions in the atmosphere, it becomes detrimental for ranging applications in rainy or foggy conditions. The light reflected, refracted, or scattered by water droplets located between the sensor and the scene causes global illumination effects which in turn affect the sensor performance in different ways. Backscattered and backreflected light will lead to an underestimation of the scene distance and increase the background noise, while single and multiple scattering or reflection events, prior or after interaction with the scene, will cause an overestimation of the scene distances. The net result will depend on the relative contribution of each effect.

4.4.2.1 Influence of Rain Droplets Between the ToF Camera and the Scene

Figures 4.13a, c, e show the simulation setup implemented to determine the effect of rain droplets on the range accuracy of ToF cameras. The camera model and the scene are identical to those described in Sect. 4.3.1. The rainfall event is modelled as an ensemble of spherical water droplets of different sizes located between the ToF camera and the scene. The droplet distribution $N(D)$ is described by a modified Gamma function of the form

$$N(D) = N_{\mathrm{G}}\, D^{\alpha} e^{-\Lambda D} \tag{4.13}$$

where D is the droplet diameter. The other parameters are determined according to the model presented in [117]. Accordingly, $\alpha = 2.160$ and the parameters N_{G} and Λ are functions of the liquid water content (LWC) expressed in g/m^3 and the rainfall rate R, expressed in mm/h:

$$\mathrm{LWC} = 0.062\, R^{0.913} \tag{4.14}$$

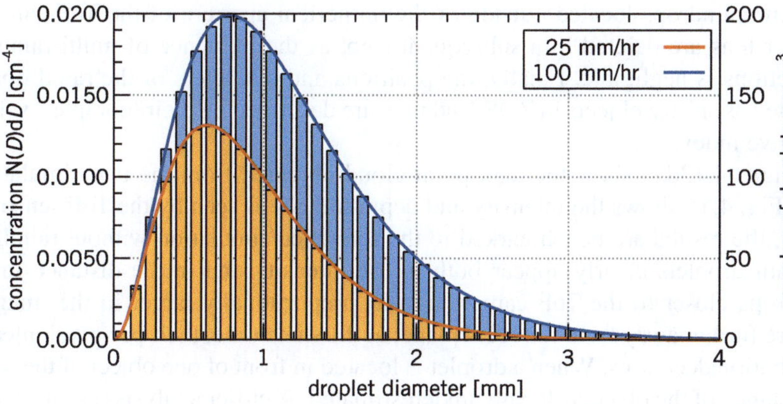

Fig. 4.12 Size distributions for the rain droplet diameter for a rainfall rates of 25 and 100 mm/h (continuous lines). Bar plots show the discretized distributions used for the ToF simulations

Defining the "average droplet volume diameter" D_0 as

$$D_0 = 0.1571 \cdot \text{LWC}^{0.1681}, \tag{4.15}$$

the other parameters in Eq. (4.13) can be expressed as

$$\begin{aligned} \Lambda &= \frac{5.5880}{D_0} \\ N_G &= \frac{512.85 \cdot \text{LWC} \cdot 10^{-6}}{D_0^4} \left(\frac{1}{D_0}\right)^{\alpha} \end{aligned} \tag{4.16}$$

In the following, we present the results obtained for two intense rainfall scenarios, corresponding to rainfall rates of 25 mm/h and 100 mm/h, respectively. As a reference, tropics in the Northern Hemisphere see a rainfall rate of 24 mm/h during 0.5% of the time in the rainiest month according to [117]. The particle diameters were discretized in 0.1 mm ranges. The continuous and discretized distributions for both investigated rainfall rates are shown in Fig. 4.12 for completeness. Note that, as the maximal particle diameter is 3.9 mm, a spherical approximation for the raindrop shape is acceptable [118, 119].

Given the raindrop distribution with n_{total} drops, the positions of the raindrops within the 3D scene are generated with a pseudo-random algorithm in Python as follows. First the cuboid volume to be filled is divided into n_{total} cubic subvolumes to fill each subvolume with exactly one drop[5]. The drop position within each subvolume is randomly generated. Then, the diameter of each subvolume's drop is randomly selected from the pool of available diameters. To speed up the simulation

[5] n_{total} and the cuboid volume borders may need to be adjusted to achieve an integer number of cubic subvolumes in each spatial direction. Otherwise, further logic has to be added handling cropped subvolume cubes at the borders of the cuboid volume.

time, the raindrops located outside of the numerical aperture of the ToF camera's receiver lens are deleted in a subsequent step, as the influence of multi-raindrops interactions is negligible. Finally, the positions and diameters of the raindrops are imported as sphere objects in ZOS, and they are described by their (complex-valued) refractive index.

Figures 4.13d, f show the data point-cloud obtained from the two simulations, while Fig. 4.14 shows the intensity and depth images as seen by the ToF sensor. As before, the results are benchmarked to the reference simulation without raindrops. The rain droplets clearly appear both in the intensity and in the distance image. Raindrops closer to the ToF camera appear proportionally larger in the image as they are further away from the focal plane; at the same time, their per-pixel intensity contribution decreases. When a droplet is located in front of one object of the scene, the distance of the object is locally underestimated. A closer analysis reveals that the major contribution to the droplet signal arises from total internal reflections within the raindrop, causing the light to be reflected back towards the ToF detection unit.

To quantify the influence of rain on the depth accuracy of the ToF measurement, we calculate the fractional error in depth as follows: From each pixel i in the reference simulation mapping an object in the scene, the relative depth error of the corresponding pixel in the simulation with rain $\varepsilon^i_{\text{depth}}$ is calculated as

$$\varepsilon^i_{\text{depth}} = \frac{d^i_{\text{rain}} - d^i_0}{d^i_0}, \tag{4.17}$$

where d^i_{rain} and d^i_0 are the distance obtained at pixel i with and without the rain droplets. The pixels that do not map an object in the reference simulation are excluded from the analysis. Figure 4.15 shows the percentage of pixels affected by a depth error as a function of the depth error magnitude for the two rain scenarios. For 25 mm/h. rainfall, about 22% of all pixels considered are affected by an error greater than 1%. This increases to 28% for a rainfall rate of 100 mm/h. If an error of 5% is considered acceptable, only about 11% of the pixels show a distance deviation larger than this threshold for 25 mm/h and 14% for 100 mm/h respectively.

Relationship Between Raindrop-Induced Distance Errors and Integration Time

Since raindrops fall at a typical speed of about 3–8 m/s [120], the simulation results shown above are accurate for integration times on the order of 0.13 ms or shorter. At longer integration times, the raindrops move during the acquisition time for lengths that are significantly larger than their size. As a result, they will appear as lines in the final image, exactly as it happens in normal photography. Commercial ToF cameras operate in the 50–150 fps regime, thus they will suffer from "smearing out" of the drop intensity over long lines. With the presented approach, it is possible to simulate the effect of long integration times by repeating the simulation phase multiple times, each time shifting the whole raindrop distribution towards the ground of a distance

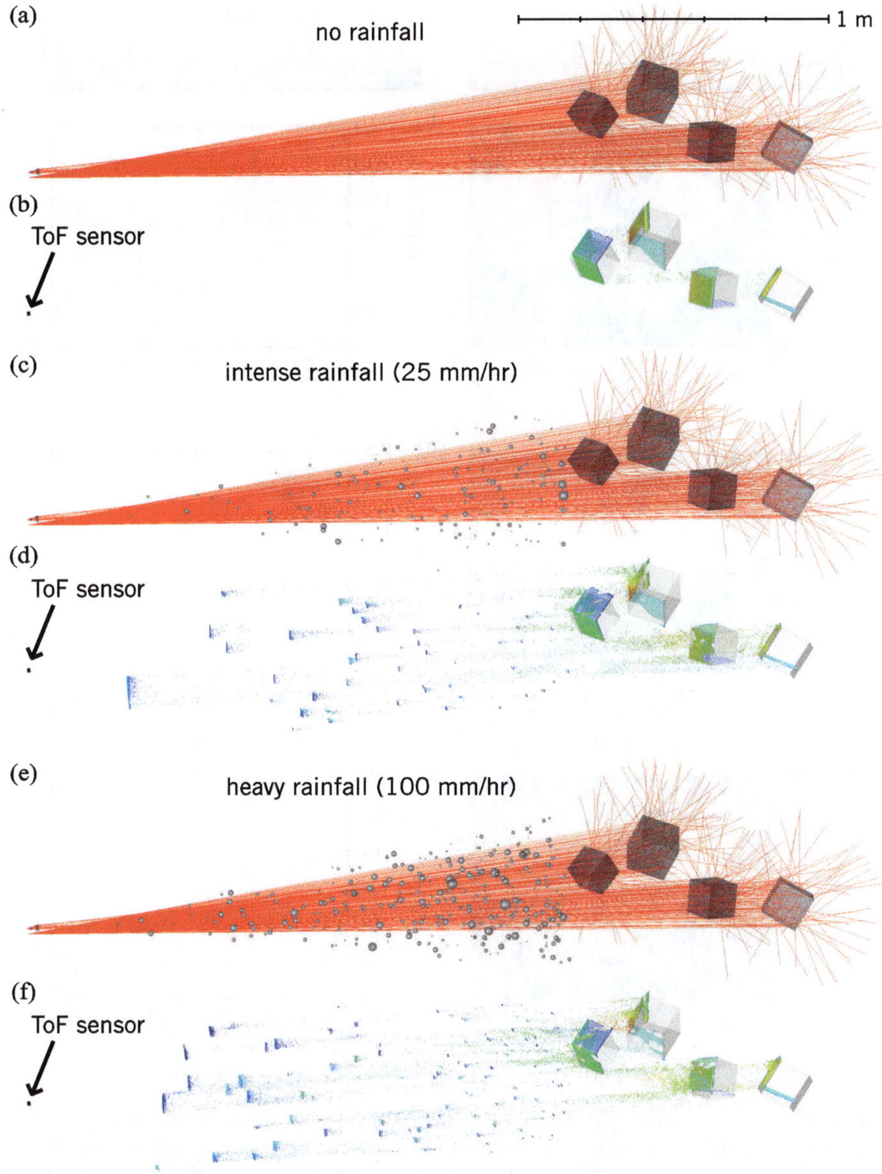

Fig. 4.13 Zemax-based simulation model and resulting ToF camera responses for a scene in the presence of raindrops. (**a**) Reference simulation model without raindrops and (**b**) corresponding modelled 3D point-cloud. (**c**, **d**) Simulation model and retrieved 3D point-cloud, respectively, for a rainfall rate of 25 mm/h and (**e**, **f**) of 100 mm/h. Raindrops in (**c**) and (**e**) are enlarged of a factor 10 for better visualization. Colors of the 3D point-clouds map the intensity reaching the sensor according to the color scale used in Fig. 4.8a

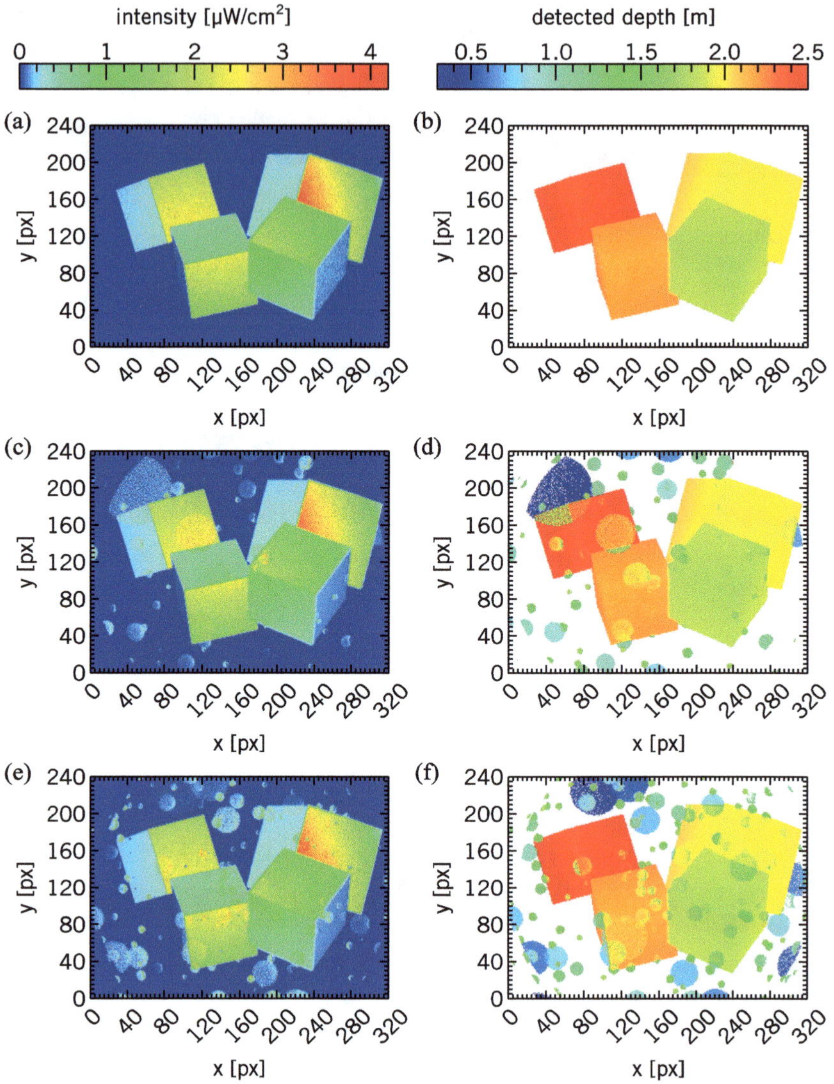

Fig. 4.14 Influence of rain droplets on ToF camera response. (**a**) Intensity and (**b**) depth for test reference test scene. (**c**) Intensity, and (**d**) depth for a rainfall rate of 25 mm/h between test scene and ToF camera. (**e**) Intensity and (**f**) depth for a rainfall rate of 100 mm/h between test scene and ToF camera

smaller than the diameter of the smallest modelled raindrop. Although we did not implement the simulation yet, it is clear that the final image will consist of a line-pattern, rather than a droplet-pattern, superimposed with the object scene. Since the objects do not move during acquisition, the range perturbation induced by a single droplet is expected to decrease with increasing integration time. This is because the

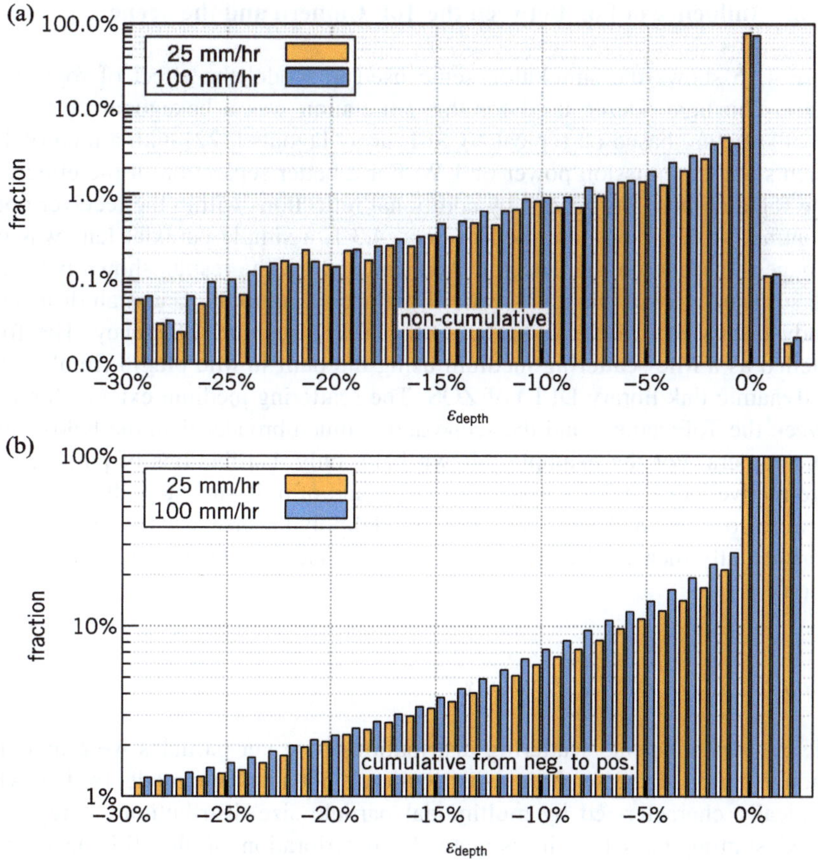

Fig. 4.15 Error on the depth accuracy of the ToF sensor caused by rain between the objects and the sensor for rainfall rates of 25 and 100 mm/h. (**a**) Non-cumulative and (**b**) cumulative distribution of the depth errors

relative contribution of the intensity reflected back from each rain droplet to the total ToF signal decreases due to the relative motion of the droplet with respect to the object behind it. A competing effect occurs at longer integration times where reflections from several droplets can be averaged at the sensor and in turn will increase the range perturbation again. Simultaneously, a larger area of the scene will be affected by a range error. Therefore, the final effect will strongly depend not just on the integration time, but also on (1) the details of the scene (e.g., the surface properties), (2) its distance (more distant scenes scatter back less light, while simultaneously the number of raindrops between camera and scene increases) and (3) the rainfall rate. As a consequence, a precise quantitative estimation of the performance of a ToF camera in a rainy scenario requires extended and detailed simulations, accounting for the typical variability in the aforementioned parameters.

4.4.2.2 Influence of Fog Between the ToF Camera and the Scene

Figure 4.16 shows the simulation setup used to model the effect of fog on ToF signals. The light source used for this simulation was a hexagonal closed-pack type of structure based on typical VCSEL array layouts [122] with a total of 1315 elements and an emission power of 1 W. For a better separation of the effects due to the fog and those introduced by additional reflections within the receiver optics, in contrast to the setup described in Sect. 4.3.1, a simple paraxial lens was used to image the scene on the detector. Additionally, for the results shown below, the cube surfaces were assumed to scatter light according to the Gaussian distribution already discussed in Eq. (4.11), but with unitary quantum efficiency. The fog is modelled as a Mie-scattering medium using the built-in Mie plugin (implemented as a dynamic link library DLL) of ZOS. The scattering medium extends for 1.6 m between the ToF camera and the scene and is much broader than the field-of-view of the camera. For the examples shown here, only distributions of particles with the refractive index and absorbance of water and of uniform (monodisperse) radial size were considered. The possibility to model polydisperse size distributions, e.g., experimentally measured distributions, will be discussed briefly at the end of this section.

Monodisperse Particle Distributions

Different monodisperse scattering distributions of water particles were modelled. Although a monodisperse particle distribution is not a realistic model for fog, which is typically characterized by multimodal particle size distributions, it represents a good starting model to investigate the deterioration of the ToF accuracy in

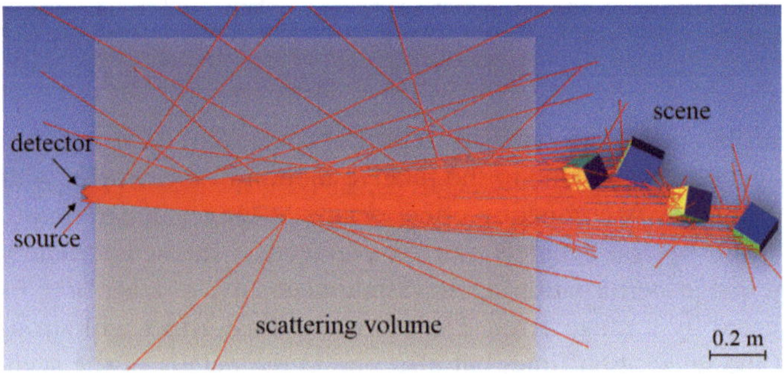

Fig. 4.16 Model setup for the investigation of ToF signals in the presence of a scattering medium. The sensor is composed of a source array, a detector array and an imaging optics which, for simplicity, was described by a paraxial lens. The scene is represented by four cubes with scattering surfaces. A volume containing a monodisperse distribution of scattering centers is located between the sensor and the scene. The interaction of light with the scattering distribution is described by a Mie-model. Reproduced from [121]

the presence of scattering centers. In particular, it allows for a straightforward correlation of the ranging errors with the characteristics of the distribution.

For monodisperse distributions of spherical particles with radius a and a particle density ρ, the single-particle scattering cross-section C_{sca} is defined as

$$C_{\mathrm{sca}} = \pi\, a^2 Q_{\mathrm{sca}} \tag{4.18}$$

where Q_{sca} is the scattering efficiency. The latter is defined as a function of the Mie-scattering coefficients a_n and b_n as

$$Q_{\mathrm{sca}} = \frac{2}{x^2} \sum_{n=1}^{\infty} (2n+1) \left(|a_n|^2 + |b_n|^2 \right) \tag{4.19}$$

where $x = ka$ and k is the wave vector of the light. The total extinction cross-section C_{ext} of the particle is obtained by summing the scattering and the absorption cross-sections: $C_{\mathrm{ext}} = C_{\mathrm{sca}} + C_{\mathrm{abs}}$. A detailed treatment of the Mie-scattering theory and derivation of these expressions can be found in the book of van de Hulst [123].

Given a particle distribution, the extinction coefficient γ determines the attenuation according to the Beer-Lambert law

$$I = I_0\, e^{-\gamma L} \tag{4.20}$$

where I and I_0 are the transmitted and initial intensities, respectively, and L is the length of the scattering medium. In general, for a scattering distribution, the total extinction coefficient γ is composed by a scattering and an absorption contribution. It can be shown that the Mie-scattering extinction coefficient of the distribution γ_{sca} is given by

$$\gamma_{\mathrm{sca}} = \rho\, C_{\mathrm{sca}} \tag{4.21}$$

For practical purposes, several online calculators or software packages offer the possibility to calculate the Mie-scattering coefficients and/or cross-sections as a function of the wavelength of the light, the particle size, and its refractive index. For the examples in this chapter, the total absorption cross-section can be considered negligible as the absorption coefficient of water is extremely small at the wavelength considered, and thus $\gamma \approx \gamma_{\mathrm{sca}}$.

Table 4.2 lists the parameters of the scattering distributions considered in this work. Figures 4.17b–e show the simulated point-cloud data for all distributions (1–4), compared to the simulated point-cloud in the absence of scattering (Fig. 4.17a). The presence of scattering particles appears as a spatially inhomogeneous, diffuse background noise which is more intense on the detector side closer to the source. This spatial inhomogeneity is caused by the combined effect of the angular distribution of Mie-scattering and the acceptance angle of the receiver optics. Besides the background noise, the distance of the scene objects is system-

Table 4.2 Parameters of the (monodisperse) scattering distributions considered in this chapter: a = radius; ρ = volumetric particle density; LWC = liquid water content; $C_{sca}{}^{particle}$ = scattering cross-section of a single particle; γ_{sca} = scattering component of the extinction coefficient of the particle distribution; I_{det}/I_0 = fraction of the total emitted intensity (I_0) reaching the detector

Distr. No.	a [μm]	ρ [part/cm^3]	LWC [g/m^3]	$C_{sca}{}^{particle}$ [cm^{-2}]	γ_{sca} [m^{-1}]	I_{det}/I_0
0[a]	–	–	0	–	–	1.09 E−5
1	1.5	497	0.007	1.22 E−7	6.1 E−3	1.17 E−5
2	17	10	0.206	1.85 E−5	1.8 E−2	1.41 E−5
3	10	40	0.168	7.10 E−6	2.8 E−2	1.27 E−5
4	5	200	0.104	1.81 E−6	3.6 E−2	1.41 E−5

[a] Distribution number 0 describes the reference simulation (i.e., without scattering medium)

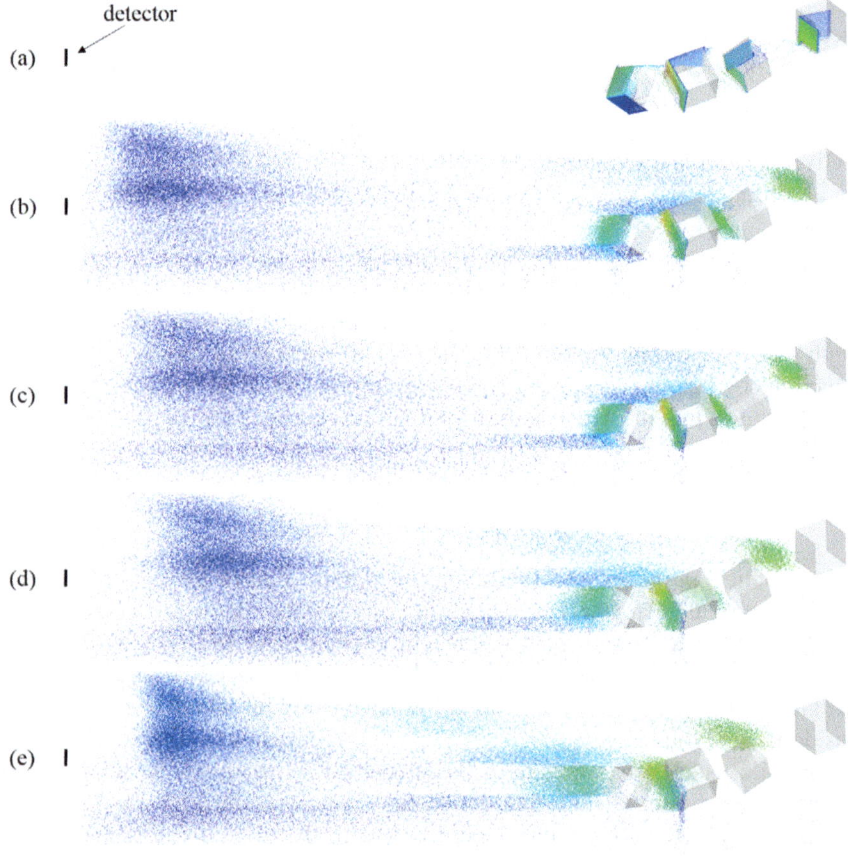

Fig. 4.17 Intensity and distance point-cloud (colors) describing the simulated ToF signal for all distributions listed in Table 4.2. Grey cubes represent the ground truth. (**a**) Reference simulation (no particles). The difference between the reconstructed image and the ground truth is caused by lens aberrations. (**b–e**) Distributions 1–4, showing the effect of an increasing γ_{sca} on the deterioration of the ranging accuracy. Adapted and extended from [121]

atically underestimated, while simultaneously the intensity at the detector (I_{det}) is systematically larger than that measured without the scattering medium. For the sake of clarity, the fraction of emitted intensity reaching the detector (I_{det}/I_0) is listed in the last column of Table 4.2.

As discussed in the beginning of Sect. 4.4.2, an underestimation of the object distance indicates the presence of a dominant multipath interference contribution from light backscattered prior to reaching the scene (direct backscattering). This hypothesis can be confirmed by analyzing the rays reaching the detector according to their intensity and the interactions they experienced on their optical path. The results of such an analysis are summarized in the two histograms of Fig. 4.18. In Fig. 4.18a, the absolute power reaching the detector is plotted for the different possible optical paths and for all scattering distributions. Figure 4.18b shows the relative power in each optical path. For each entry, the distributions are ordered from left to right with increasing scattering extinction coefficient γ_{sca}. Three major conclusions can be derived from the histogram. First, as expected the intensity arising from rays that interact exclusively with the scene decreases monotonically with rising γ_{sca} (see also inset in Fig. 4.18a). Second, the largest contribution arising from the particle distribution is indeed due to direct backscattering, which explains both the dominant underestimation of the object distance and the slight increase of intensity at the detector in the presence of scattering particles. Third, multiple scattering events are rare, and typically contribute <0.1% to the total signal.

Presenting a deep analysis of ranging errors induced by global illumination from extended scattering media and strategies for their correction are topics that go well beyond the scope of this chapter. However, having demonstrated how to generate and analyze point-cloud data for a scene in the presence of a scattering medium, a natural question is whether and how this information can be used to devise strategies to minimize the observed ranging errors. A simple example to illustrate a possible strategy is shown in Fig. 4.19. As can be seen, filtering out all signals from short distances effectively improves the ranging accuracy.

Figure 4.19a shows the mapped scene in the absence of scattering particles (reference scene), while Fig. 4.19b shows the reconstructed distance for the particle distribution number 3 assuming a 12-bit ADC (see Fig. 4.17d and Table 4.2 for the details). Figures 4.19c, d show the reconstructed depth maps obtained when neglecting all rays sampling a sensor-scene distance shorter than 61 cm and 84 cm, respectively. These values correspond to about one quarter and one third of the total scene distance and remove 36% and 52% of the fog thickness, respectively. Distant surfaces and surfaces forming the largest angles with the ToF camera are most affected by ranging errors since they scatter less intensity to the detector (see also Fig. 4.8a). Filtering out the signal from fog layers close to the sensor removes backscattered intensity overlapping with the signal coming from the scene. Therefore, the relative improvement due to this type of filtering is expected to be large for distant and tilted surfaces, while the effect on surfaces that backscatter a large amount of light to the detector is expected to be small. The reduction of the distance error by such a filtering strategy is quantified in Fig. 4.20 for the five areas marked in Fig. 4.19a, and confirms the expected trend. The most significant effect is

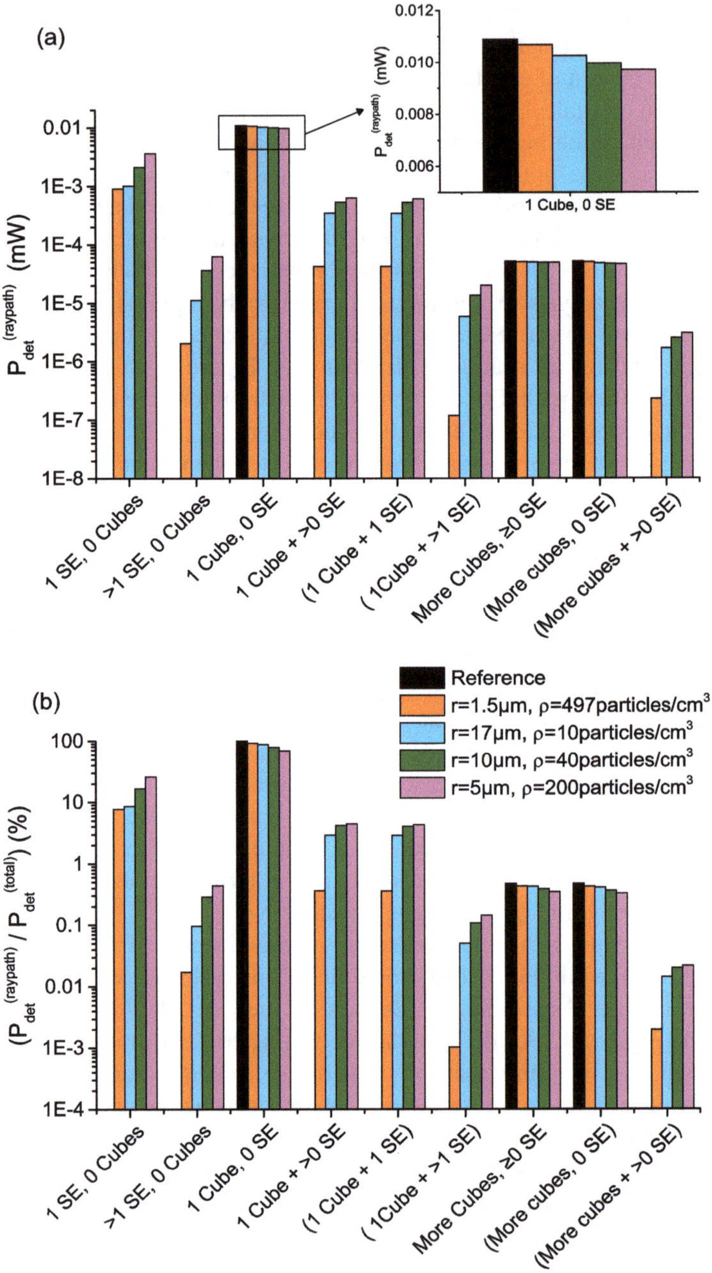

Fig. 4.18 Analysis of the ray paths and their contribution to the total power at the detector for the scattering distributions listed in Table 4.2. (**a**) Absolute power reaching the detector and (**b**) fraction of emitted power reaching the detector for the shown ray paths. *SE* = scattering event

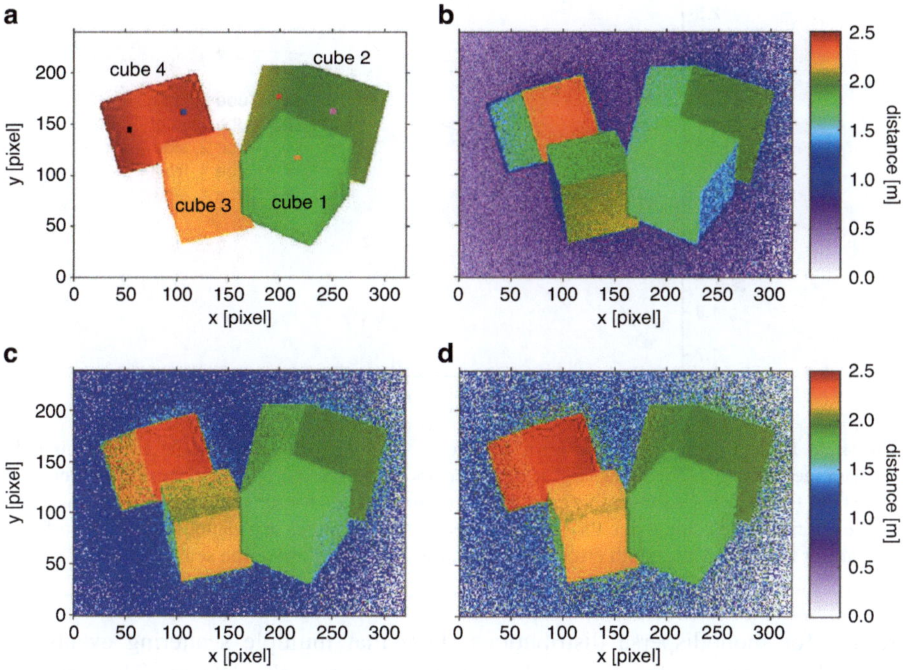

Fig. 4.19 Degradation of ToF ranging accuracy caused by a scattering medium located between the sensor and the scene and decrease of the ranging errors by removing signals from short distances. (**a**) Reference depth map. (**b**) Depth map for the fog distribution No. 3 in Table 4.2. (**c**, **d**) Reconstructed depths when filtering all signals from distances shorter than 61 cm and 84 cm, respectively

observed on the tilted side surface of cube 4, where removal of the scattering signals from the first 61 cm in front of the sensor already reduces the range error from 28 to 10%.

As a final remark, as the surfaces of the cubes have unitary scattering efficiency, the scenario modelled here represents the best possible performance for a ToF camera in the presence of the described scattering medium. Any additional absorption or losses at the surfaces will further decrease the intensity of the "direct path" channel while leaving the backscattered intensity unchanged, thus further deteriorating the range accuracy.

Modelling Polydisperse Distributions

Modelling Mie-scattering effects from polydisperse particle distributions is challenging due to the increased computational time required. The feature is not implemented in Zemax OpticStudio, but it can be included via coding of a user-defined plugin or program extension. Although this is the correct approach, the

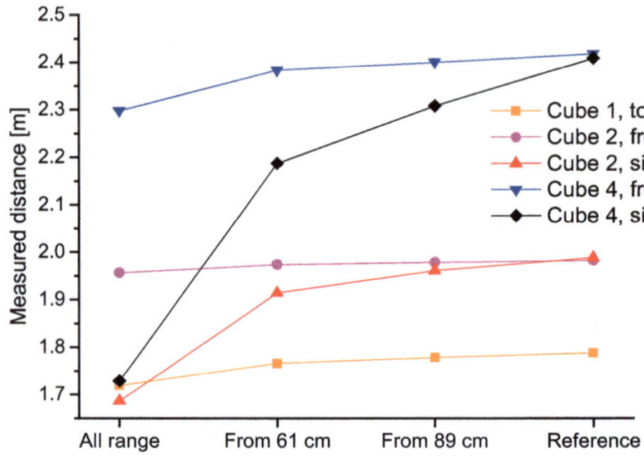

Fig. 4.20 Reduction in the depth error for the areas indicated in Fig. 4.19a by reducing the measurement depth range. The largest depth errors are observed for surfaces that scatter back little light to the ToF sensor (in this case, these are the surfaces forming the largest angles with the ToF emitter and detector, since the surface properties of all cubes are identical). For all surfaces, the largest fraction of the depth error arises from fog particles close to the sensor

results for monodisperse distributions show that multiple scattering events are typically accounting for <0.1% of the total intensity at the detector. This suggests the possibility to obtain a good approximation for a polydisperse distribution starting by proper combination of the results of multiple simulations with different monodisperse distributions mapping the total particle distribution.

4.5 Comparison with Other Methods

In the previous sections, we presented a detailed overview of a novel simulation approach for modelling an arbitrary ToF camera in front of an arbitrary scene. To be of practical relevance for testing risk scenarios and expected performances of different sensors, we believe that the modelling approach must satisfy the following requirements:

1. Provide a per-pixel intensity and distance information
2. Support an optically accurate model of the ToF sensor (especially regarding the light source properties and the lens system used)
3. Support the possibility to model an arbitrary scene
4. Support arbitrary user-defined surface properties, including specular and diffuse reflection, refraction, transparency, and translucency; support user-defined BSDF
5. Support the possibility to include multiple sources and ambient light

6. Account for global illumination (multiple scattering or reflections) and possibly allow the user to identify the origin and characteristics of the ranging errors caused by multipath interference (multipath analysis)
7. Support models for different environmental conditions (rain, fog, haze, snow)
8. Allow comparison of different hardware schemes
9. Provide simulation results "reasonably" fast
10. Be able to account for motion blur
11. Ideally, provide a good GUI.

Our approach undoubtedly fulfils points 1–7. A major advantage of the approach is that, for a given scene, a direct comparison between different hardware implementations can be performed completely separated from the ray-tracing by using one single ray-traced dataset. The major drawback of the approach is the relatively slow simulation speed due to the ray-tracing software used.

Depending on the complexity of the scene and the interactions within it, the simulation time can exceed a day (this is especially true for scattering models including the full physical description of Mie-scattering). As a consequence, motion blur effects can only be evaluated for simple scenes in a reasonable amount of time. In this regard, the CamSim toolbox which was released in the middle of 2018 is superior to our simulation approach. Using OpenGL and the full power of the GPU to trace the rays results in very short simulation times. However, it needs to be clarified if and how this platform can be extended to properly describe complex media, such as refractive objects (like raindrops) or scattering distributions (fog). Table 4.3 shows a comparison between the possibilities presented by our approach and the features of the most common (freeware or publically available) modelling environments for ToF cameras.

4.6 Conclusions

In this chapter, we have presented a novel approach to modelling ToF cameras. Our approach has several advantages with respect to current simulation environments:

- It allows retaining all optical effects (at least within the geometric optics model), including refraction and multiple reflections and/or scattering. As demonstrated in the examples of the glass window (Sect. 4.4.1) and of the rain droplets (Sect. 4.4.2), this is necessary for a correct modelling of a situation where multiple reflections/refraction events take place in the scene and give a significant contribution to the total intensity reaching the detector. Additionally, such an approach allows including a physical description of Mie-scattering from dense media.
- The approach has great versatility due to the possibility to easily modify and compare different source profiles and detector optics or to test for various scene surface properties.

Table 4.3 Comparison between the approach presented here and other advanced simulation software

Feature	Blensor	CamSim[a]	Our approach
Per-pixel intensity	No	Yes	Yes
Per-pixel distance	Yes	Yes	Yes
Lens distortions/details of imaging system	No	Yes	Yes
Specular reflection	Yes, as a binary choice (reflectivity is either 0 or 1)	Yes	Yes
Diffuse reflections	Yes	Yes	Yes
Arbitrary BSDF	No	Yes	Yes
Transparency	Not at a ToF sensor level (only rendering)	Yes	Yes
Refraction	No	No	Yes
Translucency	No	No	Yes
Scattering media	No	No	Yes
Multiple sources	No	Yes	Yes
Ambient light	No	Yes	Yes
Global illumination	Very limited (only to specular reflections)	Single bounce	Yes (full path)
Multipath analysis	No	No	Yes
Models for atmospheric effects	No	No	Yes (physical models)
Tracing speed	Fast	Very fast	Slow
Motion blur	No	Yes	Yes with restrictions (requires multiple simulations, slow)
Support for several sensor types	Yes with restrictions (to implemented systems)	Yes (except d-ToF)	Yes
GUI	Yes	No	Yes for simulation phase (ZOS[b]) Not for processing phase

[a]The list reports the information available on the official web page. Not all features have been verified
[b]Closed-source commercial product

- By separating the ray-tracing simulation from the post-processing, many features, such as different hardware implementations or different surface properties of the objects of the scene, can be systematically tested starting from the very same simulation data without the need to run a new ray-tracing. This allows for a faster and more efficient comparison between the results and can speed up ray-tracing-based optimization processes.

- The possibility to separate contributions arising from different optical paths is a powerful feature that can be used to track the origin of specific effects and devise strategies to specifically address them.

An important remark is that the implementation proposed here is not (and never will be) a real-time simulation environment. It is an approach that is meant to solve the basic objective we posed at the beginning of this chapter, that is, to help us learn about the critical phenomena we need to control (or correct for) if we want to bring ToF technology to the point of being compatible with a safe autonomous operation in any outdoor scenario we can imagine. As we have shown, our approach can be used not only to quantify the magnitude of a range error but also to identify its origin and to determine how it manifests itself in the resulting ToF image. Achieving a full mapping, quantification, and correction for such ranging errors undoubtedly requires much more work. However, many ideas and technologies are already out there. We believe that the deeper knowledge on the origin of ToF range errors can give us the key to identify the most suitable technologies to achieve the best performances in each and all possible scenarios (be it time-gating, measuring at multiple frequency, implementing post-processing correction algorithms, a mixture of these or even new ideas).

Acknowledgments This project has received funding from the Electronic Component Systems for European Leadership Joint Undertaking under grant agreement No 692480. This Joint Undertaking receives support from the European Union's Horizon 2020 research and innovation programme and Germany, Saxony, Spain, Austria, Belgium, and Slovakia. The project IoSense is co-funded by Austrian Research Promotion Agency (FFG). The authors also thank Dr. Jan Steinbrener and Dr. Gerald Auböck for helpful discussions.

References

1. J.J. Koenderink, A.J. van Doorn, A.M.L. Kappers, Depth relief. Perception **24**, 115–126 (1995)
2. P. Servos, Distance estimation in the visual and visuomotor systems. Exp. Brain Res. **130**, 35–47 (2000)
3. D. Vishwanath, P.B. Hibbard, Seeing in 3-D with just one eye: Stereopsis without binocular vision. Psychol. Sci. **24**, 1673–1685 (2013)
4. A.P. Pentland, A new sense for depth of field. IEEE Trans. Pattern Anal. Mach. Intell. **PAMI-9**, 523–531 (1987)
5. S. Weiss, D. Scaramuzza, R. Siegwart, Monocular-SLAM–based navigation for autonomous micro helicopters in GPS-denied environments. J. Field Robot. **28**, 854–874 (2011)
6. D. Antensteiner, S. Štolc, T. Pock, A review of depth and normal fusion algorithms. Sensors **18**, 431 (2018)
7. F. Santoso, M.A. Garratt, M.R. Pickering, M. Asikuzzaman, 3D mapping for visualization of rigid structures: A review and comparative study. IEEE Sensors J. **16**, 1484–1507 (2016)
8. M.T. McCann, K.H. Jin, M. Unser, Convolutional neural networks for inverse problems in imaging: A review. IEEE Signal Process. Mag. **34**, 85–95 (2017)
9. B. Girod, G. Greiner, H. Niemann (eds.), *Principles of 3D Image Analysis and Synthesis* (Springer US, New York, 2002)

10. N. Lazaros, G.C. Sirakoulis, A. Gasteratos, Review of stereo vision algorithms: From software to hardware. Int. J. Optomechatron. **2**, 435–462 (2008)
11. S. Zhang, *Handbook of 3D Machine Vision: Optical Metrology and Imaging* (CRC, Boca Raton, FL, 2013). https://doi.org/10.1201/b13856
12. P.J. Besl, Active optical range imaging sensors, in *Advances in Machine Vision*, ed. by J. L. C. Sanz, (Springer, New York, 1989), pp. 1–63
13. C. Dal Mutto, P. Zanuttigh, G.M. Cortelazzo, *Time-Of-Flight Cameras and Microsoft Kinect*™ (Springer-Verlag, New York, 2012)
14. D. Malacara, *Optical Metrology of Diffuse Surfaces* (Wiley, New York, 2007)
15. W.E.K. Middleton, A.F. Spilhaus, *Meteorological Instruments*, 3rd edn. (University of Toronto Press, Toronto, ON, 1953)
16. A. Javan, Theory of a three-level maser. Phys. Rev. **107**, 1579–1589 (1957)
17. A.L. Schawlow, C.H. Townes, Infrared and optical masers. Phys. Rev. **112**, 1940–1949 (1958)
18. T.H. Maiman, Stimulated optical radiation in ruby. Nature **187**, 493–494 (1960)
19. N.G. Basov, O.N. Krokhin, Y.M. Popov, Generation, amplification, and detection of infrared and optical radiation by quantum-mechanical systems. Sov. Phys. Usp. **3**, 702 (1961)
20. M. I. Skolnik (ed.), *Radar Handbook*, 3. [rev.] edn. (McGraw Hill, New York, NY, 2008)
21. M. Jaboyedoff, T. Oppikofer, A. Abellán, M.-H. Derron, A. Loye, R. Metzger, A. Pedrazzini, Use of LIDAR in landslide investigations: A review. Nat. Hazards **61**, 5–28 (2012)
22. D.P. Thoma, S.C. Gupta, M.E. Bauer, C.E. Kirchoff, Airborne laser scanning for riverbank erosion assessment. Remote Sens. Environ. **95**, 493–501 (2005)
23. J.S. Deems, T.H. Painter, D.C. Finnegan, Lidar measurement of snow depth: A review. J. Glaciol. **59**, 467–479 (2013)
24. W.Y. Yan, A. Shaker, N. El-Ashmawy, Urban land cover classification using airborne LiDAR data: A review. Remote Sens. Environ. **158**, 295–310 (2015)
25. C. Weitkamp (ed.), *Lidar: Range-Resolved Optical Remote Sensing of the Atmosphere* (Springer-Verlag, New York, 2005)
26. K. Sassen, The polarization lidar technique for cloud research: A review and current assessment. Bull. Am. Meteor. Soc. **72**, 1848–1866 (1991)
27. K. Fredriksson, B. Galle, K. Nyström, S. Svanberg, Lidar system applied in atmospheric pollution monitoring. Appl. Opt. **18**, 2998–3003 (1979)
28. A. Papayannis, G. Ancellet, J. Pelon, G. Mégie, Multiwavelength lidar for ozone measurements in the troposphere and the lower stratosphere. Appl. Opt. **29**, 467–476 (1990)
29. S.W. Henderson, P. Gatt, D. Rees, R.M. Huffaker, P. Gatt, D. Rees, R.M. Huffaker, Wind Lidar, in *Laser Remote Sensing*, (Taylor & Francis, Boca Raton, FL, 2005). https://doi.org/10.1201/9781420030754-10
30. A. Sathe, J. Mann, A review of turbulence measurements using ground-based wind lidars. Atmos. Meas. Tech. **6**, 3147–3167 (2013)
31. T. Fujii, T. Fukuchi, *Laser Remote Sensing* (CRC, Boca Raton, FL, 2005)
32. M.H. Conde, *Compressive Sensing for the Photonic Mixer Device: Fundamentals, Methods and Results* (Springer, Berlin, 2017)
33. Rudolf Schwarte. Method and Apparatus for Determining the Phase and/or Amplitude Information of an Electromagnetic Wave for Photomixing. (1998)
34. R. Schwarte, H. Heinol, B. Buxbaum, T. Ringbeck, Z. Xu, K. Hartmann, Principles of three-dimensional imaging techniques, in *Handbook of Computer Vision and Applications: Sensors and Imaging*, (Academic, New York, 1999), pp. 463–484
35. H.P. Stahl, Review of phase-measuring interferometry, in *Optical Testing and Metrology III: Recent Advances in Industrial Optical Inspection*, (International Society for Optics and Photonics, Bellingham, WA, 1991), pp. 704–720
36. O. Lottner, *Investigations of Optical 2D/3D-Imaging with Different Sensors and Illumination Configurations* (Shaker, Aachen, 2012)
37. T. Ringbeck, T. Möller, B. Hagebeuker, Multidimensional measurement by using 3-D PMD sensors, in *Advances in Radio Science*, (Copernicus GmbH, Göttingen, 2007), pp. 135–146

38. L. Pancheri, D. Stoppa, Sensors based on in-pixel photo-mixing devices, in *TOF Range-Imaging Cameras*, ed. by F. Remondino, D. Stoppa, (Springer, Berlin, 2013), pp. 69–89
39. B.L. Stann, W.C. Ruff, Z.G. Sztankay, Intensity-modulated diode laser radar using frequency-modulation/continuous-wave ranging techniques. Opt. Eng. **35**, 3270–3279 (1996)
40. R. Matthey, V. Mitev, Pseudo-random noise-continuous-wave laser radar for surface and cloud measurements. Opt. Lasers Eng. **43**, 557–571 (2005)
41. J.P. Godbaz, M.J. Cree, A.A. Dorrington, Understanding and ameliorating non-linear phase and amplitude responses in AMCW lidar. Remote Sens. **4**, 21–42 (2012)
42. D. Piatti, F. Remondino, D. Stoppa, State-of-the-art of TOF range-imaging sensors, in *TOF Range-Imaging Cameras*, ed. by F. Remondino, D. Stoppa, (Springer, Berlin, 2013), pp. 1–9
43. D.M. Burns, V.M. Bright, S.C. Gustafson, E.A. Watson, Optical beam steering using surface micromachined gratings and optical phased arrays, in *Optical Scanning Systems: Design and Applications*, (International Society for Optics and Photonics, Bellingham, WA, 1997), pp. 99–111
44. J.K. Doylend, M.J.R. Heck, J.T. Bovington, J.D. Peters, L.A. Coldren, J.E. Bowers, Two-dimensional free-space beam steering with an optical phased array on silicon-on-insulator. Opt Exp. **19**, 21595–21604 (2011)
45. A. Yaacobi, J. Sun, M. Moresco, G. Leake, D. Coolbaugh, M.R. Watts, Integrated phased array for wide-angle beam steering. Opt. Lett. **39**, 4575–4578 (2014)
46. C.V. Poulton, A. Yaacobi, D.B. Cole, M.J. Byrd, M. Raval, D. Vermeulen, M.R. Watts, Coherent solid-state LIDAR with silicon photonic optical phased arrays. Opt. Lett. **42**, 4091–4094 (2017)
47. R.A. Jarvis, A perspective on range finding techniques for computer vision. IEEE Trans. Pattern Anal. Mach. Intell. **PAMI-5**, 122–139 (1983)
48. J.S. Massa, G.S. Buller, A.C. Walker, S. Cova, M. Umasuthan, A.M. Wallace, Time-of-flight optical ranging system based on time-correlated single-photon counting. Appl. Opt. **37**, 7298–7304 (1998)
49. E. Charbon, M. Fishburn, R. Walker, R.K. Henderson, C. Niclass, SPAD-based sensors, in *TOF Range-Imaging Cameras*, ed. by F. Remondino, D. Stoppa, (Springer, Berlin, 2013), pp. 11–38
50. S. Bellisai, D. Bronzi, F.A. Villa, S. Tisa, A. Tosi, F. Zappa, Single-photon pulsed-light indirect time-of-flight 3D ranging. Opt. Exp. **21**, 5086–5098 (2013)
51. N.A.W. Dutton, I. Gyongy, L. Parmesan, R.K. Henderson, Single photon counting performance and noise analysis of CMOS SPAD-based image sensors. Sensors **16**, 1122 (2016)
52. J.P. Godbaz, A.A. Dorrington, M.J. Cree, Understanding and ameliorating mixed pixels and multipath interference in AMCW lidar, in *TOF Range-Imaging Cameras*, ed. by F. Remondino, D. Stoppa, (Springer, Berlin, 2013), pp. 91–116
53. M. Hebert, E. Krotkov, 3D measurements from imaging laser radars: How good are they? Image Vis. Comput. **10**, 170–178 (1992)
54. M.D. Adams, P.J. Probert, The interpretation of phase and intensity data from AMCW light detection sensors for reliable ranging. Int. J. Robot. Res. **15**, 441–458 (1996)
55. M. Schmidt, *Analysis, Modeling and Dynamic Optimization of 3D Time-Of-Flight Imaging Systems* (Ruperto-Carola University of Heidelberg, Heidelberg, 2011)
56. M. Grzegorzek, C. Theobalt, R. Koch, A. Kolb (eds.), *Time-of-Flight and Depth Imaging. Sensors, Algorithms and Applications: Dagstuhl Seminar 2012 and GCPR Workshop on Imaging New Modalities* (Springer, Berlin, 2013)
57. S.N.R. Meister, *On Creating Reference Data for Performance Analysis in Image Processing* (University of Heidelberg, Heidelberg, Germany, 2014). https://doi.org/10.11588/heidok.00016193
58. M. Baumgart, N. Druml, M. Consani, Procedure enabling simulation and in-depth analysis of optical effects in camera-based time-of-flight sensors, in *ISPRS - International Archives of the Photogrammetry, Remote Sensing and Spatial Information Sciences*, (Copernicus GmbH, Göttingen, 2018), pp. 83–89

59. R. Horaud, Hansard M, G. Evangelidis, C. Ménier, An overview of depth cameras and range scanners based on time-of-flight technologies. Mach. Vis. Appl. **27**, 1005–1020 (2016)
60. S.T.S. Holmström, U. Baran, H. Urey, MEMS laser scanners: A review. J. Microelectromech. Syst. **23**, 259–275 (2014)
61. H.W. Yoo, N. Druml, D. Brunner, C. Schwarzl, T. Thurner, M. Hennecke, G. Schitter, MEMS-based lidar for autonomous driving. Elektrotech. Infotech. **135**, 408–415 (2018)
62. S. Fuchs, Multipath interference compensation in time-of-flight camera images, in *2010 20th International Conference on Pattern Recognition*, (2010), pp. 3583–3586
63. A.A. Dorrington, J.P. Godbaz, M.J. Cree, A.D. Payne, L.V. Streeter, Separating true range measurements from multi-path and scattering interference in commercial range cameras, in *Three-Dimensional Imaging, Interaction, and Measurement*, (International Society for Optics and Photonics, Bellingham, WA, 2011), p. 786404
64. J.P. Godbaz, M.J. Cree, A.A. Dorrington, Closed-form inverses for the mixed pixel/multipath interference problem in AMCW lidar. Comput. Imag. X. **8296**, 829618 (2012). https://doi.org/10.1117/12.909778
65. A. Kadambi, R. Whyte, A. Bhandari, L. Streeter, C. Barsi, A. Dorrington, R. Raskar, Coded time of flight cameras: Sparse deconvolution to address multipath interference and recover time profiles. ACM Trans. Graph. **32**, 167:1–167:10 (2013)
66. A. Kirmani, A. Benedetti, P.A. Chou, SPUMIC: Simultaneous phase unwrapping and multipath interference cancellation in time-of-flight cameras using spectral methods, in *2013 IEEE International Conference on Multimedia and Expo (ICME)*, (2013). https://doi.org/10.1109/ICME.2013.6607553
67. F. Heide, M.B. Hullin, J. Gregson, W. Heidrich, Low-budget transient imaging using photonic mixer devices. ACM Trans. Graph. **32**, 45 (2013)
68. D. Freedman, Y. Smolin, E. Krupka, I. Leichter, M. Schmidt, *SRA: fast removal of general multipath for ToF sensors*, Computer Vision—ECCV 2014 (2014), pp. 234–249. isbn:978-3-319-10590-1
69. D. Jiménez, D. Pizarro, M. Mazo, S. Palazuelos, Modelling and correction of multipath interference in time of flight cameras, in *2012 IEEE Conference on Computer Vision and Pattern Recognition*, (2012), pp. 893–900. https://doi.org/10.1109/CVPR.2012.6247763
70. M. O'Toole, F. Heide, L. Xiao, M.B. Hullin, W. Heidrich, K.N. Kutulakos, Temporal frequency probing for 5D transient analysis of global light transport. ACM Trans. Graph. **33**, 87:1–87:11 (2014)
71. J. Lin, Y. Liu, M.B. Hullin, Q. Dai, *Fourier analysis on transient imaging with a multi-frequency time-of-flight camera. 2014 IEEE Conference on Computer Vision and Pattern Recognition* (2014), pp. 3230–3237. https://doi.org/10.1109/CVPR.2014.419
72. M. Gupta, S.K. Nayar, M.B. Hullin, J. Martin, Phasor imaging: A generalization of correlation-based time-of-flight imaging. ACM Trans Graph **34**, 156:1–156:18 (2015)
73. N. Naik, A. Kadambi, C. Rhemann, S. Izadi, R. Raskar, S.B. Kang, A light transport model for mitigating multipath interference in time-of-flight sensors, in *2015 IEEE Conference on Computer Vision and Pattern Recognition (CVPR)*, (2015), pp. 73–81
74. C. Peters, J. Klein, M.B. Hullin, R. Klein, Solving Trigonometric Moment Problems for Fast Transient Imaging. ACM Trans. Graph. **6**(34) (2015). 220:1–220:11. doi: 10.1145/2816795.2818103
75. H. Qiao, J. Lin, Y. Liu, M.B. Hullin, Q. Dai, Resolving transient time profile in ToF imaging via log-sum sparse regularization. Opt. Lett. **6**(40), 918–921 (2015). https://doi.org/10.1364/OL.40.000918
76. A. Kadambi, V. Taamazyan, S. Jayasuriya, R. Raskar. Frequency domain TOF: encoding object depth in modulation frequency. arXiv:1503.01804 [cs] (2015)
77. R. Whyte, L. Streeter, M.J. Cree, A.A. Dorrington, Resolving multiple propagation paths in time of flight range cameras using direct and global separation methods. Opt. Eng. **11**(54), 1560–2303 (2015). https://doi.org/10.1117/1.OE.54.11.113109
78. A. Jarabo, B. Masia, J. Marco, D. Gutierrez, Recent advances in transient imaging: A computer graphics and vision perspective. Vis. Inform. **1**, 65–79 (2017)

79. J. Busck, H. Heiselberg, Gated viewing and high-accuracy three-dimensional laser radar. Appl. Opt. **43**, 4705–4710 (2004)
80. A. Kirmani, T. Hutchison, J. Davis, R. Raskar, Looking around the corner using transient imaging, in *2009 IEEE 12th International Conference on Computer Vision*, (2009), pp. 159–166
81. A. Velten, T. Willwacher, O. Gupta, A. Veeraraghavan, M.G. Bawendi, R. Raskar, Recovering three-dimensional shape around a corner using ultrafast time-of-flight imaging. Nat. Commun. **3**, 745 (2012)
82. M. Laurenzis, A. Velten, Nonline-of-sight laser gated viewing of scattered photons. Opt. Eng. **53**, 023102 (2014)
83. R. Pandharkar, A. Velten, A. Bardagjy, E. Lawson, M. Bawendi, R. Raskar, Estimating motion and size of moving non-line-of-sight objects in cluttered environments. CVPR **2011**, 265–272 (2011)
84. A. Velten, D. Wu, A. Jarabo, B. Masia, C. Barsi, C. Joshi, E. Lawson, M. Bawendi, D. Gutierrez, R. Raskar, Femto-photography: capturing and visualizing the propagation of light. ACM Trans. Graph. **32**, 44 (2013)
85. B. Heshmat, G. Satat, C. Barsi, R. Raskar, Single-shot ultrafast imaging using parallax-free alignment with a tilted lenslet array, in *2014 conference on lasers and electro-optics (CLEO)—laser science to photonic applications*, (2014), pp. 1–2
86. G. Gariepy, N. Krstajić, R. Henderson, C. Li, R.R. Thomson, G.S. Buller, B. Heshmat, R. Raskar, J. Leach, D. Faccio, Single-photon sensitive light-in-fight imaging. Nat. Commun. **6**, 6021 (2015)
87. Laser Imaging Through Obscurants|LITO Technologies. https://litotechnologies.com/. Accessed 30 Jan 2019
88. Panasonic Develops Long-range TOF Image Sensor|Headquarters News. In *Panasonic Newsroom Global*. http://news.panasonic.com/global/press/data/2018/06/en180619-3/en180619-3.html. Accessed 30 Jan 2019
89. Y. Hirose, S. Koyama, M. Ishii, et al., A 250 m direct time-of-flight ranging system based on a synthesis of sub-ranging images and a vertical avalanche photo-diodes (VAPD) CMOS image sensor. Sensors **18**, 3642 (2018)
90. T. Kavli, T. Kirkhus, J.T. Thielemann, B. Jagielski, Modelling and compensating measurement errors caused by scattering in time-of-flight cameras, in *Two- and Three-Dimensional Methods for Inspection and Metrology VI*, (International Society for Optics and Photonics, Bellingham, WA, 2008), p. 706604
91. S. Jamtsho, D.D. Lichti, Modelling Scattering Distortion in 3 D Range Camera, in *Proceedings of the International Archives of Photogrammetry*, (2010), pp. 299–304
92. J. Marco, Q. Hernandez, A. Muñoz, Y. Dong, A. Jarabo, M.H. Kim, X. Tong, D. Gutierrez, DeepToF: Off-the-shelf real-time correction of multipath interference in time-of-flight imaging. ACM Trans. Graph. **36**, 219:1–219:12 (2017)
93. A. Bhandari, A. Kadambi, R. Whyte, C. Barsi, M. Feigin, A. Dorrington, R. Raskar, Resolving multipath interference in time-of-flight imaging via modulation frequency diversity and sparse regularization. Opt. Lett. **39**, 1705–1708 (2014)
94. F. Heide, L. Xiao, A. Kolb, M.B. Hullin, W. Heidrich, Imaging in scattering media using correlation image sensors and sparse convolutional coding. Opt. Exp. **22**, 26338–26350 (2014)
95. S. Achar, J.R. Bartels, W.L. Whittaker, K.N. Kutulakos, S.G. Narasimhan, Epipolar time-of-flight imaging. ACM Trans. Graph. **36**, 37:1–37:8 (2017)
96. M. Pharr, W. Jakob, G. Humphreys, *Physically Based Rendering: From Theory to Implementation*, 3rd edn. (Elsevier, Morgan Kaufmann, Amsterdam, 2017)
97. P. Shirley, R.K. Morley, P.-P. Sloan, C. Wyman, Basics of physically-based rendering, in *SIGGRAPH Asia 2012 Courses*, (ACM, New York, 2012), pp. 2:1–2:11
98. Quigley M, Conley K, Gerkey B, Faust J, Foote T, Leibs J, Wheeler R, Ng A (2009) ROS: An Open-Source Robot Operating System

99. N. Koenig, A. Howard, Design and use paradigms for gazebo, an open-source multi-robot simulator, in *2004 IEEE/RSJ International Conference on Intelligent Robots and Systems (IROS) (IEEE Cat. No.04CH37566)*, vol. 3, (2004), pp. 2149–2154
100. M. Lambers, S. Hoberg, A. Kolb, Simulation of time-of-flight sensors for evaluation of Chip layout variants. IEEE Sensors J. **15**, 4019–4026 (2015)
101. Martin Lambers/pmdsim. In *GitLab*. https://gitlab.marlam.de/marlam/pmdsim. Accessed 28 Jan 2019
102. M. Gschwandtner, R. Kwitt, A. Uhl, W. Pree, BlenSor: Blender sensor simulation toolbox, in *Advances in Visual Computing*, ed. by G. Bebis, R. Boyle, B. Parvin, et al., (Springer, Berlin, 2011), pp. 199–208
103. Martin Lambers/camsim. In *GitLab*. https://gitlab.marlam.de/marlam/camsim. Accessed 28 Jan 2019
104. D. Bulczak, M. Lambers, A. Kolb, Quantified, interactive simulation of AMCW ToF camera including multipath effects. Sensors **18**, 13 (2018)
105. H. Sarbolandi, D. Lefloch, A. Kolb, Kinect range sensing: Structured-light versus time-of-flight kinect. Comput. Vis. Image Underst. **139**, 1–20 (2015)
106. P. Fankhauser, M. Bloesch, D. Rodriguez, R. Kaestner, M. Hutter, R. Siegwart, Kinect v2 for mobile robot navigation: Evaluation and modeling, in *2015 International Conference on Advanced Robotics (ICAR)*, (2015), pp. 388–394
107. L. Pérez, ´i. Rodríguez, N. Rodríguez, R. Usamentiaga, D.F. García, Robot guidance using machine vision techniques in industrial environments: A comparative review. Sensors **16**, 335 (2016)
108. R. Bostelman, P. Russo, J. Albus, T. Hong, R. Madhavan, Applications of a 3D range camera towards healthcare mobility aids, in *2006 IEEE International Conference on Networking, Sensing and Control*, (2006), pp. 416–421
109. N. Druml, T. Pietsch, M. Dielacher, C. Steger, M. Baumgart, C. Consani, T. Herndl, G. Holweg, Virtual white cane featuring time-of-flight 3D imaging supporting visually impaired users, in *2018 21st Euromicro Conference on Digital System Design (DSD)*, (2018), pp. 450–457
110. N. Druml, B. Rutte-Vas, S. Wilfling, C. Consani, M. Baumgart, T. Herndl, G. Holweg, Safety-critical human detection featuring time-of-flight environment perception, in *2017 22nd IEEE International Conference on Emerging Technologies and Factory Automation (ETFA)*, (2017), pp. 1–7
111. N. Ziraknejad, P.D. Lawrence, D.P. Romilly, The effect of time-of-flight camera integration time on vehicle driver head pose tracking accuracy, in *2012 IEEE International Conference on Vehicular Electronics and Safety (ICVES 2012)*, (2012), pp. 247–254
112. M. Lindner, A. Kolb, Lateral and depth calibration of PMD-distance sensors, in *Advances in Visual Computing*, ed. by G. Bebis, R. Boyle, B. Parvin, et al., (Springer, Berlin, 2006), pp. 524–533
113. Schiller I, Beder C, Koch R (2008) Calibration of a PMD-camera Using a Planar Calibration Pattern Together with a Multi-camera Setup
114. Hansard M, in *Time-Of-Flight Cameras: Principles, Methods and Applications*, ed. by S. Lee, O. Choi, R. Horaud, (Springer, London, 2013)
115. R. Lange, *3D Time-Of-Flight Distance Measurement with Custom Solid-State Image Sensors in CMOS/CCD-Technology* (University of Siegen, Siegen, 2000)
116. A.D. Payne, A.A. Dorrington, M.J. Cree, D.A. Carnegie, Improved measurement linearity and precision for AMCW time-of-flight range imaging cameras. Appl. Opt. **49**, 4392–4403 (2010)
117. P.T. Willis, P. Tattelman, Drop-size distributions associated with intense rainfall. J. Appl. Meteorol. **28**, 3–15 (1989)
118. C.C. Chuang, K.V. Beard, C.C. Chuang, K.V. Beard, A numerical model for the equilibrium shape of electrified raindrops. J. Atmos. Sci. **47**, 1374–1389 (1989). Accessed 29 Jan 2019

119. M. Thurai, G.J. Huang, V.N. Bringi, W.L. Randeu, M. Schönhuber, Drop shapes, model comparisons, and calculations of polarimetric radar parameters in rain. J. Atmos. Ocean. Technol. **24**, 1019–1032 (2007)
120. G. Montero-Martínez, A.B. Kostinski, R.A. Shaw, F. García-García, Do all raindrops fall at terminal speed? Geophys. Res. Lett. **36**(11), L1181818 (2009). https://doi.org/10.1029/2008GL037111
121. C. Consani, N. Druml, M. Dielacher, M. Baumgart, Fog effects on time-of-flight imaging investigated by ray-tracing simulations. PRO **2**, 859 (2018)
122. H. Moench, R. Conrads, S. Gronenborn, X. Gu, M. Miller, P. Pekarski, J. Pollmann-Retsch, A. Pruijmboom, U. Weichmann, Integrated high power VCSEL systems, in *High-Power Diode Laser Technology and Applications XIV*, (International Society for Optics and Photonics, Bellingham, WA, 2016), p. 97330V
123. H.C. van de Hulst, *Light Scattering by Small Particles* (Dover, New York, 1981)

Chapter 5
Computational Intelligence for Simulating a LiDAR Sensor

Cyber-Physical and Internet-of-Things Automotive Applications

Fernando Castaño, Gerardo Beruvides, Alberto Villalonga, and Rodolfo E. Haber

5.1 Introduction

Recent developments demonstrate an increasing efficiency, availability and affordability of sensors, data acquisition systems, and computer networks [1]. Cyber-Physical Systems (CPS) and Internet-of-Things (IoT) applications are still growing in different engineering fields supporting applications across industries (e.g. manufacturing, electric power grids, agriculture) and even, in our personal lives (e.g. healthcare, society, mobility). CPS and IoT are growing as key strategic innovative technologies in automotive application fields, receiving in recent years a strong support and investment focused on the development of smart mobility ecosystems with several levels of autonomy and interaction between vehicles and infrastructures. Nevertheless, mainstream car manufacturers, OEMs for the automotive sector, researchers, and engineers are introducing new technological contributions as new short-term challenges have to be addressed [2, 3].

Nowadays, dozens of contributions are reported in the literature addressing key CPS issues from connecting topologies up to cognitive and self-configuration layers [4]. A detailed review of CPS-architecture functionalities is presented in [5] describing the integration of sensors, actuators, and protocols [6]. Big data analytics and cloud computing platforms are reported in [7] based on methods, software, and computer-based infrastructures. The analysis of current practices to predict future

F. Castaño (✉) · R. E. Haber
Centre for Automation and Robotics (CSIC—UPM), Arganda del Rey, Spain
e-mail: fernando.castano@car.upm-csic.es

G. Beruvides
Automotive and Industry Lab, Hitachi Europe GmbH, Schwaig-Oberding, Germany

A. Villalonga
Technical University of Madrid, UPM, Madrid, Spain

W. D. van Driel et al. (eds.), *Sensor Systems Simulations*,
https://doi.org/10.1007/978-3-030-16577-2_5

behaviour and provide security to components, machines, and infrastructures are presented in [8]. The concept to achieve greater storage for historic data to predict future trends is plausible; however, new sensory data-processing and decision-making technologies are required [9]. The computational requirements in relation to operating systems, programming languages, user interfaces, and networking technologies have become more sophisticated in relation to software managing, information flow control, error control, redundancy, reliability, and latency in heterogeneous global networks [10]. Furthermore, the knowledge acquisition, the learning [11], and its transformation into physical actions to help machines in decision-making activities [12] are priorities in the paradigm of CPSs.

Another important issue is the increase of computing power and wireless communication capabilities to expand the role of sensors from mere data collection to more demanding tasks that include sensor fusion, classification, and collaborative target tracking, that is, the inclusion of the concept of IoT sensor systems. Therefore, one of the main application fields for sensory systems and CPSs is the realistic scenarios for advanced driver-assistance systems (ADAS) and autonomous vehicles (AV) [13–15]. According to the Accenture LLP report, for the next 10 years and beyond, the key areas into the automotive vehicle industry are: (1) cyber security; (2) product liability for sensors and software and/or algorithms and (3) insuring AV infrastructure [16].

One particular challenge of autonomous driving is the accuracy and reliability estimation in vision devices such as Light Detection and Ranging (LiDAR) sensors and stereo cameras integrated in automotive driving assistance systems for pattern recognition and obstacle detection tasks [17–20]. In many scenarios, it is very difficult to certify the real topology and distance of the objects at a lower level of uncertainty, in most cases due to dead zones, object transparency, light reflection, weather conditions, and sensors failures, among others [4]. Furthermore, traditional networking devices are not designed for use in unpredictable, variable, and dynamic environments that characterize an IoT transportation ecosystem, making it necessary to develop new methodologies that characterize and estimate sensor reliability [21]. Sensor fusion is commonly applied to combine different sensors for road detection, mainly cameras and LiDARs. Current sensor fusion methods are even taking advantage of both types of sensors (cameras or LiDARs), rather than exploiting the advantages of each individual type of sensor [22]. Furthermore, the parallel processing of frames (from a camera) and (LiDAR) scans implies a high computational cost and is unnecessary in many scenarios where a sensor-based error-prediction model for assessing runtime reliability is operative [23].

Several reconstruction methods are reported in the literature to create specific geometric models of existing objects from scanned point clouds based on information obtained from LiDARs [24]. Progress with modelling techniques that simulate complex driving environments has led to realistic representations of multiple input/output variables, which have been used to determine the most influential factors in degraded reliability and to detect pedestrians, obstacles, and vehicles in real-time driving scenarios [25]. Nowadays, many classifiers rely on machine-learning approaches to exploit data redundancy and abundance to finding out patterns, trends, and relations among the input attributes and the class labels [26]. Within obstacle recognition techniques, vector support machines have been widely applied for classification and regression problems [27, 28]. An interesting application using machine learning for the pedestrian detection in autonomous vehicles based on HD 3D LiDAR is reported in [29], providing more accurate data to be successfully used in any kind of lighting conditions. Clustering techniques are also intensively used in exploratory data mining, statistical analysis, pattern recognition, image analysis, information retrieval, bioinformatics, data compression, and computer graphics [30, 31].

The main objective to simulate sensors is related to fault tolerance and reliability that perform a key role for embedded systems, such as obscured wireless sensors that are deployed in applications where physical access is difficult [32]. Throughout, this chapter will describe the basis of LiDAR sensor networks modelling and simulation techniques for CPS and IoT applications. Furthermore, the most commonly computational intelligence techniques reported in the literature for this purpose are described, doting of new real-time analytic and decision-making capabilities to the next sensor generations. Finally, a CPS use case is introduced in order to deploy a technical application scenario.

Reliable monitoring of a phenomenon (or event detection) depends on the set of data provided by the cluster of sensors rather than any individual node. The failure of one or more nodes may not cause the disconnection of operational data sources from the data sinks (command nodes or end user stations). However, node failures may increase the number of hops a data message has to go through before reaching its destination (thereby increasing the message delay), providing an estimation of the failure probabilities of both the sensors and the intermediate nodes (nodes used to relay messages between data sources and data sinks) [33].

One way to carry out these simulations is using a co-simulation framework that takes into account physical dynamics, control software, computational platforms, and communication networks, which is crucial for designing CPSs for autonomous driving [34, 35]. Co-simulation is essential for CPSs due to virtual prototyping, capable of properly emulate actor-sensor nodes with their own hardware specifications [36]. Moreover, virtual prototyping can take advantage of different modelling languages/tools and integrate them together for evaluating the behaviour of CPSs. For example, processing elements with real-time operating systems [37], communication systems, sensors, actuators, model transformations to the final virtual prototype [38], and localization error estimation and compensation [39] can be efficiently represented and modelled.

The current chapter is organized as follows. First, a brief description of the theoretical modelling of a LiDAR sensor based on the mathematical principle of operation will be described. After that, a review of the state-of-the-art of computational intelligence applications for sensor system simulations is presented. Three main issues are considered: the application of artificial intelligence techniques in IoT sensor networks, computational intelligence methods applied for pattern recognition and vision classification sensor-based analytics, and computational intelligence procedures for localization optimization and energy consumption in sensor networks. Next, a real case study of artificial intelligence techniques applied to sensor networks in a Cyber-Physical System is presented. Finally, the conclusions and future research steps are summarized.

5.2 Theoretical Modelling of a LiDAR Sensor

LiDAR, also frequently referred to as Laser Detection and Ranging (LiDAR), uses a laser to produce light pulses. LiDAR maintains some similitudes with its predecessor, Radio Detection and Ranging (RADAR), with the main deference being that it is carried out with electromagnetic radiation at a much smaller wavelength and greater frequency. Both technologies are based on Time-of-Flight (ToF) for the light pulses. In addition, light velocity allows these sensors to take numerous measurements per second [40].

Depending on the principle of operation and the ability to reconstruct or interpret a scanned scene from the data provided by these sensors can be classified in: single point and imaging. The first uses a single laser and photodetector such that the spatial information is limited to a single point in the propagation direction. Instead, the second is able to reconstruct or interpret a scanned 2D or 3D scene. This can be carried out by adding a scanning laser system to the single point LiDAR sensor so that it will rasterize an image normal to the propagation direction. Another imaging method uses an array of photodetectors to obtain the temporal and spatial information with each pulse of laser energy. Though they are able to gain spatial information using an array of photodetectors, there is often a need for scanning the scene in order to obtain a detailed image [41]. Multiple sources and multiple detectors is also an option for imaging with LiDAR. In the case of weak signals, even single photon detection can be achieved using a Geiger-mode avalanche photodiode (GmAPD) [42] for either single point or imaging systems [43]. Figure 5.1 shows the two distributions for the laser pulse travelling from the LiDAR system to the target, where Δ_r is the aperture diameter of the receiver, θ_{fov} is the angle of vertical field of view, and R is range dimension in the direction of the pulse travelling.

In the subsequent sections, mathematical models are carried out, considering automotive application's constraints.

5.2.1 *Laser Beam Model*

The intensity of the laser beam across the range is not uniform, but varies in the spatial and temporal domains [42]. It is defined as a beam profile and depends on the shape of the emitter and the technique used to generate the laser light. Generally, this profile or energy distribution is modelled as a spatial Gaussian function of radius r from the centre of the beam spot and distance z from the aperture [40]:

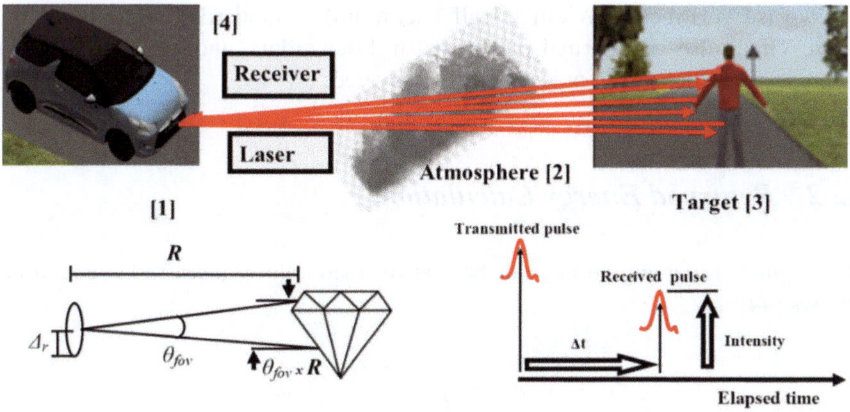

Fig. 5.1 The division of the process during simulation. Figure adapted from [40, 41, 44]

$$I\,(r,z) = I_0 \cdot \left(\frac{\omega_0}{\omega(z)}\right)^2 \cdot e^{\left(\frac{-2\cdot r^2}{\omega^2(z)}\right)} \tag{5.1}$$

where $\omega_0 = \lambda\,(\pi\;\Theta)$ is the beam width (radius) of the pulse at $z = 0$ also known as the beam waist [45].

The radial distance (in m) at which the profile value is decreased to $1/e^2$ from its peak value is $z_r = \pi\,\omega_0\,/\lambda$, where λ is the laser wavelength.

The value of I_0 is found by integrating Eq. (5.1) over r and setting the result equal to the laser power, P.

$$I_0 = \frac{2 \cdot P}{\pi \cdot \omega_0^2} \tag{5.2}$$

In addition, Eq. (5.3) represents the free space propagation of the beam results in divergence of the beam width, $\omega(z)$, where z is the propagation distance. This is true for distances that are less than the Rayleigh length [45].

$$\omega(z) = \omega_0\sqrt{1 + \left(\frac{z}{z_r}\right)^2} \tag{5.3}$$

In the temporal domain, the laser signal is modelled as a pulse. There are several pulse models with different shapes. The time propagation for a typical laser pulse can be described by:

$$p(t) = \left(\frac{t}{\tau}\right)^2 \cdot e^{-\frac{t}{\tau}}, \quad \text{where} \quad \tau = \frac{T_{1/2}}{3.5}, \tag{5.4}$$

and $T_{1/2}$ is FWHM (Full Width at Half Maximum), a standard measure of the pulse width. This pulse is a normal pulse used in laser radars, and by controlling $T_{1/2}$, different properties of the pulse can be discovered.

5.2.2 Returned Energy Calculation

The original radar equations can be transformed into a laser radar equation as follows [44]:

$$P_r = P_t \cdot \rho \cdot \frac{1}{\Omega_S R^2} \cdot \frac{\pi A_r{}^2}{4} \cdot \eta_{atm} \cdot \eta_{sys} \tag{5.5}$$

Assuming that, for an extended target, the footprint of the beam is smaller than the target surface, the returned power can be calculated using the transmitted power P_t, the travel distance of the laser beam R, the reflectance of the target surface ρ, and the aperture diameter of the receiver A_r, as represented in Eq. (5.5).

$$\eta_{atm} = T_{atm\_transmitted} \cdot T_{atm\_transmitted} = T_{atm}^2 \tag{5.6}$$

Equation (5.6) represents the round trip laser pulse, η_{atm}, as the square of the atmospheric attenuation. In addition, η_{sys} (inefficiencies and disturbances caused by the receiver system) is the product of the bandpass filter transmittance, T_{BF}, the fill factor, T_{FF}, the ND (Neutral Density) filter transmittance, T_{ND}, transmitter optics transmittance, T_T, and receiver optics transmittance, T_R.

$$\eta_{sys} = T_{BPF} \cdot T_{ND} \cdot T_{FF} \cdot T_T \cdot T_R \tag{5.7}$$

From the equations above, the received signal power at the receiver aperture P_r can be calculated using modified LADAR range equation:

$$P_r = \frac{P_t \cdot \rho \cdot A_r^2 \cdot T_{atm}^2 \cdot T_{BPF} \cdot T_{ND} \cdot T_{FF} \cdot T_T \cdot T_R}{4R^2} \tag{5.8}$$

5.2.3 Noise Energy Calculation

The problems derived from some sources of noise can be divided into four separate parts and they are all simulated individually. First, energy loose and signal fade are directly related to the aerial attenuation. Second, they the beam broadening, which makes the beam width increase with increased ranges, in addition to the beam divergence broadening. Thirdly, they the scintillation, which makes the energy intensity fluctuate randomly across the beam area [46]. Last, reflected sunlight and dark count are possibly one of the main sources of noise. Therefore, in order to simplify the mathematical LiDAR model, the incident energy of the backscattered solar radiation (see Eq. 5.9) is the unique noise modelled in this chapter.

$$E^{\text{solar}} = E_{\text{si}} \cdot \delta_\lambda \cdot \delta_{\text{t}} \cdot A \cdot \rho \cdot \frac{1}{\Omega_{\text{s}} R^2} \cdot \frac{\pi D^2}{4} \cdot \eta_{\text{atm}} \cdot \eta_{\text{sys}} \tag{5.9}$$

where E_{si} is the solar irradiance; δ_λ is the electromagnetic bandwidth of the bandpass filter; δ_{t} is the unit sampled time bin (the temporal resolution) of the system clock that measures the time; and is the area covered within the IFOV (instantaneous field of view) calculated in Eq. (5.10).

$$A = (R \cdot IFOV)^2 \tag{5.10}$$

$$\eta_{\text{atm}} = T_{\text{atm\_received}} = T_{\text{atm}} \tag{5.11}$$

On the one hand, substituting Eqs. (5.10) and (5.11) into (5.9), the incident energy of the backscattered solar radiation can be calculated by:

$$E^{\text{solar}} = \frac{E_{\text{si}} \cdot \delta_\lambda \cdot \delta_{\text{t}} \cdot \rho \cdot IFOV^2 \cdot D^2 \cdot T_{\text{atm}} \cdot T_{\text{BPF}} \cdot T_{\text{ND}} \cdot T_{\text{FF}} \cdot T_{\text{R}}}{4} \tag{5.12}$$

On the other hand, Eq. (5.13) represents the expected number of photoelectrons created by the dark count due to the thermal effects within the detector,

$$E\left[N^{\text{dc}}\right] = f_{\text{dc}} \cdot \delta_{\text{t}} \tag{5.13}$$

where f_{dc} is the dark count rate which is uniformly distributed in the time domain and every pixel in the detector has the same dark count [42].

5.2.4 Single LiDAR Model Implementation

From the mathematical formulation of the radiometric aspect of LiDAR, the next step is the implementation of a LiDAR model and its simulation, assuming the most simplified use case, that is, a single laser and a single array of photodetectors without scanning. The single LiDAR model is composed by different submodules.

On the one hand, the spatial and temporal distribution of the laser should be considered. The variable 'pulse' is the amount of energy contained in the laser pulse. The laser pulse energy is spatially distributed by Eq. (5.1) and the beam width is given by Eq. (5.3). The spatial limits of the plane that is normal to the propagation direction were defined to be three standard deviations of the Gaussian beam. The result of the spatial distribution is a 2D array that is passed to a function that temporally distributes the power according to Eq. (5.5).

On the other hand, the creation of a Dirac known as the target profile. This requires a command that describes the type of scene. The scene generator creates a 2D array that contains the range of every sample in the observed scene. A 3D Dirac matrix is then created from the 2D scene array with a length that corresponds to the TOF limits specified by the range gate. This is commonly known as the target profile.

Finally, the range equation can be applied. The final distribution of energy is calculated by convolving the distributed laser power with the target profile. The resulting 3D matrix represents all the spatial and temporal distributions in the calculation. A for-loop is then used to visit each array in the range gate to calculate the received power at the detector for each moment in time according to Eq. (5.8).

In addition, the third submodule is the selection of the attenuation coefficient based on a specific ambient effect model into the specific scenario. It creates a vector of transmission coefficients based on the range of each array in the range gate so that the spatial distribution of transmission coefficients can be easily applied to the range equation. Some of these coefficients are speckle, thermal, and background noises to the signal. Statistical methods are used to calculate noise which is followed by applying gain. Finally, the signal is passed into a decision engine for reconstructing the scene. Each sample in the photodetector array contains a time varying signal. Scene reconstruction is based on the TOF that corresponds to the maximum power for each sample in the array [43]. The structure of the LiDAR model is given in the block diagram of Fig. 5.2.

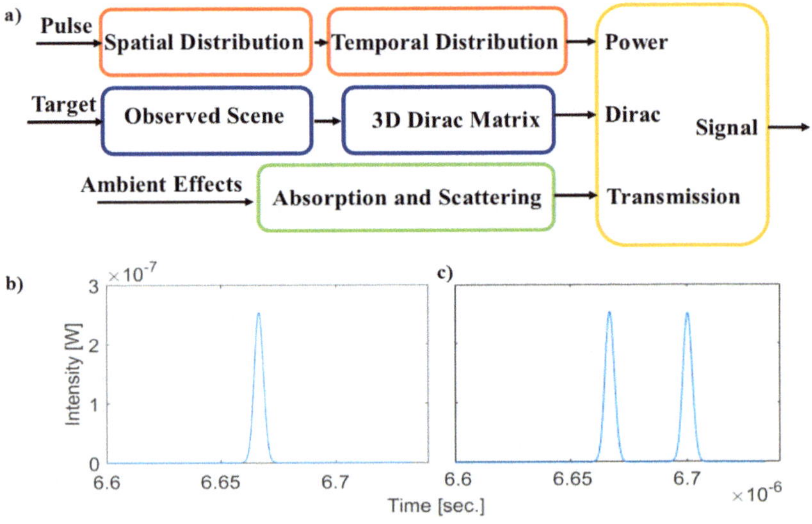

Fig. 5.2 (**a**) A block diagram showing the structure of the single LiDAR model. Signal received in the detection of one target (**b**) and (**c**) two targets

5.3 Computational Intelligence Applications in Sensor System Simulations

5.3.1 Application of Artificial Intelligence Techniques in IoT Sensor Networks

IoT sensor networks require the integration of relevant stages such as: identifying, sensing, networking, and computation. These networks enable large scale of technological innovations and value-added services which personalize user's interaction. There are numerous IoT applications that can be grouped into various domains such as health, traffic, logistics, retail, agriculture, smart cities, smart metering, remote monitoring, process automation, among others. Despite the enormous progress in different research fields including architectures, standardization, emerging technologies, security, etc. [47]. Wireless Sensor Networks (WSN) can be used to measure and track of energy consumption and production in order to optimize energy usage. In general, IoT communication is performed by producing and consuming information and by executing smart algorithms to interact intelligently with other things. The main challenges in the implementation of WSN in IoT are:

- Deployment
- Quality of service
- Localization
- Energy consumption
- Data collection and transmission
- Mobility management

In order to deal with some these challenges, new communications, protocols, and tools are reported in literature, highlighting the use of artificial intelligence techniques [48] in particular machine-learning algorithms (ML) to solve complex situation in sensor network scenarios. Artificial intelligence (AI) techniques provide to WSNs cognitive abilities in order to improve its performance by reducing power consumption, increasing network lifetime, reliability, and also in general enhancing quality of service (QoS) through the processing of great amount of data obtained from environment monitoring and applying optimization in order to determine the best parameters [49].

By the other hand, the fault detection is a key issue in autonomous systems. The main reason is that fault detection enables the systems to self-correct problems such as data package lost, determine data collision, the status of the element of the network, among others. The presence of faults in wireless sensor data may increase the network traffic and wastes the battery and power. Several techniques can be used such as nearest neighbour methods, and other clustering techniques for providing the system capabilities of self-adjust and data lost detection and correction [50]. Constructing real-time prediction models are important to maximize the network lifetime meanwhile guaranteeing connectivity and robustness [51]. The design of adaptive and intelligent WSN is essential. The adaptation and self-organization capabilities can be introduced through embedding artificial neural networks in the control systems of the WSNs providing reliability [52].

Sensor reliability is one of the main challenges found in IoT applications due to the need to ensure the correct monitoring of the process in order to ensure that the operation is within limits imposed. Since the sensor output is the only objective source of information of the status of the environment for decision-making, it is essential that the operating status of the sensors be known. The performance of online monitoring and self-correction of multiple sensors, through the application of some ML algorithms such as support vector machine (SVM) and ANNs, is very important to reduce maintenance costs, minimize the risk of using uncalibrated or faulty sensors, increase instrument reliability, and consequently increase equipment life [53, 54].

The network dynamics is another very important aspect to take into consideration into the sensor networks and IoT infrastructures. The interaction between the elements of the network can be determined by the performance of the network and it is critical to ensure its correct function. The development of scheduling strategies to ensure the coverage of points in wireless sensor networks is essential in order to optimize the sequence of states and increase the lifetime of the network, considering the resource scarcity and address the network dynamism. Techniques such as Hidden Markov Model present great performance in the scheduling design allowing the optimization of the lifetime without decrease in the network performance [55].

Protocols are also an important aspect to consider in the design of robust WSNs. The correct design of protocols has a great influence in the energy consumption, data congestion, and in general the performance of the system. As the limited power of the sensors is one of the main challenges energy save is vital to ensure a correct operation of the system. The design of protocols with low power approach along with the uses of ANNs facilitate local processing allowing faster responses as well as improve overall system efficiency and power saving [56]. In order to detect congestion the uses of algorithms like SVM to predict the level of congestion of the system allows the accurate adjustment of data transmission rate in each wireless sensor node which leads to a balanced traffic [57].

5.3.2 Computational Intelligence Applied in Pattern Recognition and Vision Classification Sensor-Based Analytics

Nowadays, several challenges arise from the sensor fusion, real-time and on-board implementation on sensor networks motivated by the need of new functionalities such as: data pre-processing, advanced analytics, pattern recognition, and classification actions based on the collection of huge amount of data points become a problem to be considered for current and future sensor manufactures, providers, and researchers [58, 59].

In particular, human activities recognition is an important research topic in the new IoT era, in order to analyse the behaviour and improve the human–computer interactions. In the last decades, different computational intelligence techniques have demonstrated their suitability to optimize industrial processes [60, 61]. In particular, artificial neural networks and fuzzy systems have been worldwide applied [62–64]. Nowadays, the increase of human–machine interactions will be one of the biggest challenges for the industry and society of the future, becoming the sensors first key element of this interaction [65]. Based on that, multiple studies are focused on the recognition of daily activities as an essential issue for maintaining healthy lifestyle, patient rehabilitation, and activity shifts among the elderly citizens, detecting and diagnosing serious illness patterns [66]. One example of that, it is the use of sensor data generated by smartphones and other wearable devices in order to dominate the research landscape in human motion analysis, activity monitoring, and detection due to their obvious advantages over other sensor modalities [67]. Recently, to overcome the challenges associated with single sensor modalities and increase generalization, many studies have proposed information fusion strategies that combine multiple sensors modalities or classifiers to increase robustness, reliabilities, derive confidence measures among different classifiers and reduce the complexity of recognition system [68]. Information fusion in human activity recognition is necessitated by increase in sensor of different modalities [69]. Information fusion techniques are prevalent in both handcrafted features and automatic feature learning using deep learning algorithms [70].

Deep learning algorithms are applied on sensor data to extract discriminative and salient features and then flattened and passed to an inference engine to recognize activities classes [71]. In this field, recently applications focused on pedestrian retrieval via part-based gradation regularization in sensor networks were reported by [72]. The solution proposes a novel part-based gradation regularization method to assign a pedestrian label for each horizontal part. For that, it has been combined the strengths of the local convolutional neural network (CNN) with an identification embedding model to extract and classify the main features. The results described in the approach not only validate the high accuracy to learn from the local featured based on CNN, they also demonstrate and improve the generalization ability of the classification model to mitigate the risk of overfitting and increase the performance of pedestrian retrieval due to rational utilization of pedestrian labels. Nevertheless, other traditional machine-learning classifiers such as: support vector machine [73], k-nearest neighbours [74], multinomial regressions [74], and Hidden Markov Models [75] are still highly used in pattern recognition and data classification applications.

On the other hand, intelligent transport systems (ITS) highly impact research topic in recent years. The possibility to provide innovative services relating to different modes of transport and traffic management and enabling various users to be better informed and make safer, more coordinated, and 'smarter' use of transport networks is very promising. ITS aim to improve transportation safety, mobility, productivity, and environmental performance for traffic planners and road users. With continuous urban road development and extensive construction of expressways, increasing interest is devoted to vehicle detection [76]. Into this transformation, the new generation of sensors, the sensor fusion, sensor network, and IoT-based solutions get an important role in function to collect real-time data from the billions of devices present in the smart mobility ecosystems [77].

One example of that is reported by [78], with the purpose to improve the real-time performance and accuracy of vehicle type recognition in complex traffic scenes. For a single vehicle, data collected from A and B nodes have a certain temporal and spatial correlation, and the signals can be enhanced and fused in the later analysis. Based on this estimation, the authors have developed a multiple sensors information fusion method, which consists of the following key components: data segmentation and filtering, vehicle feature matching, and vehicle feature waveform fusion. Finally, the experimental results obtained by the proposed method describes a 95% accuracy for small vehicles (tourism) and 90% accuracy for large vehicles (buses) approximately, becoming a key issue to provide references for multiple sensor network design and traffic data collection. Other recent example is the novel technique presented by [79] to apply a harmony search evolutionary algorithm in real-time line detection vision based on a two-wheeled robotic platform. The proposed vision-based algorithm is implemented to process the images captured from on-board camera allowed in a two-wheeled robot, follows the path via on-board microcontroller governing motors. The proposed system is able to detect and follow the path with 99% accuracy, regardless of image size and eliminates edge detection phase altogether. These are only two examples of the true potentialities of the computational intelligence-based solutions based on sensor data collection to extract the main feature of complex dynamic environments, model the potential actions and risk, and to facilitate the decision-making process for human and vehicles. The near future will demand fast solution in unknown situations, becoming the sensor reliability one of the most important attribute to be considered during the design, implemenation and performance of the future sensor networks.

5.3.3 Computational Intelligence Applied in Localization Optimization and Energy Consumption in Sensor Networks

Sensors are a crucial component in every intelligent control system [80]. Wireless sensor networks (WSNs) are one of the most rapidly developing information technologies and promise to have a variety of applications in Next Generation Networks, Internet-of-Things (IoT), becoming the sensor reliability one of the most important attribute to be considered during the design, implementation and performance of the future sensor networks [81].

Most WSNs reliability assessments are based on graph theory and probability theory. Connectivity reliability investigates the probability that the network is still connected for a given period of time under the case of some nodes or links failure.

Performance reliability analyses the end-to-end delay, packet delivery rate, and other network parameters [82, 83].

WSNs reliability is influenced by many factors such as component failure, environmental influences, task changes, and network update. These factors and network behaviour are difficult to be described or calculated by mathematical models. Network simulation becomes an important method to analyse network reliability [84]. Based on that, the objective of this subsection is to analyse computational intelligence-based implantations reported in the literature to contribute on the sensor networks reliability. Following, four main topics were selected based on node localization, energy management, network multi-objective optimization, and self-capabilities approach to describe the impact of the computational intelligence techniques into the present and future sensor networks [85, 86].

The emergence IoT-based applications are revolutionizing the current concept of industry, transportation systems, energy grids, cities, and society [87]. To archive this transformation, many daily applications based on IoT require an efficient node position and localization mechanism [88, 89]. Localization technology has been a core component for Internet-of-Things, especially for Wireless Sensor Network.

On the one hand, the localization techniques range-based are based on measuring the distance or the angle between the directions of reference nodes called anchors. Several technologies allow the nodes distance measure such as: Received Signal Strength Indicator [90], Time of Arrival [91], Time Difference of Arrival [92], and Angle of Arrival [93], among others. On the other hand, range-free-based localization solutions can avoid the increase of the energy consumption and the inappropriate cost of the equipment. Range-free technologies introduce an independence between the hardware devices and the distance computing [94]. Hence, this independence allows the adaption technology of different wireless transmission types based on locations as centroid, Approximate Point in Triangle Test [95] and Distance Vector-Hop (DV-Hop) [95]. Although these algorithms provide more economic and simple estimations, their results are not as accurate as tracking error level. A precise detection of the transmitter node position allows an effective routing of information with minimum energy consumption. Another important parameter is the localization time which is the necessary time for the network localization [96]. Nevertheless, range-free algorithms have relatively lower localization accuracy, but it is more widespread to be used because of the inexpensive cost and simplicity, especially in the large-scale network [97].

In this topic, an interesting result is introduced by [98], developing a novel high accurate localization algorithm based on DV-Hop and differential evolution (DECHDV-Hop) for WSN. In particular, DECHDV-Hop improves the positioning accuracy without adding extra equipment to the sensor network. In order to further reduce the localization error, DE algorithm is used to locate unknown nodes by formulating the location estimation process as an optimization problem. The obtained results show a reduction in the average localization error about 70% compared with traditional DV-Hop techniques in the simulation of four different network situations (Random, Grid, C-shaped random, and C-shaped grid). The average localization error was estimated based on the following equation:

$$\Delta L = \frac{\sum_{i=1}^{(1-p)\cdot N} \sqrt{(x_i - \tilde{x}_i)^2 + (y_i - \tilde{y}_i)^2}}{R\cdot (1-p)\cdot N} \tag{5.14}$$

where (x_i, y_i) and $(\tilde{x}_i, \tilde{y}_i)$ are the estimated location and actual location, respectively; N is the sensor nodes count, p is the percentage of anchor nodes, and R is the communication range. In all cases, the smaller the average localization error value, the higher the localization accuracy.

Other interesting result based on an adaptive neural fuzzy inference system (ANFIS) for accurate localization of WSNs is introduced by [99]. Figure 5.3 describes the ANFIS topology selected to investigate the propagation model for track cycling in indoor and outdoor localization application.

The obtained results demonstrate the divergence between the ANFIS and theoretical models when the distance increased. A convergence between the theoretical and measured plots has been observed for the outdoor velodrome. On the contrary, there is a big divergence among the theoretical and measured plots for the indoor velodrome relative to the outdoor velodrome. This is because of the multipath effect, due to the presence of reflections, scatters, and diffractions from indoor objects such as furniture, doors, windows, and walls in the sports hall. The results showed that the distance estimation accuracy was improved by 84% and 99% for indoor and outdoor velodromes, respectively, after applying the ANFIS optimization.

Multi-objective heuristic optimization algorithms become an important tool [100, 101] to solve localization in wireless sensor networks, especially those without the aid of a Global Positioning System (GPS). Pareto-based solutions contribute to find the better performance in WSNs, considering several aspects or key issues

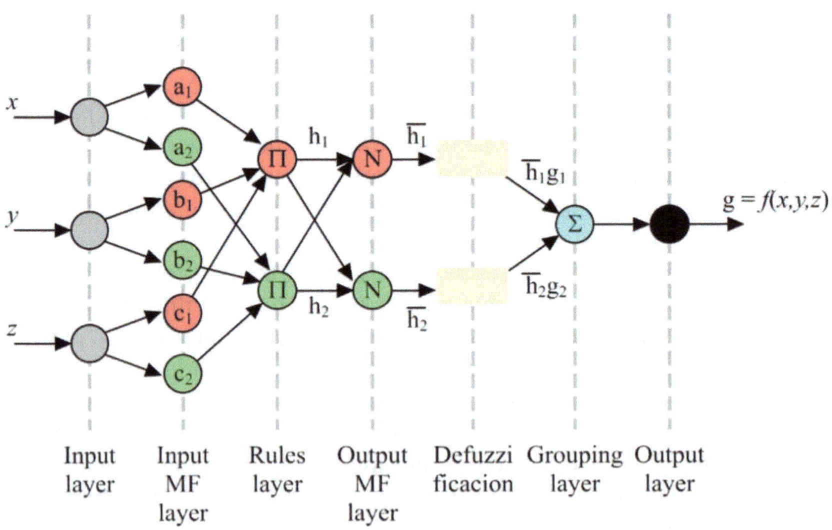

Fig. 5.3 ANFIS model representation based on the model proposed by [99]

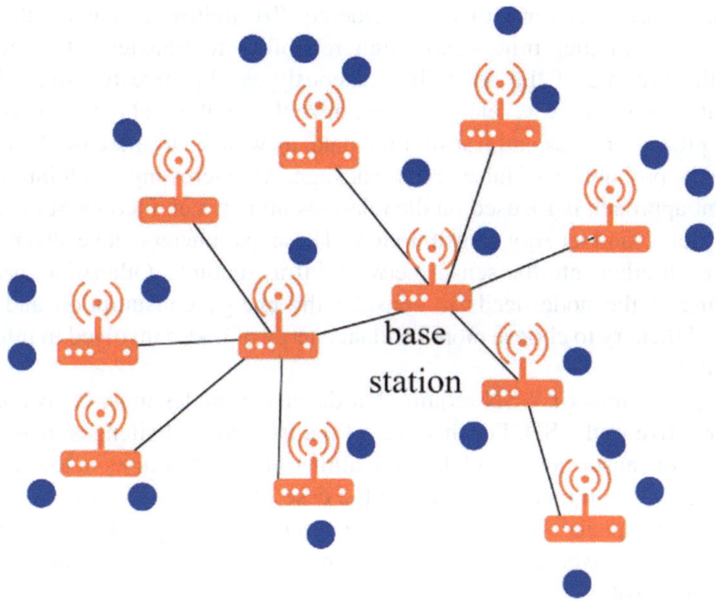

Fig. 5.4 Problem instance where the blue points represent locations to be sensed for WSN nodes

to find the optimal balance for each particular situation. In this direction, a hybrid multi-objective evolutionary algorithm Sensor Placement Optimizer (M-SPOT) is introduced by [102], with the objectives to minimize the total number of devices deployed into the network and the total energy dissipated by the sensors (see Fig. 5.4).

Based on the approach contributions and considering that the initial monetary cost to deploy any sensor network is associated with the number of deployed sensors, it can conclude that the heuristic optimization algorithms are a powerful tool to increase the overall performance sensor networks. Besides, more complex problem can be addressed, extending the number of objectives to include reliability to perform placement in three dimensions' objective function. A similar approach is introduced by [103] in order to improve the location estimation precision in non-uniform deployments. The proposed solution combines two techniques based on a hybrid model using fuzzy logic and extreme learning machine with vector particle swarm optimization, which adjusts the estimation accuracy to be closer to the actual node locations in relation with the diversity of signal strengths. The current results validate the performance of the proposed algorithm on irregular topology, outstanding (as determined by the lowest location estimation error with a comparatively low complexity trade-off) in comparison with other soft-computing-based localization approaches.

Another particular application of the heuristic techniques is reported by [104]. In a medical sensor network, the users' information needs to be transmitted to the doctor workstation with smaller time delay. Meanwhile, the reliability and energy consumption are necessary to be considered. To archive this goal, the authors considered the smaller time delay, high reliability to transfer information, and prolong the lifetime of the node simultaneously as objective functions. For that, a differential evolution was selected to optimize the smaller delay time needs sensor network quickly process and transmit the data, as well as, to increase the reliability to reduce the probability of false or loss package. An interesting result introduced by the current approach is focused on the relationship between increment of reliability with the delay and energy consumption. These parameters have clear conflicts between each other into the sensor network infrastructures. Otherwise, prolonging the lifetime of the node needs to consider the energy consumption and residual energy, and then try to choose more residual energy of nodes involved in information transmission.

Most applications of WSN require that the sensor nodes must be aware of their location relative to the SN. For this, one of the important challenges in WSN is the localization of sensor nodes and determination of nodes positions in the sensor field [105]. As it was commented before, in the case of sensor networks, the energy is other key performance indicator of efficiency and reliability values. For this, GPS devices are not convenient as valid solution of localization problem in WSN for many reasons [106]:

- Once the expensive GPS devices are added to every sensor node in the network, this solution becomes unfeasible due to the increase of the deployment cost.
- With the inclusion of GPS device the sensor nodes size becomes quite large, limiting the required properties of WSN where nodes should be designed in small size.
- To work property, GPS devices always depend on satellites information. In cases of failure with satellite link, GPS ceases to function. It is the case of many real applications as indoor and explorative applications.
- Finally, GPS in WSN depend on many factors such as: power consumption, cost, and size, being not the case of other localization techniques.

In order to sustain long-term sensing capability over a large coverage set, the energy conservation becomes an important factor to extend the battery life in sensor networks. Replacement of a node battery is infeasible due to the harsh environment in which the sensor networks are usually deployed in [107], becoming an imperative issue to prioritize energy efficiency in order to maximize network lifetime and performance. In time-critical applications, routing delay also becomes a pressing factor [108]. In order to address the energy consumption issue effectively, the energy-efficient grid-based routing algorithm introduced by [109] overcomes the hotspot problem and utilizes the energy of nodes more efficiently. This algorithm consists of three different phases, namely, grid formation, grid coordinator election, and grid-based routing phases. For that, each grid is elected by applying fuzzy logic where the fuzzy variables are used for grid coordination selection based on the residual energy of the nodes, their mobility pattern, and the distance of the nodes from the sink node. As the transmission consumes energy based on the distance, the distance to the sink is considered as an important issue to select the grid coordinator, monitoring the energy utilization of the network continuously.

In terms of sensor networks energy consumption optimization topics, the clustering techniques take an important role. Several researchers have introduced interesting contribution in the field during the last years. One example of that is the energy aware routing for throughput optimization in clustered IoT-wireless sensor networks introduced by [110]. The extensive devices connected through IoT bring forth many challenges such as high battery power consumption, designing for smart transportation, health and home control system devices, among others. To solve the above problems, an optimal clustering mechanism based on Fuzzy-C means clustering is designed to divide the sensor nodes into different clusters [111]. The results introduced in the approach demonstrate the potentialities of the clustering-based algorithms to optimize the energy consumption in WSNs.

Nowadays, one of the biggest challenges of the computational intelligence-based solutions is the dotation of new self-capabilities to become the current devices, machines, processes and ecosystems in smart-solutions. The sensors as a key element in the whole system chain are not exempt from this new digitalization era. In this direction, several researchers and component manufactures start to consider the possibility to dote the SNs with new functionalities based on self-optimization, self-configuration, and even self-healing skills. One example of that are the contributions reported by [112], based on self-maintenance model for WSNs. The approach introduces a distributed self-healing approach for both node and cluster head levels, archiving a detection accuracy of sensor circuit fault tolerate up to 76.9%; 52% of battery fault; and 71.96% of receiver faults. The Pseudocode 1 describes the algorithm proposed by [112] for self-fault detection in sensor networks.

Pseudocode 1 Fault Detection Algorithm at CH Level Extracted from Elsayed et al. [112]

```
Begin
 N_fault = [0]
 N_controller = [0]
 N_k = [0]
 Sensor node transfers HB_msg to CH
 IF HB_msg periodically sensed in CH:
      Node is healthy; CH checks its aggregated data
      IF the aggregated data was closed:
            N_k = N_k + N
      ELSE:
            N_controller = N_controller + N
 ELSE:
    Node is malfunction: N_fault = N_fault + N
    Isolate it and update the topology to replace this node
 Sensor node periodically transfers introduction message to CH.
 IF CH received sleeping notification from sensor node:
      Remove this sensor and assign its function to posterior
      Select new path
 ELSE:
      N_k = N_k + N
End
```

where N is the number of nodes, N_k is the active vertices of the network, CH is the cluster head, and HB_{msg} is periodic heartbeat message.

For the same topic, but this time focused on the self-configuration algorithm, an efficient neural network model to minimize the weakly connected dominating sets in WSNs is introduced by [113]. The experimental results developed by the authors demonstrate that the proposed neural network with direct convergence achieves much better performance in the quality of solutions and efficiency. Furthermore, the results highlight the importance of a correct selection of a suitable transmission radius to make the network more stable and having a long lifespan. As it can be appreciated along this section, the computational intelligence-based solutions have become a key issue to optimize localization, energy, and accuracy in current and future sensor networks, improving higher levels of efficiency, reliability, and self-healing capacities.

5.4 Use Case: Sensor System Application

As demonstrated in the previous section, computational intelligence can be used in a multitude of applications. In this chapter, as an example, a use case will be presented, that is based on the methodologies described in [25, 114, 115]. The CI methods will be applied in a 3D scenario simulation software, in which has been used the mathematical LiDAR model set in the previous section. This simulation software is Webots R2018b for automobiles [116].

The purpose of this use case is apply a CI model library of object classification using the LiDAR data in a virtual sensor network in a Cyber-Physical System to obstacle recognition in driving assistance. For this, four artificial intelligence-based methods and a probabilistic method for obstacle recognition are explored, such as a multi-layer perceptron, a self-organized map, a support vector machine, k-nearest neighbours and naive Bayes.

5.4.1 LiDAR Model Simulation

A simulation procedure is the first step before applying the CI into a CPS or IoT scenario in order to create a training data set with the information collected from the sensor network in order to integrate a conjunction of CI methods with the mathematical four-layer LiDAR model (described in Sect. 5.2.4). The aim of this first simulation is to get data acquired by the LiDAR model in the simulation tool to generate classification models based on CI methods. The classification models will be applied to the LiDAR sensors to be able to compare the results when recognizing obstacles.

A virtual driving traffic scenario that emulates a real test track was generated in Webots. The virtual scenario emulates the real setup available in the Centre for

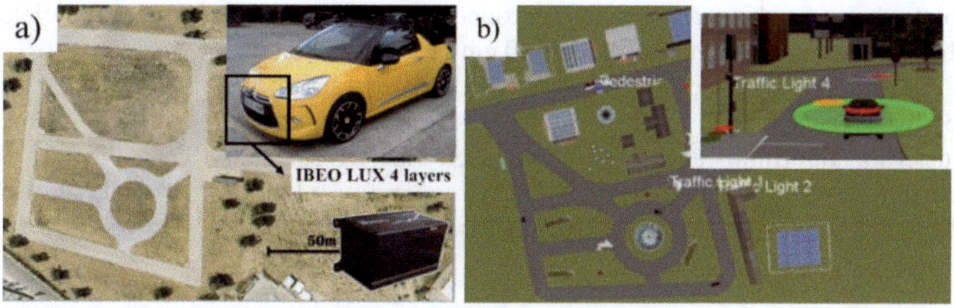

Fig. 5.5 Aerial view of the (**a**) real and (**b**) 3D simulated scenario, with a real and modelled vehicle with sensors incorporated

Automation and Robotics (CAR-CSIC), Madrid, Spain[1]. This consists of a grid form, simulating city blocks, along with some additional elements: a roundabout in one of the quadrants, traffic lights in the central crossing, and additional curves on the main straight. Based on that an urban environment was achieved, with a broader set of curves, a small diagonal segment; every street has a minimum width of 6 meters, allowing traffic in both directions (see Fig. 5.5a).

Although each vehicle has been equipped with different specific sensors and actuators, the control architecture is, in essence, the same for all of them. Its backbone is an on-board unit that handles sensor inputs, and a fuzzy logic controller that manages individual actions on throttle, brake, and steering wheel [117]. The last vehicle listed (Citroën DS3) is the one with a higher number of sensors: DGPS (20 Hz) receiver Trimble BD960, 3D Stereo vision camera Bumblebee X2, and a Lidar 4 layers Ibeo Lux.

Therefore, in order to emulate this real testing scenario, a virtual driving traffic scenario was created in Webots Automobile simulation tool based on the state-of-the-art of LiDAR sensor model. Figure 5.5b also illustrates the aerial view of some of these 3D scenarios in Webots for automobiles for driving assistance. The scenario for simulation was set up with three fully automated vehicles with the on-board sensors.

The next step once the scenario has been created is to simulate all the models configured in Webots, including the theoretical model of the LiDAR. Four hours of sensors data provided by the virtual sensors data were recorded. The data obtained from these sensors are camera images with objects recognized and the LiDAR point clouds. In total, a benchmark with 1031 scenes was available with the same number of LiDAR scans, captured images, and annotation files with the localization of each recognized obstacles.

Subsequently, in order to facilitate the training of the classification models and determine the error-prediction models, the raw data from the LiDAR sensors required filtering and pre-processing. For example, in this particular use case, the

[1] www.car.upm-csic.es

points that map the road asphalt at ground level and the vegetation were deleted. These points were mainly located at 20 cm above ground level.

The next data-processing step consisted of extracting the points by mapping nearby obstacles that correspond to a specific point-cloud sequence, using a clustering algorithm, and to compare each LiDAR calculated centroid with the actual location generated by the object recognition algorithm of each obstacle. Therefore, a training data set of 1031 samples was generated.

5.4.2 Cyber-Physical Application

The aim of this use case is the LiDAR model simulation implemented in the last section for obstacle detection in a transportation Cyber-Physical System. During the simulation, the library initially contains some classification models by default, and later, the content will be enriched in the runtime during the process simulation of the scenario. In this particular use case, three techniques are then considered, i.e. multi-layer perceptron neural network, self-organization map, support vector machine, k-nearest neighbour and naive Bayes. The main rationale for their selection is the solid mathematical foundations, demonstrated ability in modelling in complex scenarios, and a wide range of successful applications.

The first technique was a multi-layer perceptron (MLP) neural network with an input layer with 18 neurons, a hidden layer of 40 neurons, and an output layer with a single neuron and linear activation function [118–120]. For training, the method used was gradient descent with momentum and adaptive learning rate backpropagation. The initial values of learning rate and performance goal were 10^{-7} and 10^{-8}, respectively. The network was trained during 50,000 iterations, after which it reached a best performance of 0.0216 and a gradient of 0.0021. Using the validation set, values of mean square error (MSE) of 0.0409 and correctly classified instances (CCI) equal to 95.91% was reached.

The second method for the obstacle recognition library is a self-organizing map (SOM) [121]. Specifically, a topology function that creates neurons in an N-dimensional random pattern was used, and the dimensions were 22×2. Finally, the *Manhattan* function was applied as distance function. In addition, an input weight equal to the number of observations in the training set was set, i.e. $w = 1050$. The network was trained during a cover step of 10000 and an initial neighbour size of 4, after which it reached a MSE of 0.132 and a CCI equal to 89.55% was reached using the validation set.

The third, a support vector machine (SVM) was also implemented in the library [122]. This nonlinear classifier uses a Gaussian Kernel function with a kernel scale $\sigma = 0.94$ and a box constrain of $9.78e^4$. The supervised learning method was trained during 1255 iterations, until its reason for convergence gradient reached a $\Delta < 0.001$. The results obtained during validation were a MSE of 0.0636 and a CCI equal to 93.64%.

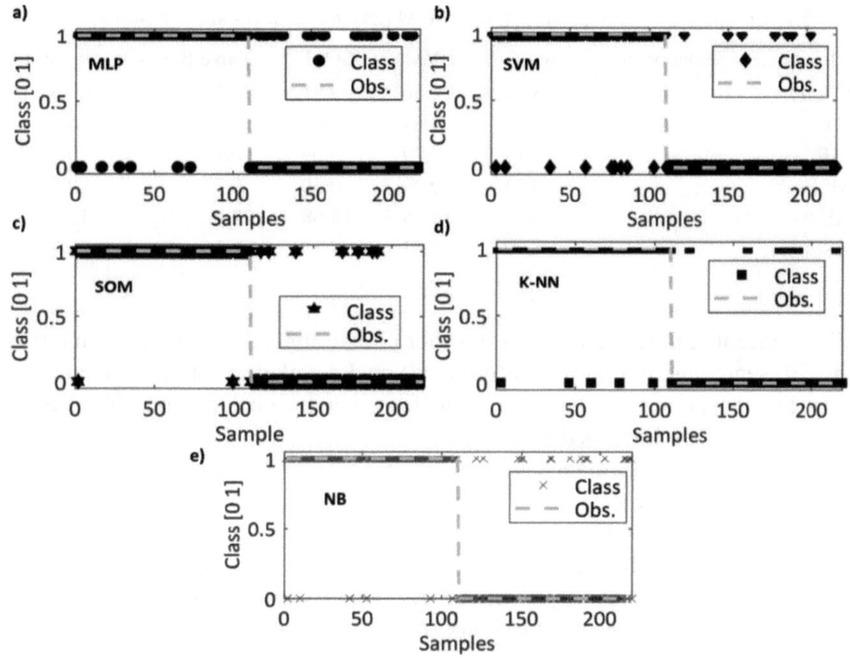

Fig. 5.6 Validation results in pedestrian detection (**a**) MLP, (**b**) SVM, (**c**) SOM, (**d**) k-NN, and (**e**) NB

The fourth, selected technique was a k-Nearest Neighbours (kNN) clustering algorithm. K-NN is one of the simplest non-parametric classification methods with an easily interpretable output, low calculation times, and high predictive power [123–126]. The setting parameters used was a Euclidean distance and the number of neighbours equal to 5. The results obtained during validation step were a MSE of 0.1388 and a CCI equal to 86.82%.

Finally, a simple 'probabilistic classifier' based on applying Bayes theorem with strong (naive) independence assumptions between the features [122]. The Naive Bayes (NB) classifier is set using a Gaussian distribution, normal distribution names, and number of observations of 1050. A MSE of 0.1182 and a CCI equal to 88.18% was reached using the validation set.

Figure 5.6 shows the classifiers outputs of the five models vs. observed classes using the validation set. The classifiers outputs indicate whether the detected object is a pedestrian (class 1) or not (class 0). MLP, followed very closely by SVM, has the smallest error and the highest number of correctly classify instances. Instead, SOM and k-NN have also obtained acceptable success rates but below 90%. Even the worst result that corresponds to k-NN has very good performance indices. This study is not conclusive and therefore a validation with unknown data set is required using more performance indices to make a more complete comparative study among the five classifiers.

Table 5.1 Comparative study of MLP, SVM, SOM [25], naive Bayes and k-NN [19]

Performance index\approach	MLP	SVM	SOM	Naive Bayes	k-NN
CCI (%)	88.19	91.36	90.91	88.18	93.64
ICI (%)	11.81	8.64	9.09	11.82	6.36
MAE	0.12	0.09	0.09	0.11	0.0636
RMSE	0.34	0.29	0.30	0.33	0.25
RAE (%)	23.64	17.29	18.68	23.62	12.74
RRSE (%)	9.27	7.93	8.36	9.26	6.81

The current testing set in good weather climatological conditions (sunny) contains 230 segments (not known beforehand), and detailed annotations regarding the pedestrian appearances (in terms of occlusion), namely, occluded/partial pedestrians (class-0) and entire body pedestrians (class-1).

A total of six performance indices were considered in the validation test on the basis of experimental run as it is described as follows: number of correctly classified instances (CCI), the number of incorrectly classified instances (ICI), the mean absolute error (MAE), the root mean squared error (RMSE), the relative absolute error (RAE), and the root relative squared error (RRSE). The results of the comparative study of the classifiers (MLP, SVM, and SOM) were summarized in Table 5.1.

The application of the previously unknown test data set on the MLP classification model that obtained the best results in validation studies yielded 23.64% of RAE. On the contrary, SVM and SOM achieved an excellent accuracy, for instances 17.29% and 18.68% in RAE, respectively, although not too much lower in percentage than the MLP error. The excellent behaviour is also endorsed with high correct classified instances of 93.64%, achieved by k-NN. The worst result occurred with NB although it is the model that less varies its merit figures regardless of the input data set.

5.5 Conclusions

In this chapter, the modelling and simulation techniques for LiDAR sensor networks are presented. Highlighting the most commonly computational intelligence techniques reported to add new functionalities and decision-making capabilities into LiDAR sensor networks for Cyber-Physical Systems and Internet-of-Things environments. Nowadays, the main objective to simulate the sensor behaviours is related with the ability to increase the fault prediction and reliability, defining as a key performance role far from the current physical testing operations. In addition, it was demonstrated that on-chip light detection and ranging concept has driven a great technological challenge into sensor networks.

In this chapter, with the aim of a more exhaustive compression of the behaviour of a LiDAR sensor, a brief description of the theoretical modelling based on mathematical principles of operation is outlined. The mathematical modelling is focused on the radiometric aspect of single LiDAR that is composed by different submodules, such as laser beam, returned energy, and energy noise. As an example, this sensor model is implemented, assuming the most simplified use case which is a single laser and a single array of photodetectors without scanning and will be used in the later simulation steps.

Subsequently, the chapter describes a general overview of state-of-the-art computational intelligence techniques commonly used in sensor system simulation applications. Some of these applications are AI-based techniques for CPS and IoT sensor network solutions, computational intelligence for pattern recognition and vision classification sensor-based analytics, localization optimization, and energy consumption in WSN. In summary, the technical review draws the following key concluding remarks. Firstly, due the great advance of informatics and communication technologies, which has allowed the rebirth of the AI-based solutions, the applications of machine-learning algorithms in engineering have risen especially in fields of recent emergence such as CPS, and IoT. Furthermore, the WSN, considered one particular field into the IoT-based solutions, has also received a lot of attention on the application of ML algorithms, becoming one of the main tools to address the main challenges for implementation in industrial and services scenarios. Secondly, the near future will demand fast solution in unknown situations, becoming both the sensor and computational intelligence fusion key issues to develop more robust, efficient, and safety human–machine ecosystems. Finally, the computational intelligence-based solutions have become relevant to optimize localization, energy, and accuracy in current and future sensor networks, improving higher levels of efficiency, reliability, and self-healing capacities.

In order to corroborate the potential of the computational intelligence for adding new capabilities to sensor networks in Cyber-Physical Systems and Internet-of-Things environments, a use case is presented. CI methods were applied to a 3D scenario simulation software (Webots) for automobiles, using the mathematical LiDAR model defined in this chapter. The application of a model library for object classification tasks, using the LiDAR data in a virtual sensor network for obstacle recognition actions in driving assistance in a Cyber-Physical System scenario showed very promising results. Four artificial intelligence-based methods and a probabilistic method for obstacle recognition were applied: a multi-layer perceptron, a self-organized map, a support vector machine, k-nearest neighbours and naive Bayesian network. The results demonstrated the suitability of applying computational intelligence methods to increase the reliability of sensor networks, addressing the key challenges to increase the safety and security in automotive applications.

Acknowledgments The authors wish to thank the support given by the European project IoSENSE: Flexible FE/BE Sensor Pilot Line for the Internet of Everything. This project has received funding from the Electronic Component Systems for European Leadership Joint Undertaking under grant agreement No 692480. This Joint Undertaking receives support from the European Union's Horizon 2020 research and innovation programme and Germany, Saxony, Austria, Belgium, the Netherlands, Slovakia, and Spain.

References

1. A.J.C. Trappey, C.V. Trappey, U.H. Govindarajan, J.J. Sun, A.C. Chuang, A review of technology standards and patent portfolios for enabling cyber-physical systems in advanced manufacturing. IEEE Access **4**, 7356–7382 (2016). https://doi.org/10.1109/ACCESS.2016.2619360
2. X. Krasniqi, E. Hajrizi, Use of IoT technology to drive the automotive industry from connected to full autonomous vehicles. IFAC-PapersOnLine **49**(29), 269–274 (2016). https://doi.org/10.1016/j.ifacol.2016.11.078
3. J. Pajares Redondo, L. Prieto González, J. García Guzman, B.L. Boada, V. Díaz, VEHIOT: Design and evaluation of an IoT architecture based on low-cost devices to be embedded in production vehicles. Sensors **18**(2), 486 (2018)
4. T. Bures, D. Weyns, M. Klein, R.E. Haber, 1st International Workshop on Software Engineering for Smart Cyber-Physical Systems (SEsCPS 2015), in *2015 IEEE/ACM 37th IEEE International Conference on Software Engineering, 16-24 May 2015*, (2015), pp. 1009–1010. https://doi.org/10.1109/ICSE.2015.326
5. J. Lee, B. Bagheri, H.-A. Kao, A cyber-physical systems architecture for industry 4.0-based manufacturing systems. Manufact. Lett. **3**, 18–23 (2015b)
6. T. Sanislav, L. Miclea, Cyber-physical systems-concept, challenges and research areas. J. Contr. Eng. Appl. Inform. **14**(2), 28–33 (2012)
7. A. Bhave, B.H. Krogh, D. Garlan, B. Schmerl, View consistency in architectures for cyber-physical systems, in *Proceedings of the 2011 IEEE/ACM Second International Conference on Cyber-Physical Systems*, (IEEE Computer Society, 2011), pp. 151–160
8. P. Gölzer, P. Cato, M. Amberg, Data Processing Requirements of Industry 4.0-Use Cases for Big Data Applications, in *ECIS*, (2015)
9. X. Yuan, C.J. Anumba, K.M. Parfitt, Review of the potential for a cyber-physical system approach to temporary structures monitoring. Int. J. Architect. Res. ArchNet-IJAR **9**(3), 26–44 (2015)
10. T. Watteyne, V. Handziski, X. Vilajosana, S. Duquennoy, O. Hahm, E. Baccelli, A. Wolisz, Industrial wireless ip-based cyber–physical systems. Proc. IEEE **104**(5), 1025–1038 (2016)
11. J. Wan, D. Zhang, S. Zhao, L. Yang, J. Lloret, Context-aware vehicular cyber-physical systems with cloud support: architecture, challenges, and solutions. IEEE Commun. Mag. **52**(8), 106–113 (2014)
12. J. Ruan, W. Yu, Y. Yang, J. Hu, Design and realize of tire production process monitoring system based on cyber-physical systems, in *2015 International Conference on Computer Science and Mechanical Automation (CSMA)*, (IEEE, 2015), pp. 175–179
13. A. Iovine, F. Valentini, E. De Santis, M.D. Di Benedetto, M. Pratesi, Safe human-inspired mesoscopic hybrid automaton for autonomous vehicles. Nonlinear Anal. Hybrid Syst. **25**, 192–210 (2017). https://doi.org/10.1016/j.nahs.2016.08.008
14. D. Reiser, D. Paraforos, M. Khan, H. Griepentrog, M. Vázquez-Arellano, Autonomous field navigation, data acquisition and node location in wireless sensor networks. Precis. Agric., 1–14 (2016)
15. X. Zhao, K. Mu, F. Hui, C. Prehofer, A cooperative vehicle-infrastructure based urban driving environment perception method using a D-S theory-based credibility map. Optik **138**, 407–415 (2017). https://doi.org/10.1016/j.ijleo.2017.03.102
16. W. Rapberger, The emergence of autonomous vehicles. Accenture (2017)
17. C.-I. Chen, R. Koseluk, C. Buchanan, A. Duerner, B. Jeppesen, H. Laux, Autonomous aerial refueling ground test demonstration—A sensor-in-the-loop, non-tracking method. Sensors **15**(5), 10948–10972 (2015)
18. G. Pandey, J.R. McBride, S. Savarese, R.M. Eustice, Automatic extrinsic calibration of vision and lidar by maximizing mutual information. J. Field Robot. **32**(5), 696–722 (2015)
19. S. Sim, J. Sock, K. Kwak, Indirect correspondence-based robust extrinsic calibration of LiDAR and camera. Sensors **16**(6), 933 (2016)

20. R. Vivacqua, R. Vassallo, F. Martins, A low cost sensors approach for accurate vehicle localization and autonomous driving application. Sensors **17**(10), 2359 (2017)
21. M. Ahmad, Reliability models for the internet of things: a paradigm shift, in *Paper Presented at the 2014 IEEE International Symposium on Software Reliability Engineering Workshops, 3-6 Nov 2014*, (2014)
22. L. Xiao, R. Wang, B. Dai, Y. Fang, D. Liu, T. Wu, Hybrid conditional random field based camera-LIDAR fusion for road detection. Inform. Sci. **432**, 543–558 (2018b). https://doi.org/10.1016/j.ins.2017.04.048
23. Y. Zeng, H. Yu, H. Dai, S. Song, M. Lin, B. Sun, W. Jiang, M. Meng, An Improved Calibration Method for a Rotating 2D LIDAR System. Sensors **18**(2), 497 (2018)
24. S. Hu, Z. Li, Z. Zhang, D. He, M. Wimmer, Efficient tree modeling from airborne LiDAR point clouds. Comput. Graph. **67**, 1–13 (2017). https://doi.org/10.1016/j.cag.2017.04.004
25. F. Castaño, G. Beruvides, R. Haber, A. Artuñedo, Obstacle recognition based on machine learning for On-Chip LiDAR sensors in a cyber-physical system. Sensors **17**(9), 2109 (2017a)
26. R. Kala, 3—Perception in autonomous vehicles, in *On-road intelligent vehicles*, (Butterworth-Heinemann, Oxford, 2016), pp. 36–58. https://doi.org/10.1016/B978-0-12-803729-4.00003-9
27. A. Apatean, A. Rogozan, A. Bensrhair, Visible-infrared fusion schemes for road obstacle classification. Transport. Res. Pt. C Emerg. Technol. **35**, 180–192 (2013). https://doi.org/10.1016/j.trc.2013.07.003
28. H. Taghavifar, A. Mardani, H. Karim Maslak, A comparative study between artificial neural networks and support vector regression for modeling of the dissipated energy through tire-obstacle collision dynamics. Energy **89**, 358–364 (2015). https://doi.org/10.1016/j.energy.2015.05.122
29. P.J. Navarro, C. Fernández, R. Borraz, D. Alonso, A machine learning approach to pedestrian detection for autonomous vehicles using high-definition 3D range data. Sensors **17**(1), 18 (2016)
30. B. Shi, L. Han, H. Yan, Adaptive clustering algorithm based on kNN and density. Pattern Recogn. Lett. **104**, 37–44 (2018). https://doi.org/10.1016/j.patrec.2018.01.020
31. S. Zhang, D. Cheng, Z. Deng, M. Zong, X. Deng, A novel kNN algorithm with data-driven k parameter computation. Pattern Recogn. Lett. (2017). https://doi.org/10.1016/j.patrec.2017.09.036
32. D. Bein, V. Jolly, B. Kumar, S. Latifi, Reliability modeling in wireless sensor networks. Int. J. Inform. Technol. **11**(2), 1–8 (2005)
33. H.M.F. AboElFotoh, S.S. Iyengar, K. Chakrabarty, Computing reliability and message delay for Cooperative wireless distributed sensor networks subject to random failures. IEEE Trans. Reliabil. **54**(1), 145–155 (2005). https://doi.org/10.1109/TR.2004.842540
34. A. Artuñedo, J. Godoy, R. Haber, J. Villagrá, Toro RMd Advanced Co-simulation Framework for Cooperative Maneuvers Among Vehicles, in *2015 IEEE 18th International Conference on Intelligent Transportation Systems, 15-18 Sept. 2015*, (2015), pp. 1436–1441. https://doi.org/10.1109/ITSC.2015.235
35. Z. Zhang, E. Eyisi, X. Koutsoukos, J. Porter, G. Karsai, J. Sztipanovits, A co-simulation framework for design of time-triggered automotive cyber physical systems. Simul. Model. Pract. Theor. **43**, 16–33 (2014). https://doi.org/10.1016/j.simpat.2014.01.001
36. F. Ferracuti, A. Freddi, A. Monteriù, M. Prist, An integrated simulation module for cyber-physical automation systems. Sensors **16**(5), 645 (2016)
37. I.D. de Souza, S.N. Silva, R.M. Teles, M.A. Fernandes, Platform for real-time simulation of dynamic systems and hardware-in-the-loop for control algorithms. Sensors **14**(10), 19176–19199 (2014)
38. M. Mozumdar, Z.Y. Song, L. Lavagno, A.L. Sangiovanni-Vincentelli, A model-based approach for bridging virtual and physical sensor nodes in a hybrid simulation framework. Sensors **14**(6), 11070–11096 (2014)
39. B.-H. Lee, J.-H. Song, J.-H. Im, S.-H. Im, M.-B. Heo, G.-I. Jee, GPS/DR error estimation for autonomous vehicle localization. Sensors **15**(8), 20779–20798 (2015a)

40. A.A. Al-Temeemy, The development of a 3D LADAR simulator based on a fast target impulse response generation approach. 3D Res. **8**, 31 (2017). https://doi.org/10.1007/s13319-017-0142-y
41. R. Richmond, S. Cain, *Direct-detection LADAR systems* (Society of Photo-Optical Instrumentation Engineers, Bellingham, WA, 2010)
42. S. Kim, I. Lee, Y. Kwon, Simulation of a Geiger-mode imaging LADAR system for performance assessment. Sensors **13**(7), 8461 (2013)
43. T.R. Morris Ii, *Radiometric simulation of LADAR imaging for underwater environments* (Universitat Politècnica de Catalunya, Barcelona, 2014)
44. T. Carlsson, O. Steinvall, D. Letalick, *Signature simulation and signal analysis for 3-D laser radar* (Swedish Defence Research Agency, Sweden, 2001), pp. 7–8
45. O. Svelto, D.C. Hanna, *Principles of lasers*, vol 4 (Springer, Berlin, 1998)
46. S. Ove, C. Tomas, Three-dimensional laser radar modeling, in *Proc. SPIE*, (2001), pp. 23–34
47. A. Čolaković, M. Hadžialić, Internet of Things (IoT): A review of enabling technologies, challenges, and open research issues. Comput. Netw. **144**, 17–39 (2018). https://doi.org/10.1016/j.comnet.2018.07.017
48. A. Gajate, R.E. Haber, J.R. Alique, P.I. Vega, Transductive-weighted neuro-fuzzy inference system for tool wear prediction in a turning process, in *Lecture Notes in Computer Science (including subseries Lecture Notes in Artificial Intelligence and Lecture Notes in Bioinformatics)*, vol. 5572 LNAI, (2009). https://doi.org/10.1007/978-3-642-02319-4_14
49. S. Gheisari, M.R. Meybodi, A new reasoning and learning model for Cognitive Wireless Sensor Networks based on Bayesian networks and learning automata cooperation. Comput. Netw. **124**, 11–26 (2017). https://doi.org/10.1016/j.comnet.2017.05.031
50. H. Zhang, J. Liu, A.-C. Pang, A Bayesian network model for data losses and faults in medical body sensor networks. Comput. Netw. **143**, 166–175 (2018). https://doi.org/10.1016/j.comnet.2018.07.009
51. X. Hao, L. Wang, N. Yao, D. Geng, B. Chen, Topology control game algorithm based on Markov lifetime prediction model for wireless sensor network. Ad Hoc Netw. **78**, 13–23 (2018). https://doi.org/10.1016/j.adhoc.2018.05.006
52. G. Serpen, J. Li, L. Liu, AI-WSN: Adaptive and intelligent wireless sensor network. Proc. Comput. Sci. **20**, 406–413 (2013). https://doi.org/10.1016/j.procs.2013.09.294
53. Y. Chuan, L. Chen, The application of support vector machine in the hysteresis modeling of silicon pressure sensor. IEEE Sens. J. **11**(9), 2022–2026 (2011). https://doi.org/10.1109/JSEN.2011.2109706
54. M. Nsabagwa, I. Mugume, R. Kasumba, J. Muhumuza, S. Byarugaba, E. Tumwesigye, J.S. Otim, Condition monitoring and reporting framework for wireless sensor network-based automatic weather stations, in *2018 IST-Africa Week Conference (IST-Africa), 9-11 May 2018*, (2018), pp. 1–8
55. P. Chaturvedi, A.K. Daniel, Hidden markov model based node status prediction technique for target coverage in wireless sensor networks, in *2017 International Conference on Intelligent Communication and Computational Techniques (ICCT), 22-23 Dec. 2017*, (2017), pp. 223–227. https://doi.org/10.1109/INTELCCT.2017.8324049
56. I. Lobachev, R. Maleryk, S. Antoschuk, D. Filiahin, M. Lobachev, Integration of neural networks into smart sensor networks, in *2018 IEEE 9th International Conference on Dependable Systems, Services and Technologies (DESSERT), 24-27 May 2018*, (2018), pp. 544–548. https://doi.org/10.1109/DESSERT.2018.8409192
57. M. Gholipour, A.T. Haghighat, M.R. Meybodi, Hop-by-Hop congestion avoidance in wireless sensor networks based on genetic support vector machine. Neurocomputing **223**, 63–76 (2017). https://doi.org/10.1016/j.neucom.2016.10.035
58. C. Mammadova, H. Ben Hmida, A. Braun, A. Kuijper, New approach for optimizing the usage of situation recognition algorithms within IoT domains, in *Ambient intelligence*, (Springer International, Cham, 2017), pp. 183–196

59. P.K.D. Pramanik, S. Pal, P. Choudhury, Beyond automation: The cognitive IoT. Artificial intelligence brings sense to the internet of things, in *Cognitive Computing for Big Data Systems Over IoT: Frameworks, Tools and Applications*, ed. by A. K. Sangaiah, A. Thangavelu, V. Meenakshi Sundaram, (Springer International, Cham, 2018), pp. 1–37. https://doi.org/10.1007/978-3-319-70688-7_1
60. R. Haber-Guerra, S.Y. Liang, J.R. Alique, R. Haber-Haber, Fuzzy control of spindle torque in high-speed milling processes. J. Manuf. Sci. Eng. Trans. ASME **128**(4), 1014–1018 (2006). https://doi.org/10.1115/1.2194063
61. R.E. Haber, J.R. Alique, A. Alique, J. Hernández, R. Uribe-Etxebarria, Embedded fuzzy-control system for machining processes: Results of a case study. Comput. Indus. **50**(3), 353–366 (2003). https://doi.org/10.1016/s0166-3615(03)00022-8
62. A.G. Martin, R.E.H. Guerra, Internal model control based on a neurofuzzy system for network applications. a case study on the high-performance drilling process. IEEE Trans. Automat. Sci. Eng. **6**(2), 367–372 (2009). https://doi.org/10.1109/tase.2008.2006686
63. R.E. Precup, M.B. Radac, R.C. Roman, E.M. Petriu, Model-free sliding mode control of nonlinear systems: Algorithms and experiments. Inform. Sci. **381**, 176–192 (2017). https://doi.org/10.1016/j.ins.2016.11.026
64. M.B. R\v{a}dac, R.E. Precup, E.M. Petriu, S. Preitl, C.A. Drago\c{s}, Data-driven reference trajectory tracking algorithm and experimental validation. IEEE Trans. Indus. Inform. **9**(4), 2327–2336 (2013). https://doi.org/10.1109/tii.2012.2220973
65. T. Pflanzner, M. Fidrich, A. Kertesz, Simulating sensor devices for experimenting with IoT cloud systems, in *Connected Environments for the Internet of Things: Challenges and Solutions*, ed. by Z. Mahmood, (Springer International, Cham, 2017), pp. 105–126. https://doi.org/10.1007/978-3-319-70102-8_6
66. M. Rantz, L.J. Phillips, C. Galambos, K. Lane, G.L. Alexander, L. Despins, R.J. Koopman, M. Skubic, L. Hicks, S. Miller, A. Craver, B.H. Harris, C.B. Deroche, Randomized trial of intelligent sensor system for early illness alerts in senior housing. J. Am. Med. Dir. Assoc. **18**(10), 860–870 (2017). https://doi.org/10.1016/j.jamda.2017.05.012
67. M. Cornacchia, K. Ozcan, Y. Zheng, S. Velipasalar, A survey on activity detection and classification using wearable sensors. IEEE Sens. J. **17**(2), 386–403 (2017). https://doi.org/10.1109/JSEN.2016.2628346
68. I.M. Pires, N.M. Garcia, N. Pombo, F. Flórez-Revuelta, From data acquisition to data fusion: a comprehensive review and a roadmap for the identification of activities of daily living using mobile devices. Sensors **16**(2), 184 (2016)
69. R. Gravina, P. Alinia, H. Ghasemzadeh, G. Fortino, Multi-sensor fusion in body sensor networks: State-of-the-art and research challenges. Inform. Fusion **35**, 68–80 (2017). https://doi.org/10.1016/j.inffus.2016.09.005
70. K. Xiao, R. Wang, H. Deng, L. Zhang, C. Yang, Energy-aware scheduling for information fusion in wireless sensor network surveillance. Inform. Fusion (2018a). https://doi.org/10.1016/j.inffus.2018.08.005
71. H.F. Nweke, Y.W. Teh, M.A. Al-Garadi, U.R. Alo, Deep learning algorithms for human activity recognition using mobile and wearable sensor networks: State of the art and research challenges. Exp. Syst. Appl. (2018)
72. S. Liu, X. Hao, Z. Zhang, Pedestrian retrieval via part-based gradation regularization in sensor networks. IEEE Access **6**, 38171–38178 (2018)
73. H. Qiao, F.-Y. Wang, X. Cao, Application of a decomposed support vector machine algorithm in pedestrian detection from a moving vehicle, in *Intelligence and Security Informatics*, (Springer, Berlin, 2005), pp. 662–663
74. R. Madarshahian, J.M. Caicedo, Human activity recognition using multinomial logistic regression, in *Model Validation and Uncertainty Quantification*, vol. 3, (Springer International, Cham, 2015), pp. 363–372
75. X. Niu, M. Li, X. Cui, J. Liu, S. Liu, K.R. Chowdhury, WTrack: HMM-based walk pattern recognition and indoor pedestrian tracking using phone inertial sensors. Pers. Ubiquit. Comput. **18**(8), 1901–1915 (2014). https://doi.org/10.1007/s00779-014-0796-x

76. Z. Yang, L.S.C. Pun-Cheng, Vehicle detection in intelligent transportation systems and its applications under varying environments: A review. Image Vision Comput. **69**, 143–154 (2018). https://doi.org/10.1016/j.imavis.2017.09.008
77. W. Shi, M.B. Alawieh, X. Li, H. Yu, Algorithm and hardware implementation for visual perception system in autonomous vehicle: A survey. Integration **59**, 148–156 (2017). https://doi.org/10.1016/j.vlsi.2017.07.007
78. F. Li, Z. Lv, Reliable vehicle type recognition based on information fusion in multiple sensor networks. Comput. Netw. **117**, 76–84 (2017)
79. M.B.B. Mahaleh, S.A. Mirroshandel, Harmony search path detection for vision based automated guided vehicle. Robot. Auton. Syst. **107**, 156–166 (2018). https://doi.org/10.1016/j.robot.2018.06.008
80. G. Beruvides, R. Quiza, R. Del Toro, R.E. Haber, Sensoring systems and signal analysis to monitor tool wear in microdrilling operations on a sintered tungsten-copper composite material. Sens. Actuat. A. Phys. **199**, 165–175 (2013). https://doi.org/10.1016/j.sna.2013.05.021
81. I. Kabashkin, J. Kundler, Reliability of sensor nodes in wireless sensor networks of cyber physical systems. Proc. Comput. Sci. **104**, 380–384 (2017)
82. L.A. Laranjeira, G.N. Rodrigues, Border effect analysis for reliability assurance and continuous connectivity of wireless sensor networks in the presence of sensor failures. IEEE Trans. Wirel. Commun. **13**(8), 4232–4246 (2014)
83. M. Yan, K.-Y. Lam, S. Han, E. Chan, Q. Chen, P. Fan, D. Chen, M. Nixon, Hypergraph-based data link layer scheduling for reliable packet delivery in wireless sensing and control networks with end-to-end delay constraints. Inform. Sci. **278**, 34–55 (2014)
84. M.-T. Vo, T.T. Thanh Nghi, V.-S. Tran, L. Mai, C.-T. Le, Wireless sensor network for real time healthcare monitoring: Network design and performance evaluation simulation, in *5th International Conference on Biomedical Engineering in Vietnam*, (Springer International, Cham, 2015), pp. 87–91
85. D. Cacciagrano, R. Culmone, M. Micheletti, L. Mostarda, Energy-efficient clustering for wireless sensor devices in internet of things, in *Performability in Internet of Things*, (Springer, Cham, 2019), pp. 59–80
86. S. Hossain, A.R. Fayjie, O. Doukhi, D.-J. Lee, CAIAS Simulator: Self-driving Vehicle Simulator for AI Research, in *International Conference on Intelligent Computing & Optimization*, (Springer, Berlin, 2018), pp. 187–195
87. M. Mukherjee, I. Adhikary, S. Mondal, A.K. Mondal, M. Pundir, V.A. Chowdary, Vision of IoT: Applications, Challenges, and Opportunities with Dehradun Perspective, in *Proceeding of International Conference on Intelligent Communication, Control and Devices*, (Springer, Singapore, 2017), pp. 553–559
88. A. Bradai, A. Benslimane, K.D. Singh, Dynamic anchor points selection for mobility management in software defined networks. J. Netw. Comput. Appl. **57**, 1–11 (2015)
89. S. Kumar, D. Lobiyal, Novel DV-Hop localization algorithm for wireless sensor networks. Telecommun. Syst. **64**(3), 509–524 (2017)
90. J. Hightower, G. Borriello, Location systems for ubiquitous computing. Computer **34**(8), 57–66 (2001). https://doi.org/10.1109/2.940014
91. G.J. Pottie, W.J. Kaiser, Wireless integrated network sensors. Commun. ACM **43**(5), 51–58 (2000). https://doi.org/10.1145/332833.332838
92. N.B. Priyantha, A. Chakraborty, H. Balakrishnan, The Cricket location-support system, in *Paper presented at the Proceedings of the 6th annual international conference on Mobile computing and networking, Boston, Massachusetts, USA*, (2000)
93. A. Savvides, C.-C. Han, M.B. Strivastava, Dynamic fine-grained localization in Ad-Hoc networks of sensors, in *Paper presented at the Proceedings of the 7th annual international conference on Mobile computing and networking, Rome, Italy*, (2001)
94. J. Sheu, P. Chen, C. Hsu, A distributed localization scheme for wireless sensor networks with improved grid-scan and vector-based refinement. IEEE Trans. Mob. Comput. **7**(9), 1110–1123 (2008). https://doi.org/10.1109/TMC.2008.35

95. T. He, C. Huang, B.M. Blum, J.A. Stankovic, T. Abdelzaher, Range-free localization schemes for large scale sensor networks, in *Paper presented at the Proceedings of the 9th annual international conference on Mobile computing and networking, San Diego, CA, USA*, (2003)
96. F. Tlili, A. Rachedi, A. Benslimane, Time-bounded localization algorithm based on distributed multidimensional scaling for wireless sensor networks, in *2014 IEEE International Conference on Communications (ICC), 10-14 June 2014*, (2014), pp. 233–238. https://doi.org/10.1109/ICC.2014.6883324
97. A. Khan, I. Ali, H.A. Mahmood, Localization free variable transmit power routing protocol for underwater wireless sensor networks, in *Advances in Network-Based Information Systems*, (Springer International, Cham, 2018), pp. 136–147
98. L. Cui, C. Xu, G. Li, Z. Ming, Y. Feng, N. Lu, A high accurate localization algorithm with DV-Hop and differential evolution for wireless sensor network. Appl. Soft Comput. **68**, 39–52 (2018)
99. S.K. Gharghan, R. Nordin, A.M. Jawad, H.M. Jawad, M. Ismail, Adaptive neural fuzzy inference system for accurate localization of wireless sensor network in outdoor and indoor cycling applications. IEEE Access **6**, 38475–38489 (2018)
100. G. Beruvides, R. Quiza, R.E. Haber, Multi-objective optimization based on an improved cross-entropy method. A case study of a micro-scale manufacturing process. Inform. Sci. **334-335**, 161–173 (2016). https://doi.org/10.1016/j.ins.2015.11.040
101. R.E. Haber, G. Beruvides, R. Quiza, A. Hernandez, A simple multi-objective optimization based on the cross-entropy method. IEEE Access **5**, 22272–22281 (2017). https://doi.org/10.1109/access.2017.2764047
102. A.J. Perez, M-SPOT: A hybrid multiobjective evolutionary algorithm for node placement in wireless sensor networks, in *2018 32nd International Conference on Advanced Information Networking and Applications Workshops (WAINA), 16-18 May 2018*, (2018), pp. 264–269. https://doi.org/10.1109/WAINA.2018.00096
103. S. Phoemphon, C. So-In, D.T. Niyato, A hybrid model using fuzzy logic and an extreme learning machine with vector particle swarm optimization for wireless sensor network localization. Appl. Soft Comput. **65**, 101–120 (2018)
104. Y. Xu, L. Wang, Z. Wang, Y. Lv, Y. Zhang, W. Chen, Using differential evolution to research the multi-objective optimization of medical sensor networks: A brief discussion, in *2018 Chinese Control And Decision Conference (CCDC)*, (IEEE, 2018)
105. K. Singh, K. Singh, A. Aziz, Congestion control in wireless sensor networks by hybrid multi-objective optimization algorithm. Comput. Netw. **138**, 90–107 (2018)
106. S. Amri, F. Khelifi, A. Bradai, A. Rachedi, M.L. Kaddachi, M. Atri, A new fuzzy logic based node localization mechanism for wireless sensor networks. Future Gener. Comput. Syst. **93**, 799–813 (2017)
107. N.A. Pantazis, S.A. Nikolidakis, D.D. Vergados, Energy-efficient routing protocols in wireless sensor networks: A survey. IEEE Commun. Surv. Tutor. **15**(2), 551–591 (2013). https://doi.org/10.1109/SURV.2012.062612.00084
108. N. Ramluckun, V. Bassoo, Energy-efficient chain-cluster based intelligent routing technique for wireless sensor networks. Appl. Comput. Inform. (2018)
109. R. Logambigai, S. Ganapathy, A. Kannan, Energy–efficient grid–based routing algorithm using intelligent fuzzy rules for wireless sensor networks. Comput. Electr. Eng. **68**, 62–75 (2018)
110. S.B. Shah, Z. Chen, F. Yin, I.U. Khan, N. Ahmad, Energy and interoperable aware routing for throughput optimization in clustered IoT-wireless sensor networks. Future Gener. Comput. Syst. **81**, 372–381 (2018)
111. S. Su, S. Zhao, An optimal clustering mechanism based on Fuzzy-C means for wireless sensor networks. Sustain. Comput. Inform. Syst. **18**, 127–134 (2018)
112. W. Elsayed, M. Elhoseny, S. Sabbeh, A. Riad, Self-maintenance model for wireless sensor networks. Comput. Electr. Eng. **70**, 799–812 (2017)

113. H. He, Z. Zhu, E. Makinen, A neural network model to minimize the connected dominating set for self-configuration of wireless sensor networks. IEEE Trans. Neural Netw. **20**(6), 973–982 (2009)
114. F. Castaño, G. Beruvides, R.E. Haber, A. Villalonga, Time-To-failure modelling in On-Chip LiDAR sensors for automotive applications. Proceedings **1**(8), 809 (2017b)
115. F. Castaño, G. Beruvides, A. Villalonga, R.E. Haber, Self-tuning method for increased obstacle detection reliability based on internet of things LiDAR sensor models. Sensors (Switzerland) **18**(5) (2018). https://doi.org/10.3390/s18051508
116. O. Michel, Cyberbotics Ltd. Webots™: professional mobile robot simulation. Int. J. Adv. Robot. Syst. **1**(1), 5 (2004)
117. J. Godoy, J. Pérez, E. Onieva, J. Villagrá, V. Milanés, R. Haber, A driverless vehicle demonstration on motorways and in urban environments. Transport **30**(3), 253–263 (2015). https://doi.org/10.3846/16484142.2014.1003406
118. A. Alique, R.E. Haber, R.H. Haber, S. Ros, C. Gonzalez, Neural network-based model for the prediction of cutting force in milling process. A progress study on a real case, in *IEEE International Symposium on Intelligent Control—Proceedings*, (2000), pp. 121–125
119. I.L. Fe, G. Beruvides, R. Quiza, R. Haber, M. Rivas, Automatic selection of optimal parameters based on simple soft computing methods. A case study on micro-milling processes. IEEE Trans. Indus. Inform. **99**, 1-1 (2018). https://doi.org/10.1109/TII.2018.2816971
120. R.E. Haber, J.R. Alique, Nonlinear internal model control using neural networks: An application for machining processes. Neural Comput. Appl. **13**(1), 47–55 (2004). https://doi.org/10.1007/s00521-003-0394-8
121. E.M. Voumvoulakis, N.D. Hatziargyriou, Decision trees-aided self-organized maps for corrective dynamic security. IEEE Trans. Power Syst. **23**(2), 622–630 (2008)
122. M.N. Murty, R. Raghava, Kernel-based SVM, in *Support Vector Machines and Perceptrons: Learning, Optimization, Classification, and Application to Social Networks*, (Springer International, Cham, 2016), pp. 57–67. https://doi.org/10.1007/978-3-319-41063-0_5
123. S. Aziz, E.A. Mohamed, F. Youssef, Traffic sign recognition based on multi-feature fusion and ELM classifier. Proc. Comput. Sci. **127**, 146–153 (2018). https://doi.org/10.1016/j.procs.2018.01.109
124. G. Beruvides, F. Castaño, R.E. Haber, R. Quiza, A. Villalonga, Coping with complexity when predicting surface roughness in milling processes: Hybrid incremental model with optimal parametrization. Complexity (2017). https://doi.org/10.1155/2017/7317254
125. B. Khaldi, F. Harrou, F. Cherif, Y. Sun, Self-organization in aggregating robot swarms: A DW-KNN topological approach. Biosystems **165**, 106–121 (2018). https://doi.org/10.1016/j.biosystems.2018.01.005
126. F. Penedo, R.E. Haber, A. Gajate, R.M. Del Toro, Hybrid incremental modeling based on least squares and fuzzy K-NN for monitoring tool wear in turning processes. IEEE Trans. Indus. Inform. **8**(4), 811–818 (2012). https://doi.org/10.1109/TII.2012.2205699

Chapter 6
A Smartphone-Based Virtual White Cane Prototype Featuring Time-of-Flight 3D Imaging

Norbert Druml, Thomas Pietsch, Marcus Baumgart, Cristina Consani, Thomas Herndl, and Gerald Holweg

6.1 Introduction

In 2014, the World Health Organization (WHO) estimated that there were 285 million visually impaired people worldwide, with 246 million having low vision and 39 million being blind. It is a challenging task for the visually impaired to safely and quickly navigate through an unknown or unfamiliar environment. Still the traditional white cane and the guide dog are the most common and reliable travel aids available. However, using the traditional white cane, collisions can only be detected once they already occurred. While guide dogs are able to guide the user and to avoid collisions, they are rather expensive, require massive amounts of training, and the visually impaired might not want to or be able to care for an animal.

Time-of-Flight 3D imaging is a depth perception technology. By emitting continuously modulated laser light and by measuring the time it takes for the light to travel into the scenery and back to the sensor, depth information can be gathered. Thanks to recent advances in this field of technology, this perception technology can be regarded as an enabler for new applications, in particular for aiding people while travelling.

In this work, we particularly evaluate the Time-of-Flight 3D imaging technology as an electronic travel aid, i.e. virtual white cane. For this purpose, the Lenovo Phab 2 Pro is employed. The Phab 2 Pro is an affordable smartphone with an integrated Time-of-Flight 3D imaging sensor and OpenCL capabilities. These capabilities are exploited in order to detect the ground floor and the free space in front of the user.

N. Druml (✉) · T. Pietsch · T. Herndl · G. Holweg
Infineon Technologies Austria AG, Graz, Austria
e-mail: norbert.druml@infineon.com

M. Baumgart · C. Consani
Photonic Systems, CTR Carinthian Tech Research AG, Villach, Austria

W. D. van Driel et al. (eds.), *Sensor Systems Simulations*,
https://doi.org/10.1007/978-3-030-16577-2_6

Fig. 6.1 Obstacle warning implementation of an electronic travel aid, cf. [1], (© IEEE 2018), reprinted with permission

In order to detect the ground floor robustly, a new approach is introduced that combines the well-known v disparity and RANSAC algorithms. Once the ground floor is detected, all remaining points are considered to be obstacles. The resulting information is then represented via the so-called Conservative Polar Histogram. With the Conservative Polar Histogram the closest obstacles in each direction are detected and a tactile and/or acoustic warning is given, see also Fig. 6.1. At its current implementation level, the virtual white cane is able to scan the environment with 7 FPS, while the software stack of the Lenovo Phab 2 Pro provides depth data only at 5 FPS.

Summarizing, this chapter makes the following contributions:

- It presents a virtual white cane that exploits the Time-of-Flight 3D imaging technology.
- Two novel concepts are introduced, the combined v disparity/RANSAC ground plane detection and the Conservative Polar Histogram, both improving detection and tracking capabilities of obstacles.
- It proofs the feasibility of our virtual white cane concept by means of several test scenes and provides performance data.

6.2 Related Work

This section summarizes the current state-of-the-art of depth-based techniques for aiding visually impaired people. The authors in [2] divide the electronic devices for navigation into three categories:

- Vision Enhancement: This helps users with some remaining residual vision by providing enhanced visual output.
- Vision Replacement: This approach replaces the human vision and provides output directly to the optic nerve or the human brain.
- Vision Substitution: This uses other sensory means, such as auditory or tactile feedback, in order to provide a mental image of the scene.

Vision substitution can be further subdivided in the following three main categories:

- Electronic Travel Aid (ETA): provides information normally perceived visually by transforming them into a form which can be communicated through other sensory modality.
- Electronic Orientation Aid (EOA): supports the user by providing orientation aid.
- Position Locator Device (PLD): provides the user with positional information, for example, by means of GPS.

RGB-Depth (RGB-D) camera solutions are nowadays widely available, thanks to the introduction of Google Tango, Google ARCore, and Microsoft Kinect devices. These devices either operate directly on the point cloud, for example, by using RANSAC-based algorithms, or use more complex computer vision-aided approaches, such as marching cubes or region growing. As an example, the system presented by the authors in [3] directly operates in a time-critical manner on the point cloud. Their algorithm is therefore interruptible and shall provide conservative and meaningful results. In [4], the authors introduced a RANSAC-based approach that uses the point cloud data which is directly obtained from a RGB-D sensor. Aladren et al. use the RANSAC algorithm in order to fuse RGB-D data for extending the range of the depth sensor. They exploit a voxel-based filter in order to down-sample the point cloud, which significantly reduces the work load (7,000 points are considered instead of 300,000 without losing crucial information). Then, the authors use a Manhattan world model (three main directions which are orthogonal between each other [5]) in order to find the most prominent planes. In [6], the authors presented a fast plane detection solution. In their work, a Swissranger SR3000 3D range camera was used to obtain the point cloud data. Furthermore, region growing was employed in order to detect the planes. The authors of [7] used a different approach for obstacle detection: a depth map and the marching squares algorithm were used, instead of finding structures in the point cloud. In addition, the authors used sound output to inform the user of obstacles. While the pitch of the sound represents the height (y-axis), the volume represents the distance to the user, and the x position is encoded using stereo. A computer vision-based approach was presented by the authors of [8] using Microsoft's Kinect in combination with Speeded Up Robust Feature (SURF) local descriptors that are clustered via a Bag-of-Visual-Words (BOVW) model. The BOVW constructs feature vectors that serve as input for a multi-class Support Vector Machine (SVM) in order to classify common office objects like laptops, chairs, stairs, etc. In [9], the authors used Google Tango in order to help visually impaired users with the detection and recognition of objects and faces. While face and object recognition were carried out via deep neural networks (NVIDIA's CUDA Deep Neural Network Library cuDNN), the depth relation of these objects and faces was carried out with Google Tango's depth data.

A revolutionary and completely new algorithm to detect the ground plane in stereo images, in particular for autonomous and assisted driving, was presented in [10]. By using the disparity map of a stereo camera, they calculated a v disparity map. In this particular v disparity map, a ground plane is mapped to a straight line. Therefore, in order to detect such a straight line in this map, a standard Hough transformation is used. If a non-flat road geometry is given, then a succession of parts of planes is assumed. These parts of planes are given as a piecewise linear curve. Obstacles and objects are represented as vertical lines in the v disparity image. The authors of [11] enhanced the v disparity approach in order to handle also off-road scenarios. First, they used v disparity to find ground plane assumption. Then, static calibration data was employed to compute the expected ground correlation line on a flat surface and the ground correlation line for different pitch angles. Afterwards, the v disparity along these candidate lines was accumulated and the line with the best correlation was chosen allowing obtaining pitch estimation. Further algorithm advancements were achieved by the authors in [12]. They combined v disparity's ability to detect the road [10] with u disparity, thus improving the ability to detect objects and obstacles. The authors proposed a propagation algorithm in order to cope with difficulties such as roll, yaw, pitch, or road slant. In [13], the authors applied for the first time a threshold to the u disparity and thus obtained an obstacle map. By using a threshold, only obstacles with a certain height in pixels are kept. The v disparity is calculated on the free map. The free map is defined as every pixel that is not given as an obstacle in the obstacle map. As a consequence, removing the pixels that are part of obstacles increases the chance of the road surface to be the most prominent line in the v disparity map. In [14], the authors demonstrated that the approach of [12] can be feasibly applied to 3D cameras as well. They calculated the disparity by assuming a baseline of a stereo-vision camera. In order to enhance the segmentation of obstacles on the road, the authors exploited both the v disparity and the u disparity. Another approach was presented in [15]. The authors assumed that a robot/vehicle moves on the ground plane and all points detected above the ground plane can be assumed to be obstacles. Furthermore, this approach required an initial space directly in front of the robot that is free, which is used for calibration purposes. Whenever there are more than 40% invalid depth readings, the authors assumed that it is not safe to proceed any further.

Summarizing, even though this research area is highly active, there are still lots of gaps. Therefore, this chapter provides an innovative contribution to the ongoing discussion in this important field of Electronic Travel Aids.

6.3 Virtual White Cane for Vulnerable Road Users

6.3.1 Requirements

In the following, the fundamental requirements for a virtual white cane, which shall use the Lenovo Phab 2 Pro smartphone (see also Fig. 6.2), are summarized. The most crucial requirement of this system is to warn the user whether a possible obstacle is in his way. In order to robustly achieve this goal, it is required to detect and to remove the ground floor from the obstacle list because the ground floor should not be considered to be an obstacle. The remaining data points are then considered to be not traversable and therefore are considered as obstacles. Furthermore, the user shall be warned of objects in his way. This shall be achieved with two diverse warning methods, the acoustic warning and the vibration warning. The acoustic warning shall provide the user an acoustic image/representation of the identified obstacles by playing a sound that originates from the object. By doing so, the user is able to create a mental image of the scenery in front of the device. Given the design of the virtual white cane, further constraints are introduced in order to optimize its performance: the ground floor is considered to be between 0.8 and 1.3 m below the virtual white cane at any point in time. These constraints coincide with the average height a mobile device is held at while the user is standing.

Fig. 6.2 The Lenovo Phab 2 Pro smartphone implements not only a RGB color camera, but also a Time-of-Flight 3D imaging camera and a fish-eye camera, cf. [1], (© IEEE 2018), reprinted with permission

6.3.2 Lenovo Phab 2 Pro Hardware

The Lenovo Phab 2 Pro is a state-of-the-art smartphone that integrates a Qualcomm Snapdragon 652 processor featuring an octa-core processor architecture and the Adreno 510 GPU. Its 64 bit CPU comes with the following specification: 4× 1.8 GHz Coretex-A72, 4× 1.4 GHz Cortex-A53, and 4 GB RAM. In order to implement a high-performance virtual white cane, a GPU is of high importance, because it enables the usage of Khronos OpenCL for improved computational performance. The Adreno GPU runs at 600 MHz and supports 128 ALUs, a unified memory, and a unified shader model. The unified memory is of particular importance because memory can be shared between CPU and GPU which improves the performance of memory accesses without the need of dedicated memory copies. Typically, these memory copies represent a bottleneck of GPGPU programs. The Lenovo Phab 2 Pro's Adreno 510 GPU supports the following APIs:

- OpenCL 2.0 Full Profile.
- OpenGL ES 3.2.
- OpenGL 3.2.
- Vulkan 1.0.

Although Vulkan is supported by the Adreno GPU, it is not supported by Android version 6.0 Marshmallow that is available on the Lenovo Phab 2 Pro. Once Android 7.0 Nougat will be available for the Lenovo Phab 2 Pro, also Vulkan will be fully supported. Furthermore, there is no official support of OpenCL via Android. Yet, the Lenovo Phab 2 Pro comes with the Open CL 2.0 libraries that can be employed by developers, and which was also employed for the presented Virtual White Cane project.

With regard to sensors, the Lenovo Phab 2 Pro integrates a 224 × 172 pixel Time-of-Flight depth camera (which was jointly developed by Infineon Technologies and PMDTechnologies), a 640 × 480 fish-eye lens camera using the software stack for motion tracking, and a high resolution 1920 × 1080 pixel RGB color camera. In addition, also a compass/magnetometer, proximity sensor, accelerometer, ambient light sensor, gyroscope, and barometer are available. The Time-of-Flight camera, which represents the most crucial sensor for the Virtual White Cane, integrates a patented Suppression of Background Illumination technology. This technology ensures a robust operation not only during dark environment condition, but more importantly also during harsh and bright sunlight enabling outdoor operation.

6.3.3 Tilt Angle Considerations

The tilt angle of our virtual white cane device is mostly irrelevant. This is due to the fact that the angle of the line that has to be found within the v disparity map directly correlates directly, in a linear way, with the device's tilt angle. By employing the

Lenovo Phab 2 Pro's internal gravity sensor, the orientation of the device can be detected and thus the angle of the line in the v disparity map can be calculated (according to Eq. (6.1) where k and d represent the parameters of the line) and found.

$$\text{angle} = k * \text{tilt} + d \tag{6.1}$$

The parameters k and d can be found by following the following procedure:

- Record the v disparity map of an empty room without any obstacles
- Employ Hough transformation in order to find lines
- Record angle and tilt angle and perform a linear regression on these data points

As a result, a k value of 0.1478 and a d value of 163.223 can be found. These parameters together with the tilt angle enable the calculation of the line angle with a fairly tight bound. As a consequence, it is possible to allow for a deviation of only 1°.

6.3.4 Basic Flow of Operations

The high-level flow of operations of the virtual white cane is depicted by Fig. 6.3. First, the *VirtualWhiteCaneService* connects to Android's point cloud provider and forwards the point cloud to the *GroundDetectionModule*. This module calculates the v disparity map and runs RANSAC on this input data. The result is then forwarded to the *WarningModule* that generates the *ConservativePolarHistogram* and warns the user in case of a detected obstacle.

6.3.5 Ground Detection Module

The ground detection module does not solve the problem of finding obstacles but solves instead the inverse problem of finding the free space in front of the user. For this task, the ground plane has to be detected and filtered. Any remaining points that are not part of the ground plane are considered as obstacles. Given the application scenario of the virtual white cane (smartphone being held in various heights and distances to the floor), a combination of two algorithms is used in order to improve the robustness of detecting the ground plane. At first, the v disparity is calculated and the line, which represents a candidate for ground plane, is found via Hough transformation. A filter then separates the points belonging to the ground plane and the points representing obstacles. The ground plane candidate is then fed into the RANSAC algorithm. The output of the RANSAC algorithm is then used to crosscheck the point cloud data. In case the v disparity algorithm is not able to find a ground plane at all, then all points are obstacles. The basic flow of this ground

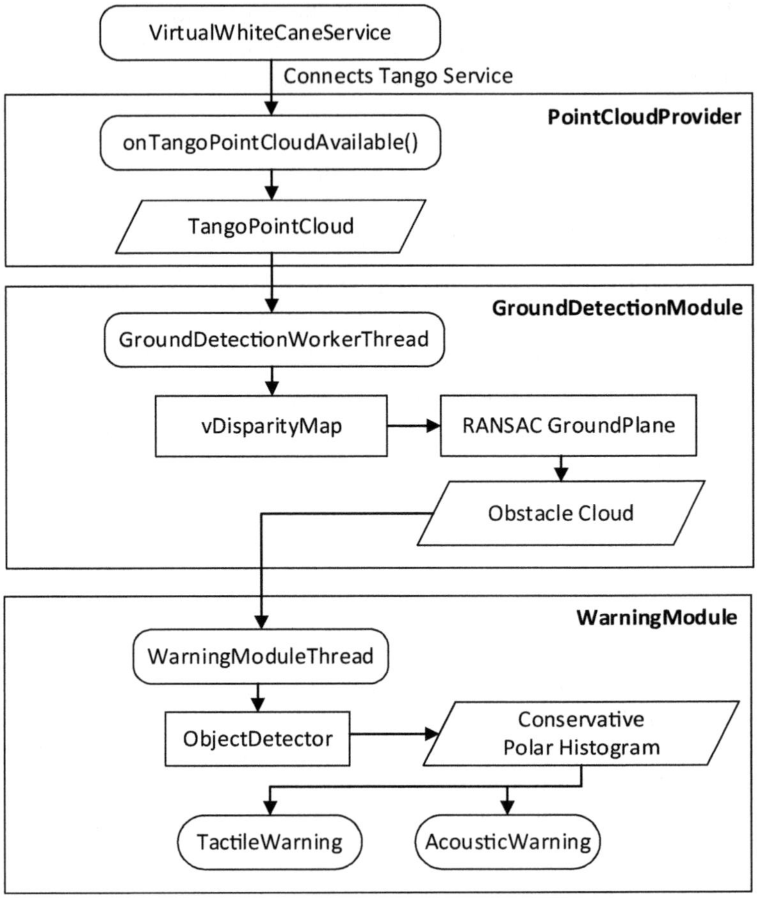

Fig. 6.3 Basic flow of operations of the virtual white cane, cf. [1], (© IEEE 2018), reprinted with permission

plane detection approach is depicted in Fig. 6.4. In the following, the ground plane detection will be explained in detail.

6.3.6 v Disparity Map

The v disparity map calculation requires the point cloud to be represented in an ordered (u, v) grid. After an input point cloud is ordered, the v disparity can be calculated according to the following procedure. Given disparity map I_{Δ}, the v disparity map $I_{v\Delta}$ is calculated by summing up the points of same disparity in I_{Δ} along the $\overrightarrow{v}$ axis. An example of such a v disparity map calculation is illustrated in Fig. 6.5.

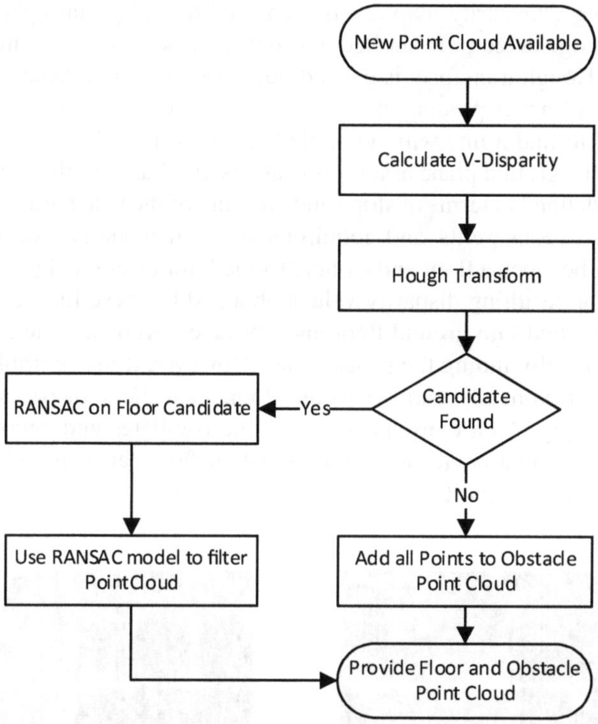

Fig. 6.4 Ground plane detection: v disparity map is calculated from point cloud. The most relevant lines are detected with Hough transformation. The floor candidate is fed into RANSAC and the candidate is refined. If no candidate is found, then all points are considered as obstacles. Image obtained from [1], (© IEEE 2018), reprinted with permission

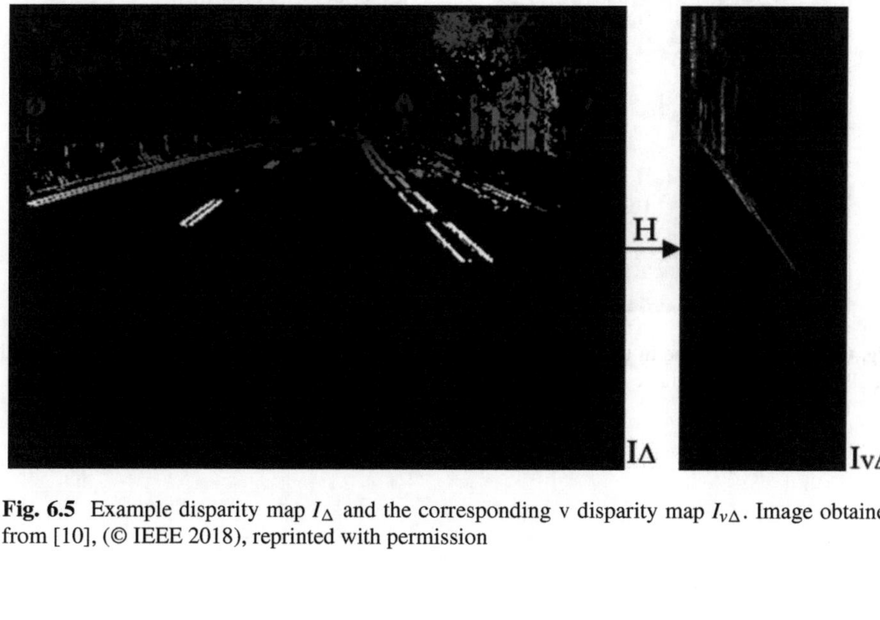

Fig. 6.5 Example disparity map I_{Δ} and the corresponding v disparity map $I_{v\Delta}$. Image obtained from [10], (© IEEE 2018), reprinted with permission

The resulting v disparity map $I_{v\Delta}$ is then used to find ground plane candidates. First, a canny edge detector is applied in order to preserve only the most prominent lines. Then, a Hough transform is carried out in order to find these lines. Since the ground floor will never produce a line at a non-acute angle, the Hough transform can be limited to find a line with an angle θ between $\left[0, \frac{\pi}{2}\right]$ and only lines which correspond to the ground plane assumption are used. Thanks to these constraints, the lines are well-defined in terms of slope and crossing of the v disparity image borders. Following these constraints and requirements is mandatory because otherwise discrimination between a floor and a plane formed, for example, by a table wouldn't be possible. The resulting disparity values obtained by these lines are then used to filter the point cloud into ground floor and obstacles. Knowing the tilt angle of the smartphone can help finding the proper lines. However, if the smartphone's bottom edge is not approximately parallel to the floor, then the v disparity map will be altered accordingly. As a consequence, the line oscillates and becomes wider, as shown in Fig. 6.6, violating the assumption that the floor can be found as a piecewise linear curve in the v disparity map.

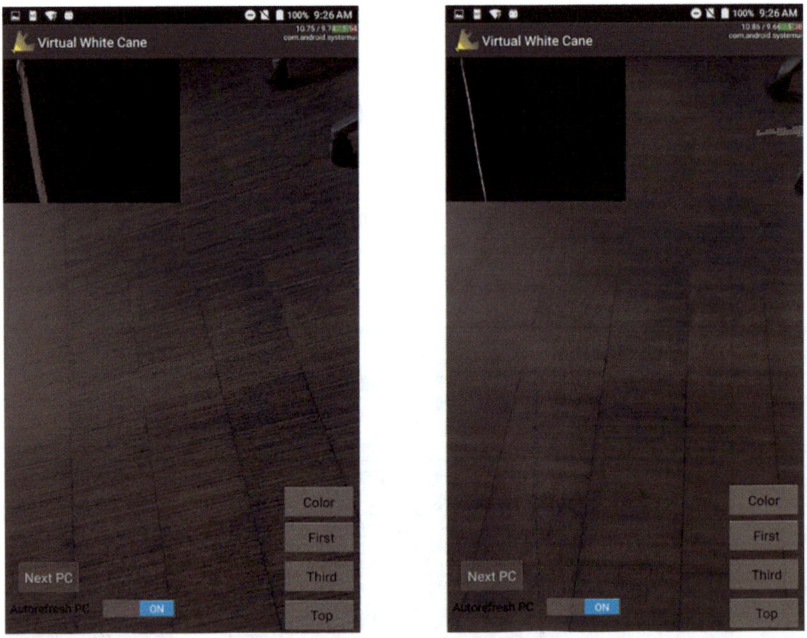

Fig. 6.6 v disparity (line in the top left corner) of the same scene, in case the device is not held parallel to the ground. Image obtained from [1], (© IEEE 2018), reprinted with permission

6.3.7 RANSAC

In order to improve the robustness of detecting the ground plane and detecting obstacles, a combined v disparity/RANSAC approach is employed. Ground plane candidates that were found with the v disparity filtering are fed into the RANSAC algorithm. To ensure the found plane has the correct orientation in space, the normal must be verified. This also allows compensating the point cloud up to a certain degree if the user is holding the lower edge of the device not parallel to the ground. Even though only fractions of the ground plane are used as input to the RANSAC, it is still capable of finding a good fit. As a final step, the result of the RANSAC is used to filter the point cloud and to remove the ground plane.

6.3.8 Obstacle Detection

This work introduces a novel approach, the so-called Conservative Polar Histogram, which detects obstacle in the point cloud in such a way that meaningful warnings can be provided to the user. The novel algorithm assigns each point to a polar grid and creates a disparity histogram for each bin of this grid. Remaining outliers are filtered, and the nearest remaining bin is considered to be an obstacle.

After this processing step, the filtered point cloud should contain only obstacles. Obstacle detection is performed now in a conservative way by prioritizing the closest obstacle in the path. Then, the point cloud is transformed into the OpenGL world coordinate system that defines the smartphone's camera as the origin. The points in the OpenGL coordinate system can be projected easily onto the ground plane by omitting the y axis. This approach results in a 2D point pt_2. Each point in the obstacle point cloud is associated with its respective bin k according to their angle in polar coordinates that are calculated as defined by (6.2) and (6.3) with an $\alpha = 5^\circ$. In addition, the Euclidean distance d from the origin (which is the camera) is calculated according to (6.4) and (6.5).

$$\beta_{i,j} = \tan^{-1}\frac{y_i - y_0}{x_j - x_0} \tag{6.2}$$

$$k = \left\lceil \frac{\beta_{i,j}}{\alpha} \right\rceil \tag{6.3}$$

$$pt_2 = \left(pt_{\text{ogl}}.x,\ pt_{\text{ogl}}.z\right) \tag{6.4}$$

$$d = \text{dot}\left(pt_{2,}\,pt_2\right) \tag{6.5}$$

As a next step, the Euclidean distance d is inverted, scaled, and discretized in order to fit into a range from [0, 255]. Given the fact that the exposure time of the

Time-of-Flight camera is fixed in the Lenovo Phab 2 Pro, there is a limit with regard to the smallest valid depth value. This minimum valid distance is 0.2 m and is due to overexposure at short distances. Therefore, a depth reading closer to 0.2 is very unlikely to be valid and the numerator of d_{pseudo} can be set to 0.2. Because depth values are not gained via stereo imaging but via Time-of-Flight 3D imaging, the values are interpreted as pseudo disparity d_{pseudo} as defined in (6.6).

$$d_{\text{pseudo}} = \left\lceil \frac{0.2}{d} * 255 \right\rceil \tag{6.6}$$

Finally, a disparity histogram is calculated for each bin of the grid by counting the number of occurrences for each disparity value d_{pseudo}. If we assume a k value of 72, then an array size of 72*256 is required. The disparity histogram enables further filtering of outliers from the obstacle point cloud. Thus, bins that contain only a few points are removed. The biggest remaining non-zero disparity represents the nearest obstacle in this very direction. An example of a Conservative Polar Histogram is depicted in Fig. 6.7. A pseudo code listing of the calculation of Conservative Polar Histogram is given an Annex A. The area around the user/smartphone is divided into angular bins. The blue areas represent obstacles. A disparity histogram is calculated for each bin. The red line indicates the distance from the camera to the objects. Figure 6.8 depicts the flow of operations carried out by the obstacle detection algorithm. The largest disparity has to be converted back to the 3D space

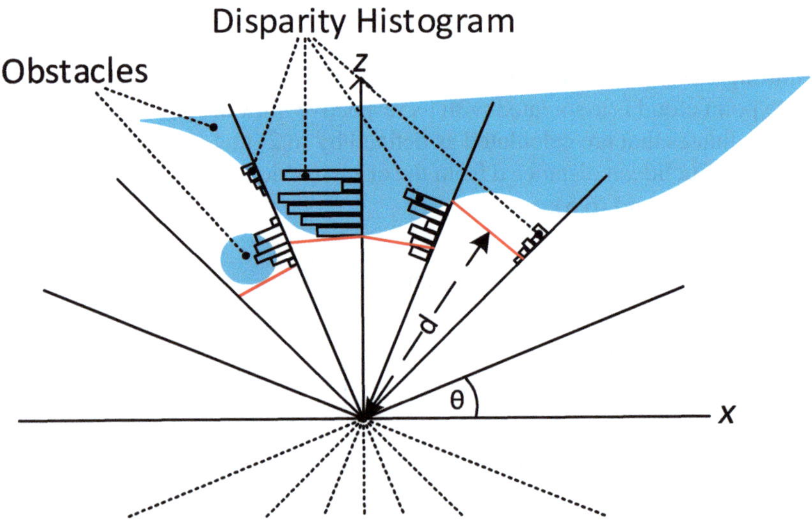

Fig. 6.7 Example of a Conservative Polar Histogram. The blue objects represent obstacles, while the red line indicates distance to the object. Image obtained from [1], (© IEEE 2018), reprinted with permission

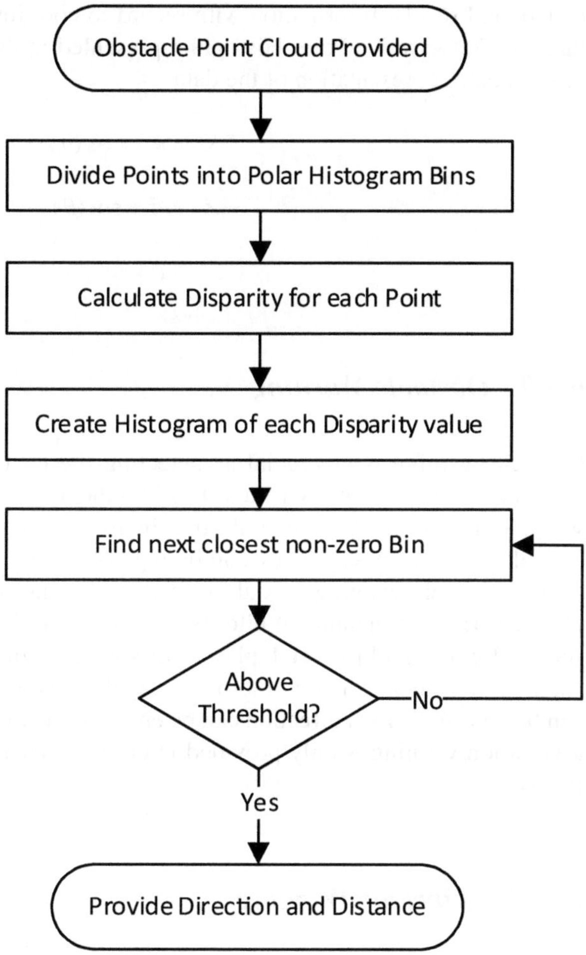

Fig. 6.8 Flow of operations of the obstacle detection approach. Image obtained from [1], (© IEEE 2018), reprinted with permission

that is defined by *X*, *Y*, and *Z* coordinates. The bin k is converted back to an angle θ according to (6.7).

$$\theta = \frac{k * 5 + 2.5}{180} * \pi \tag{6.7}$$

Furthermore, the disparity value is converted back into a depth value d by inverting (6.6) according to (6.8)

$$d = \frac{0.2}{d_{\text{pseudo}}/255} \tag{6.8}$$

As a final step, the angle phi and distance d (which are represented as polar coordinates) are converted into the Euclidean coordinate system according to (6.9),

(6.10), and (6.11). Information with regard to coordinate Y are lost due to the fact that the Conservative Polar Histogram presented in this work only supports a two-dimensional representation of the data.

$$X = d * \cos(\theta) \tag{6.9}$$

$$Z = d * \sin(\theta) \tag{6.10}$$

$$Y = 1 \tag{6.11}$$

6.3.9 Obstacle Warning

Obstacle warning is as crucial as detecting the obstacles because if the warning cannot be easily and quickly interpreted by the user in a robust way, then a warning will not provide any benefit at all. In principle, warnings can be generated either in a tactile or acoustic way, or by combining both methods.

The obstacle warning module receives the nearest detected obstacles from the obstacle detector module. While the acoustic warning submodule places a sound source for each object and plays the sounds sequentially, the tactile warning submodule is more limited in terms of capabilities: neither the vibration's intensity can be modified, nor a direction can be encoded in any way. Given these constraints, a vibration warning is only provided in case an obstacle is immediately in front of the user.

6.3.10 Vibrator Warning

Given the highlighted constraint that the vibrator warnings in Android cannot be altered in terms of intensity, it is only possible to provide a warning once an obstacle is within a certain distance, defined as 1 m, of the user. Furthermore, it is only possible to communicate a direction to the user if the virtual white cane is moved in a scanning way. Therefore, only the front facing bins of the Conservative Polar Histogram are regarded for triggering a vibrator warning. This is feasible because the Conservative Polar Histogram resides in a world coordinate system with the device rotating at the origin and the front facing bins are calculated in relation to the device's orientation.

6.3.11 Acoustic Warning

Acoustic sounds represent an essential approach in order to warn the user of approaching obstacles. Thanks to stereo sound and the Head-Related Transfer Function (HRTF), a source of sound can be positioned in a certain direction to the user. Therefore, an approaching obstacle can be located by the user in its direction.

The Acoustic Warning module positions a source of sound for each bin of the Conservative Polar Histogram if a certain distance limit is reached. These sounds are played sequentially like sonar/radar, which enables the visually impaired person to create a mental image of the surroundings with its obstacles. Experiments with acoustic warnings showed that choosing the right sound to play is not a trivial selection. While such a sound shall not only be easily differentiable from any natural sound source, it shall also contain various frequencies instead of only a simple sine wave. This is due to the fact that the more complex a sound is, the easier is the localization of such a sound. However, a sound shall not be played too long because several sound sources will overlap due to the sonar/radar like playstyle. The Acoustic Warning module plays the sounds in a repeating and round robin fashion, which creates an animated impression (the sound moves along the obstacles). Repeating and animated sounds help the visually impaired person to pinpoint obstacles much faster.

6.4 Results

In this section, the results and the performance of the Virtual White Cane are discussed. In the following figures, the color of the pixels defines their classification after the ground detection algorithm. While green pixels represent the ground plane, red pixels indicate detected obstacles.

Our proposed combination of v Disparity and RANSAC algorithms improves significantly the accuracy of a pure RANSAC approach. Due to RANSAC's random approach and fitting process, the best resulting fit is not always reflecting the real scenery. This behavior is shown in Fig. 6.9. While the RANSAC shows a poorly fitting ground plane, our approach (which combines RANSAC with v Disparity) detects the ground plane and the obstacles properly. Our combined RANSAC/v Disparity approach also overcomes problems with cluttered scenes, as demonstrated in Fig. 6.10.

The v disparity segmentation algorithm often classifies too many points as ground plane, while RANSAC is unable to find a properly fitting plane. This is the case for example if a large obstacle is covering most of the floor. Figure 6.11 illustrates this situation. While RANSAC is not able to detect the correct floor at all, v disparity can clearly identify the floor. Again, our approach that combines both algorithms can robustly detect the floor and the scenery's obstacles.

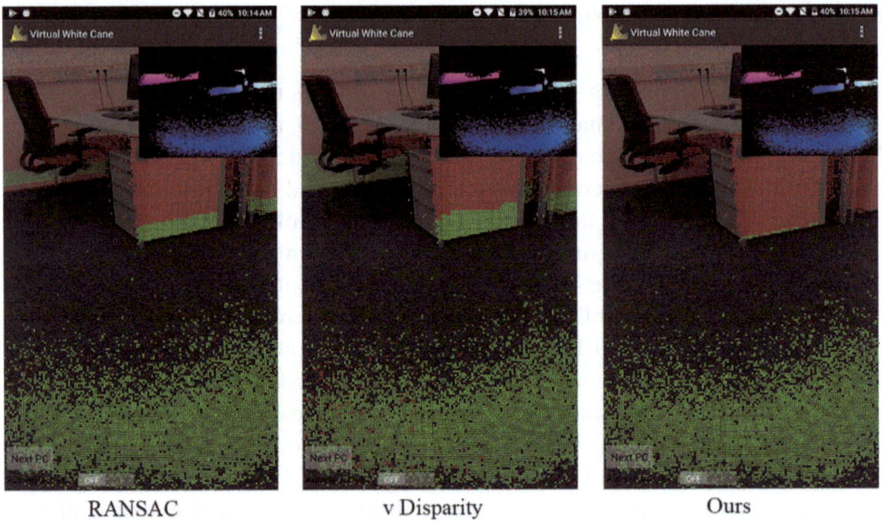

Fig. 6.9 Comparison of floor segmentation results. Images obtained from [1], (© IEEE 2018), reprinted with permission

Fig. 6.10 Our method outperforms in cluttered scenes. Images obtained from [1], (© IEEE 2018), reprinted with permission

Figure 6.12 demonstrates the results of another critical situation. In case the device is held in such a way that the bottom line of the device is not parallel to the ground, then v disparity shows reduced performance. This degraded performance is caused by the line representing the ground floor in the v disparity image. This

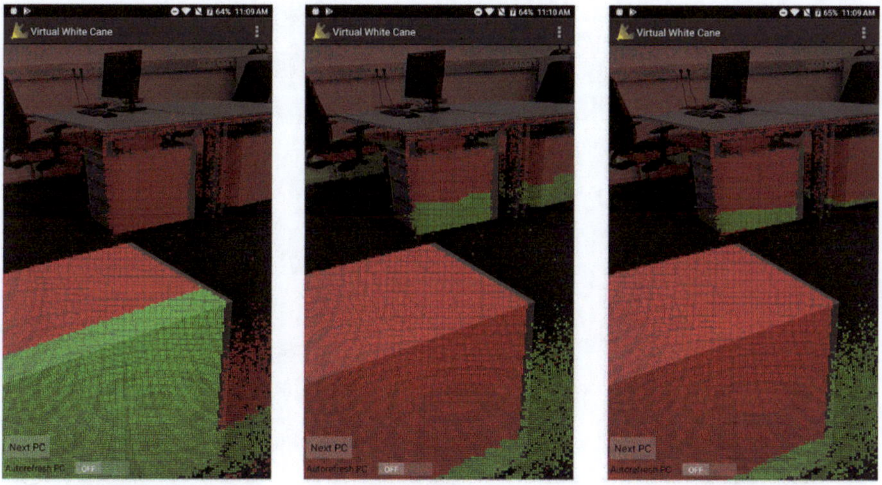

Fig. 6.11 Our method outperforms with a large object in front of the user. Images obtained from [1], (© IEEE 2018), reprinted with permission

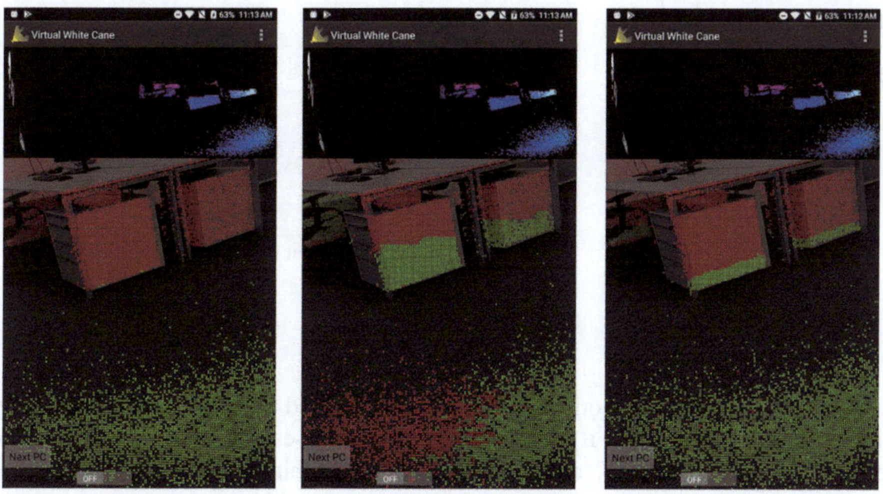

Fig. 6.12 Our implementation fixes the problems with pure v disparity. Images obtained from [1], (© IEEE 2018), reprinted with permission

line oscillates in case the device is held not parallel to the ground and therefore the ground plane is not found properly. Our approach, which combines RANSAC and v disparity, is capable of solving this degraded v disparity performance and thus provides a robust ground plane detection.

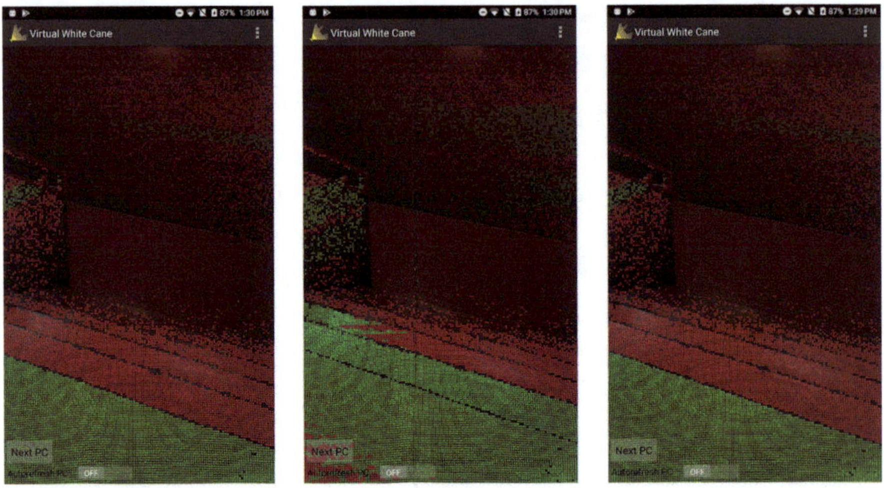

Fig. 6.13 Downstairs scene, the stairs are considered as obstacles. Images obtained from [1], (© IEEE 2018), reprinted with permission

Table 6.1 Performance data of combined RANSAC/v disparity ground detection

	Scenery 1	Scenery 2
#Points [1]	13,621	12,869
v disparity [ms]	2.89	2.66
Hough [ms]	25.12	24.64
Filtering [ms]	4.32	4.42
v disparity total [ms]	71.59	70.62
RANSAC [ms]	8.73	9.50
RANSAC + transformation [ms]	55.55	57.34
Total [ms]	141.70	142.56
FPS [Hz]	7.05	7.01

Another crucial situation that needs to be tackled robustly and properly by an electronic white cane is the detection of staircases. Figure 6.13 depicts the performance of RANSAC and v disparity. By combining both approaches, the staircases are identified correctly, and the user is warned in time.

Finally, Fig. 6.14 shows a visualization of the Conservative Polar Histogram (bottom right corner). The Conservative Polar Histogram triggers a vibrator or acoustic warning in case an obstacle reaches a defined distance-threshold.

Table 6.1 summarizes performance data of the most crucial steps of the ground plan detection. According to our measurements, a frame rate of 7 FPS can be achieved. Furthermore, the Conservative Polar Histogram algorithm runs an update rate of more than 50 FPS.

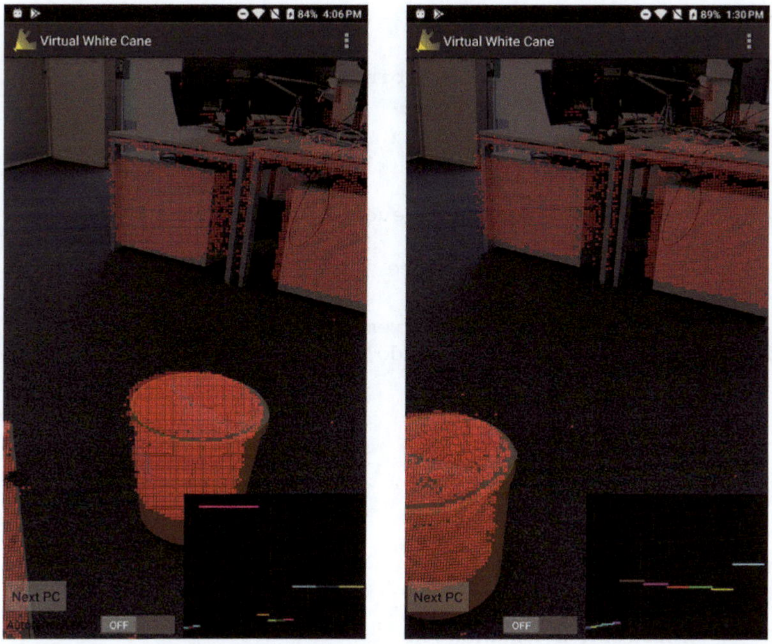

Fig. 6.14 Two scenes and their visualized Conservative Polar Histogram in the bottom right corner. Images obtained from [1], (© IEEE 2018), reprinted with permission

6.5 Conclusions

This chapter presents the results of a virtual white cane demonstration that exploits the innovative features of the Time-of-Flight 3D imaging technology and its integration into recent smartphones.

Our virtual white cane prototype runs with 7 FPS, implements two novel concepts, the combined v disparity/RANSAC ground plane detection and the Conservative Polar Histogram, and thus considerably improves state-of-the-art: real-life tests showed that our virtual white cane prototype robustly detects obstacles and warns the visually impaired person in time.

A.1 Appendix: Calculation of the Conservative Polar Histogram

The following pseudo code depicts the calculation of the Conservative Polar Histogram.

```
foreach point pt_ogl
   //project to 2D
```

```
pt_2 = (pt_ogl.x, pt_ogl.z);

// calculate angle beta and bin number
// e.g. ALPHA = 5 degrees
beta = atan(pt_2.x / pt_2.z);
k = beta/ALPHA;

// calculate distance and pseudo disparity
d = dot(pt_2, pt_2);
d_pseudo = floor(0.2/distance * 255);

// use k and d_pseudo to increment the histogram value
disparityHistogram[k][d_pseudo]++;
```

Glossary

BOVW	Bag-of-Visual-Words
ETA	Electronic Travel Aid
EOA	Electronic Orientation Aid
FOV	Field of View
GPGPU	General Purpose Graphics Processing Unit
HRTF	Head-Related Transfer Function
PLD	Position Locator Device
RANSAC	RANdom SAmple Consensus
RGB-D	Color and Depth sensing Camera
SURF	Speeded Up Robust Feature
SVM	Support Vector Machine
ToF	Time-of-Flight

Acknowledgements This project has received funding from the Electronic Component Systems for European Leadership Joint Undertaking under grant agreement No 692480. This Joint Undertaking receives support from the European Union's Horizon 2020 research and innovation program and Germany, Netherlands, Spain, Austria, Belgium, and Slovakia.

References

1. N. Druml, T. Pietsch, M. Dielacher, C. Steger, M. Baumgart, C. Consani, T. Herndl, G. Holweg, Virtual white cane featuring time-of-flight 3D imaging supporting visually impaired users, in *2018 21st Euromicro Conference on Digital System Design (DSD)*, (2018), pp. 450–457
2. D. Dakopoulos, N.G. Bourbakis, Wearable obstacle avoidance electronic travel aids for blind: A survey. IEEE Trans. Syst. Man Cybern. Part C Appl. Rev. **40**(1), 25–35 (2010)
3. D. Bernabei, F. Ganovelli, M.D. Benedetto, M. Dellepiane, R. Scopigno, A low-cost time-critical obstacle avoidance system for the visually impaired, in *International Conference on Indoor Positioning and Indoor Navigation*, (2011), pp. 21–23

4. M. Vlaminck, L. Jovanov, P. Van Hese, B. Goossens, W. Philips, A. Pizurica, Obstacle detection for pedestrians with a visual impairment based on 3D imaging, in *2013 International Conference on 3D Imaging*, (2013)
5. J.M. Coughlan, A. Yuille, Manhattan world: Compass direction from a single image by Bayesian inference, in *Proceedings of the Seventh IEEE International Conference on Computer Vision*, vol. 2, (1999)
6. J. Poppinga, N. Vaskevicius, A. Birk, K. Pathak, Fast plane detection and polygonalization in noisy 3D range images, in *2008 IEEE/RSJ International Conference on Intelligent Robots and Systems, IROS*, (2008), pp. 3378–3383
7. M. Brock, P.O. Kristensson, Supporting blind navigation using depth sensing and sonification, in *Proceedings of the 2013 ACM Conference on Pervasive and Ubiquitous Computing Adjunct Publication—UbiComp '13 Adjunct*, (2013)
8. A. Bhowmick, S. Prakash, R. Bhagat, V. Prasad, S. M. Hazarika, IntelliNavi: Navigation for blind based on kinect and machine learning, in *Multi-disciplinary Trends in Artificial Intelligence* (2014)
9. R. Jafri, *A GPU-accelerated real-time contextual awareness application for the visually impaired on Google's project Tango device* (2016)
10. R. Labayrade, D. Aubert, J.-P. Tarel, *Real Time Obstacle Detection in Stereovision on Non Flat Road Geometry Through "v-disparity" Representation*, vol 2 (2002), pp. 646–651
11. A. Broggi, C. Caraffi, R.I. Fedriga, P. Grisleri, Obstacle detection with stereo vision for off-road vehicle navigation, in *2005 IEEE Computer Society Conference on Computer Vision and Pattern Recognition (CVPR'05)-Workshops*, (2005), p. 65
12. N. Soquet, D. Aubert, N. Hautiere, *Road Segmentation Supervised by an Extended V-Disparity Algorithm for Autonomous Navigation* (2007), pp. 160–165
13. B. Musleh, A. De La Escalera, J.M. Armingol, U-V disparity analysis in urban environments, in *Lecture Notes in Computer Science*, vol. 6928, (2012), pp. 426–432
14. Y. Gao, X. Ai, J. Rarity, N. Dahnoun, Obstacle detection with 3D camera using U-V-disparity, in *International Workshop on Systems, Signal Processing and Their Applications, WOSSPA*, (2011)
15. B. Peasley, S.T. Birchfield, Real-time obstacle detection and avoidance in the presence of specular surfaces using an active 3D sensor, in *2013 IEEE Workshop on Robot Vision (WORV)*, (2013)

Chapter 7
Power Consumption Minimization of Wireless Sensor Networks in the Internet of Things Era

Jordi Serra, David Pubill, and Christos Verikoukis

7.1 Introduction

WSN are key components of the upcoming IoT revolution, where billions of devices with sensing, communications, and actuating capabilities will set up a huge geo-distributed perception system to interact with the background. Thereby, IoT creates a plethora of applications such as air quality monitoring in smart cities and self-driving cars. However, IoT benefits may be impaired by the limited energy budget of battery-powered WSN devices. Energy constraints limit the WSN devices lifetime, which leads to a performance loss in an IoT application. Moreover, the limited energy budget provokes the battery replenishment of a huge number of WSN devices and increases the maintenance cost.

Therefore, this chapter proposes methods for an efficient management of the WSN nodes' power consumption. In the first part, two techniques are jointly considered to reduce the transmitted power of the WSN nodes to achieve a desired quality of service at the receiver side, which is measured in terms of the Mean Square Error (MSE) in the estimation of the Signal of Interest (SoI). The first of these techniques is a power allocation method at each WSN node, i.e., a weight that controls the transmitted power. Namely, this weight takes into account whether there is redundant information among WSN nodes, the quality of measurement, and the quality of the wireless link between the WSN node and the access point (AP). The second technique considers the use of multiple distributed APs at the receiver side to decrease the transmitted power. Numerical simulations have been carried out to validate the proposed methods. These simulations highlight that if power allocation at each WSN node is jointly used with multiple APs, then it is possible to reduce

J. Serra (✉) · D. Pubill · C. Verikoukis
Telecommunications Technological Center of Catalonia (CTTC), Castelldefels, Spain
e-mail: jserra@cttc.es; dpubill@cttc.es; cveri@cttc.es

W. D. van Driel et al. (eds.), *Sensor Systems Simulations*,
https://doi.org/10.1007/978-3-030-16577-2_7

drastically the WSN nodes' transmitted power required to achieve a target quality of service.

In the second part, we evaluate the use of artificial light EH to power perpetually the WSN nodes. Namely, we propose an experimental setup for the EH of a WSN node. The EH module consists of a Photovoltaic (PV) cell that harvests artificial light from a nearby lamp, a boost converter that transforms the gathered energy into the proper electrical features of the WSN node and a 3 V rechargeable coin battery that stores the electrical energy. This EH module is the power supply for a commercial WSN node, a Zolertia Z1 WSN node. A proper configuration was found to power perpetually the WSN node in three different scenarios: (1) no activity in the mote, (2) performing a simple computation, and (3) transmitting a temperature reading over the air to another WSN node. To achieve this aim, several transmission time periods have been considered and it has been concluded that it is possible to extend the WSN node lifetime without bound, when the transmission time period is above a certain threshold. Moreover, it has been studied which is the most appropriate rechargeable coin battery for the setup.

7.2 Employing Multiple Access Points and a Power Allocation to Reduce the Transmit Power in WSN

7.2.1 Introduction

WSN have been widely used for decentralized inference tasks. Moreover, they are of paramount importance in disruptive technologies such as IoT. WSN are composed of sensor nodes, which have a limited energy budget, as they are usually battery powered. Thus, the design of inference problems in WSN incorporates the energy efficiency as an objective. Herein, the energy efficient estimation of a source vector observed by a WSN is considered. Several algorithms have been proposed to tackle the energy efficient estimation problem. The authors in [1] studied the quantization of the WSN data measurements. A method based on sensor selection was introduced in [2]. It minimizes the estimation error subject to an energy budget constraint. Data reduction algorithms, which rely on a linear compression of sensor nodes' data, were proposed in [3, 4] bearing in mind its influence on the estimation error. The authors in [5] considered a multiple antenna Fusion Centre (FC) scenario and proposed a power allocation at each sensor node to minimize the MSE subject to power constraints.

Actually, Energy Efficiency (EE) is a cornerstone in future wireless communications systems [6]. Namely, the capacity needs to increase dramatically without increasing the current power consumption substantially to avoid economical and societal issues. Thus, new system architectures were proposed recently to increase the EE through a network densification, i.e., increasing the number of access points or communication links per unit area [6, 7]. One of them is collocated massive

Multiple Input Multiple Output (CM-MIMO), which relies on a large number of antennas at the base station (BS) [8]. This system architecture was recently studied in WSN to reduce the transmitted power in estimation tasks. Namely, considering a coherent Multiple Access Channel (MAC), [9, 10] dealt with the Maximum Likelihood (ML) and Linear Minimum Mean Square Error (LMMSE) estimation of a scalar parameter at the FC, respectively. Within this setting [11] studied the LMMSE estimation of a correlated source vector.

Herein, multiple distributed access points and a power allocation technique are jointly used to reduce dramatically the WSN transmit power for a given estimation quality at the receiver side. The idea is that increasing the number of virtual antennas at the fusion center leads to reduce the transmit power of the WSN. Thereby, herein the problem of estimating a correlated source vector in a Distributed Massive-MIMO (DM-MIMO) architecture is tackled and an LMMSE estimator is proposed to achieve this aim. Namely, a WSN obtains noisy measurements of the source vector and sends them through a coherent MAC channel to the FC. The FC is made of a large number of single antenna APs, which are connected to a Central Processing Unit (CPU) by means of a backhaul network. The LMMSE estimation is performed at the CPU. Moreover, a power allocation technique is considered at each WSN node to minimize the total transmitted power subject to an estimation error constraint. Thereby unlike the related works in [9–11] a DM-MIMO system is proposed herein to carry out the estimation. These type of systems, a.k.a. large distributed antenna systems (L-DAS), have been considered recently to increase dramatically the EE in 5G networks, see e.g., [7] or [12]. The contributions are as follows. First, the MSE associated to the LMMSE estimation is obtained as well as an approximation of it, bearing in mind the DM-MIMO context. This paves the way to obtain the optimal power allocation that minimizes the total transmitted power for a given estimation error as an SDP problem. The numerical simulations show that in comparison to [11], which considers a CM-MIMO, the total WSN transmitted power is reduced for a given MSE constraint, both in the uniform and the optimal power allocation problems. Moreover, the work herein is more general than the one in [9, 10] as a vector source is tackled in the estimation rather than a scalar. Last but not least, it is important to say that the proposed power allocation method can still be used in a distributed MIMO system, with a moderate number of virtual antennas. The difference is that in this case the power allocation is suboptimal, as it is derived assuming a large number of virtual antennas.

7.2.2 System Model

The proposed system model is depicted in Fig. 7.1. More specifically, a WSN composed of L sensor nodes that takes noisy observations of a source vector $\boldsymbol{\theta}$ is considered. The vector of measurements $\mathbf{x}$ is linearly modelled as

$$\mathbf{x} = \boldsymbol{\theta} + \mathbf{n},$$

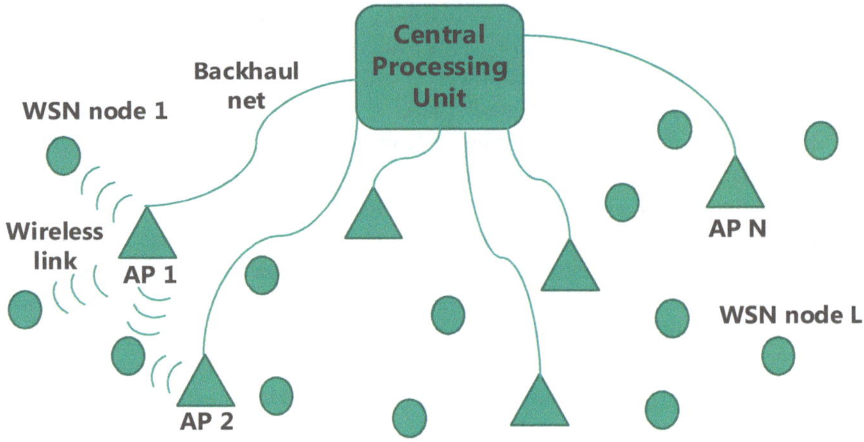

Fig. 7.1 Proposed DM-MIMO system for decentralized estimation of a source. © IEEE 2018, reprinted with permission

where $\mathbf{n}$ is the sensing noise. Both $\boldsymbol{\theta}, \mathbf{n} \in \mathbb{C}^L$ are modeled as circularly symmetric complex Gaussian random vectors, i.e., $\boldsymbol{\theta} \sim (\mathbf{0}, \mathbf{C}_{\boldsymbol{\theta}})$ and $\mathbf{n} \sim \mathcal{CN}(\mathbf{0}, \mathbf{C_n})$, with $\mathbf{C_n} = \text{diag}\left(\sigma_{n1}^2, \ldots, \sigma_{nL}^2\right)$.[1] Power allocation is implemented at each sensor node, i.e., the l-th node multiplies its observation by a complex gain α_l and sends it through a coherent MAC channel to the FC.

The FC consists of a large number N of geographically distributed single antenna APs, which are assumed to serve simultaneously all the sensor nodes. The APs are connected to a CPU through a backhaul network. Then, the CPU performs the estimation of $\boldsymbol{\theta}$ given the available WSN measurements stacked in $\mathbf{y}$. Herein it is assumed that $N \gg L$, thereby this is a DM-MIMO system, a.k.a. L-DAS or cloud radio access network (C-RAN) if the CPU is implemented within the cloud computing paradigm. As it is demonstrated below, the assumption $N \gg L$ permits to derive an approximation of the MSE in the estimation of the parameter of interest. That approximation permits to derive an optimal power allocation to minimize the transmit power for a given estimation error constraint in the DM-MIMO setting. As it will be highlighted below, in a distributed MIMO scenario with a moderate number of virtual antennas the power allocation could still be used although the power allocation would be suboptimal. DM-MIMO has been recently proposed to increase the EE in future wireless communications systems, see e.g., [7] or [12]. Herein, DM-MIMO is considered within the context of wireless sensing to assess its potential to reduce the transmitted power of the WSN for a given estimation error constraint. The next linear model is considered for the received signal at the CPU,

[1]From now on $\text{diag}(x_1, \ldots, x_n)$ expresses a diagonal matrix with elements $x_1, \ldots, x_n$ in its main diagonal.

$$\mathbf{y} = \mathbf{HDx} + \mathbf{v} \tag{7.1}$$

where $\mathbf{D} = \mathrm{diag}(\alpha_1, \ldots, \alpha_L)$ is the matrix of gains to be designed, i.e., the one implementing the power allocation. $\mathbf{v} \sim \mathcal{CN}(\mathbf{0}, {\sigma_v}^2\mathbf{I})$ is an Additive White Gaussian Noise (AWGN) and $\mathbf{H} \in \mathbb{C}^{N \times L}$ is the channel between the sensor nodes and the APs. Note that the l-th column of $\mathbf{H}$ is the channel between the l-th sensor node and each of the APs and it is modeled as follows:

$$\mathbf{H} = (\mathbf{h}_1, \ldots, \mathbf{h}_L)\,, \quad \mathbf{h}_l = {\boldsymbol{\Gamma}_l}^{1/2}\mathbf{f}_l \tag{7.2}$$

where ${\boldsymbol{\Gamma}_l}^{1/2}$ is a diagonal matrix modeling the large-scale fading, i.e., $\boldsymbol{\Gamma}_l \triangleq \mathrm{diag}(\gamma_{1l}, \ldots, \gamma_{Nl})$, $\gamma_{nl} = {d_{nl}}^{-2\beta}$, d_{nl} is the distance between the l-th sensor node and the n-th AP and β is the pathloss exponent. The small-scale fading is modeled by $\mathbf{f}_l$ whose components are i.i.d random variables, namely $[\mathbf{f}_l]_i \sim \mathcal{CN}(0, 1)\ \forall i = 1, \ldots, N$.

Remark 1 A set of assumptions are made in the proposed DM-MIMO system. Firstly, as stated in, e.g., [12], a perfect backhaul network is presumed between the APs and the CPU. Secondly, perfect synchronization is assumed between the WSN nodes and the APs. This allows a coherent reception at the FC, similarly than in [11]. Thirdly, the FC has perfect knowledge of the channel and noise statistics.

Remark 2 The channel model in Eq. (7.2) highlights that the proposed DM-MIMO generalizes the CM-MIMO system in [11]. That is, when all the antennas are collocated at a given AP, then the distance between the l-th sensor node and any of the antennas is approximately the same, i.e., the elements of ${\boldsymbol{\Gamma}_l}^{1/2}$ in Eq. (7.2) can be approximated as $\gamma_{nl} = \gamma_l\ \forall n = 1, \ldots, N$. This leads to obtain the channel model for a CM-MIMO proposed in Eq. (3) of [11].

7.2.3 *Optimal Power Allocation in a DM-MIMO System*

7.2.3.1 LMMSE Estimation and Asymptotic MSE Approximation

This section treats the LMMSE estimation of the SoI $\boldsymbol{\theta}$ for the system architecture proposed in Sect. 7.2.2. That is, the signal received at the CPU has the model stated in Eq. (7.1) with the assumptions described in *Remark* 1. Furthermore, recall that the LMMSE estimation of $\boldsymbol{\theta}$ is given by the conditional expectation $\hat{\boldsymbol{\theta}} = \mathrm{E}\,[\boldsymbol{\theta}|\mathbf{y}, \mathbf{H}]$ see Chap. 15 in [13]. Thereby, in a DM-MIMO setting $\hat{\boldsymbol{\theta}}$ has the next expression

$$\hat{\boldsymbol{\theta}} = \mathbf{E}\,[\boldsymbol{\theta}|\mathbf{y}, \mathbf{H}] = \left(\mathbf{C}_{\boldsymbol{\theta}}^{-1} + \mathbf{D}^H\mathbf{H}^H\mathbf{C}_w^{-1}\mathbf{H}\mathbf{D}\right)^{-1}\mathbf{D}^H\mathbf{H}^H\mathbf{C}_w^{-1}\mathbf{y} \tag{7.3}$$

After easy manipulations involving the matrix inversion lemma, one concludes that the MSE achieved by $\hat{\boldsymbol{\theta}}$ has the next expression,

$$\mathrm{MSE}=\mathrm{Tr}\left[\left(\mathbf{C}_{\theta}^{-1}+\sigma_{v}^{-2}\mathbf{D}^{H}\mathbf{H}^{H}\mathbf{H}\mathbf{D}-\sigma_{v}^{-4}\mathbf{D}^{H}\mathbf{H}^{H}\mathbf{H}\mathbf{D}\times\left(\mathbf{C}_{n}^{-1}+\sigma_{v}^{-2}\mathbf{D}^{H}\mathbf{H}^{H}\mathbf{H}\mathbf{D}\right)^{-1}\mathbf{D}^{H}\mathbf{H}^{H}\mathbf{H}\mathbf{D}\right)^{-1}\right]. \tag{7.4}$$

Next, an asymptotic approximation of the MSE is obtained. This has several benefits. Firstly, a more compact MSE expression is obtained. Secondly, the randomness of the channel due to the small-scale fading vanishes. This is importan t for the channel estimation task, as the estimation must be updated less frequently. This is because the large-scale fading varies much more slowly than the small-scale fading. Thirdly, the MSE approximation paves the way for the design of the optimal power allocation within a convex optimization problem, see Sect. 7.2.3.2. To achieve the desired MSE approximation, let us consider the next lemma,

Lemma 1 *Consider the asymptotic regime where $N \to \infty$ and L remains fixed. Then, $\mathbf{H}^H\mathbf{H}$, with $\mathbf{H}$ defined in (7.2), converges in probability to the next expression,*

$$\mathbf{H}^{H}\mathbf{H} \xrightarrow{p} \mathrm{diag}\left(\sum_{n=1}^{N}\gamma_{n1},\ldots,\sum_{n=1}^{N}\gamma_{nL}\right) \tag{7.5}$$

Proof First, the expression (7.2) leads to the next equality, $[\mathbf{H}^H\mathbf{H}]_{ll} = \mathbf{f}_l^{\mathrm{H}}\boldsymbol{\Gamma}_l\mathbf{f}_l$. Bearing in mind the expression of $\boldsymbol{\Gamma}_l$ in (7.2) and defining $\mathbf{f}_l = (f_{1l}, \ldots, f_{Nl})^T$, $[\mathbf{H}^H\mathbf{H}]_{ll}$ can be reformulated as

$$\left[\mathbf{H}^{H}\mathbf{H}\right]_{ll} = \gamma_{1l}|f_{1l}|^2 + \cdots + \gamma_{Nl}|f_{Nl}|^2 \tag{7.6}$$

Next, to ease the notation, we introduce S_{ll}, x_n, and μ_n, which have the next definitions,

$$S_{ll} = \left[\mathbf{H}^{H}\mathbf{H}\right]_{ll}, \quad S_{ll} = \sum_{n=1}^{N} x_n, \quad \mu_n = E\left[x_n\right].$$

Thereby, assuming that $\mathrm{var}(x_n) \le c < \infty$ and x_n are independent $\forall n$, by the Chebyshev's inequality the next inequalities hold for any real number $\epsilon > 0$,

$$\Pr\left[|N^{-1}S_{ll} - N^{-1}\sum_{n=1}^{N}\mu_n| \ge \varepsilon\right] \le \frac{\sum_{n=1}^{N}\mathrm{var}\left(x_n\right)}{N^2\varepsilon^2} \le \frac{c}{N\varepsilon^2}$$

Note that $\mu_n = [\gamma_{nl}|f_{nl}|^2] = \gamma_{nl}$, as $f_{nl} \sim (0, 1)$. Thereby, taking $N \to \infty$ in both sides of the last expression, the next convergence in probability is obtained,

$$\lim_{N\to\infty} \Pr\left[|N^{-1}S_{ll} - N^{-1}\sum_{n=1}^{N}\gamma_{nl}| \geq \varepsilon\right] = 0 \tag{7.7}$$

The same procedure can be followed to demonstrate that the off diagonal elements of $\mathbf{H}^H\mathbf{H}$ converge to 0. To this end, observe that $[\mathbf{H}^H\mathbf{H}]_{lk} = \mathbf{f}_l^H{\mathbf{\Gamma}_l}^{1/2}{\mathbf{\Gamma}_k}^{1/2}\mathbf{f}_k$ and bear in mind that $\mathbf{f}_l$, $\mathbf{f}_k$ are independent with zero mean elements.

Lemma 1 paves the way to obtain the desired approximation for the MSE. This is formally stated in the next theorem.

Theorem 1 Let define the diagonal matrix $\mathbf{B}$ with diagonal elements $b_l = |\alpha_l|^2\sum_{n=1}^{N}\gamma_{nl}$ for $l = 1, \ldots,$ L. Then, when $N \to \infty$ for a fixed L, the MSE in (7.4) converges in probability to $\overline{\text{MSE}}$, whose expression is given next,

$$\text{MSE} \xrightarrow{\text{p}} \overline{\text{MSE}} = \text{Tr}\left[\left(\mathbf{C}_\theta^{-1} + \sigma_v^{-2}\mathbf{B} - \sigma_v^{-4}\mathbf{B} \times \left(\mathbf{C}_n^{-1} + \sigma_v^{-2}\mathbf{B}\right)^{-1}\mathbf{B}\right)^{-1}\right]. \tag{7.8}$$

Proof Equation (7.8) is obtained by applying Lemma 1 into Eq. (7.4).

7.2.3.2 Optimal Power Allocation

The energy is a limited resource in a WSN and the proposed DM-MIMO architecture permits to reduce notably the WSN power consumption by having a large number of virtual antennas at the FC. Thus, this section faces the problem of minimizing the total transmitted power P_t of the WSN nodes for a given performance constraint in the estimation procedure, within the DM-MIMO context proposed herein.[2] This problem is stated as,

$$\begin{aligned} &\underset{\{\text{b}_\text{l}\geq 0\}_{\text{l}=1}^{\text{L}}}{\text{minimize}} && P_t\left(b_l\right) \\ &\text{subject to} && \overline{\text{MSE}}\left(b_l\right) \leq \delta \end{aligned} \tag{7.9}$$

where δ is the maximum allowable estimation error in the estimation task at hand and the MSE approximation, i.e., $\overline{\text{MSE}}\left(b_l\right)$ in Eq. (7.8), is considered. This will pave the way to reformulate Eq. (7.9) as a convex optimization problem. Furthermore,

[2] It is worth saying that in a distributed MIMO scenario with a moderate number of APs, the transmitted power of the WSN can still be reduced. Actually, the power allocation proposed in this section still holds in that scenario, but it is suboptimal, as the MSE approximation in Eq. (7.8) requires a large number of APs.

according to Sect. 7.2.2, $P_t(b_l)$ has the expression,

$$P_t\left(b_l\right) = \mathbb{E}\left[\|\mathbf{D}\left(\boldsymbol{\theta}+\mathbf{n}\right)\|^2\right] Tr\left[\mathbf{D}^H\mathbf{D}\left(\mathbf{C}_{\boldsymbol{\theta}}+\mathbf{C}_{\mathbf{n}}\right)\right] \tag{7.10}$$

Now, let us define $\tilde{\boldsymbol{\Gamma}}$ as

$$\tilde{\boldsymbol{\Gamma}} = \text{diag}\left(\sum_{n=1}^{N}\gamma_{n1},\ldots,\sum_{n=1}^{N}\gamma_{nL}\right)$$

Given this definition of $\tilde{\boldsymbol{\Gamma}}$ and bearing in mind the expression of b_l in Theorem 1 leads to express $P_t(b_l)$ as follows:

$$P_t\left(b_l\right) = \text{Tr}\left[\tilde{\boldsymbol{\Gamma}}^{-1}\mathbf{B}\left(\mathbf{C}_\theta+\mathbf{C}_n\right)\right] \tag{7.11}$$

Thus, the optimization problem Eq. (7.9) can be reformulated as follows after substituting Eqs. (7.11) and (7.8) into it,

$$\begin{array}{ll}\underset{\mathbf{B}\geq 0}{\text{minimize}} & \text{Tr}\left[\tilde{\boldsymbol{\Gamma}}^{-1}\mathbf{B}\left(\mathbf{C}_\theta+\mathbf{C}_n\right)\right] \\ \text{subject to Tr}\left[\left(\mathbf{C}_\theta^{-1}+\sigma_v^{-2}\mathbf{B}-\sigma_v^{-4}\mathbf{B}\times\left(\mathbf{C}_n^{-1}+\sigma_v^{-2}\mathbf{B}\right)^{-1}\mathbf{B}\right)^{-1}\right]\leq\delta & \end{array}\cdot \tag{7.12}$$

Next, after introducing slack variables **X**, **Z** and by considering the Schur complement of a block matrix, (7.12) can be rewritten as follows:

$$\begin{array}{ll}\underset{\mathbf{B},\mathbf{X},\mathbf{Z}\geq 0}{\text{minimize}} & \text{Tr}\left[\tilde{\boldsymbol{\Gamma}}^{-1}\mathbf{B}\left(\mathbf{C}_\theta+\mathbf{C}_n\right)\right] \\ \text{subject to} & \text{Tr}\left[\mathbf{Z}\right]\leq\delta \\ & \begin{pmatrix}\mathbf{C}_\theta^{-1}+\sigma_v^{-2}\mathbf{B}-\sigma_v^{-4}\mathbf{X} & \mathbf{I}\\ \mathbf{I} & \mathbf{Z}\end{pmatrix}\geq 0\cdot \\ & \begin{pmatrix}\mathbf{X} & \mathbf{B}\\ \mathbf{B} & \mathbf{C}_n^{-1}+\sigma_v^{-2}\mathbf{B}\end{pmatrix}\geq 0\end{array} \tag{7.13}$$

This is an SDP problem and thereby can be solved efficiently, i.e., in polynomial time, by software packages relying on interior point methods.

Remark 3 The comparison of the proposed algorithm in eq. (7.13) to a uniform power allocation algorithm is of paramount importance. This is presented in Sect. 7.2.4. In the latter case, $\mathbf{D} = \alpha\mathbf{I}$ and thus $\mathbf{B} = |\alpha|^2\tilde{\boldsymbol{\Gamma}}$ and $P_t = \text{Tr}[|\alpha|^2(\mathbf{C}_{\boldsymbol{\theta}}+\mathbf{C}_{\mathbf{n}})]$ in Eq. (7.13).

7.2.4 Numerical Results

Next, we evaluate the power allocation algorithm obtained in Eq. (7.13), which seeks to minimize the total transmitted power of a WSN in a DM-MIMO scenario, for a given estimation error constraint. The simulation parameters considered for the experiments are explained next. An exponential covariance matrix model is assumed for the source covariance, i.e., $[\mathbf{C}_{\boldsymbol{\theta}}]_{ij} = \rho_\theta{}^{|i-j|}$ with $0 \leq \rho_\theta \leq 1$. The sensing and AWGN noise variances, i.e., $\sigma_v{}^2$, $\sigma_{nl}{}^2$, are detailed below. The number of WSN nodes is $L = 15$. In the coefficients modeling, the large-scale fading, i.e., $\gamma_{nl} = d_{nl}{}^{-2\beta}$, the pathloss exponent $\beta = 1$. Both the WSN nodes and the APs, in the DM-MIMO system, are within a square area with side length $a = 160$, and their coordinates are generated randomly according to a uniform distribution between 0 and a at each iteration of the simulation, which leads to obtain d_{nl}. For the CM-MIMO architecture, the same scenario is considered, though the antennas of the FC are fixed at the center of the square area. Thereby, $\gamma_l = d_l{}^{-2\beta}$, see *Remark* 2 above for further insights. Finally, δ in Eq. (7.13) is set to the middle of the interval $(\text{Tr}[(\mathbf{C}_{\boldsymbol{\theta}}{}^{-1} + \mathbf{C}_{\mathbf{n}}{}^{-1})^{-1}], \text{Tr}[\mathbf{C}_{\boldsymbol{\theta}}]]$, which characterizes its possible values according to [11].

In Fig. 7.2, the asymptotic approximation of the MSE proposed in Eq. (7.8) is evaluated. We consider a WSN with a total transmitted power of $P_t = 10^{-2}$ W, $\rho_\theta = 0.95$, $\sigma_{nl}{}^2 = 10^{-3}$ W $\forall l$, and $\sigma_v{}^2 = 10^{-6}$ W. Furthermore, a uniform power

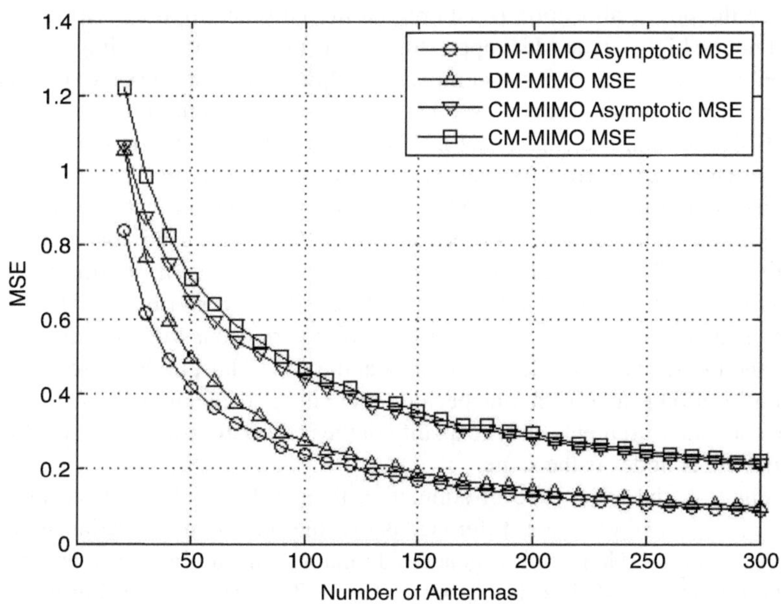

Fig. 7.2 Comparison between MSE in and asymptotic MSE in of a DM-MIMO and a CM-MIMO.

allocation (UPA) is implemented, i.e., $\mathbf{D} = \alpha\mathbf{I}$. Thus, according to Eq. (7.10), $|\alpha|^2 = P_t/(L([\mathbf{C}_\theta + \mathbf{C}_n]_{ll}))\ \forall\ l$. Furthermore, a Monte Carlo simulation with 100 iterations is considered. At each iteration the coordinates of the WSN nodes and APs are generated randomly as explained above, this obtains a given $\mathbf{H}$. With these simulation conditions, we plot the exact MSE in Eq. (7.4) and its approximation in Eq. (7.8) obtained by the LMMSE in a DM-MIMO setting. For comparison purposes, we display the exact and approximated MSE obtained by the LMMSE in a CM-MIMO context as well. These MSE expressions were obtained in [11] and are a particular case of Eqs. (7.4) and (7.8) due to the *Remark* 2 explained above. Figure 7.2 highlights that $\overline{\text{MSE}}$ tends to MSE when N increases for a fixed L as predicted in Theorem 1 and for $N > 100$ on the approximation is tight. Observe that for a UPA setting, the performance of the proposed LMMSE in a DM-MIMO outperforms the one of the LMMSE in a CM-MIMO proposed in [11]. The rationale relies on its distributed nature, as a DM-MIMO setting permits to shorten the distance between the WSN nodes and the virtual antennas of the FC compared to a CM-MIMO scheme. Moreover, there is more diversity in the links as it is highlighted in the corresponding channel models, see Eq. (7.2) and *Remark* 2.

Figure 7.2 also shows that for a given transmitted power of the WSN, the estimation quality improves as the number of virtual antennas or APs increases. Furthermore, the MSE approximation is rather tight even for a moderate number of antennas. Thus, the proposed power allocation algorithm in Sect. 7.2.3.2 could be applied in a distributed MIMO scenario where the number of virtual antennas or APs is moderate, compared to the number of WSN transmitting nodes. That is, although the power allocation is suboptimal in that scenario, the performance loss should not be high, as the MSE approximation is rather tight according to Fig. 7.2.

In Figs. 7.3, 7.4, and 7.5, we compare the Optimal Power Allocation (OPA) and UPA in a DM-MIMO scenario (see [8] and *Remark* 3) to the OPA and UPA in a CM-MIMO setting, see Eq. (7.12) in [11]. In Fig. 7.3, the SoI is an uncorrelated source vector, ${\sigma_v}^2 = 10^{-3}$ W and ${\sigma_{nl}}^2 = 10^{-6}$ W $\forall l$. Figure 7.3 shows that the total transmitted power of the WSN, for a given estimation error constraint, can be drastically reduced in the proposed DM-MIMO setting compared to a CM-MIMO setting. The rationale is similar to the one of Fig. 7.2. Furthermore, Fig. 7.3 highlights that both the UPA and OPA obtain similar performance. The reason is that the source is uncorrelated, the sensing noise variance is the same among WSN nodes and the AWGN variance is the same at the APs. Thus, in this case the optimal power allocation is a uniform one because there is no redundant information among the sensors' measurements and the quality of the information arriving at the FC from different WSN nodes is the same.

In Fig. 7.4, the same type of simulation than in Fig. 7.3 is carried out. Now, ${\sigma_v}^2 = 10^{-3}$ W and the main difference is that the sensing noise variance is varied in the WSN nodes. Namely, it is generated randomly according to the next uniform distribution $\sigma_{n_l}^2 \sim \mathrm{U}(0.5 \times 10^{-3}, 0.9 \times 10^{-3})$. This variable sensing noise leads the OPA to obtain better performance than the UPA both for the DM-MIMO and CM-MIMO settings, as a higher weight is assigned to the nodes with better sensing information. Furthermore, in this setting the OPA DM-MIMO and UPA DM-MIMO

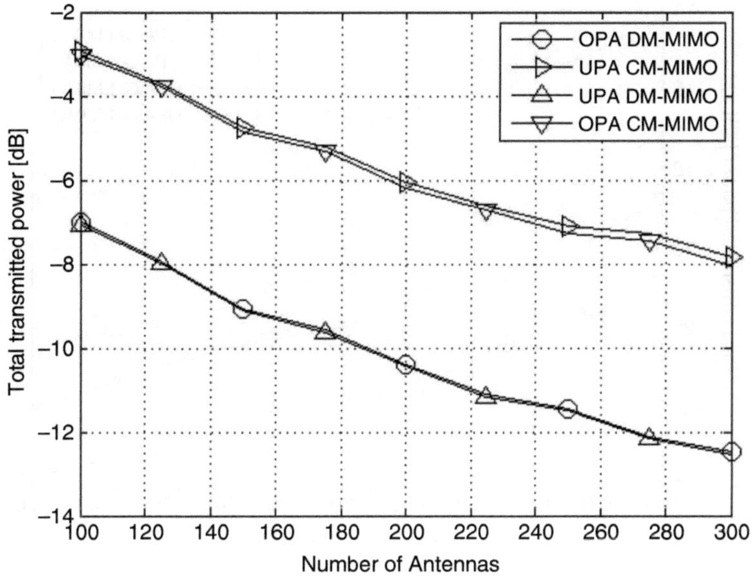

Fig. 7.3 Comparison for uncorrelated source vector (i.e., $\rho_\theta = 0$). © IEEE 2018, reprinted with permission

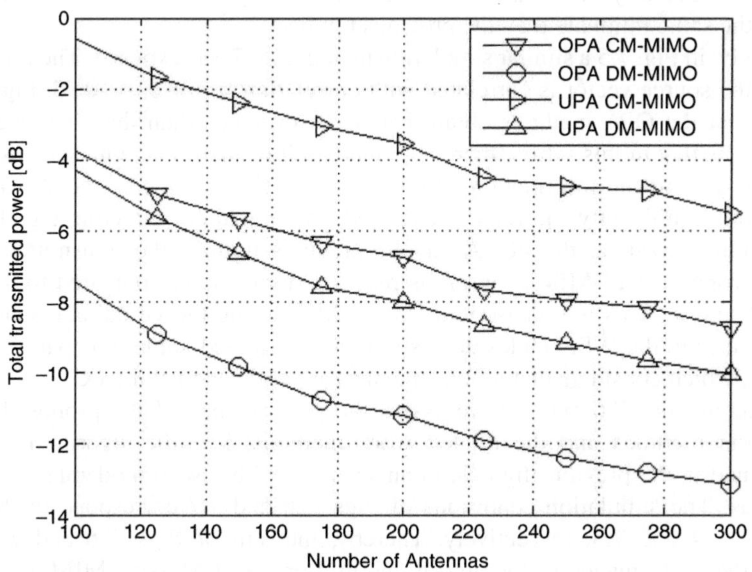

Fig. 7.4 Comparison for uncorrelated source vector (i.e., $\rho_\theta = 0$) and variable sensing noise variance ($\sigma_{n_l}^2 \sim U(0.5 \times 10^{-3}, 0.9 \times 10^{-3})$). © IEEE 2018, reprinted with permission

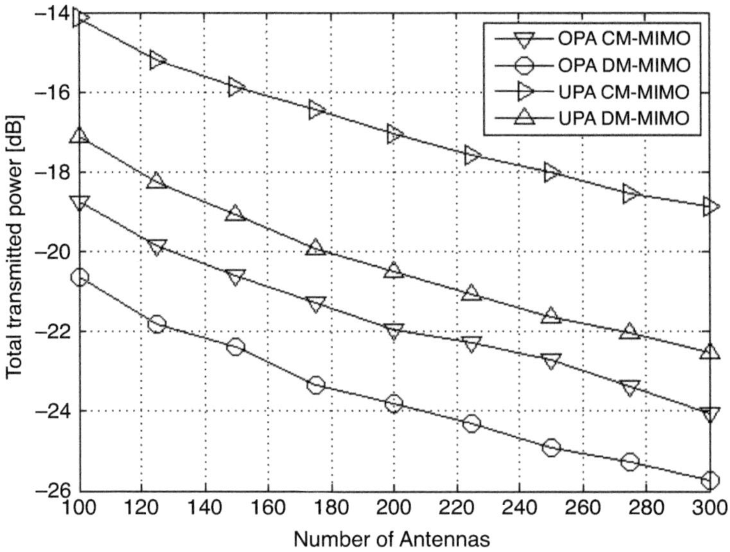

Fig. 7.5 Comparison for correlated source vector with $\rho_\theta = 0.95$. © IEEE 2018, reprinted with permission

reduce significantly the total transmitted power compared to the CM-MIMO case due to the same rationale than the previous figures.

Finally, in Fig. 7.5 a similar simulation than in Fig. 7.3 is exposed. The difference is that the source vector is correlated with correlation factor $\rho_\theta = 0.95$. Figure 7.5 shows that the OPA methods obtain better performance than the UPA ones. The rationale is that in this setting there is redundant information among sensors. Thus, one can discard the information or lower the weight at some of the WSN nodes to reduce significantly the total transmitted power without affecting significantly the resulting MSE at the FC. As in the previous figures, the transmitted power in the proposed DM-MIMO setting is reduced significantly compared to the CM-MIMO setting. This is because in a DM-MIMO setting the virtual antennas lower the distance to the WSN nodes thanks to their distributed nature compared to CM-MIMO, which considers that all the antennas are collocated at the FC.

Next, in Figs. 7.6 and 7.7 we assess the performance of the proposed power allocation methods and the related work methods, i.e., the algorithms already mentioned in the previous figures, in an scenario with low to moderate number of antennas. The simulation conditions of Figs. 7.6 and 7.7 correspond to the ones of Figs. 7.4 and 7.5, respectively. Thereby, the aim of Figs. 7.6 and 7.7 is to assess the performance of the algorithms in a practical Massive MIMO scenario, where the number of receiving antennas is not so large, i.e., moderate, and in a classical MIMO architecture situation where the number of antennas is low. Note that in these situations, the presented methods are suboptimal, as all of them are based on an asymptotic approximation of the MSE that assumes the number of

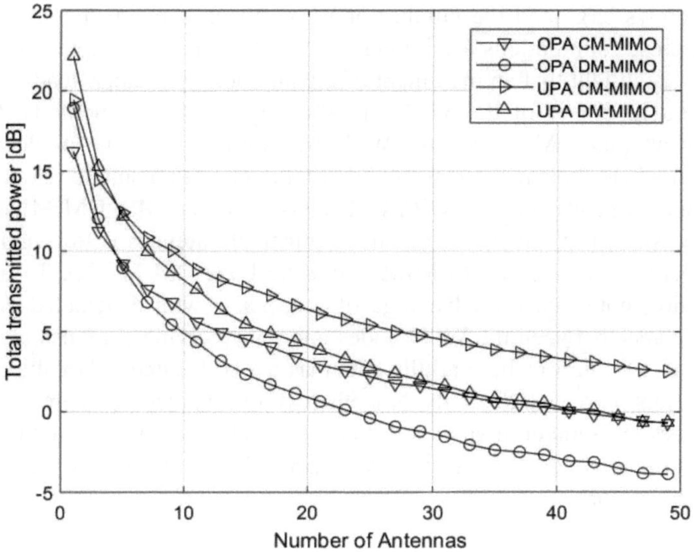

Fig. 7.6 Comparison for uncorrelated source vector (i.e., $\rho_\theta = 0$) and variable sensing noise variance ($\sigma_{n_l}^2 \sim U(0.5 \times 10^{-3}, 0.9 \times 10^{-3})$), for a small to medium number of antennas

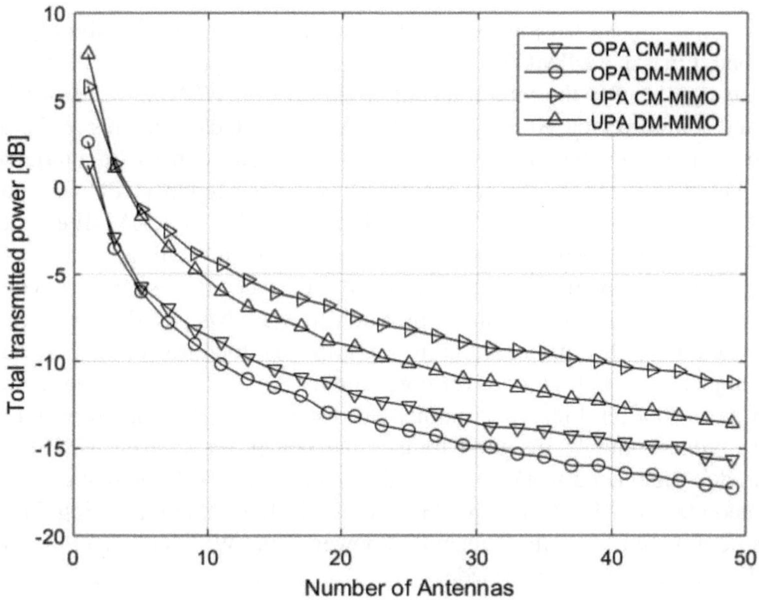

Fig. 7.7 Comparison for correlated source vector with $\rho_\theta = 0.95$, for a small to medium number of antennas

antennas grows large and the number of WSN nodes is fixed, see e.g., Eq. (7.8). However, theses simulations are interesting for the sake of completeness. Figures 7.6 and 7.7 highlight that an increment of the number of antennas leads to decrease the transmitted power of the WSN significantly in all the methods, for a low number of antennas. Moreover, the WSN transmitted power keeps decreasing in the interval where their number is moderate, though with a smaller slope. Another interesting comment is that the OPA CM-MIMO and the OPA DM-MIMO tend to provide the same performance when the number of antennas is low, and the same happens between the UPA CM-MIMO and the UPA DM-MIMO. To understand this behavior, note that the advantage of a scenario with distributed antennas is that the distance between the WSN nodes and the receiving antennas is shortened, compared to a scenario where all the antennas are collocated. Also, the distributed scenario offers more link diversity. Nonetheless, those advantages are noticeable as one increases the number of antennas, thereby for a low number of antennas one can expect that the methods developed for a distributed scenario converge to the ones of a collocated scenario.

7.2.5 Conclusions

This work has considered jointly the use of a large number of distributed APs, i.e., a distributed massive MIMO scenario, and the use of a power allocation technique to reduce significantly the total transmitted power of a WSN in the estimation of a correlated source vector. Namely, a WSN takes noisy measurements of the SoI and transmit them via a coherent MAC channel to a FC. This consists of a large number of distributed APs, or virtual antennas, which are connected to a CPU via a backhaul network. We have presented the exact MSE associated to an LMMSE estimation at the CPU of the FC. Then, leveraging the massive MIMO setting, an approximated MSE has been obtained, which converges to the exact MSE when L remains fixed and N grows without bound. This approximated MSE has paved the way to design an optimal power allocation method to minimize the total transmitted power of the WSN subject to an estimation error constraint at the FC. The obtained method can be cast as an SDP problem. The numerical results show that the proposed method leads to reduce notably the total transmitted power for a given MSE constraint compared to the related work where all the antennas are collocated at the FC, i.e., a collocated massive MIMO setting. Also our work shows that increasing the number of APs permits to reduce the transmitted power of the WSN and maintain the same estimation error at the receiver side.

7.3 Harvesting Artificial Light Indoors to Power Perpetually a WSN Node

7.3.1 Introduction

EH systems capture energy from different environmental sources. A variety of EH sources exist depending on the scenario and the application, e.g., light, motion, vibration, wind, heat, or electromagnetic waves [14]. Moreover, EH systems transform the harvested energy into electrical energy and store it in a battery. A power management subsystem, within the EH module, defines a proper electrical interface to power a WSN device with the harvested energy. Thereby, EH is the cornerstone to extend the WSN devices' lifetime. This work investigates EH for indoors and, in particular, we focus on EH from artificial light. EH in indoor environments is a challenging scenario since the average power that can be harvested from ambient light, is 10 to 100 times lower than in outdoor environments [15].

Several works deal with EH from ambient light in the literature. In [15], a hybrid EH module is proposed, which gathers energy from both ambient and thermal light. They propose a single power management unit for both sources of energy. The authors in [16] characterize the indoors light energy availability and they develop energy allocation algorithms for EH devices. The work in [17] considers WSN devices that harvest ambient light energy, both from solar and artificial light sources. Each WSN device is equipped with presence and light sensors, and their measurements are used by a central controller to adapt the illumination of the indoor light system. In [18], the authors characterize the indoor light conditions not only in terms of light intensity, but also in terms of its spectral information. This leads to a precise estimation of the output power that a photovoltaic panel can obtain from the given ambient light conditions.

Herein we demonstrate experimentally energy neutrality in WSN devices based on energy harvested from artificial light. The proposed solution consists of an EH module that powers a WSN source node which is transmitting data to a WSN sink node. The EH module consists of a PV cell that harvests artificial light from a nearby lamp, a boost converter that transforms the gathered energy into the proper electrical features of the WSN node and a 3 V rechargeable coin-type battery that stores the electrical energy.

7.3.2 Related Work

Several related works have been presented in the literature though some of them focus on harvesting solar light in outdoor scenarios. In [19], the authors present a low cost solar harvesting system constituted by WSN MICAz nodes and the MAX1724EZK33 boost converter, by Maxim Integrated, which achieves a perpetual power supply. Unlike in [19], this work focuses in an indoor environment.

In [20], the authors study the performance of different PV cells. To test the cells, they design a prototype based on the Texas Instruments bq25504 boost converter. In our work, the use of the same boost converter is different, here it is used to design an EH module to power a WSN node. Finally, in [21] this device is used to build an EH WSN device. Their conclusion is that the consumed power is higher than the harvested one. Herein we present a way of powering perpetually a WSN node in three different scenarios: (1) no activity in the data acquisition device, (2) performing a simple computation, and (3) periodically transmitting temperature measurements to another device. We consider several transmission periods and we conclude that it is possible to extend the device's lifetime when the frequency of the transmissions is above a certain threshold. The rest of the sections are organized as follows. Section 7.3.3 describes two setups, the proposed photovoltaic cell setup and a baseline one, which is used as a reference to compare the results obtained. Section 7.3.4 presents in detail the components of the setups and Sect. 7.3.5 shows the experimental results. Finally, Sect. 7.3.6 concludes our work.

7.3.3 Artificial Light EH Setup

In this section, the block diagram of the proposed EH system and two setups based on it are presented. Figure 7.8 shows the EH block diagram, it consists of a Raspberry Pi, a WSN node, a PV cell, a boost converter, and a 3 V coin battery. As it can be seen in the figure, the EH module is constituted by the cell, the boost converter, and the battery.

Figure 7.9 displays the photovoltaic cell setup, which is constituted by three WSN nodes (source, sink, and auxiliary), the EH module, and a Raspberry Pi. The function of the WSN source node is to send temperature measurements to the WSN

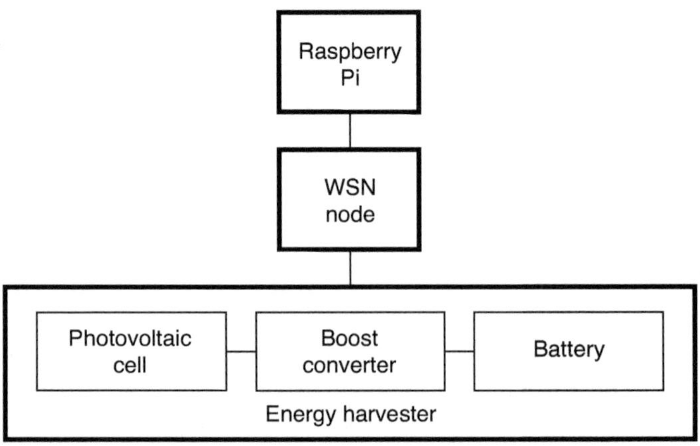

Fig. 7.8 Energy harvesting block diagram

Fig. 7.9 Photovoltaic cell setup

sink node while the WSN auxiliary node collects the battery level and, finally, the Raspberry Pi gathers the data for later analysis. All the experiments of this work have been carried out indoors.

The light generated by a common desk lamp is the energy source used to recharge the battery. According to the manufacturer, the power of the lamp is around 50 watts and 400 lumens. Taking into account that a lux is the unit of illuminance equal to 1 lumen per square meter and that the distance between the lamp and the photovoltaic cell was around 25 cm for all the experiments, Eq. (7.1) can be used to calculate the illuminance in lux, $E_v(lx)$

$$E_v(lx) = \frac{\varphi_v(lm)}{4\pi r^2} \tag{7.14}$$

where $\varphi_v(lm)$ is the illuminance in lumens and r is the spherical radius which corresponds to the distance between the lamp and the PV cell. Therefore, the illuminance is around 500 lux (Fig. 7.10).

In order to compare the results of the experiments with the photovoltaic cell setup, a baseline setup (see Fig. 7.11) has been defined. It consists of three WSN nodes, a 3 V coin battery, and a Raspberry Pi. In this case, the 3 V battery is connected directly to the WSN source node without considering EH. The objective of the baseline scenario is to characterize the discharge behavior of the battery for different operation modes of the WSN source node:

1. **No activity:** No activity in the WSN source node. This scenario represents a lower bound on the energy consumption of the WSN device.

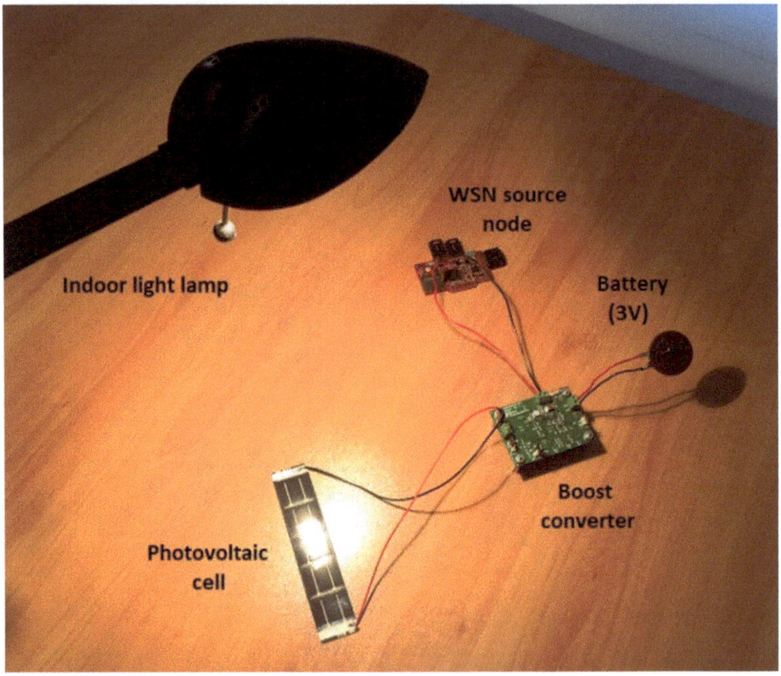

Fig. 7.10 Indoor light lamp and photovoltaic cell

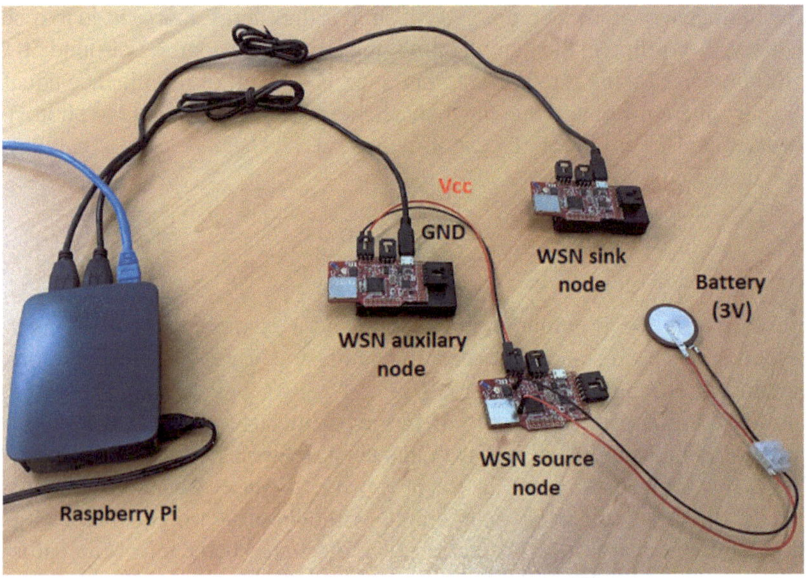

Fig. 7.11 Baseline setup

2. **Computing**: A simple computation is running in the WSN source node; it increments a sequence number in an infinite loop.
3. **Computing + transmission (Computing + Tx)**: The RF transceiver of the WSN source node is active. The WSN source node transmits data to the WSN sink node every certain period.

For both setups, the voltage output pin of the WSN source node (V_{cc}) is connected to the data input pin of the WSN auxiliary node in order to collect the battery level. It should be noted that the maximum battery level, according to the battery used, is around 3 V. However, when it is discharging with the time, the voltage on the output V_{cc} pin drops. These data is transferred and stored on the Raspberry Pi. Note that the WSN sink node is only used for the "Computing+Tx" scenario since it is used as a receiver of the temperature measurements transmitted by the WSN source node.

7.3.4 Setup Components

In this section, we explain more in detail the components of the setups proposed in Section 7.3.3 and their features.

7.3.4.1 WSN Node

The WSN nodes used are the Zolertia Z1 devices (see Fig. 7.12) [22]. The Z1 is a low power wireless module compliant with IEEE 802.15.4 and Zigbee protocols. It is equipped with a second-generation low-power microcontroller, which features a 16-bit RISC CPU @16 MHz clock speed, a built-in clock factory calibration, 8 KB RAM, and 92 KB flash memory. It includes the CC2420 transceiver, operating at

Fig. 7.12 Z1 WSN node

2.4 GHz frequency band with a data rate of 250 Kbps. It supports Contiki OS, an open-source operating system for the IoT, which connects tiny, low-cost, low-power microcontrollers to the Internet. Each Z1 device can operate as either a source or a sink node. According to [22], the consumption in transmission mode is around 17.4 mA for a transmission power of 0 dBm, 18.8 mA in reception mode, less than 10 mA when the microcontroller is active and the consumption of the memory circuit is up to 15 mA. The Z1 can be powered with two AA batteries (1.5 V each one). However, in this work the EH module has been used to power the device.

7.3.4.2 Raspberry Pi 3

Figure 7.13 shows a Raspberry Pi 3, which is a single-board computer with wireless LAN and Bluetooth connectivity. It is equipped with a Quad Core 1.2 GHz Broadcom BCM2837 64 bit CPU and 1 GB RAM. Its role is mainly to gather data from the WSN source node's lifetime for the different scenarios and setups.

7.3.4.3 Rechargeable Coin Batteries

Two 3 V coin-type rechargeable lithium batteries of 50 mAh (VL2330) and 100 mAh (VL3032) capacity have been chosen, they are provided by Panasonic [23] (see Fig. 7.14). Although there are smaller capacity batteries in the market, they are not considered because they cannot assume the power consumption of the WSN node and the boost converter.

7.3.4.4 Boost Converter

Figure 7.15 shows the boost converter used for the photovoltaic cell setup; it is the bq25504 evaluation module by Texas Instruments. According to [24], the

Fig. 7.13 Raspberry Pi 3 Model B

Fig. 7.14 Rechargeable coin batteries

Fig. 7.15 Boost converter

VBAT_OK parameter defines high/low thresholds programmed at 2.8 and 2.4 V, respectively. A VBAT_OK high signal would typically indicate that the battery is ready to be used. If it were low, it would indicate that the battery is discharged. Therefore, the system load should be reduced or disabled.

7.3.4.5 Photovoltaic Cell

Figure 7.16 shows the MP3-25 photovoltaic cell by PowerFilm Solar [25]. It is created by depositing amorphous silicon on a thin plastic substrate, operates at 3 V and generates a current of 31 mA. Its dimensions are 24 × 114 mm. This module is lightweight, paper-thin, and durable. To connect it to the boost converter, a cable has been welded to every tin-coated copper tape.

Fig. 7.16 Photovoltaic cell

7.3.5 Experimental Results

This section is structured in two subsections. The first subsection considers the baseline setup and shows the performance of both 50 and 100 mAh capacity batteries. The objective of these experiments is to evaluate both batteries and decide whether the 50 mAh capacity battery is enough to achieve a perpetual power supply of the source WSN node or the 100 mAh capacity battery is needed. Then, once decided the most appropriate battery, the second set of experiments demonstrates how we achieve the goal of this work, evaluating the battery for the different scenarios described in Sect. 7.3.3.

7.3.5.1 Baseline Setup Experiments

These experiments have been performed using the baseline setup described in Sect. 7.3.3 (see Fig. 7.11). The objective here is to study the behavior of the batteries without considering the EH module; therefore, connecting directly the battery to the WSN source node. The operating range of the WSN device microcontroller ranges from 1.8 to 3.6 V, this means that below 1.8 V the WSN source node may not work properly.

Figures 7.17 and 7.18 shows the baseline results for the 50 and 100 mAh capacity battery for “No activity,” “Computing,” and “Computing + Tx” (with a transmission period of 5 s) scenarios. As it can be seen if the activity in the WSN source node increases, the operating time of the battery decreases. Namely, the power consumption due to the communication resources and computational usage is greater in the “Computing+Tx” scenario than in the “Computing” one and the same happens in the “Computing” scenario compared to the “No activity” one. In fact, Fig. 7.17 shows the performance of the 50mAh battery, it can be appreciated

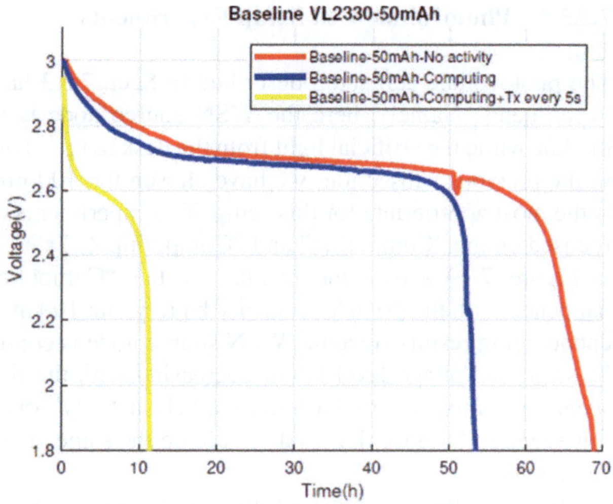

Fig. 7.17 Baseline setup, 50 mAh, 3 scenarios

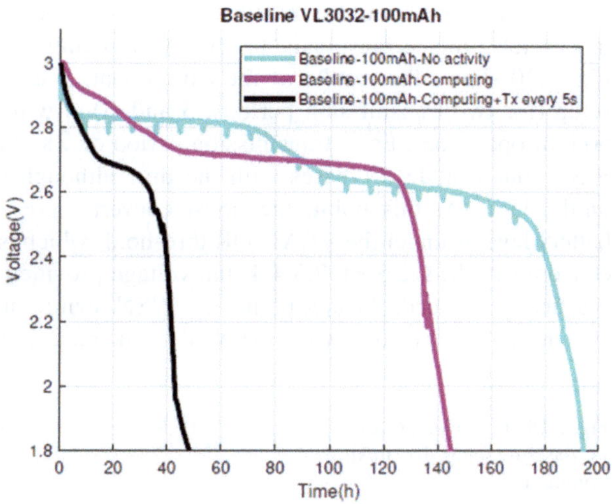

Fig. 7.18 Baseline setup, 100 mAh, 3 scenarios

that there is an important difference between the "No activity" scenario (69 h of operation) and the "Computing + Tx" (11 h). The last result makes us glimpse that with the 50 mAh battery it will be very difficult to obtain an infinite power supply for the WSN source node; as it can be seen in Fig. 7.17, the voltage drops very quickly for the "Computing+Tx" scenario. Therefore, the 100 mAh capacity battery is considered as the best candidate for the photovoltaic cell setup. Figure 7.18 shows that the operating time using the 100 mAh capacity battery is significantly better for all the scenarios. Thus, the next set of experiments considers only the 100 mAh capacity battery and compares the results obtained for the photovoltaic cell setup with the ones obtained for the baseline one.

7.3.5.2 Photovoltaic Cell Setup Experiments

The photovoltaic cell setup described in Sect. 7.3.3 has been considered for these experiments. Namely, here the WSN source node is powered, thanks to the EH module while the artificial light from the desk lamp is being harvested. As concluded in the previous subsection, we have chosen the 100 mAh capacity battery since it is the most appropriate for this setup. The experiments of this subsection have been focused on the "Computing" and "Computing + Tx" scenarios.

Figure 7.19 shows the results for the "Computing" scenario. Unlike what happened with the 50 mAh capacity battery, the 100 mAh capacity battery presents encouraging results since the WSN source node is continuously powered. As it can be seen the voltage level is not decreasing with the time. Since the "Computing" scenario is more restrictive than the "No activity" one, it is not necessary to test this scenario because the result would be the same, a perpetual supply of the WSN source node.

However, the objective of this work was to find the scenario in which a WSN source node could transmit data to a WSN sink node every certain periods indefinitely, this corresponds to the "Computing+Tx" scenario. In this sense, Fig. 7.20 shows the results of the experiment. It compares the photovoltaic cell setup (for two transmission periods: 5 and 10 s) with the baseline setup (around 50 h of operation). For a transmission period of 5 s, the voltage at the input of the WSN source node decreases with the time although its operating time increases until 135 h. At this point, the boost converter presents a constraint. Once the battery level is under the VBAT_OK threshold (which is a functionality of the boost converter itself, see Sect. 7.3.4.4) the voltage provided to the WSN source node is drastically reduced. At this point, the WSN source node stops working properly and the communication with the WSN sink node is interrupted. However, for a

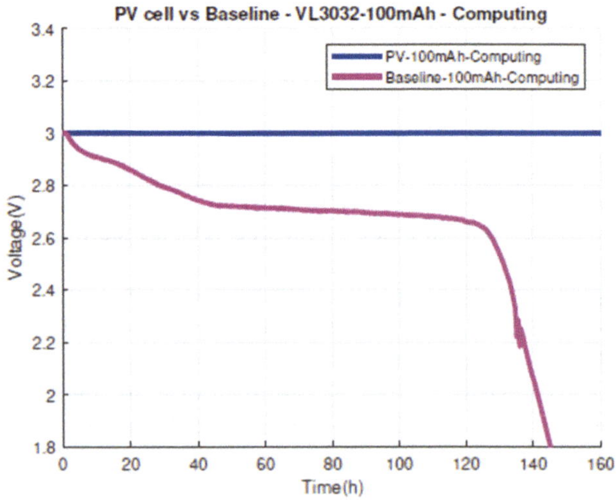

Fig. 7.19 Photovoltaic cell vs. Baseline setup, 100 mAh, Computing

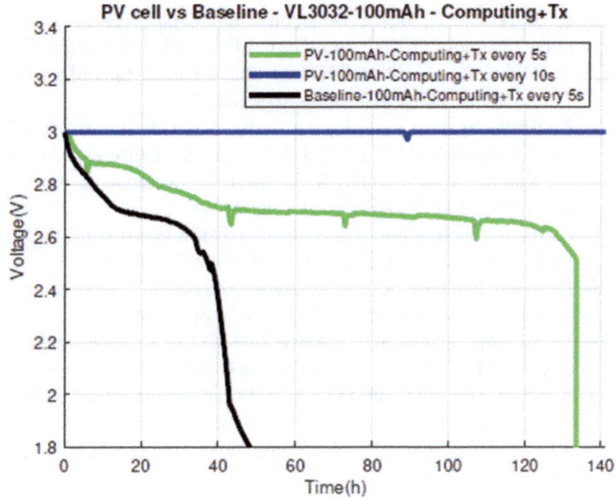

Fig. 7.20 Photovoltaic cell vs. Baseline setup, 100 mAh, Computing + Tx

transmission period of 10 s the WSN source node is powered perpetually. Other experiments with higher transmissions periods have been performed with the same result. We can conclude that the threshold (i.e., the transmission period between two consecutive transmissions) above which the WSN source node can be perpetually powered is around 10 s.

7.3.6 Conclusions

This work presents a setup able to power perpetually a WSN node using an EH module with a 100 mAh capacity battery in an indoor environment for a transmission period equal or greater than 10 s. The energy source for the energy harvester is a common desk lamp, which is 25 cm from the PV cell generating an illuminance of 500 lux. In future work, more tests can be done increasing the distance between the PV cell and the source of light, for instance, using other sources of energy such as fluorescent lights.

References

1. Z. Luo, Universal decentralized estimation in a bandwidth constrained sensor network. IEEE Trans. Inform. Theor. **51**, 2210–2219 (2005)
2. S. Liu et al., Sensor selection for estimation with correlated measurement noise. IEEE Trans. Signal Proc. **64**, 3509–3522 (2016)
3. I. Schizas et al., Distributed estimation using reduced-dimensionality sensor observations. IEEE Trans. Signal Proc. **55**, 4284–4299 (2007)

4. A. Shirazinia, S. Dey, Power-constrained sparse gaussian linear dimensionality reduction over noisy channels. IEEE Trans. Signal Proc. **63**, 5837–5852 (2015)
5. A. Behbahani, A. Eltawil, H. Jafarkhani, Decentralized estimation under correlated noise. IEEE Trans. Signal Proc. **62**, 5603–5614 (2014)
6. E. Bjronson et al., Deploying dense networks for maximal energy efficiency: Smmall cells meet massive MIMO. IEEE J. Select. Areas Commun. **34**, 832–847 (2016)
7. M. Kamel, W. Hamouda, A. Youssef, Ultra-dense networks: A survey. IEEE Commun. Surv. Tutor. **18**, 2522–2545., Fourth quarter (2016)
8. T. Marzetta, Noncooperative cellular wireless with unlimited numbers of base station antennas. IEEE Trans. Wirel. Commun. **9**, 3590–3600 (2010)
9. F. Jiang, J. Chen, A. Swindlehurst, Phase-only analog encoding for a multi-antenna fusion center, in *Proc. IEEE ICASSP*, (2012), pp. 2645–2648
10. A. Swindlehurst et al., Massive MIMO for wireless sensing with a coherent multiple access channel. IEEE Trans. Signal Proc. **63**, 3005–3017 (2015)
11. A. Shirazinia et al., Massive MIMO for decentralized estimation of a correlated source. IEEE Trans. Signal Proc. **64**, 2499–2512 (2016)
12. T.L. Marzetta et al., Cell-free massive MIMO versus small cells. IEEE Trans. Wirel. Commun. **16**, 1834–1850 (2017)
13. S. Kay, *Fundamentals of Statistical Signal Processing: Estimation Theory* (Prentice-Hall, Upper Saddle River, NJ, 1993)
14. O. Cetinkaya, O. Akan, Electric-field energy harvesting from lighting elements for battery-less internet of things. IEEE Access **5**, 7423–7434 (2017)
15. Y. Tan, S. Panda, Energy harvesting from hybrid indoor ambient light and thermal energy sources for enhanced performance of wireless sensor nodes. IEEE Trans. Indus. Electr. **58**, 4424–4435 (2011)
16. M. Gorlatova et al., Networking low-power energy harvesting devices: Measurements and algorithms. IEEE Trans. Mob. Comput. **12**, 1853–1865 (2013)
17. S. Li, A. Pandharipande, Networked illumination control with distributed light-harvesting wireless sensors. IEEE Sens. J. **15**, 1662–1669 (2015)
18. X. Ma, S. Bader, B. Oelmann, Characterization of indoor light conditions by light source classification. IEEE Sens. J. **17**, 3884–3891 (2017)
19. L.J. Chien et al., A simple solar energy harvester for wireless sensor networks, in *6th International Conference on Intelligent and Advanced Systems (ICIAS)*, (2016), pp. 1–6
20. M. Rasheduzzaman et al., A study of the performance of solar cells for indoor autonomous wireless sensors, in *10th International Symposium on Communication Systems, Networks and Digital Signal Processing (CSNDSP)*, (2016), pp. 1–6
21. A.E. Xhafa et al., Towards a perpetual wireless sensor node. IEEE Sens. J., 1–4 (2013)
22. "Z1 datasheet, Zolertia." http://zolertia.sourceforge.net/wiki/images/e/e8/Z1_RevC_Datasheet.pdf
23. "Coin type rechargeable lithium batteries (VL series), Panasonic." https://industrial.panasonic.com/ww/products/batteries/primary-batteries/lithium-batteries/coin-type-rechargeable-lithium-batteries-vl-series?reset=1
24. "bq25504 EVM—Ultra Low Power Boost Converter with Battery Management for Energy Harvester Applications, Texas Instruments." http://www.ti.com/lit/ug/sluu654a/sluu654a.pdf
25. "MP3–25 Photovoltaic cell, PowerFilm." http://www.powerfilmsolar.com/products/?&show=product&productID=271534&productCategoryIDs=6573

Chapter 8
Model-Based Design of Secured Power Aware Smart Sensors

Thomas Wolfgang Pieber, Thomas Ulz, and Christian Steger

8.1 Introduction

Sensors can be found everywhere in our society. They measure the environment, gather useful data to navigate robots through the environment, and enable machines to sense the environment they want to interact with and record the changes they cause. To get better at these versatile tasks, new sensors are developed steadily.

To develop new sensors efficiently, simulations of the intended use case are created and model simulations of the new sensor are placed in the simulated environment. Research prototypes can then be used to verify the simulation results. If the simulation results do not resemble the real measurements, the prototypes can be used to search for errors in the simulation and gain new insights in the processes of the use case.

This model-based design approach of sensors can be used to efficiently create optimized sensors for any use case.

The act of sensing, as well as the processing, storing, securing, and transmitting of the gathered data requires energy in the form of electric power. This energy can be provided by various methods. For many sensors this energy is provided by electro-chemical batteries or is gathered from the environment by energy harvesting methods. These forms of energy provisioning have in common that the sensor needs to manage its available power carefully as either the energy is limited or only a limited amount of power can be extracted from the environment. When designing a sensor that should be operated with one of these methods, it is necessary to know the characteristics of the component parts of the sensor and the possibilities of the energy provisioning system with high accuracy. These parameters can influence the

T. W. Pieber (✉) · T. Ulz · C. Steger
Institute for Technical Informatics, Graz, Austria
e-mail: thomas.pieber@tugraz.at; thomas.ulz@tugraz.at; steger@tugraz.at

W. D. van Driel et al. (eds.), *Sensor Systems Simulations*,
https://doi.org/10.1007/978-3-030-16577-2_8

design of the sensor massively. That means that the components of the sensor need to match the requirements for the intended use case and the energy provisioning needs to be capable of providing sufficient energy to operate the sensor for the intended lifetime and operations. To work efficiently the sensor has to be aware of its power usage and the current capabilities of the energy provisioning system.

The data that is generated by the sensors is mostly seen as not harmful to the process they monitor. This can be seen as the most common approach to data security of sensor data is "*What can be measured can be seen by anyone*". This notion of data security is now seen as a fallacy as the *STUXNET* [9] and *HAVEX* [4] attacks have been noted. Furthermore, as the *Mirai botnet* [2] shows, unsecured sensors also pose a threat to other targets. To circumvent these threats, measures to secure the system and the generated data need to be taken.

To prevent the new sensor to be exploited in such way, the security of the system and data is of utmost importance, and thus, must not be neglected during the design of the new sensor system.

The paragraphs above describe a sensor that can not only generate data about the environment, but also about itself and take informed decisions. It can furthermore modify the data and perform security relevant operations on it. Beyond that, it can receive and send such data to form a network. It also knows its current power status and can perform actions to prolong its lifetime or let the network know if the remaining energy gets too low.

Such a sensor that is able to perform informed measures on itself, the data, or the network can be called a smart sensor.

This chapter follows the design process shown in Fig. 8.1.

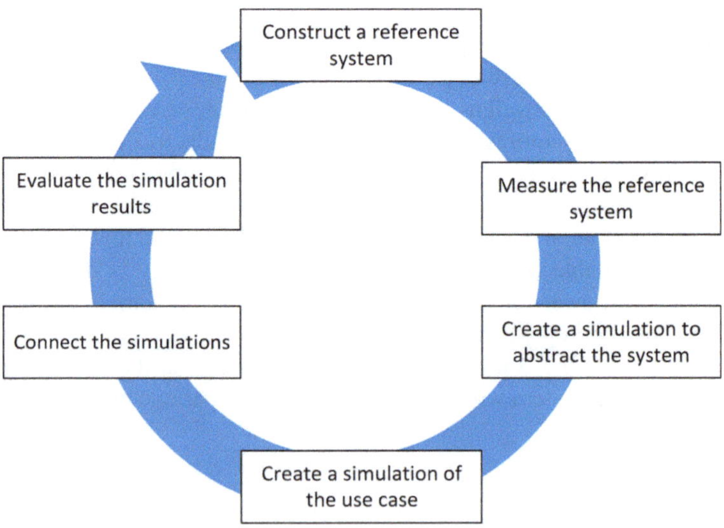

Fig. 8.1 Concept of a model based design process for a secured power aware smart sensor

- A reference system is created that shows all processes we want to model.
- The reference system is measured to understand the processes.
- A simulation of the system is created to abstract the system.
- The use case is modelled and simulated.
- The simulation of the new system is connected to the simulation of the use case.
- The results of the simulations are evaluated. If the results are not as expected, the process is restarted.

In this chapter we use this design approach to create a sensor that is read and charged by using NFC technology. To do so autonomous robots in an *Industry 4.0* inspired setting will communicate to the sensors and connect them to a network.

8.2 Related Publications

Simulation Background
The use of simulation to predict sensor functionality and the methods to get the parameters of the sensor components have been described in many publications.

In this publication we want to use the robotic simulation tool Gazebo [14] as a core system. The connection of a Matlab environment to Gazebo has been published on the official website [12]. This tutorial shows how to connect Matlab to Gazebo via the ROS (Robot Operating System) interface to send and receive data from the Gazebo simulation.

Further publications connect the Gazebo simulator with different software for machine learning [26]. Also here the software is connecting to the operating system controlling the robot and influences the path the robot takes to reach a goal.

SystemC [1] is a hardware description language based on C++. It is capable of describing the hardware on different levels of abstraction. This is especially useful as a detailed description of all components results in a very slow, albeit accurate, simulation. This is furthermore useful if developing a simulation from scratch as the components can be described in an abstract way and be defined in later steps. The publication of Panda shows design processes to create SystemC simulations [15].

The connection of SystemC to different simulation tools has been described in many publications [3, 7, 11, 13, 18]. In these works the SystemC simulation is primarily used as a kernel to provide the functionality of the quasi parallel execution.

The connection of SystemC to Matlab was designed by Bouchhima et al. [3] As Matlab works with discrete events, they created this cosimulation to additionally simulate continuous events.

To speed up the SystemC simulations, Huang et al. [7] describe a possibility to spread the simulation across a network of computers.

SystemC can also be connected to a simulation tool designed for integrated circuit simulation [11]. Martin et al. connected it to a SPICE (Simulation Program with Integrated Circuit Emphasis) simulator as well as to a VHDL (Very high speed integrated circuit Hardware Description Language) simulation.

Mueller-Gritschneder et al. [13] used SystemC to compute physical processes that affect the robot while it is moving. This simulation therefore generates data in the SystemC simulation that is then used to model the robots behaviour. In contrast to that, our approach generates data in the environment and uses this as input for the SystemC simulation.

Pieber et al. [16] describe methods to connect a SystemC simulation to the Gazebo environment in one of their publications. This paper describes in detail how the connection between the simulations can be formed and what requirements need to be fulfilled to create a successful simulation. This publication furthermore announces possible advancements to increase the simulation speed if multiple SystemC simulations need to be run at the same time.

Power Awareness for Sensors

The idea that sensors are aware of their own power levels and power consumption has been discussed in many publications [5, 8, 10, 19, 20, 25]. Using this information, the sensors can make decisions about the routing of information inside the network, reduce the quality of the measurements to prolong the lifetime, or inform the network that the energy level is too low to actively take part in the task of the network.

Chen et al. [5] developed a small energy harvesting sensor node that is capable of providing its own energy. This sensor node relies on solar cells to charge a battery. If the charge in the battery is sufficient, the sensor wakes and performs its operations until the voltage drops below a predefined value. Then the sensor switches to a ultra-low power sleep mode. In this scenario the sensors duty cycle is provided by the capabilities of the energy harvesting system.

To calculate a duty cycle for a energy harvesting sensor system the energy usage and harvesting capabilities must be known. Kansal et al. [8] formalized these calculations in their work. These configurations can be used to initialize such system. If the observed parameters at the position of the sensor differ from the predicted ones, the team of researchers propose a dynamically changing duty cycle. Using this dynamically changing system the researchers noted a significantly improved performance of their research system.

Additionally Rahimi et al. [19] explore the possibility to expand the sensors lifetime using energy harvesting methods. In their research they try to exploit autonomously moving nodes. These nodes search in a so-called Energy Cell for the optimal spot to harvest solar energy. Afterwards the moving nodes provide this energy to the sensor nodes in their assigned cell.

Another possible energy source that can be used by the sensor nodes is RF energy from broadcasting stations. Sogorb et al. [20] use sensor nodes with two antennas to research this possibility. In their research prototype one antenna is used to harvest the energy from broadcasting stations, the other one to transmit the gathered data to a base station.

Another strategy that uses RF energy harvesting has been explored by Lee et al. [10]. This research team focused on the possibility to harvest energy using antiparallel resonant loops. The researchers explore RFID's (Radio Frequency

IDentification) capability to transmit energy. They stated that antiparallel loops can improve the efficiency of energy transmission to 87% from about 50%.

The researchers around Yan [25] explored methods to construct more efficient sensor networks. This is done by making the sensors aware of their own energy state. In this research Yan et al. implemented two levels of energy saving. "*Node-level energy saving*" adaptively regulates the transmission power of the sensor node. "*Network-level energy saving*" adaptively reconfigures the sensor networks configuration.

Security for Sensors
Security for sensor devices has been researched in the literature [6, 17, 22–24]. Most researchers however see security as a side topic to be added later to an existing sensor hardware.

The work by Ulz et al. [24] describes multiple methods how sensor data can be secured. The researchers pointed towards problems that arise if sensor data is encrypted and how these problems can be mitigated. In their publication the researchers use authenticated encryption (AE) to securely transmit data between a sensor of a control system and the controller. The researchers state that using only this technique a DoS attack is easier to perform. To counteract this problem forward error correction techniques have been proposed. Furthermore, the researchers state that a security controller can be used to perform these operations as this processing unit is specialized to perform the task of encrypting more efficiently than a regular microcontroller.

In addition to securing the data of a sensor against adversaries Ulz et al. [22] proposed methods to secure a sensor network against misuse and misconfigured sensors using a security controller. This paper presents the idea of a two-layer attestation system that first checks the validity of a sensors firmware, and in a second step validates the version of this firmware. Another publication by this researchers describes an update mechanism for such sensor system [23]. Here the researchers state that the interface to update the sensor needs to be separated of the main communication interface.

To prevent the misuse of single sensors, authentication to see and modify data is necessary. Pieber et al. [17] describe a method to use a password-based authentication method that is lightweight enough to be run on a single sensor.

Haase et al. [6] propose a system to (re-)configure a sensor system via NFC. They furthermore use cryptography to authenticate the reconfiguration device.

8.3 Obtaining Data for the Models

To get a decent simulation of any system, the processes and components of the system need to be understood. A smart sensor consists at least of the following parts:

- The sensor frontend: The part of the sensor that does the conversion from an external stimulus to digital information.

- The controller.
- The memory.
- A communication interface.
- The energy provisioning system.
- (Optional) A security controller to perform security relevant operations.

For these components data sheets are available that specify the peak current at an optimal voltage source. This is sufficient information to build the sensor, as overestimation of the components energy usage leads to a more powerful energy provisioning system. To get a better estimation of the energy demands of the system, and therefore be able to design a more optimal energy provisioning system, measurements have to be performed. This requires the development of a research prototype.

To create an efficient sensor, a model of the environment and the intended use case is useful. Such simulation need not be very detailed as this allows the simulation to be used in various ways and thus create a multitude of different stimuli for the sensor. The simulation of the environment needs at least these components:

- A representation of the new sensor.
- Objects influencing the sensor or the communication with the sensor.
- Communication partners for the smart sensor to test the used communication protocols.
- Things for the sensor to measure.

8.3.1 Measuring Data for the New Sensor

The prototype that is used to measure the sensor components should be able to vary the input voltages to emulate the energy provisioning system. This concept is shown in Fig. 8.2. Here a control program gets the values of the current consumption of the components and calculates how the voltage of the energy provisioning system reacts to this current flow. The updated settings are transmitted to the testbench where the voltage is generated. The smart sensor can now be operated with this updated voltage and the current flow changes. To have an optimal measurement of the drawn current a gain control unit can be used to set the sensor gains of the current measurement units. The measured values are given to the control program to be stored and to calculate the new settings for the testbench. The data gained from the measurements are used to construct models of the sensor components.

The data that is generated can then be used to create models of the sensor components energy usage at different voltages. This data can furthermore be used to specify the requirements for the energy provisioning system of the final sensor.

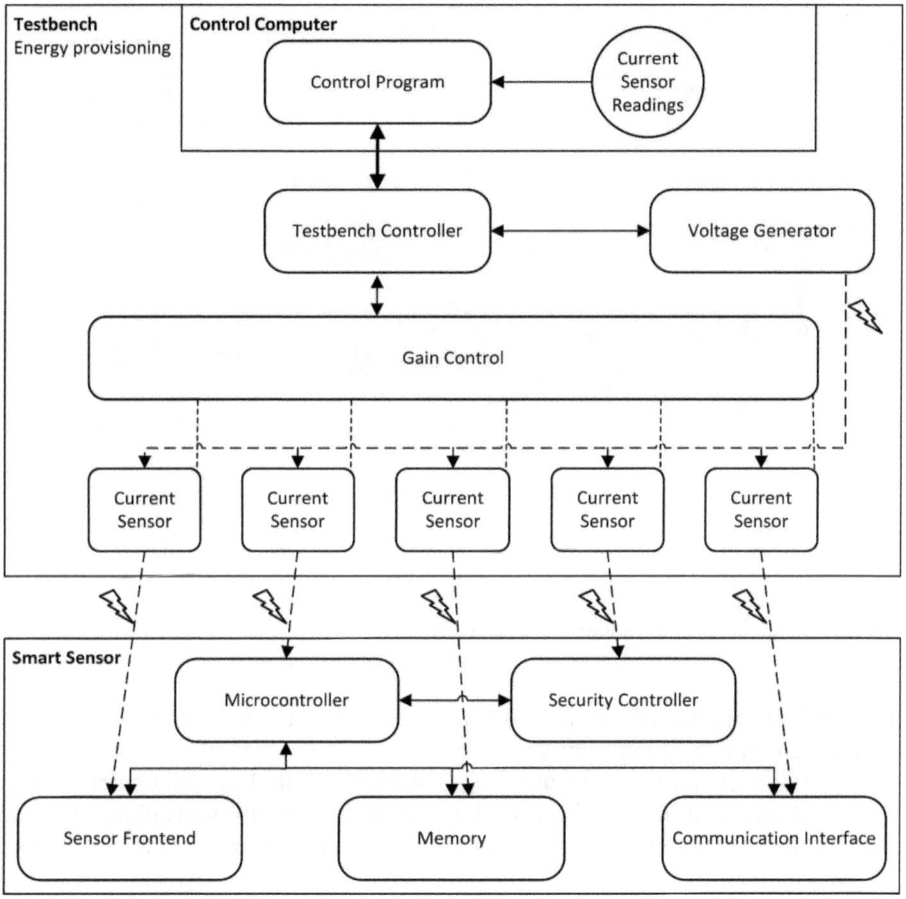

Fig. 8.2 Concept of the measurement system for a smart sensor

8.3.1.1 An Example

To create an energy efficient sensor, all components should be optimized for low energy usage. Furthermore, components that are not needed still consume energy. To counteract that, load switches can be used to cut the components off the power supply. In Fig. 8.3 the same sensor device is connected to the channels *EXT-1* and *EXT-2*. *EXT-1* is cut from the power supply with a load switch. This sensor uses 33 μW. If it is switched on, the sensor consumes 75 μW.

Such simple method can reduce the energy consumption drastically and therefore prolong the sensors lifetime.

The only component that cannot be cut from the power supply is the microcontroller. There the energy consumption can be reduced by switching into a low-power state. Figure 8.3 shows a measurement of a smart sensor. Here the sensor connected

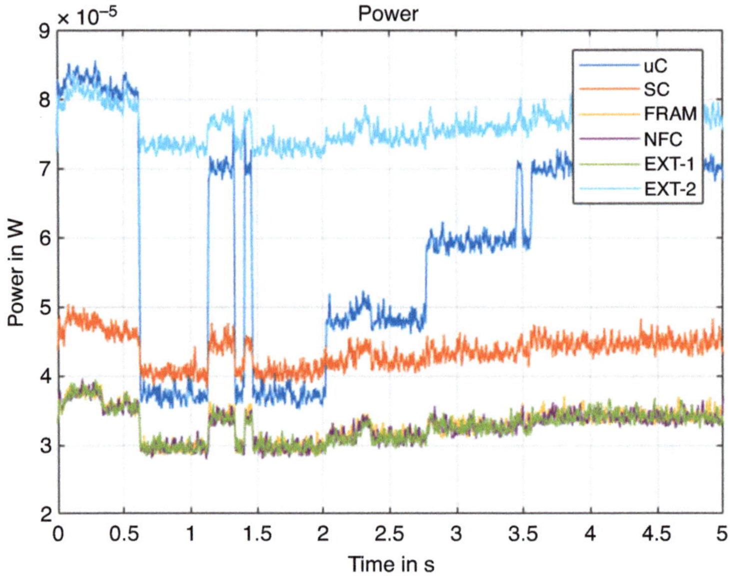

Fig. 8.3 Example of a measurement of sensor components

to *EXT-2* is continuously active. LEDs, connected to the microcontroller, switch according to the measured value.

Additional memory may be useful for a sensor system. Here configurations can be stored and measurements can be buffered until they are transmitted. For the FRAM memory module in this design the energy consumption in standby is around 35 μW.

To communicate with other devices and to harvest energy, this example design features an NFC interface. This interface consumes about 33 μW in idle mode.

Stated as an optional component is a security controller. The energy consumption of such element is around 45 μW in an idle phase. Also here a load switch can reduce the energy consumption and prolong the sensors lifetime.

Using this approach, the energy provisioning system is replaced by the testbench. That implies that the energy provisioning needs either to be measured using a different method or an existing simulation can be used to get to the needed data. There are numerous simulations of energy harvesting methods and batteries available that can be used for this purpose.

To simplify the simulation of the electric system of the sensor, an intermediate simulation tool can be used. Figure 8.4 shows how the energy consuming parts of the sensor can be modelled in LTSpice.

The top part represents the energy harvesting and energy storing of the sensor system. Here, the input voltage (*V1*) represents the capabilities of the energy harvesting device. If this voltage is higher than the sensor's supply voltage, the voltage from the harvesting device is switched to the supply voltage. The sensor

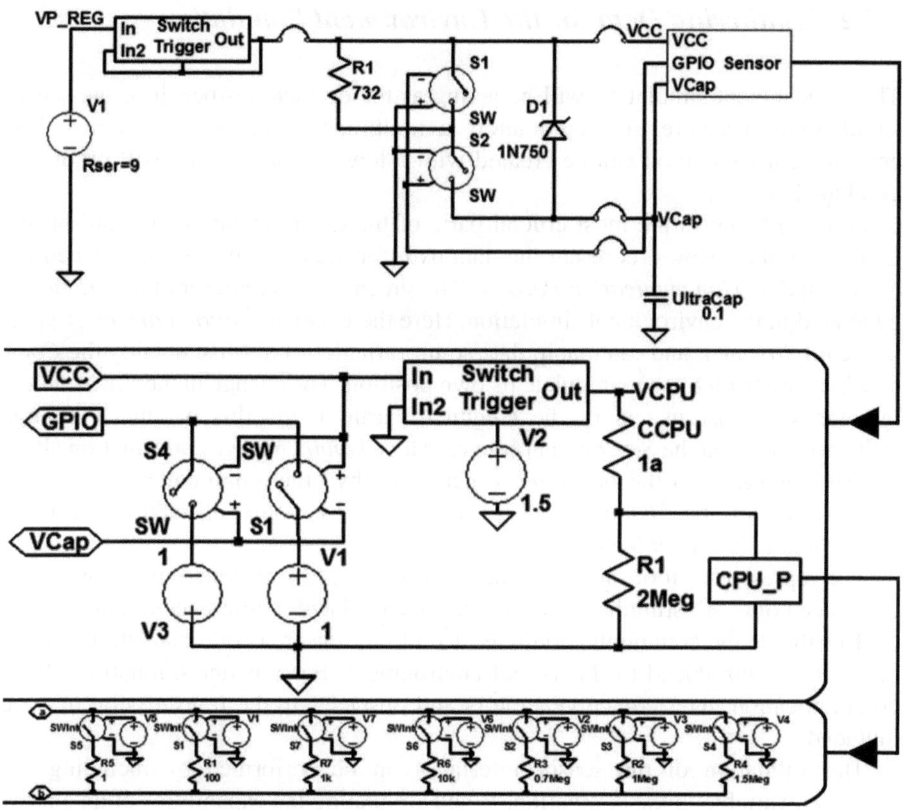

Fig. 8.4 Electrical model of a smart sensor

is then driven by the harvested energy. When the sensor system (*Sensor*) detects that the supply voltage is larger than the voltage at the energy storage (*UltraCap*) the switches *S1* and *S2* are flipped and the energy storage is charged. If the harvested voltage gets below the voltage from the capacitor, the switches are flipped again and the sensor is not charged any longer. If the harvested voltage drops below the voltage that can be supported by the energy storage, the sensor is again driven by the stored energy.

The middle section of Fig. 8.4 shows the mechanism that flips the switches in the top section (*S1, S4, V1*, and *V3*). Additionally, the mechanism that completely switches off the sensor if the voltage gets too low (*Switch* and connected *V2*) is implemented in this section. The operational part of the smart sensor (*CPU_C*) is then driven with that voltage.

In the bottom section of this figure various electrical networks (in this case resistors) can be switched on and off. These networks are tuned to represent the different components and power states of the sensors components.

This reduced electrical model of the sensor can be used to model the energy consumption of the sensor components in the final simulation.

8.3.2 Gathering Data for the Environment Simulation

The environment simulation will be acting as the de-facto testbench of the sensor simulation. Therefore, it should allow a multitude of different scenarios. The environment simulation can be created with a low complexity and still produce a good quality result.

Figure 8.5 shows the most crucial parts of the environment simulation. In this figure, the blue arrows represent the data path for the communication between the *Sensor* and its *Communication Partner*. The green arrows represent the data that is generated in the environment simulation. Here the *Communication Partner* is a part of the *Environment* and can manipulate some variables of it. Most notably the *Communication Partner* can manipulate its own position. The change in the *Environment* influences the parameters of the *Channel*. Furthermore, this parameter change influences the data the *Sensor* can observe. The *Channel* receives information about the *Environment* and the *Interference* generated by other communications. Using this information the *Channel* modifies the data that is communicated between the *Sensor* and its *Communication Partner*.

The representation of the sensor needs to be placed somewhere in the world as reference point. All stimuli for the sensor are calculated in reference to that point.

To validate the communication interface of the sensor, a communication partner needs to be introduced to the virtual environment. For a better simulation of the communication, interference generators and obstacles to the transmission may be included.

The validation of the sensor interface can be performed by including the measured variable in the simulation. When simulating the capabilities of an energy harvesting device, also this stimulus needs to be generated in the simulation.

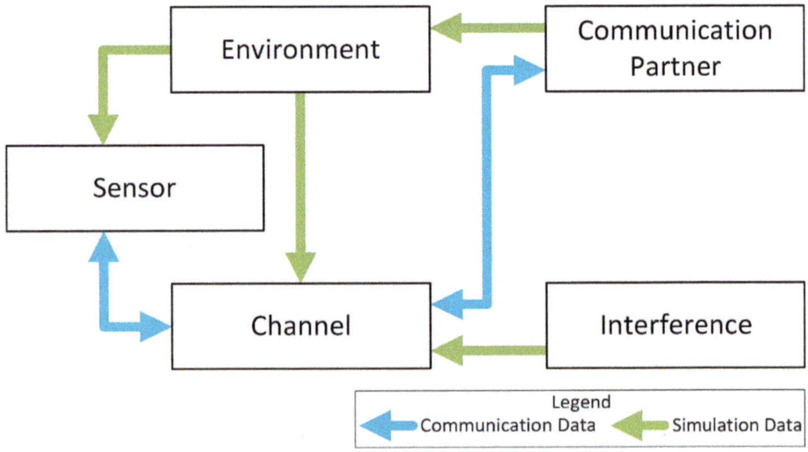

Fig. 8.5 Example of an environment of a sensor

This virtual environment allows the testing of the sensors use case. The environment can be modified to test different use case scenarios, boundary conditions of the communication, and reaction of the sensor to faulty signals.

8.4 Creating the Simulations

When designing a new sensor, simulations play an essential role. Not only the simulation of the sensor system, but also a simulation of the intended use case is useful.

8.4.1 Designing the Sensor Model

Using the data gathered in Sect. 8.3.1 a model of a generic sensor can be created using SystemC. This models needs to represent all features of a smart sensor. These features are:

- **The sensor frontend:** Depending on the use case the smart sensor can be equipped with different frontends. Typical frontends can just be switched on or off, thus it can be modelled as a static energy consumer.
- **The control unit:** The microcontroller can be represented as a timed state machine. Different power states that the microcontroller operates in can be defined (e.g. calculating, memory access, idle, sleep).
- **The memory:** If the sensor includes an additional memory unit, a detailed description is needed. Typical memory modules have very few operational modes (data access, data write, sleep). These can be modelled very accurately without having an overly complex model.
- **The communication unit(s):** The communication is typically slow and energy intensive. The model for the communication needs to be as accurate as possible. The description of the hardware itself can again be abstracted as a timed state machine operating with different power levels (sending, receiving, sleep, energy harvesting).
- **The energy provisioning system:** Many sensors are either connected to a bigger machine or to the power grid. For those sensors we can assume the voltage to be constant as it should be possible to provide sufficient energy to the sensor. If the sensor is operated by a battery or using an energy harvesting system and a capacitor, this system needs to be modelled with high accuracy. Voltage drops due to high energy demand can cause the sensor to stop operating. To counteract this, the sensor can use measures to decrease the energy consumption by temporarily disabling some functionality. In addition to these radical changes to the energy consumption, the non-linear behaviour of capacitors and batteries is important to the accurate simulation of the voltage levels of the sensor.
- **A security coprocessor:** The modelling of a security coprocessor can be tricky as the detailed description of it is most of the times classified. Nevertheless, a model

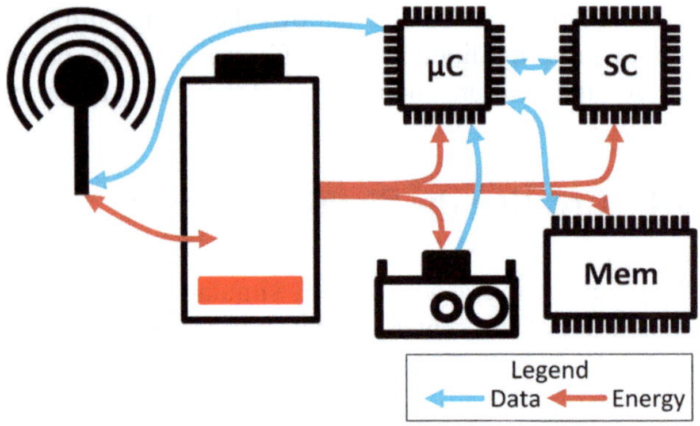

Fig. 8.6 The overall structure of a smart sensor

of it needs to be constructed, as this coprocessors typically consume much energy to conceal their behaviour. The security coprocessor can be modelled as a static energy consumer when switched on. This overestimates the real consumption and leads to a more robust design.

After these components have been created, they need to be connected to each other. Figure 8.6 shows how these connections can be set. The red arrows represent the energy transfer between the components. In the case that NFC is used as means of communication, energy can be harvested from the RF field. This energy is stored in the sensor's capacitor or accumulator. The antenna system also requires some energy if it is not supplied by an external field. All other components have to be connected to the energy supply. The blue arrows in this figure represent the data connections. All components are connected to the microcontroller. Either, as represented here directly, or via a bus system.

When creating this simulation a tradeoff between accuracy of the model and the simulation speed needs to be made. The more complex the model of the smart sensor becomes, the slower the simulation will be. In addition to reducing the complexity of some of the sensor components models, optimizations for longer simulation periods where the sensor is idle need to be made. These optimizations require the knowledge of the sensors non-linearities, the stimuli that wake the sensor from sleep, and the actions the sensor takes when stimulated.

Such optimizations can stop the execution of the full simulation and switch to a reduced version, or just estimate how the transient variables would behave during these periods. When estimating the variables, non-linearities in the sensor system need to be accounted for.

8.4.2 Designing the Environment

To test if the new sensor will be able to perform all tasks of the intended use case, an environment needs to be created that can influence the simulation for the sensor.

The environment can be simulated using the "Gazebo" simulator. To describe the different actions an entity can take, plugins need to be created. This simulation can then be used as a testbench for the sensor simulation. Using the data from Sect. 8.3.2 a model of the environment can be created. Its vital components are:

- **The sensor:** A representation of the sensor in the environment. This component implements the interface to the sensor simulation. All interactions of the environment with the sensor are relative to this representation. Additionally, all parameters the simulation calculates for the sensor to measure are calculated with this reference model.
- **Something to measure:** A simulation tool representing 3D objects primarily computes the relative motion and size between objects. Other parameters such as communication signals, air humidity, temperature, or light might not be calculated. Depending on the use case of the sensor these variables need to be calculated and passed to the sensor model.
- **A communication partner:** To test the communication unit of the sensor a counterpart needs to be implemented in the environment. This can be performed by another instance of the new sensor or an already existing device. To be able to test different scenarios it can be useful to allow the communication partner to be moved.
- **Interference and obstacles:** Any communication is influenced by the environment it is performed in. In the case of a wireless communication interference from other machinery, nearby communication, signal attenuation, and the scattering and reflecting of the signal produce communication errors. To see how the sensor reacts to faulty messages an implementation of the communication channel, including some interferences, is needed.

These components are then placed in a 3D virtual environment and logically connected.

The sensor receives information about the communication from the implementation of the channel and information about the measured variable directly from the environment. This data is then forwarded to the sensor simulation. The data that is produced by the sensor simulation is filtered and forwarded to the intended destination. This destination can be an actuator or the communication channel.

The communication partner sends the information that is intended for the sensor to the communication channel and receives the information from the sensor via this channel. If the communication partner is movable, the information about the changing position is sent to the environment.

The channel receives data from all communication partners and noise sources. Additionally, information about the 3D environment is gathered. Using this information the received data is modified (bit errors are introduced and the transmitted energy is calculated) and sent to the receiver.

The environment gathers data about the location of all objects. The information about the measured variable is then calculated and sent to the sensor.

8.4.3 Inserting the Sensor in the Environment

The simulation of the sensor and the simulation of the environment are very different simulation types. On the one hand, a simulation of a hardware and the processes that occur in the hardware during the operation. On the other hand, a simulation of some physical processes, movements, and communication. The major difference in these simulations are the time steps in which the simulation operates. The simulation of the hardware can occur at a very low level of abstraction and thus the time steps are small. SystemC, for example, supports time steps as small as 1 fs. The movement of robotic appliances is simulated in larger time steps. Gazebo, a robotic simulation tool, operates in time steps of 1 ms. This difference in simulation speed requires a complex *Interface* between the simulations.

Figure 8.7 shows the overall structure of the connection between the Gazebo simulation of the environment and the SystemC simulation of the sensor. The red arrows in this figure represent the flow of energy in the simulations. The blue and green arrows show how data is transmitted between the modules. Here, the blue arrows represent information that is transferred to or from the sensor, the green ones information that is required by the environment.

To create a connection between the 3D representation of the sensor in Gazebo and the simulation of the sensor in SystemC, a plugin is connected to the Gazebo model. This plugin forms the *Interface* between the simulations. To perform the task of connecting the two simulations, all the gathered data is packed and transferred to the SystemC process and the returning data is distributed in the Gazebo simulation. This plugin furthermore handles the synchronization between the simulations. This is done by adding additional information about the time step and a signal that the SystemC simulation should perform the calculations for this step.

In the SystemC simulation the original testbench is replaced by the counterpart of the *Interface*. It receives the information and forwards it to the destination. The

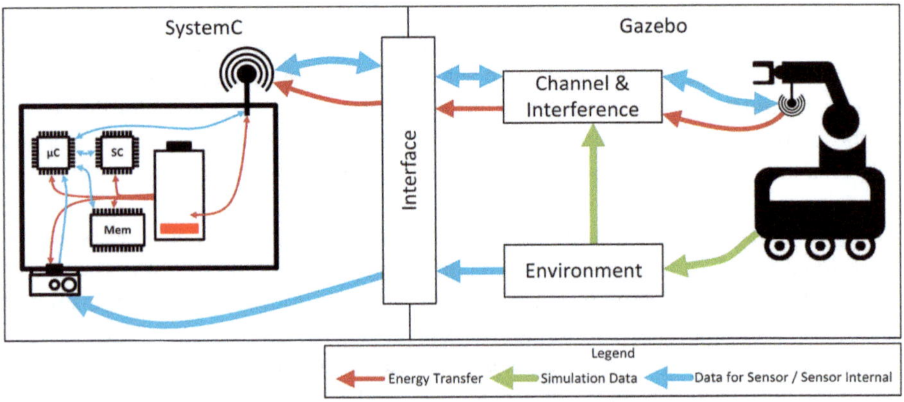

Fig. 8.7 The connection of the sensor and environment simulation

destination for the measured variable is the model of the sensor frontend. The data that is sent via the communication channel is forwarded to the communication interface. The additional information about the time step is kept at the testbench. With the information about the received values the testbench can calculate whether the full simulation needs to be run or if the changing parameters can be estimated. During the calculations of the simulation messages may be created that are then sent to the Gazebo simulator. When all calculations for this time step are finished the testbench sends a signal that the Gazebo simulation can perform another step.

Beyond the Immediate Neighbourhood

Using the data of the sensor system and the immediate interaction with the environment, more simulations can be created. These simulations can model the behaviour of a network of such sensors. The sensors in this network can be placed in a larger scale environment with moving communication partners transporting data and delivering energy. The approach of Ulz et al. [21] to form a network using mobile communication partners can here be used to also distribute the required energy to the sensor nodes.

8.5 The Simulation in Detail

In the Gazebo simulator all operations are performed in plugins that are connected to entities in the simulation. These plugins are executed sequentially such that the plugin that is being executed can already work with the results of the plugins executed before.

To synchronize the two simulations Gazebo needs to be halted until SystemC has finished and SystemC needs to wait for Gazebo to provide the parameters for the new time step.

The simulation sequence is shown in Fig. 8.8. This figure shows the initialization phase, simulation steps, and the termination of the cosimulation.

Initialization
When the simulation is started all Gazebo plugins run an initialization routine. In these routines the communication structures between the plugins are created and initial values are loaded. In the case of the environment description, all initial positions are gathered. The implementation of the channel loads the relative positions of the antennas and obstacles. The interface implementation forks a new process that will execute the SystemC simulation. It furthermore spawns a new thread that will be listening to the forked process. Finally, initial configuration parameters are sent via a communication channel.

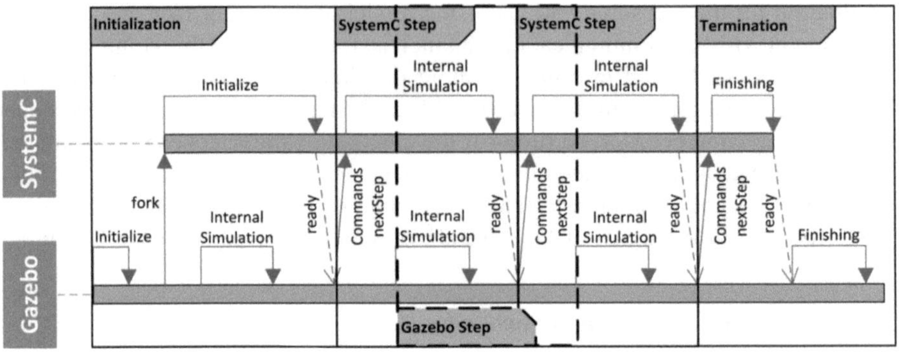

Fig. 8.8 Sequence for the Gazebo-SystemC cosimulation

The newly created process starts the SystemC simulation, loads the initial configuration from the provided communication channel and informs its parent process, the Gazebo simulation, about its status and waits for instructions.

Simulation Steps

The Gazebo plugins that control the communication partner check if inputs from a user or the robot's operating system are available. These commands are then translated to actions the communication partner will take. These actions include moving itself or sending some messages.

The plugin controlling the environment receives the information about movements and calculates new input for the sensor.

If messages are sent, the channel plugin gathers the information about the relative distance and orientation between the communicating parties and obstacles. Using this information the messages are altered to include bit errors. Additionally the transmitted energy gets changed to account for the channel properties. The modified messages are then forwarded to their destination.

The interface implementation waits for the SystemC simulation to have finished its last simulation step. When this signal is received the gathered information is forwarded towards the intended destination. After that all information that is sent to the interface plugin is packed and sent via the communication channel to the other process. This information is appended by information regarding the time step and a signal that informs the SystemC simulation that all information is sent. The SystemC simulation can now execute the needed operations while the Gazebo simulator computes the information needed for the next simulation step. Parallel to the computation of the next step, the thread of the interface listens to the channel between the processes for information the SystemC simulation sends to the Gazebo environment.

The SystemC simulation receives all information sent to it by the Gazebo simulator. The testbench can now adapt the simulation parameters and start the simulation.

While the simulation is being executed messages can be sent to the Gazebo simulator. These messages can originate from the communication interface module. If this sensor also includes an actuator or some visual status indicator, these modules can also send information to the environment simulation.

When the simulation step is finished the testbench sends the signal that the simulation step is finished.

Termination

After the simulation has been performed the plugins receive a signal that the simulation is about to be ended.

The interface plugin relays this information and waits for the remaining messages from SystemC. When it receives a signal that the SystemC simulation is terminated the plugin cleans its data structures and terminates.

When the SystemC simulation receives the signal to terminate, a post-processing step for the gathered data is initiated. This post-processing is a part of the runtime optimizations and performs tasks to correctly display the gathered data. The data is furthermore compressed by the post-processing step. Finally the data is saved and the Gazebo simulation is informed of thc termination of the simulation.

8.5.1 Optimizations

As the SystemC simulation operates in smaller time steps, each step of the Gazebo simulation causes thousands of SystemC steps. This results in the SystemC simulation being much slower than the Gazebo one, and thus slowing down the whole process. To speed up the simulation, the accuracy of the results can be reduced at certain times. A possible method to reduce the accuracy is shown in Fig. 8.9.

To take advantages of this possibility, the testbench needs to estimate the state after the simulation step. If the sensor is in idle mode and the input parameters do not change, the accuracy can be reduced by skipping the simulation and keeping a number that describes the amount of time that has been skipped since the last

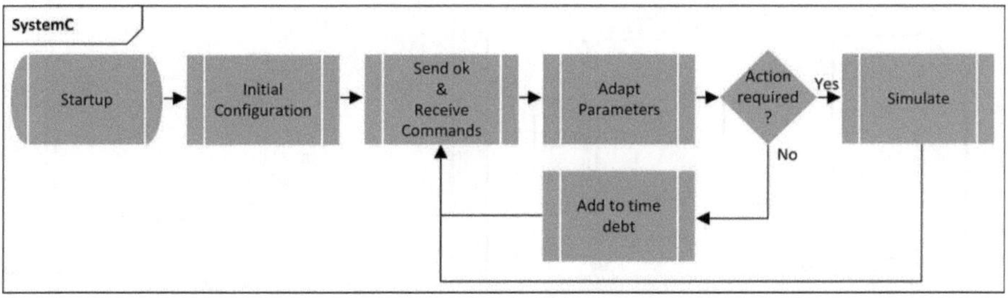

Fig. 8.9 The testbench of the SystemC simulation is adapted to skip the simulation if it is likely to yield results that can be estimated

simulation step. If the estimator evaluates that a simulation is needed this *time debt* is used to calculate changing parameters. To account for non-linearities in the simulation that have not been included in the estimation process a simulation step is needed if the *time debt* exceeds a defined time interval.

As SystemC is not created to support this behaviour, the output files of the simulation need the post-processing to be correctly displayed.

8.6 Evaluating the Simulations

The first step in the evaluation of the simulation results is the verification that the components of the simulations show the same characteristics as the measured ones.

The results of the LTSpice simulation are compared in Fig. 8.10. Here the left image shows the measured variables of a research prototype. In this scenario the sensor was charged and performed measurements every 60 s. The right-hand side shows the same operation simulated using LTSpice. The direct comparison shows that the LTSpice simulation approximate the real system to a high degree.

There exist some possibilities to optimize the energy consumption of the sensor itself.

- The duty cycle can be adapted to reduce the number of measurements. The spikes in the sensors current indicate the measurement.
- Some of the sensor components can cut off the energy supply. For example, the sensor can take many measurements and keep the values in its internal storage. If this storage gets too full, all data is transferred to the external memory module. To do so, the module is then switched on and all data is transferred. This reduces the amount of time the memory module spends in its startup phase.
- The communication module only needs to be switched on if some communication is about to happen. The same principle applies to the security controller of the sensor.

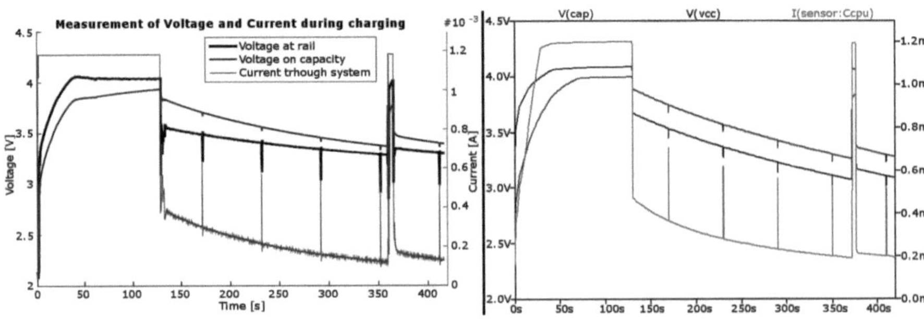

Fig. 8.10 Comparison between the measured values during charging and the electrical LTSpice simulation

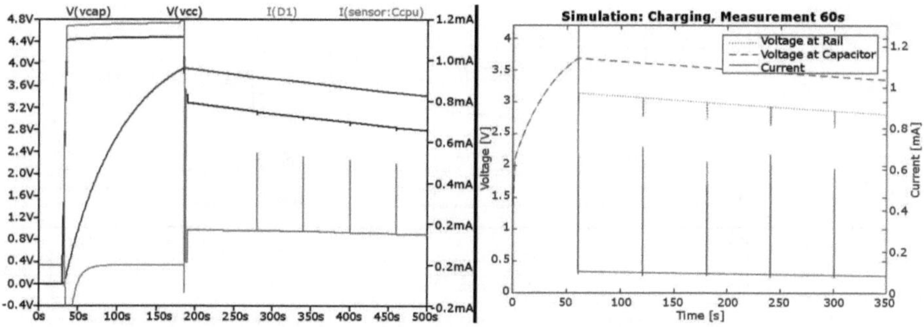

Fig. 8.11 Comparison between the measured values and the results of the SystemC simulation

Figure 8.11 shows a comparison between the simulation in LTSpice and the simulation performed in SystemC. The left-hand side of this figure shows the results of another LTSpice simulation. The right-hand side shows the same process in the SystemC simulation. Also here the capacitor has been charged and after that a simulated measurement has been performed every 60 s. The results show that the two simulations behave similarly. To optimize the energy consumption of the complete sensor system, some more options exist.

- The more valuable the measured variables are, the more often the sensor needs to be looked after. If the significance of the variable changes, the sample rate can be adapted to reduce the energy consumption of the sensor. Additionally, the sensor can change the interval in which the measured values are transmitted. This can be changed through configuration of the sensor.
- If the data is transmitted using mobile nodes, the moving partners can optimize their routes to gather the data while they are on the way to perform some tasks.

Examining the Effects of the Optimizations
The runtime optimization that can skip simulation steps introduces some problems. Figure 8.12 shows two traces to examine this effect. The top trace is taken from the optimized simulation, in the bottom trace the effects are removed. In this figure two of three operation modes of the operation can be seen clearly.

The first one is in the time span from 47.4 to 69.7 ms of optimized trace that corresponds to the time span from 24.8 to 26.3 s of the last trace. This equals to a time reduction of 98.51%. During this time the input from the Gazebo simulation was only the changing field of the removing NFC reader. This value is constant for any time step. As the rest of the sensor is idle, the simulation is only run for a short period of time to check if any values are changing in an unexpected way, after that the simulation is skipped and the rest of the changes are approximated. This can be seen in a thin line in the *advanceTime* line.

In contrast to that, the first 1.2 ms of the optimized trace corresponds to the first 6.3 s of the last trace. This equals to a time reduction of 99.98%. In this time no RF

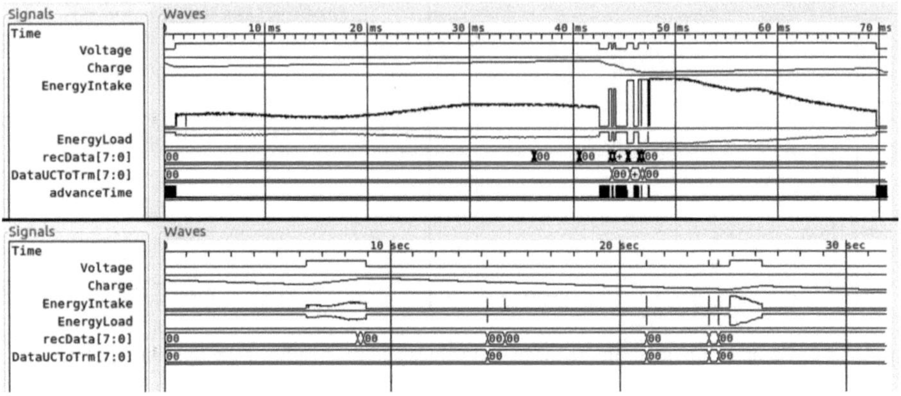

Fig. 8.12 Effects that result from the runtime optimization by skipping simulation steps. The simulation time is compressed if the sensor is idle and the effects can be estimated

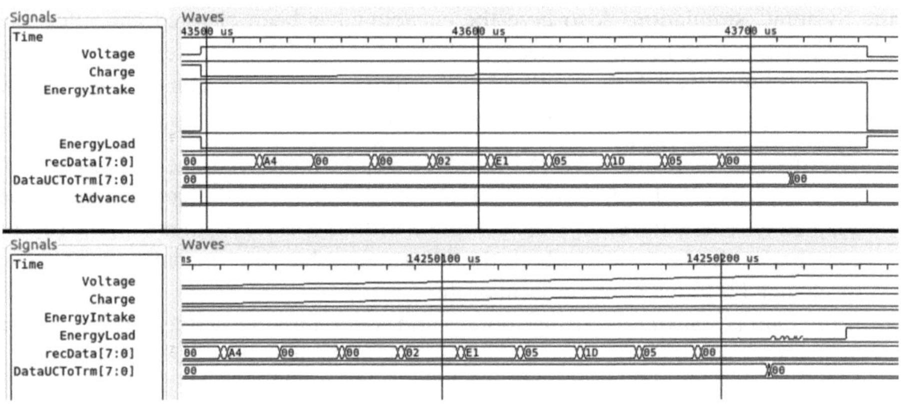

Fig. 8.13 Third operation mode of the optimization. If the sensor is not idle, no compression occurs

field is detected and the sensor is idle. In this phase the simulation is only started if the *time debt* of Fig. 8.9 reaches a predefined value to update the change rates due to non-linear effects. During this phase most changes are approximated. The *advanceTime* line shows a thick line while this phase lasts.

The third operation mode is seen if the *advanceTime* line is not elevated. Figure 8.13 shows a section of Fig. 8.12 where the sensor is not idle. This phase starts from 43.5 ms of the optimized trace and has a duration until 43.74 ms. This corresponds to the time span from 14,250 to 14,250.24 ms. In this time the sensor is receiving data that needs to be processed. Therefore the sensor is not idle and the effects cannot be estimated. Here the full simulation is performed and no time reduction occurs.

The values of the simulation time reduction highly depend on the processor state and the simulated use case.

8.7 Security of Smart Sensors

When developing new sensors, not only the efficiency of them is important, they furthermore should be secure. The security should not only care about the data that is generated by the sensor, but also the sensors should not endanger systems they are connected to.

Any security solution for smart sensors needs to consider the following six points:

- The sensor has limited resources in terms of energy and processing power.
- The overhead in memory usage, computational effort, and data transmission sizes that are imposed due to security considerations can be large in comparison with the payload data.
- To provide security to a sensor network every sensor needs to perform the security relevant operations.
- The security features should be easy to use by untrained end users.
- The end user needs to trust the security features of the sensor.
- Dedicated hardware that performs the security operations is better, but needs more energy and can be costly.

8.7.1 Data Security

The sensor data needs to be protected from unauthorized access. Otherwise this data can disclose company secrets. Furthermore, if the sensor data can be replicated, control mechanisms can be fooled to perform harmful operations that can endanger the facility or human lives. In the case of the STUXNET attack [9] the attacker has learned the behaviour of the system by observing the sensor data. After the learning phase the attacker manipulated the sensor's data to turbines of a nuclear enrichment plant while it was hiding its doings by displaying normal looking data to the next layer of controls.

This attack showed that the sensor data needs to be protected. To be able to read (or guess) the sensor data can enable an attacker to manipulate the processes they observe.

To protect the gathered data in a sensor network from unauthorized access, the data needs to be encrypted. As Ulz et al. showed, such measures increase the severity of introduced bit errors [24]. To still be able to communicate correctly, a forward error correction scheme can be used on the encrypted data. Using this method, bit errors can be detected before decrypting, thus increasing the resilience against bit errors of the communication. Ulz et al. also suggested a solution to this problem which they called Joint Encryption and Error Correction (JEEC).

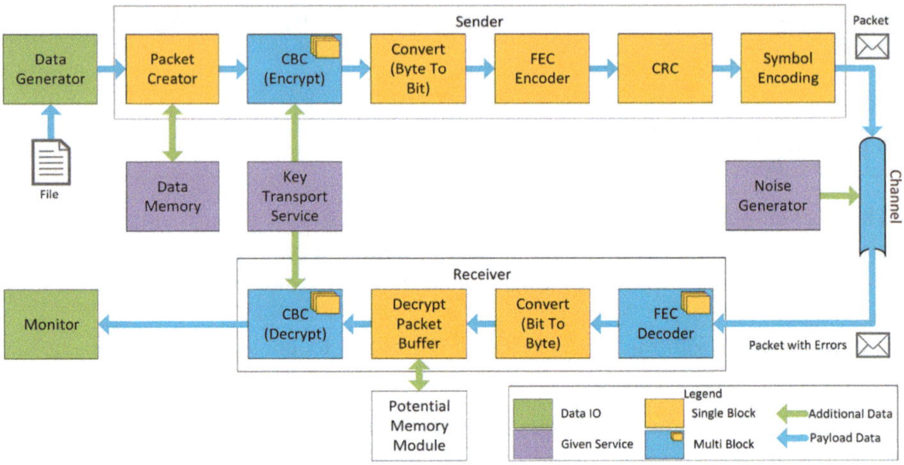

Fig. 8.14 Concept to simulate the JEEC approach by Ulz et al. [24]

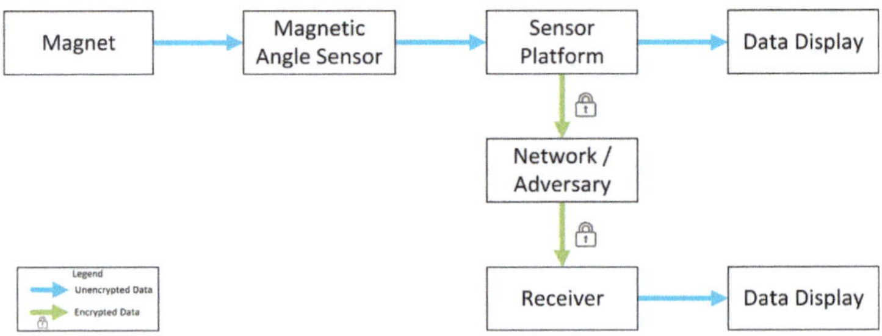

Fig. 8.15 Concept for testing the secured data transmission of Ulz et al. [24]

To simulate the security of the data during transmission, another simulation has been created. Figure 8.14 shows the concept of this simulation. Here the green fields represent blocks where the data is not secured. The yellow and dark blue fields represent steps for the data processing, where the dark blue describes that this process consists of many subprocesses. The light blue fields are considered given in this simulation.

The data to be transmitted is provided by a file. The top row shows the process the sender takes before the data is sent. During this process the data is chunked into packets, encrypted, a forward error correcting (FEC) code is added, and a checksum is calculated. This packet is then sent via a channel. This channel can see the transmitted data and introduce bit errors. Before the receiver can use the data, it needs to be processed again. This process, shown in the bottom row, consists of the recovering of flipped bits using the FEC and decrypting.

Figure 8.15 shows a concept to test the JEEC approach. Here a magnetic angle sensor is stimulated by a magnet. The sensor data (blue arrows) is then sent to the smart sensor platform where it is processed. The original data gets displayed for later comparison. Additionally, the sensor data gets encrypted and encoded using a FEC scheme. This packed data is then sent via a network to the receiver (green arrows with locks). The adversary is in the network and can manipulate the data. The receiver tries to decode the data. If this succeeds the plain data is displayed to be compared against the original one.

8.7.2 System Security

Recently another kind of attack using sensors has been reported. In this scenario a large number of sensors connected to a network are used to perform a DDoS (Distributed Denial of Service) attack. In this type of attack the sensors firmware is altered to send messages through the network to a common destination. Using a large number of manipulated sensors, the common destination receives more data than it can handle. This blocks the recipients capabilities to perform its normal operations.

To use the sensors for such malicious purposes, their firmware needs to be changed. Ulz et al. presented an approach to exclude sensors from the sensor network that use unauthorized firmware [22]. This is done in a two-layer system. In the first step the firmware is checked by an on-board security module. If this check verifies that the software is from a trusted source, the sensor's network stack is released. Now the firmware version can be checked at a trusted backend server. This verifies that the sensor is running the newest version of the firmware.

8.8 Conclusions

This chapter presents an approach to design secured power aware smart sensors using hardware models. To be able to use this model-based design, the model parameters need to be known. Therefore, we discussed the methods for gathering such data and how the data can be simplified to be used in an abstracted hardware model. In addition to the model, use cases for the new sensor need to be known. Here a simple simulation can be performed that generates the inputs necessary for the hardware simulation. If the use case extends beyond one sensor, the gathered data can be abstracted again and included in higher layer simulations. Additionally, we presented the main phases of such simulations and mentioned possibilities for runtime optimizations. Finally, we discussed the possibilities of attacks on and with the sensors and mentioned methods to increase the difficulty of a successful attack. Additionally, the concepts to secure the data against adversaries has also been simulated and tested.

As the simulation should resemble a generic smart sensor, some details are lost at every layer of abstraction of the models. This loss in detail is necessary to create usable simulations.

The main focus of this chapter is on the creation of the simulation of such sensor systems. A detailed description on the model generation and the simulation steps is given.

Using these simulations it is possible to analyse the effects of changes to the sensor system, the communication protocols, or the routines in the environment. This can be used to optimize the energy consumption of each individual components as well as the energy consumption of the complete environment.

Acknowledgements This project has received funding from the Electronic Component Systems for European Leadership Joint Undertaking under grant agreement No 692480. This Joint Undertaking receives support from the European Union's Horizon 2020 research and innovation programme and Germany, Netherlands, Spain, Austria, Belgium, Slovakia.

IoSense is funded under the agreement number 853326 by the Austrian Federal Ministry of Transport, Innovation and Technology (BMVIT) under the program "ICT of the Future" between May 2016 and May 2019. More information https://iktderzukunft.at/en/.

We would like to thank Infineon Technologies and especially Rainer Matischek for providing us the security controllers used in the system and for their support that helped creating the prototypes and simulations.

References

1. Accelera: SystemC. (2000). http://accellera.org/downloads/standards/systemc. Accessed 17 Jan 2017
2. M. Antonakakis, T. April, M. Bailey, M. Bernhard, E. Bursztein, J. Cochran, Z. Durumeric, J.A. Halderman, L. Invernizzi, M. Kallitsis, et al., Understanding the Mirai botnet, in *USENIX Security Symposium* (2017)
3. F. Bouchhima, M. Briere, G. Nicolescu, M. Abid, E. Aboulhamid, A SystemC/simulink co-simulation framework for continuous/discrete-events simulation, in *2006 IEEE International Behavioral Modeling and Simulation Workshop* (Institute of Electrical and Electronics Engineers (IEEE), Piscataway, 2006). https://doi.org/10.1109/bmas.2006.283461
4. R.J. Campbell, Cybersecurity issues for the bulk power system. Electr. Deliv. Secur Fed Oversight Activities Funding, 63–108 (2016). https://www.hsdl.org/?abstract&did=767255
5. G. Chen, M. Fojtik, D. Kim, D. Fick, J. Park, M. Seok, M.T. Chen, Z. Foo, D. Sylvester, D. Blaauw, Millimeter-scale nearly perpetual sensor system with stacked battery and solar cells, in *2010 IEEE International Solid-State Circuits Conference - (ISSCC)* (IEEE, Piscataway, 2010). https://doi.org/10.1109/isscc.2010.5433921
6. J. Haase, D. Meyer, M. Eckert, B. Klauer, Wireless sensor/actuator device configuration by NFC, in *2016 IEEE International Conference on Industrial Technology (ICIT)* (IEEE, Piscataway, 2016), pp. 1336–1340
7. K. Huang, I. Bacivarov, F. Hugelshofer, L. Thiele, Scalably distributed SystemC simulation for embedded applications, in *2008 International Symposium on Industrial Embedded Systems* (Institute of Electrical and Electronics Engineers (IEEE), Piscataway, 2008). https://doi.org/10.1109/sies.2008.4577715
8. A. Kansal, J. Hsu, S. Zahedi, M.B. Srivastava, Power management in energy harvesting sensor networks. ACM Trans. Embed. Comput. Syst. **6**(4), 32 (2007). https://doi.org/10.1145/1274858.1274870

9. R. Langner, Stuxnet: dissecting a cyberwarfare weapon. IEEE Secur. Priv. **9**(3), 49–51 (2011)
10. W.S. Lee, W.I. Son, K.S. Oh, J.W. Yu, Contactless energy transfer systems using antiparallel resonant loops. IEEE Trans. Ind. Electron. **60**(1), 350–359 (2013). https://doi.org/10.1109/tie.2011.2177611
11. D. Martin, P. Wilsey, R. Hoekstra, E. Keiter, S. Hutchinson, T. Russo, L. Waters, Integrating multiple parallel simulation engines for mixed-technology parallel simulation, in *Proceedings 35th Annual Simulation Symposium. SS 2002* (Institute of Electrical and Electronics Engineers (IEEE), Piscataway, 2002). https://doi.org/10.1109/simsym.2002.1000082
12. Mathworks: get started with gazebo and a simulated turtlebot (2016). https://de.mathworks.com/help/robotics/examples/get-started-with-gazebo-and-a-simulated-turtlebot.html. Accessed 03 Jan 2017
13. D. Mueller-Gritschneder, K. Lu, E. Wallander, M. Greim, U. Schlichtmann, A virtual prototyping platform for real-time systems with a case study for a two-wheeled robot, in *Design, Automation & Test in Europe Conference & Exhibition (DATE), 2013* (EDAA, 2013). https://doi.org/10.7873/date.2013.274
14. Open Source Robotics Foundation: Gazebo simulator (2004). http://www.gazebosim.org. Accessed 03 Jan 2017
15. P.R. Panda, SystemC - A modelling platform supporting multiple design abstractions, in *Proceedings of the 14th International Symposium on Systems Synthesis - ISSS* (Association for Computing Machinery (ACM), New York, 2001). https://doi.org/10.1145/500001.500018
16. T.W. Pieber, T. Ulz, C. Steger, Systemic test case generation with the gazebo simulator, in *Proceedings of the 7th International Conference on Simulation and Modeling Methodologies, Technologies and Applications - Volume 1: SIMULTECH*. INSTICC, (SciTePress, Setúbal, 2017). https://doi.org/10.5220/0006404800650072
17. T.W. Pieber, T. Ulz, C. Steger, R. Matischek, Hardware secured, password-based authentication for smart sensors for the industrial internet of things, in *International Conference on Network and System Security* (Springer, Berlin, 2017), pp. 632–642
18. H. Possadas, J.A. Adamez, E. Villar, F. Blasco, F. Escuder, RTOS modeling in SystemC for real-time embedded SW simulation: a POSIX model. Des. Autom. Embed. Syst. (2005). https://doi.org/10.1007/s10617-006-9725-1
19. M. Rahimi, H. Shah, G. Sukhatme, J. Heideman, D. Estrin, Studying the feasibility of energy harvesting in a mobile sensor network, in *2003 IEEE International Conference on Robotics and Automation (Cat. No.03CH37422)* (IEEE, Piscataway, 2003). https://doi.org/10.1109/robot.2003.1241567
20. T. Sogorb, J.V. Llario, J. Pelegri, R. Lajara, J. Alberola, Studying the feasibility of energy harvesting from broadcast RF station for WSN, in *Instrumentation and Measurement Technology Conference Proceedings, 2008, IMTC'08* (IEEE, Piscataway, 2008). pp. 1360–1363
21. T. Ulz, T. Pieber, C. Steger, S. Haas, R. Matischek, Sneakernet on wheels: trustworthy NFC-based robot to machine communication, in *2017 IEEE International Conference on RFID Technology & Application (RFID-TA)* (IEEE, Piscataway, 2017), pp. 260–265
22. T. Ulz, T. Pieber, C. Steger, S. Haas, R. Matischek, H. Bock, Hardware-secured configuration and two-layer attestation architecture for smart sensors, in *2017 Euromicro Conference on Digital System Design (DSD)* (IEEE, Piscataway, 2017), pp. 229–236
23. T. Ulz, T. Pieber, C. Steger, C. Lesjak, H. Bock, R. Matischek, Secureconfig: NFC and QR-code based hybrid approach for smart sensor configuration, in *2017 IEEE International Conference on RFID* (IEEE, Piscataway, 2017), pp. 41–46
24. T. Ulz, T. Pieber, C. Steger, R. Matischek, H. Bock, Towards trustworthy data in networked control systems: a hardware-based approach, in *2017 22nd IEEE International Conference on Emerging Technologies and Factory Automation (ETFA)* (IEEE, Piscataway, 2017), pp. 1–8
25. R. Yan, H. Sun, Y. Qian, Energy-aware sensor node design with its application in wireless sensor networks. IEEE Trans. Instrum. Meas. **62**(5), 1183–1191 (2013). https://doi.org/10.1109/tim.2013.2245181
26. I. Zamora, N.G. Lopez, V.M. Vilches, A.H. Cordero, Extending the OpenAI Gym for robotics: a toolkit for reinforcement learning using ROS and Gazebo. arXiv preprint arXiv:1608.05742 (2016)

Chapter 9
A Software Toolkit for Complex Sensor Systems in Fog Environments

Dominik Grzelak, Carl Mai, René Schöne, Jan Falkenberg, and Uwe Aßmann

9.1 Introduction

The Software Toolkit (STK) is a component-oriented framework, which facilitates the integration of single sensors, sensor arrays, and complex sensor systems in the fog. It is developed with the vision of prospective use for different application domains (e.g., manufacturing or smart home), where the properties of the STK are designed in such a way to follow domain-specific standards and consider supplementary conditions for particular domains. The goal of the STK is to provide the functional coupling between the system environment and the sensor elements as well as realizing the efficient and secure data aspects between sub-systems and the application's main system. From low-level sensor signals to a complex IoT system, multiple connected sensors are the key elements of emerging smart rooms, which require a unified software framework to create value-adding applications (apps) for a future smart room software ecosystem. With the STK, a focus on aspects regarding reliability, weight, cost, performance, interoperability, and standardization was set. Furthermore, the software toolkit facilitates the application of new developments by third-party users or system developers.

9.1.1 Motivation

In the next years more and more things and information will be connected to the internet resulting in a complex network, which becomes denser with every single

D. Grzelak (✉) · C. Mai · R. Schöne · J. Falkenberg · U. Aßmann
TU Dresden, Fakultät Informatik—Institut für Software- und Multimediatechnik/Lehrstuhl Softwaretechnologie, Dresden, Germany
e-mail: dominik.grzelak@tu-dresden.de

W. D. van Driel et al. (eds.), *Sensor Systems Simulations*,
https://doi.org/10.1007/978-3-030-16577-2_9

device—this is referred to as the *Internet of Things* (IoT). A huge variety and amount of data are generated from various devices and sensors, which need to be processed and responded to in near real-time with the cloud becoming an indispensable part of that process. It is expected that 50 billion devices will be connected to the Internet by 2020 [1].

Because IoT is becoming more prominent and the number of devices is steadily increasing, smart objects and devices equipped with sensors will become an important source of data as they can gather different information not only from their internal states but also about the environment and entities they interact with. This data is used as a basis for various kinds of decision algorithms, machine learning, and artificial intelligence in general. Hence, the importance of collecting and distributing this data to the right services. On the flip side, this will also produce more and more traffic in the network and become a burden for low-bandwidth and high-latency networks.

A purely cloud-based solution is not able to overcome these issues as the physical distance between the user, edge devices, and the cloud services increase with the transmission latency and response times. Therefore, complex real-time applications are not possible anymore. The next logical step is to push the cloud services to the edge of the network—to the devices (e.g., sensors, actuators, and embedded computers) gathering the actual data, moving the processing of data closer to the edge as well. Deriving higher-level data and performing decisions locally on those devices instead of a physically distant cloud server will significantly reduce the amount of data that is being sent through the network, reduce the required bandwidth and also increase data security. These are the main principles of *Fog Computing* [2], which we will use as the basic paradigm for developing a software architecture and the *IoSense Software Toolkit* to facilitate the integration of newly developed sensors from IoSense with the existing IoT infrastructure.

The goal of the software toolkit is to bridge the gap between the cloud and connected sensors (e.g., the newly developed light sources), as shown in Fig. 9.1. To demonstrate complex scenarios with an installation of sensors, actuators, or sensor arrays in combination with the software system, we present a component-based software architecture, which copes with each of the challenges mentioned above. Also, the architecture intends to facilitate the integration and interaction of new and existing IoT devices of different vendors (e.g., sensors and actuators).

To overcome above problems, the STK provides a solution to the issues above: a decentralized system architecture based on fog computing paradigms and a software framework providing a feasible solution for integrating a wide range of sensors in complex systems and systems of systems. The STK uses recently developed concepts to leverage the separation of important functionality into small components, which can be reused and distributed to various kinds of heterogeneous devices (such as single-board computers, and sensor systems). The STK makes the integration of and communication among all IoT components as easy as possible. The processing of data and execution of the workload directly on a device closer to the user's network is speeding up applications that require low-latency responses. Even if they are not continuously connected to the network, devices can make smart

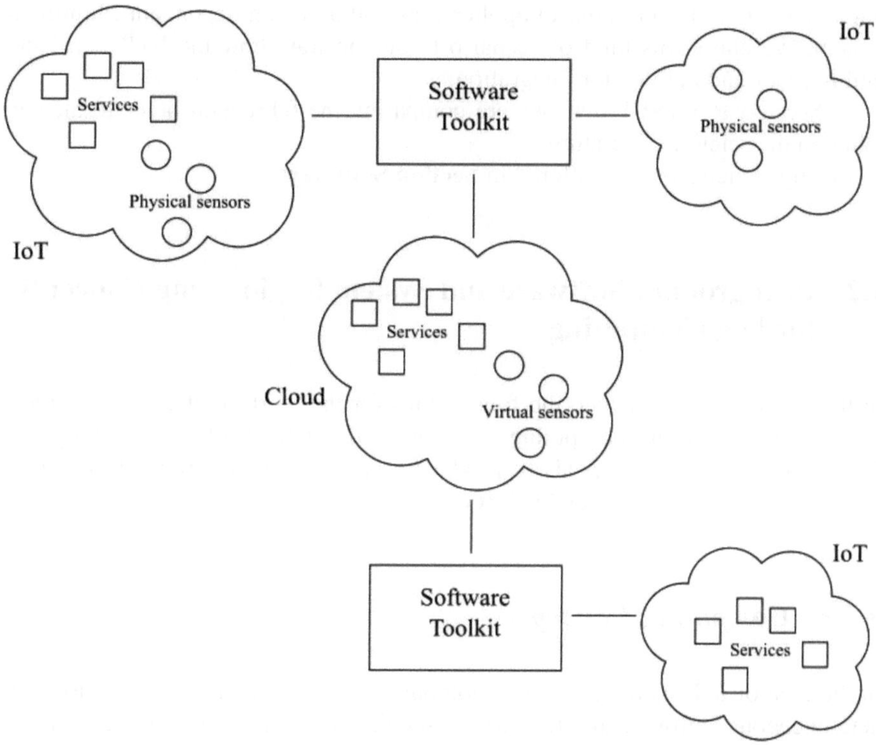

Fig. 9.1 The software architecture bridging the gap between a connected sensor system in the IoT layer and the cloud services

decisions and send the data at a later point in time. That is a crucial requirement when developing complex sensor system applications in IoT environments with resource constraints and dynamic, mobile devices.

9.1.2 Structure

Fog can handle the diversity of different industry standards, sensors, devices, and so on. Therefore, **Section Background** describes the underlying computing concepts and properties that are vital for the development of the fog-based framework.

The key element of a fog computing framework can bring its elements regarding communication and self-adaption in conformity with the fog paradigms. In **Section Architecture,** the architecture of the STK is introduced where the presented concepts of the previous section are adopted.

The development of the STK is based on several demonstrators and use cases. In **Section Evaluation,** the evaluation is guided by three case studies, involving the implementation of robot control applications and a person-dependent illumination system. We chose this kind of scenario to demonstrate how the STK can handle multi-sensor and multi-actor integration.

In **Section Related Work**, we are comparing the STK with other frameworks found in the scientific literature.

Finally, conclusions are offered in **Section Summary**.

9.2 Background: Software and System Engineering Concepts for Fog Computing

In this section, we introduce the foundations for the work on the software toolkit by examining various computing concepts and methods of sensor integration. Combining this with our goals to produce a versatile and distributed system, we explain our choice in **Section Architecture**.

9.2.1 Computing Concepts

In the area of IoT, there are various computing concepts employed because of the heterogeneous hardware and the various requirements on an IoT system. Here we will explain the most important concepts, which all are part of the considerations to implement the STK.

9.2.1.1 Embedded

Embedded systems purely integrate the entire processing capability into a microcontroller. Potential collaborations with other embedded systems are defined at design time due to their inability to change the firmware code during runtime. However, modern microcontrollers use standard communication protocols (e.g., TCP/IP) to communicate with others. The computational performance of microcontrollers is usually very limited. Therefore, only simple tasks such as non-intensive data processing and data analysis can be performed.

9.2.1.2 On-Premise

In many cases, an embedded system is accompanied by a computer (e.g., local computer or server) with higher performance and more storage on premise (in the building) of the person or organization. The computer usually uses an $\times 86$ architecture and a high-level operating system (e.g., Linux). In this way, the long-term sensor data storage is possible. Hence, it allows for implementing higher-level data processing and data analysis strategies with high-level programming languages (e.g., Java, R). The usage of object-oriented programming languages, such as Java, enables the developer to implement more complicated and flexible systems. For example, the system supports runtime adaptation and runtime extension. It is very important for IoT systems where IoT device operations are heterogeneous and dynamic.

9.2.1.3 Cloud

When an embedded device collects sensor data and implements simple data fusion or data analysis strategies, it can then send the pre-processed data to a cloud. The cloud again collects and re-processes the pre-processed data through its data fusion and data analysis strategies.

If the embedded device is also an actuator, then it can receive control messages from the cloud for a specific task. The cloud uses virtualization technologies with many available resources to provide significant storage and processing capability. Therefore, some well-known data technologies such as Big Data and Deep Learning can be used to provide completely new features for the user. Besides the advantages, the cloud architecture still has some disadvantages such as deployment costs and high communication latency between IoT devices and the cloud. These problems require a more efficient architecture.

9.2.1.4 Edge and Fog

Edge computing is pushing data from computing applications and services away from centralized server nodes to the logical extremes of a user's network, e.g., routers. It allows physical things (such as lamps, doors, fridges, and sensors) to process the data they gather locally together with additional domain knowledge leveraging artificial intelligence and machine learning algorithms to make sense of the data.

This enables some application processing to be performed by a small edge server positioned between the cloud and the sensor in a location physically closer to the user. The workload from the cloud will be offloaded to a user's device for processing while at the same time speeding up applications that require a low-latency response.

Figure 9.2 shows a multi-layer fog architecture. This example combines the benefits of the on-premise solution with the cloud solution. It includes a number of high-performance local computation and storage nodes located within the edge network and near the IoT devices. Main data processing strategies such as data fusion and data analysis can be executed on these fog nodes. For the long-term data storage and higher processing power, a cloud layer is added on top. The characteristics and advantages of fog computing with respect to [3] are as follows:

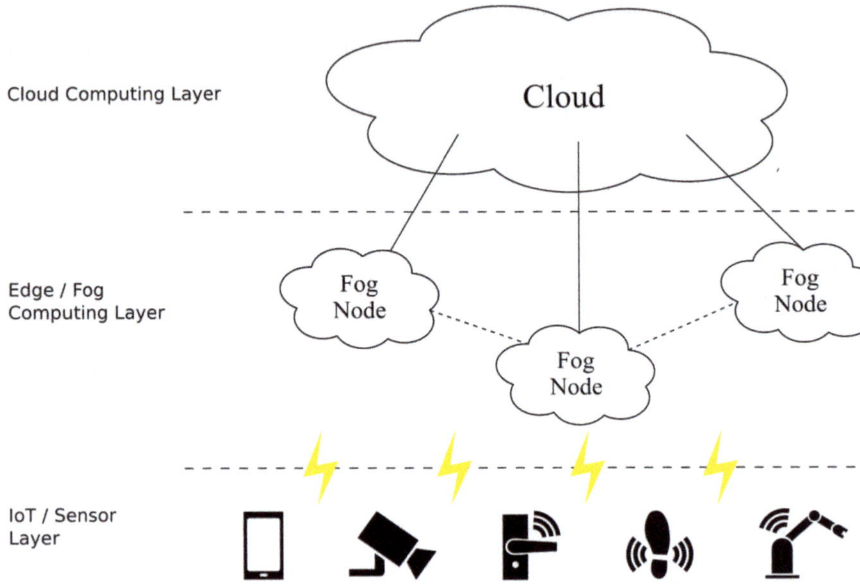

Fig. 9.2 Example architecture of a multi-layer fog framework

1. **Location**: Better delay performance.
2. **Distribution**: Deploy many "micro" data centers in contrast to few cloud data centers.
3. **Scalability**: Increase the number of "micro" centers to a growing number of users.
4. **Density of devices**: Resilience and replication.
5. **Mobility support**: "Mobile" cloud near the user.
6. **Standardization**: Incorporation of different cloud vendors.
7. **On the fly analysis**: Compute partial aggregation results and big data on the cloud.

The concept fog computing, in general, creates value by addressing the following limitations of cloud-centric architectures:

1. **Volume**: There will not always be sufficient bandwidth to push all IoT data to cloud services.
2. **Connectivity**: The data sources (objects–devices) will not always have a cloud connection.
3. **Velocity**: The latency and jitter between things, sensors, and cloud services will not always be low enough for critical applications (e.g., autonomous decision-making through analytics).
4. **Cost**: The cost of connectivity to a cloud service (especially via public clouds) can be unacceptable for the data volumes required.

5. **Security**: For some system owners, it will not be appropriate (because of privacy or security concerns) to push operational data to a cloud service.
6. **Interoperability**: Peer-to-peer device is often preferred or required for cloud interoperability for latency, security, risk, or distributed computing reasons.

9.2.2 Sensor Integration

A sensor must be physically and logically integrated into an IoT system. The communication can either be wired or wireless. In both cases, different combinations of communication protocols across the ISO-OSI layers exist (e.g., I2C, CAN, TCP, UDP). Depending on the available interfaces and protocols, the embedded software must provide software components for establishing a connection to the individual sensor. Once the system can connect sensors, other new challenges comprise the storage and efficient sensor data processing. The STK provides a standard component model for sensor integration, sensor data collection, sensor data fusion, and sensor data analysis. The aim is to overcome the problems above as well as to build a system with re-usable components.

9.2.2.1 Data Collection

The collection of sensor data encompasses the transient collection and storage of sensed data. Transient storage is usually formed by a set of triple of parameters (*Sensor–Time–Data*). In which, *Sensor* indicates the information of the corresponding sensor, *Time* specifies time information, and *Data* is the payload of the sensor data. Depending on the available storage and computational performance of a device, it can only store the captured data for a limited time.

9.2.2.2 Data Fusion

Sensor data fusion is the process of integrating data from multi-sensors to produce more accurate and useful information. Many contextual facts of interest cannot be determined by using only one type of sensor. A typical example is *depth* information that is only specified by the difference between two stereoscopically aligned RGB-cameras. In order to acquire the *depth* information, the data streams of both cameras have to be synchronized and combined. The implementation of sensor data fusion solutions on IoSense STK is one of the main goals.

9.2.2.3 Data Analysis

IoT systems use sensor data to provide their services and adapt themselves according to changing conditions. Sensors are essential to measure the environmental state and provide accessible services from software backend. Sensor data analysis is performed on sensor data by using inspection, transformation, and modeling tools to discover useful information as well as support decision-making.

9.3 Reference Architecture

The following section introduces the detailed concepts of the IoSense STK. We describe the reference architecture and breaking it down into the various components of fog including orchestration, distribution and communication and what are the applicable building blocks in an architectural perspective.

9.3.1 System Infrastructure

A high-level overview of STK and its architecture are depicted in Fig. 9.3. The STK aims to create a software ecosystem, also, providing the core functionalities for sensor integration and communication. The STK also defines interfaces that developers can use to program their app (e.g., application) using the core functionality of the platform. In this way, the app can connect with the STK directly through the interfaces.

Furthermore, it may contain new functionalities (i.e., *Value Added Functionality*) for processing specific tasks. The STK is also designed to integrate easily into other frameworks. Therefore, the flexibility of the STK is one of the most considered advantages.

Inspired by the openHAB architecture, the STK also uses OSGi as a core framework. Each bundle is a service which is developed separately. It runs independently or communicates with other bundles to carry out a specific task. Therefore, services can have dependencies on other services allowing them to request additional functionality among themselves.

Figure 9.4 shows the software architecture of the STK and various open-source components it exploits. The open-source software platforms are described briefly in Table 9.1. A service (we use the term "service" for an OSGi bundle) is a component which provides a specific functionality of STK such as data analysis, data processing, or connector functionality. The design of a bundle is developed by using the well-known *Concierge* open-source software that is based on OSGi model R5 (see Table 9.1). Either they are deployed on fog nodes or fog nodes can communicate with services through well-defined interfaces (see Fig. 9.3). This even allows low-end devices that do not support this service architecture due to resource constraint

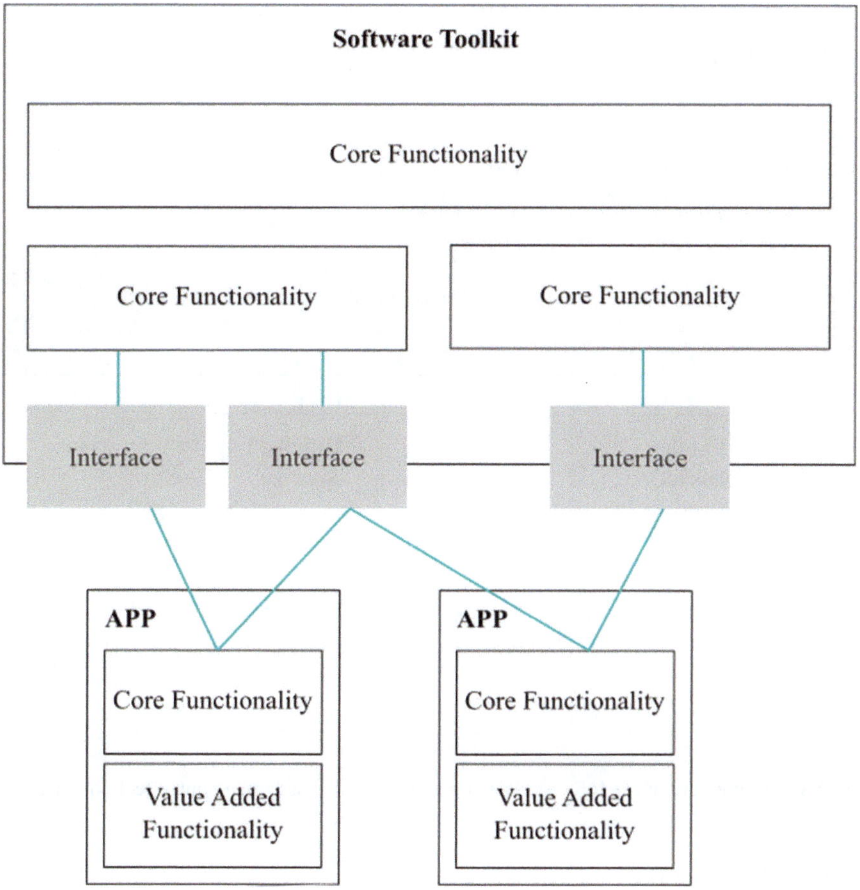

Fig. 9.3 High-level overview of the IoSense STK architecture

to interact with an existing system. Services can communicate through a message bus. Exemplary, an MQTT Broker like Mosquitto, or a data streaming pipeline like Kafka can be incorporated to transfer data and messages. IoT devices can connect to a service or the message bus directly. To solve the connection problem between various IoT devices, open-source tools (e.g., Cap'n Proto, Protobuf) are used to provide the flexible automated mechanism for serializing and deserializing structured data. In this way, a unified structured data is written and read to and from a variety of data streams with different languages from many kinds of IoT hardware platform.

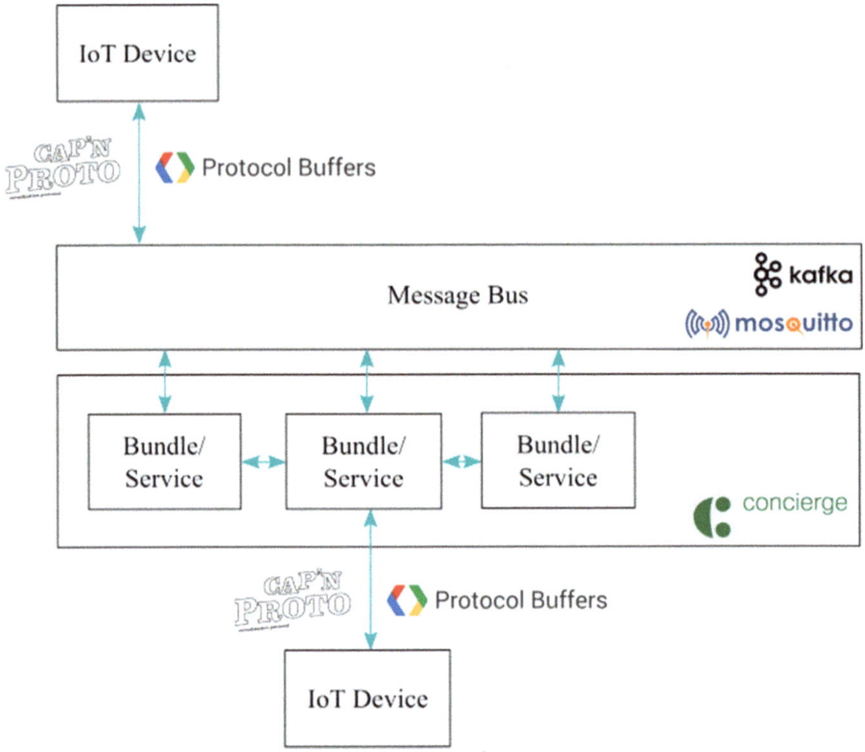

Fig. 9.4 Architecture of the IoSense STK together with the used open-source technologies

9.3.1.1 Message Bus

The STK implements a generic message bus, thus offering a broader approach by using, for example, Apache Kafka or Mosquitto as a broker. The STK is not limited to a special message queue technology. This architectural decision in the STK can be much better exploited.

Especially when using, for example, Kafka, real-time data analysis of sensor data is possible with very high throughput. On the other hand, MQTT is preferably used in low-bandwidth and high-latency environments which makes it an optimal candidate for single-board computers with CPU limitations. This makes the STK very versatile: For every individual environment, the optimal message bus technology can be used, as it abstracts the whole message queue layer. Furthermore, a robust system is realized which can be easily scaled. For instance, many MQTT brokers can be distributed on many nodes providing failsafe operations. They are forming an MQTT cluster and logically act as one broker. The cluster is accessed via a load balancer providing a single point of entry. This analogously applies, for example, to Apache Kafka.

Table 9.1 Open-source frameworks used by IoSense STK

Name	Description
Kafka	Apache Kafka is a distributed streaming platform. It is used for building real-time data pipelines and streaming apps. Kafka provides low-latency, high-throughput, fault-tolerant publish and subscribe pipelines. It can process streams of events. (https://kafka.apache.org/)
Mosquitto	Eclipse Mosquitto is an open-source message broker based on the MQTT protocol. It is designed to guarantee a lightweight specification. Therefore, Mosquitto is suitable for all devices from low power devices to full servers. (https://mosquitto.org/)
Concierge	OSGi is a popular framework for dynamic modules for the Java language. Eclipse Concierge provides a small and lightweight implementation optimized for mobile, embedded systems and especially IoT. (https://www.eclipse.org/concierge/)
Protobuf	Protobuf uses an open-source license and is patent-free, well-tested (used by Google since 2001) and language-independent as well as platform-neutral. Protocol buffers (short for Protobuf) are a method of serializing structured data, established by Google. Protobuf provides a code generator for multiple languages under an open-source license. It is useful in developing programs to communicate with each other over a wire or for storing data. (https://developers.google.com/protocol-buffers/)
Cap'n Proto	Cap'n Proto is a lightweight framework and provides a fast data interchange format. Also, it also supports a capability-based RPC system with better performance compared to traditional RPC. (https://capnproto.org/)

9.3.1.2 Connectors

An essential feature of every IoT framework is the ability to connect to sensors, actuators, and other similar devices. In openHAB, the notion of sensor bindings is used which is completely different from the STK's way to interact with such sensors.

The core questions are: What kind of protocol is used in the device and what kind of data format is used for exchange? Moreover, there are many ways how data can be transferred. The STK abstracts from these details and uses, for example, Google's Protobuf. This technology makes it very easy to transfer data from different devices into the STK and then to other components through the message bus.

Usually, data processing is performed on fog nodes which improves the overall application performance when the computation is done locally directly on the fog devices where the data is produced, or actuators must be executed. Thus, minimizing latency and shrink response-time.

The communication of the bundle for the distributed processing is carried out over the generic message bus, as described in the previous section, **Section Message Bus**. The abstract message bus layer is capable of handling different implementation of message queue technologies and different data format for the information exchange of sensor and actuator data as well.

The STK is not limited to a single message format technology when connecting sensors. However, for a running application built with the STK, it must be prior defined which data interchange format is facilitated. Usually, any data format can be used, e.g., JSON or XML. However, as consistency is preferred over mixed formats, the STK already utilizes existing message formats, like Cap'n Proto, Google Protobuf, or Apache Avro which can be used according to the actual implementation requirements. These message formats have well-defined notations to define datatypes, protocols and offer serialization and deserialization mechanism for the data, thus, providing a consistent format across the whole application. In the following, the integration of Protobuf is illustrated to demonstrate the STK's capabilities.

Protobuf is very lightweight, and one of its properties is its efficient encoding which allows faster transmissions through a network. This is valuable regarding low-bandwidth and high-latency environments. Also, it is well suited for single-board computers and other devices that have limited CPU, memory, and bandwidth, but it can also be used in large-scale applications. Reading arbitrary sensor information requires to define the structure of the Protobuf messages by plain text files before. Those files are then compiled using the Protobuf compiler. After the compilation process, the data accessors are automatically generated, as well as simple accessors to read and write fields, methods to serialize and deserialize structures from and to raw bytes. This is the basic workflow when working with Protobuf, which is very well integrated into the STK. Since the STK provides an abstract layer for connecting sensors and other IoT devices, an implementation based on the Cap'n Proto data interchange format and Apache Avro support are provided.

9.3.1.3 Bundle Architecture

The STK uses *concierge* [4] as a lightweight version of the OSGi framework. This makes it preferably in IoT environments. In contrast, openHAB uses Equinox (http://www.eclipse.org/equinox/) as a reference implementation of the OSGi Core 4.x specification. Both frameworks, the STK and openHAB, offer the key features of OSGi, including modularity through the bundle concept, runtime dynamics for managing components at runtime and the possibility of inter-component communication through services. To explain, functionality can be easily maintained and extended on-demand. Thus, the flexibility and scalability are guaranteed.

The STK applies the MAPE-K [5] reference model inside a bundle to model an autonomous system with decentralized components.

Figure 9.5 shows the complete detailed architecture of a service. Input data from the message bus is received first and then deserialized by the deserializer module. Here, any message format can be used, e.g., Protobuf (see Table 9.1). The core functionality of a service is the data processing which is realized by four steps: *Monitor*, *Analyze*, *Plan*, and *Execute*, finally generating output data. The service architecture conforms to the MAPE-K [5] feedback loop, a widely applied reference control model used for self-adapting systems. After that, the output data combined

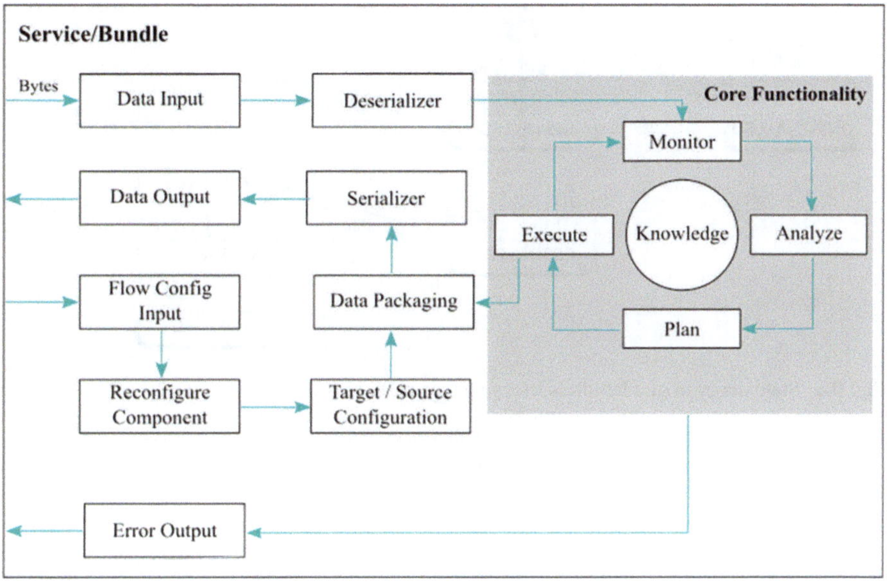

Fig. 9.5 Service architecture

with user configurations is serialized and sent to arbitrary IoT devices or a bundle via the message bus. Furthermore, the bundle also provides a logging service to monitor and report computation errors and exceptions.

The Lifecycle of a Bundle

One key feature of OSGi includes modularity through the bundle concept, which allows runtime dynamics for managing components at runtime. Components can be installed, updated, or removed without shutting down the running system. An associated `Bundle` object exists for every installed bundle in the OSGi framework ([6], p. 89). It can be used to manage the lifecycle of a bundle.

The dynamic nature of the OSGi platform transitions a bundle through different states in their lifecycle (see Fig. 9.6). A bundle can be in various states, e.g., it can be installed, resolved, stopped, or uninstalled. The `BundleListener` class provides a callback method `bundleChanged (BundleEvent event)` that allows the system to get notified of the different states.

A bundle can be started by calling the start method of a `Bundle` object, or the OSGi framework can automatically start a bundle if it is in the ready state and the autostart setting of a bundle is set accordingly. After the OSGi framework is launched and passes start level zero, it will enter start level one. All bundles which are assigned to this start level are started then.

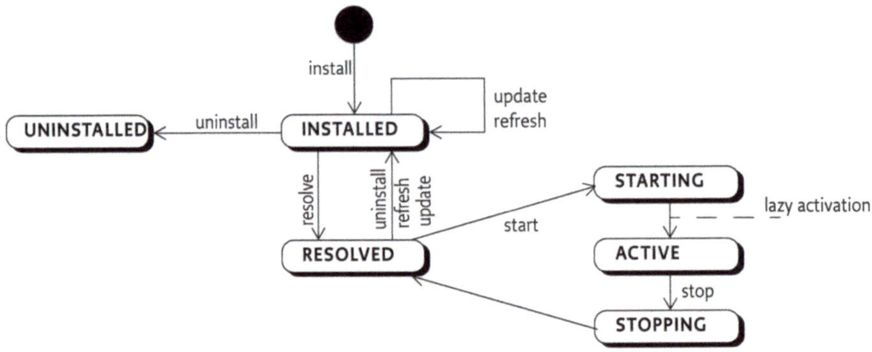

Fig. 9.6 State diagram of a bundle's lifecycle (see p. 90 in [6])

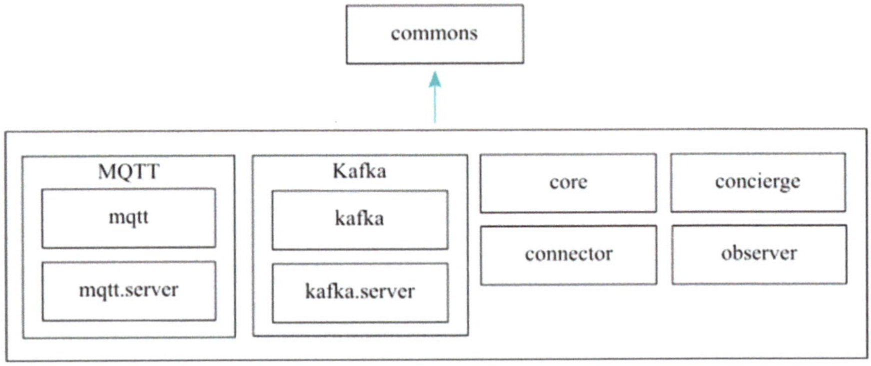

Fig. 9.7 Overview of all core bundles of the STK

This is utilized by a configuration file which is passed when starting the IoSense STK indicating which bundle should be started at which start level and in which order.

System Bundles

After presenting the high-level architecture and functionality, the specific software bundles that are incorporated in the STK are highlighted, which is depicted in Fig. 9.7. This will give a detailed view on the STK what bundle is responsible for the concrete functionality of the architecture.

The commons bundle is a system bundle which defines all available data types that can be used as data exchange formats between the bundles. Moreover, it is a base bundle for all other bundles. The connector bundle is the "glue code" for connecting sensor interfaces, grepping data from them and input them in the queue, like Kafka or MQTT. The server bundles are complete implementations for message queues. They include implementation of Apache Kafka Server in Java and contain an internal implementation of the Mosquitto broker, respectively. The MQTT bundle can be launched if no external MQTT broker is present. In the observer bundle, the common observer pattern is implemented for the STK. This package is optional, but it can be used for tasks, where it is important to listen to a specific number of connected sensor types and print them to the console only for visualization. The concierge bundle manages the complete lifecycle and holds all configuration files, for instance, which bundles should be loaded and when. The bundle is responsible for starting the OSGi framework. The core bundle is an empty module that can be used to include functionality to provide the algorithms throughout all bundles.

Bundle Composition

A configuration file defines the order in which the services are started. This is regarded as the first level of bundle orchestration of an application. This file is passed as program argument to an STK application.

Depending on the actual requirements, the services are deployed on different devices. The interaction and communication of the services are carried out over the message bus. Thus, services can make requests to other services in their current system boundary among themselves in order to execute certain functionalities provided by these services. The problem here is that OSGi bundles are bound to the scope in their currently running JVM.

Therefore, a distribution of services across all devices that depend on each other in the fog network is performed. Single bundles are deployed on-demand which enables services to request dependencies from remote devices as if they were running in the same JVM. Services can be deployed and run on devices depending on the device's performance and locality in the fog computing environment. Meaning that the deployment of a service on a device near the physical location where the action takes place (e.g., actuator movement of a robotic arm). Therefore, increasing the execution speed through short network runtimes and decreased network delay.

9.3.2 Fog Computing Application

Being able to harmonize the elements regarding communication and self-adaption is the key element of a fog computing framework. Thus, after having detailed the inner workings of the architecture, this section introduces how the STK and its components are deployed in the fog.

An application inside the STK comprises a set of so-called bundles, where each bundle defines a specific set of functions and is regarded as a component. Hence, the application itself can be divided into small separate components. A fully working application can run on a single device or can operate as an application, which will pull the necessary bundles from other devices to perform a task.

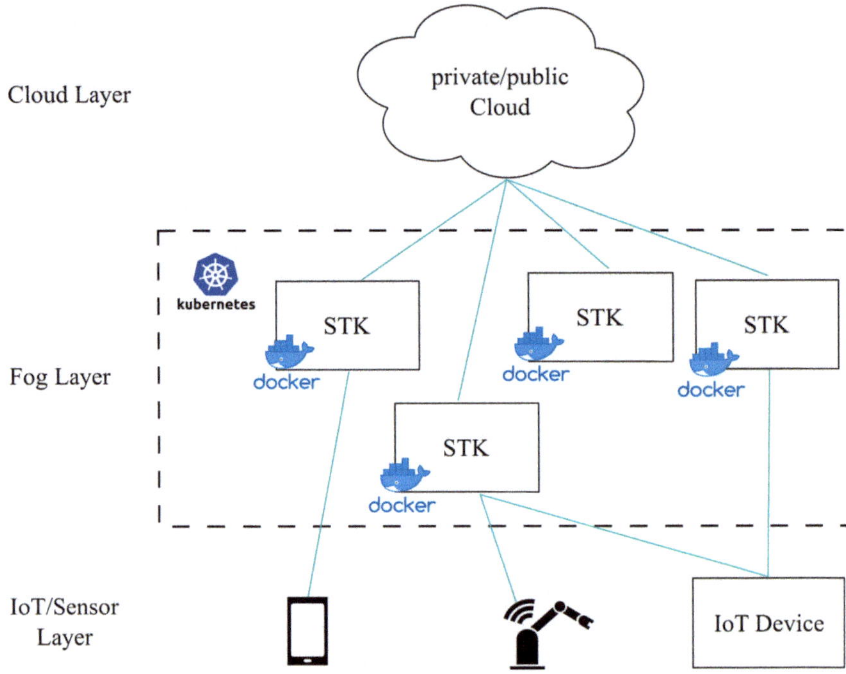

Fig. 9.8 Deployment of the IoSense STK for a fog computing environment

Figure 9.8 describes the practical deployment of the IoSense STK in a fog computing environment. The IoSense STK runs on each fog node for a specific task. Using the bundle architecture, direct communication with IoT devices and the cloud is possible at the same time. The architecture of the STK makes the deployment of either whole applications or individual components possible due to the OSGi bundles. In Fig. 9.8 this deployment process is illustrated using Kubernetes and Docker, as both technologies can be incorporated for monitoring and orchestrating STK operations on fog nodes.

Also, they provide tools to deploy the individual instances of the STK efficiently and simply when a large number of fog nodes need to be set up. Kubernetes can be facilitated for that very task allowing the deployment and scaling of bundles in the network. Using Kubernetes in conjunction with OSGi enables to maintain and extend services on-demand, thus guaranteeing flexibility and scalability of the system.

As described in **Section Message Bus,** different variants of the message bus technology can be used. The instances of the message bus must be managed as well. The scalability of the system is achieved by ensuring to provide redundant brokers, thus, achieving the necessary load balancing. Two examples are given and are intended to illustrate the scalability of the message bus—supported by the STK—using different technologies.

MQTT brokers are used for the communication via the MQTT protocol. Eliminating the Single Point of Failure (SPF) is achieved by creating an MQTT broker cluster making the communication fault-tolerant. Regarding the scalability, MQTT brokers are deployed in the cluster depending on the network load and other measurements provided by default from Kubernetes. This approach applies to Apache Kafka. In the second case, Apache Zookeeper is used as a load balancer to coordinate a Kafka cluster consisting of many Kafka brokers. A Kafka broker is publishing and receiving messages for several topics from several consumers and producers, respectively. IoT devices communicate with Kafka through a REST proxy. Since nearly every embedded device can send HTTP requests, it makes a REST proxy as an interface to a Kafka cluster a valid alternative.

9.3.3 Summary

To sum up, an overview of the IoSense STK was given; our key driving factors for the development of the STK were multiple IoT-centered demonstrators that will be explained in the next section.

This section explained the underlying reference architecture and compared openHAB to the STK to identify and highlight the differences. It turned out that the STK makes it as easy as possible to connect the most common sensor interfaces. It was shown that the STK's architecture supports integrating sensors in complex systems or building them because the used technologies make it possible to integrate into nearly every environment seamlessly—either in a fog or a cloud. For example, Google's Protobuf and Cap'n Proto can be used on almost every microcontroller. The STK provides an abstraction layer for binding sensors and IoT sensors to a system and handling the communication and control. Therefore, the STK makes it very easy to read the data out from a generic serial port or a general-purpose input/output (GPIO) serial port—it implements both data interchange formats which is possible due to the abstraction layer. Protobuf is an efficient technology which is beneficial in low-bandwidth and high-latency environments as in a fog. Using this approach makes the STK very versatile to communicate to almost any sensor or similar IoT devices and be prepared for future formats and protocols.

After getting the data, it must be written to some storage so that it can be used by other applications or components—depending on the specific use case. This is an answer to the issue of data distribution, which is a strength of the STK—it is so powerful as it also abstracts the message queue layer completely in this regard. Kafka and MQTT as ready to use applications are implemented, and the STK is open to extensions for additional message queue technologies. The STK implements the MAPE-K feedback loop concept inside a bundle. This enables the modeling of distributed self-adaptive systems with decentralized components which lays the foundation for the use in a fog environment. Furthermore, a high-level overview as well as an in-depth look at the incorporated bundles was described to explain the intrinsic workings.

9.4 Evaluation

The STK is evaluated by implementations of practical IoT scenarios. In this section, we demonstrate the STK with three scenarios. On the one hand, one basic robotic application, which can be controlled by an Arduino-based input device. One complex robotic application, where a wearable controls the movement of a robot arm, on the other. Finally, a context-dependent lighting demonstration which can be utilized for example in a distributed smart home scenario.

9.4.1 Robotic Demonstrators

In the field of robotics most industrial applications are used for steering robot arms, like LBR iiwa manufactured by KUKA AG, or the UR10 developed by Universal Robots. Those applications have a monolithic architecture. The developers write cumbersome software with a complex configuration that cannot cope with the increasing innovation of new sensors and other complex systems, thus, making the handling and integration very difficult. Specifically, it is not possible to inject routines during the execution of the software or to integrate external signals (e.g., from a force-momentum sensor) into the application's flow.

Since until now, elementary systems of industrial robots could not handle the flood of information and take the necessary reactions resulting from the data. In consequence, companies like KUKA AG re-organized their whole system and developed their application from the beginning for the KUKA LBR iiwa robot for a better human-machine-collaboration experience.

Recently, considering robotics in the industry, only external information was available to interact with the robot when harm was occurring. That was an emergency stop button which could be activated by a worker only if some unsupervised person entered the protected area or when an irregular event occurred. Such an event leads most of the time to a complete downtime of the whole process. Certainly, a cumbersome restart of the application is necessary. As a result, the workpiece could be made unusable and causing additional costs.

For the new generation of robots regarding human-machine-collaboration, those systems are not anymore economically justified and can seriously harm humans. A robot should not be shut down immediately when a human is entering the room where the robot is working. Those robots should be equipped with various sensors, for instance, force-torque sensors and environmental scanners, to detect the presence of a human, and thus react accordingly. For example, the system should minimize the overall velocity, take short breaks, or change and adapt the work process. The only reaction for sensor information should not be the activation of an emergency stop button to stop the whole work process.

All those reactions need complex event processing systems that were not included in the development and implementation of such robots. The robot's adaption of the work process with multiple collaborating human workers is a further area. Without a doubt, every human has different body proportions which make the adaption of the robot necessary.

Specifically, KUKA AG was taking this event processing approach with their Sunrise-OS. A robot arm gets different events from other robots which can be processed in parallel, and therefore react accordingly and adapt automatically at execution time/runtime of the system. This implementation approach is very appropriate for the above-mentioned use case. One drawback still exists: this solution is not scalable with increasing sensors, as in other areas like currently evolving for smart homes. Therefore, different approaches are indeed necessary—where the STK is one of those.

9.4.1.1 Controlling a Robotic Arm with a 3D Mouse

For the first evaluation, a three-dimensional mouse is implemented to control any three-link robot with revolute joints. An arbitrary IoT device is utilized to steer a robotic arm. This first example demonstrates some basic features of the STK regarding the deployment of single services on different devices and how these services communicate through a message bus.

The robot we used here is a UR10, produced by Universal Robots. The input controller representing the three-dimensional mouse is described in a moment. The input controller generates the necessary data that is generated by the motions and transmits it to the robot, finally, reproducing the results of the received data by imitating the motion of the controller. This process is depicted in Fig. 9.9.

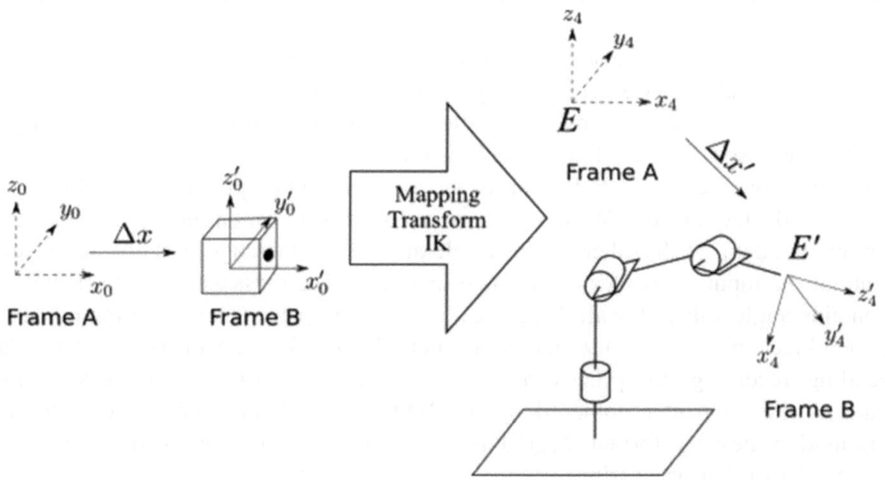

Fig. 9.9 The translation of the input controller (that is, the cube symbol) from a frame A to frame B results in Δx which is translated to $\Delta x'$. The input controller's position corresponds to the end effector orientation

In the figure two frames A and B are shown; between these two frames, a motion of the controller is executed resulting in a change, denoted by Δx. Several steps including mapping, transformation, and inverse kinematics computation are performed to get the corresponding $\Delta x'$ for the robot manipulator which changes the position of the end effector E accordingly. This change represents the same motion as performed with the controller.

Input Controller

The controller is regarded as a three-dimensional mouse, intended to be held by one hand. It has the form of a cube where the housing is made from cardboard with a dimension of 100 mm × 100 mm × 100 mm (H × W × D). A Feather M0 from Adafruit is used as portable microcontroller board with an ATSAMD21G18 processor running at 48 MHz with 3.3 V power. Connected to the Feather M0 is the BNO055, a 9-DOF sensor to acquire the necessary orientation value along two axes, and additionally it is used to measure the displacement with respect to the device's initial position. The sensors deliver orientation information and acceleration data which are used to build the spatial vector representing the current position of this input controller.

Service Components

Software bundles contain certain parts of the whole functionality. These bundles are distributed on different fog nodes in the local network allowing the local processing near the devices producing the data or actuators, thus connecting the fog and IoT layer.

First, the input controller sends data using a client to communicate via MQTT. Therefore, we created the MQTT topics `o/orientation` and `o/acceleration` to send the angles and acceleration values of the input controller, respectively. After the broker receives the message, it sends it to the *Connector* bundle. The data format for the transmission is specified using Google's Protocol Buffers Format. We defined two message formats, on the one hand for the orientation data, on the other for the acceleration. Each message contains the current state of the input controller for a single time step. The message for the orientation contains angle values for all three axes, a "face" property (indicating the upwards pointed face, perpendicular to the floor), and a 3-tupel for time attributes storing the sending, receiving, and processing time. The message format for the acceleration data of the controller gathered by the BNO055 is defined analogously to the orientation message format. MQTT is used to transmit the data to the *Connector* bundle for further processing.

Secondly, as the *Connector* receives the data packages, it sends these to the *Analyzer*. Before, the connector validates the data and performs some initial pre-processing. The *Analyzer* bundle contains the whole functionality of the motion translation process which includes value mapping, coordinate system transforma-

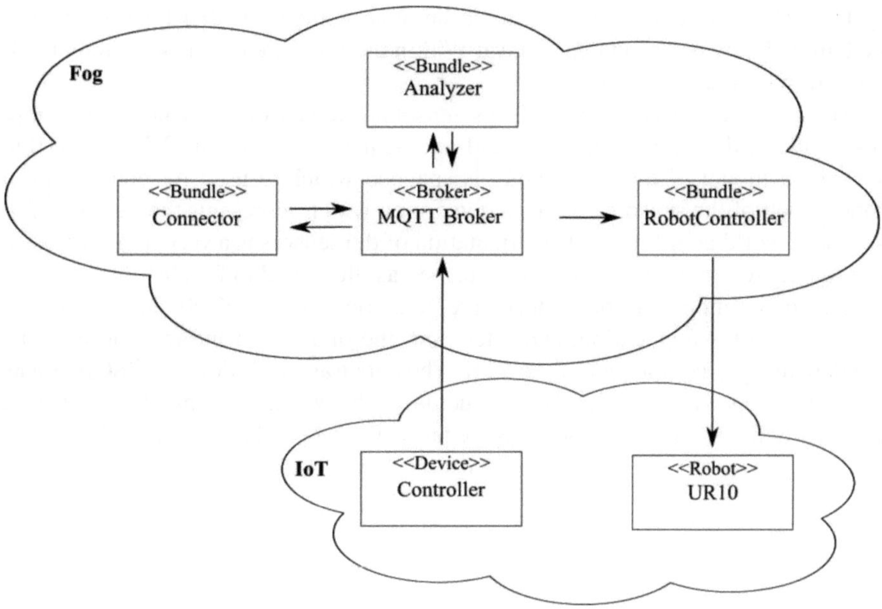

Fig. 9.10 The whole workflow is separated into many services, implemented as bundles, and deployed on several devices. Processing and execution take place near the device executing the actual task

tion, and inverse kinematics computation. The input controller data (orientation and position) are used as parameters for the inverse kinematics model of the UR10.

Finally, the results of the computations are sent to the *RobotController* bundle. This bundle contains the control classes to interact with the UR10 via TCP/IP over the network. Before, we specified the IP address of the robot. Now, a simulation can be run, or the actions can be transferred to the physical instance of the UR10, as depicted in Fig. 9.10.

9.4.1.2 Sensor Jacket and Mr. T

In this section, the steering of a robotic arm with the help of a sensor jacket is described as a practical example of the implementation. The goal is to control two robot arms, the UR05 and UR10, developed by the company Universal Robots, with sensors attached to the human body by a wearable jacket. Therefore, 9-axis position sensors from Xsens (https://www.xsens.com/) were used. Those sensors gather information about their spatial orientation and movement via a proprietary interface. With a C# application, a connection to this sensor interface can be established. This C# application feeds the sensor information into the toolbox—and in this special case, over a server socket. As an afterthought for a later development step, the C# application could implement an MQTT client to transfer the data directly into the system. The information transfer could be done without another interpreter in the system which is implemented in Java. Within the STK the data is gathered and fed into the system as a data stream.

The jacket consists of three Xsens sensors: one sensor is located at the upper body, one at the right upper arm, and one at the right forearm. With these three sensors, a model of the upper body is spanned which is used for controlling the robot. Each of the sensors outputs a quaternion which represents the rotation of the sensor in world coordinates. The orientation of the sensors is a very important issue. For this reason, the sensors must be attached as illustrated in Fig. 9.12.

The quaternions of the sensors are read out by the WPF application (see Fig. 9.11), a C# application, and afterward, the quaternion information is sent to the data mapper component of the STK. The data transfer is done via JSON-Strings which is mapped to the internal data structure of the wearable within the data mapper and finally dispatched over the message bus (refer to **Section Message Bus**).

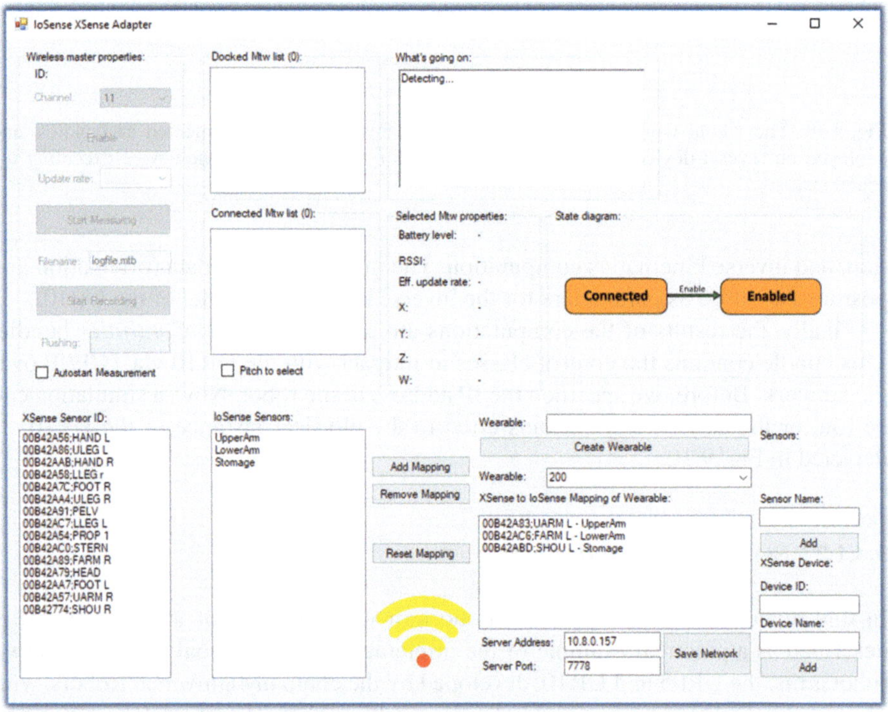

Fig. 9.11 WPF application for reading the Xsens sensor information

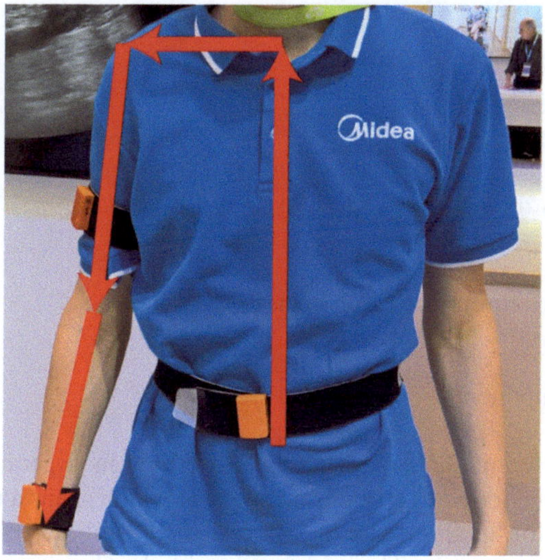

Fig. 9.12 The wearable jacket—positioning of the sensors

IoSense Xsens Adapter

The IoSense Xsens adapter is a Windows™ WPF application that establishes a connection to the IoSense receiver (see Fig. 9.11). The application's task is to assign the several Xsens sensors to the individual limbs of the body.

In this use case, the list box "IoSense Sensors" shows the three available limbs of the persons, and the list box "XSense Sensor ID" shows a list of all possible Xsens sensors where the first 8 letters represent the HEX ID of the sensors and the last ones represent the description of the sensor's nameplate. The list can be extended with other sensors using this GUI. For the available sensors, a mapping is then created, assigning the corresponding wearable to the sensors.

To automatically transfer the sensor data to the STK of all available sensors, the button "Start Measurement" must be pressed—a WLAN symbol shows the connection status to the STK.

Data Mapper Component

The data mapper converts the quaternion data from the IoSense adapter (see **Section IoSense Xsens Adapter**) in the internal data structure of the "Wearables" which consists of an upper arm, forearm, upper body, and the hand. This data is forwarded through the OSGi bus to the Goal Mapper component.

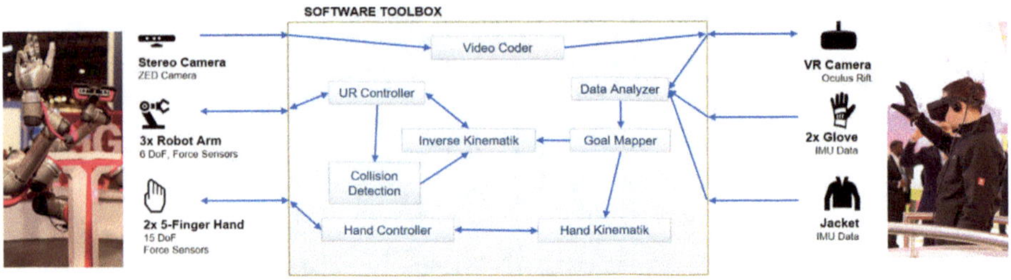

Fig. 9.13 Detailed implementation details of the demonstrator

Goal Mapper Component

The *Goal Mapper* component converts the quaternion in target coordinates. Therefore, the data is transferred to an external package which spans a model of the body. By that, the position in space can be calculated (in millimeters) where the upper body is the coordinate origin. The settings file with the Concierge module stores the body dimensions of all limbs of the operating person. They are used as a reference for the calculation of the target point. In Fig. 9.12, the vector lengths are displayed in a simplified scheme. Those values must be adjusted for different users.

Besides the normal computation of the spatial positions, a calibration must be done. For that, the user must position himself/herself looking at the actual target of the operation. After that, the calibration can be started. Thereupon the user is calibrated regarding the working direction. A complete rotation on the user's axis is the maximum range of movement.

In case, another mapping of those goals should be executed, for instance, limiting the working area to a pre-defined space, the mapping can be implemented in the same component, or a new component can be created.

The result of the "goal mappers" is so-called goal directions—spatial vectors which show in the direction of the robot's end effector. The corresponding orientation of the vector is responsible for the rotation of the last joint.

Inverse Kinematic Component

The *Inverse Kinematic* component converts the result of the *Goal Mapper*—which are points in space to a millimeter accuracy and two vectors that are describing the orientation and the rotation of a hand into a target axis position of the robot. Hence, an external IK algorithm is executed where the previously calculated results are passed as arguments which in turn output the target joints. Attention should be paid that this result can also be an approximation if the target cannot be reached effectively.

Finally, the calculated joint values are sent through the OSGi bus as radian values, where the *UR Controller* (see Fig. 9.13) can interpret the data and send it to the robot afterward.

9.4.2 Smart Lighting

The last evaluation of the STK is the implementation of a person-dependent lighting control within a building. The goal is to illuminate the room to consume as little energy as possible, however, sufficiently provide as much light as needed. Currently, the whole building or complete parts of it are illuminated, whether a person is in there or not, or whole corridors when only one person is present. However, using movement sensors, person tracking technologies and light sensors luminaires for buildings can be optimized regarding the energy efficiency with the use of the IoSense STK.

Regarding the technical implementation, three components are involved (see Fig. 9.14):

- Light controller: A link component for the light control
- Intensity analyzer: A sensor link for the brightness sensor
- Component detecting the presence of a person: A sensor link for the movement sensor

The *Intensity Analyzer* component analyzes the brightness values and creates the needed light intensity. The other component analyzes the presence of a person and adapts the brightness of the lights. It deactivates all lights automatically when no person is near the lamps.

Additionally, a *User Config Evaluation* component takes person-dependent configuration settings, for instance, to consider different personal light settings. Finally, the *Light Controller* is in charge to turn the data into actions. Therefore, the light sources have to be controlled with the STK, depending on the room lighting and additional knowledge about the presence of other persons. In the first example, two lamps are controlled based on brightness sensors and a signal which simulates the presence of a human being. In the following, the implemented STK components are described.

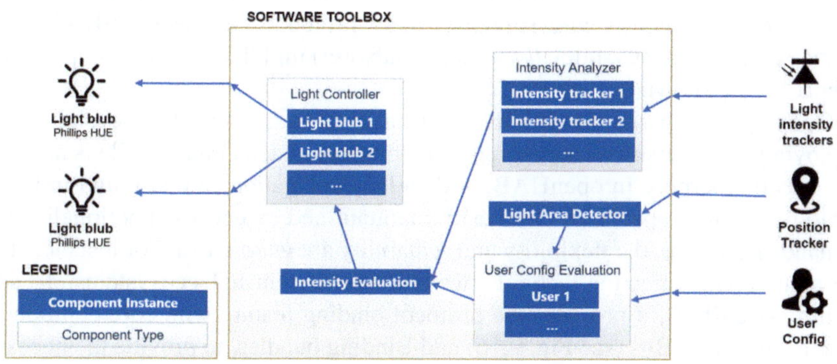

Fig. 9.14 Services for the lighting application

The STK allows getting the different information from all those IoT sensor devices to implement a self-adapting light control system, considering the user preferences.

9.4.3 Summary

This section presented the various IoSense demonstrators and how they contributed to the final stage of the STK's implementation. The different use cases were key driving factors in the development of the STK.

Controlling a robot with the use of a sensor jacket is a demanding and sophisticated use case which was implemented with the Mr. T demonstrator. It solves limitations regarding real-time capabilities. The 3D mouse had shown an alternative approach of steering a robot arm. The STK was used to develop a robust real-time system suitable for this scenario within fog computing environments.

Extending the application spectrum even further, another demonstrator in the field of smart homes was developed. The *Smart Light* demonstrator implemented a person-dependent illumination control system.

9.5 Related Work

Within the last years, several software systems were proposed to tackle the new challenges from the IoT. These systems range from full-fledged solutions for smart home over research prototypes, which only focus on a few concerns. Our proposed STK can be seen as an architectural foundation to all of these. While we do not deliver specific sensor implementations, we proposed an architecture which can be adjusted to fit most scenarios.

One of the most famous frameworks in the area of home automation, which has a large community and supports many protocols, is the **openHAB** (Open Home Automation Bus, https://www.openhab.org/) middleware. All components of openHAB are shown in Fig. 9.15.

OpenHAB is based on OSGi that is a Java framework for developing and deploying modular software programs and libraries. Each OSGi bundle is dedicated to a specific purpose in openHAB. A detailed explanation can be found in [8]. By exploiting OSGi, openHAB can easily maintain and extend its functionalities on-demand. Therefore, the flexibility and scalability are guaranteed. For instance, there are many communication technologies among different IoT ecosystems. To adapt them to openHAB, it provides the protocol binding feature with core components, such as the Event Bus (see Fig. 9.16) and Binding bundles, to provide the necessary communication abilities for different kinds of underlying technologies, shown in Fig. 9.16. In openHAB, *Bindings* integrate small devices, physical hardware, and external systems into the system. They are a special type of add-on for openHAB.

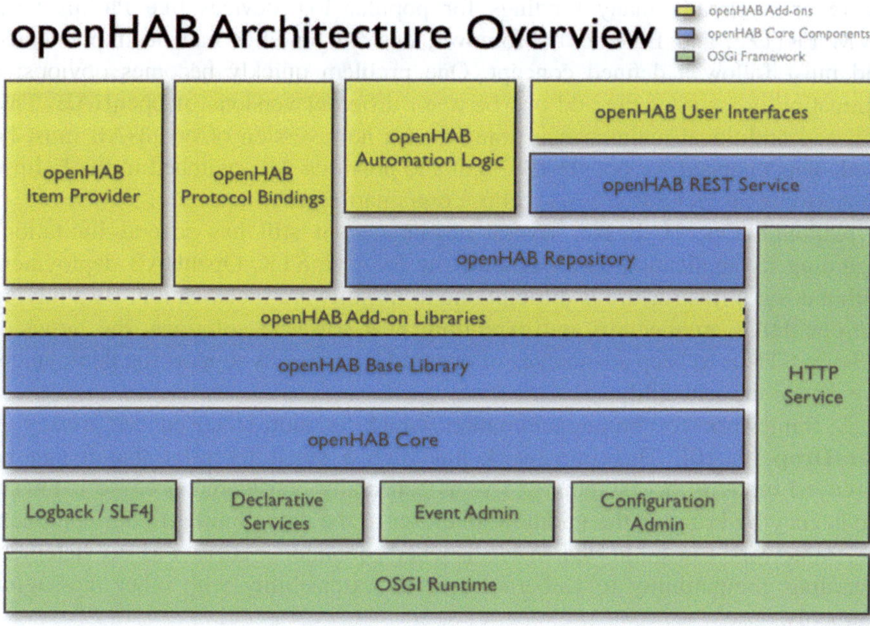

Fig. 9.15 The openHAB architecture (see [7])

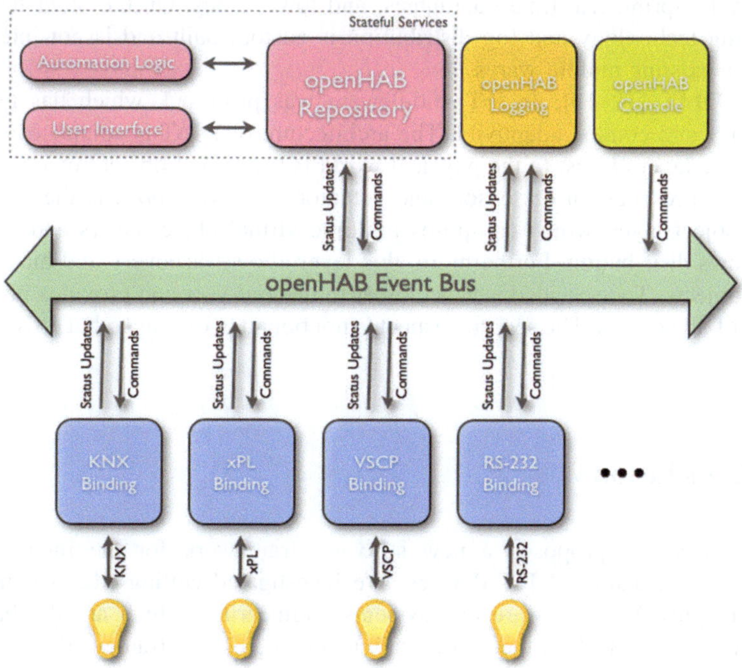

Fig. 9.16 The openHAB Event Bus (see [8])

There exist already many bindings for popular IoT devices like Philips Hue, AVM FRITZ!, and Bluetooth. Moreover, custom bindings can be implemented and must follow a defined concept. One problem quickly becomes obvious: a compatibility issue of the add-ons between different versions of openHAB. This will add additional maintenance time when a new version of openHAB must be used. This changed for openHAB 2.0 as the previous API switched to the Eclipse SmartHome project (http://www.eclipse.org/smarthome/).

Although OpenHAB has several advantages, it still has several limitations regarding its application as a basis for the IoSense STK. OpenHAB deployment within a fog infrastructure is difficult due to its software specification. By studying the openHAB architecture and exploiting new software solutions, the target of IoSense STK is to keep advantages of openHAB as well as support fog deployment to enhance the scalability.

A framework for home automation which is more focused on privacy is **ParaDrop** [9, 10]. This framework has such a small footprint that it can be deployed on WiFi access points. This is especially considered because it lowers the latency by bringing the middleware closer to the data. Similar to our approach, the framework can be hosted inside Docker containers. There was no statement regarding compatibility to Kubernetes or its extensibility with other messaging protocols.

Lisa 2.0 was proposed in [11] and is a low-level IoT framework based directly on a real-time enabled micro kernel RIOT. Because of that, Lisa has advantages of a low footprint, real-time guarantees, and low latency. On the other hand, the messaging technology and fog distribution is custom built and is not integrating well with existing middle-wares.

With **DIAT** [12–14], an IoT middleware was proposed, which has its focus on privacy and context adaptivity. The architecture of DIAT is composed of three layers. The layer closest to the physical world is the virtual object layer, providing a virtual representation of sensors and actuators. A layer above is the composite virtual object layer, which composes multiple virtual object layers and therefore also allows distribution. Furthermore, this layer allows discovery and matching of virtual objects. The service layer is closest to the end-user and provides high-level control of all devices. The security aspect is not bound to one layer, but cross-cutting across all.

9.6 Conclusion

In this work, we proposed a new software framework for the integration of sensors and actuators of IoT devices. We investigated cutting-edge technologies and principles from software and systems engineering, which are the basis for the IoSense STK. In IoT, there are many technologies and frameworks to support application development. A central concept and requirement for IoT applications is the distribution of data and computation, which is why computing models and system architectures such as fog computing, cloud computing, edge computing, and dew computing [15] are emerging as new paradigms and becoming more important in this area. With the STK, we focus on the application of future-proof technologies and standards.

Within the course of our investigation and evaluation of technologies, frameworks, applications, and architectures related to IoT and sensors with respect to their advantages and disadvantages to being used within our sensor framework, we identified various concepts that will form the basis of the IoSense Software Toolkit. The new paradigm of *fog computing* will serve as the central system architecture principle as it fulfills the requirements of fast data processing, achieving low latency and high bandwidth as well as security and fault-tolerance for sensor applications close to the computing edge with real-time demands. The principle of *Bundles*, taken from the OSGi standard, permits a component-oriented software development approach and flexible deployment including runtime updates. The reference architecture of an IoT middleware for smart homes described by openHAB shows promising concepts regarding the combination of the aforementioned technologies. We will use this architecture as a basis and guide for our developments (e.g., having an *event bus* for communication and adapters to integrate new sensors). To remedy the lack of support for the fog computing infrastructure, we will use state-of-the-art and future-proof *containerization* and *orchestration* techniques based on Docker and Kubernetes.

In our ongoing work, we are combining the system of bundles with formal methods of Petri nets and context adaptation. With an adaptive Petri net [16], the bundles and its order of execution are configured based on a context (e.g., a sensor value). With this, the designed architecture can be verified according to the specification with temporal logic. Furthermore, we are working on an improved data handling by combining the STK with the statistics tool R. To cope with the huge amount of data in our distributed fog architecture, the data handling has to be managed in a distributed manner by utilizing R Container [17].

The IoSense STK was evaluated on three use cases, ranging from medium complexity to highly complex robotic applications. They demonstrate how the components of the STK can be utilized and formed in a way that they fit modern IoT-centered projects.

References

1. 50 billion devices online by 2020, https://www.telegraph.co.uk/technology/internet/9051590/50-billion-devices-online-by-2020.html. Accessed 21 Apr 2018
2. F. Bonomi, R. Milito, J. Zhu, S. Addepalli, Fog computing and its role in the internet of things, in *Proceedings of the First Edition of the MCC Workshop on Mobile Cloud Computing*, pp. 13–16. MCC '12
3. A. Al-Fuqaha, M. Guizani, M. Mohammadi, M. Aledhari, M. Ayyash, Internet of things: a survey on enabling technologies, protocols, and applications. IEEE Commun. Surv. Tutor. **17**(4), 2347–2376 (2015)

4. Rellermeyer und Alonso, "Concierge", in *Proceedings of the 2nd ACM SIGOPS/EuroSys European Conference on Computer Systems* (2007). https://doi.org/10.1145/1272996.1273022
5. "An Architectural Blueprint for Autonomic Computing.", White Paper, http://www-03.ibm.com/autonomic/pdfs/AC%20Blueprint%20White%20Paper%20V7.pdf. Accessed 9 Apr 2018
6. OSGi, "OSGi Core Release 5", https://osgi.org/download/r5/osgi.core-5.0.0.pdf. Accessed 10 Apr 2018
7. https://commons.wikimedia.org/wiki/File:OpenHAB_Architecture.png. Accessed 12 Apr 2018
8. "openHAB—empowering the smart home", http://www.openhab.org/. Accessed 15 Aug 2018
9. P. Liu, D. Willis, S. Banerjee, Paradrop: enabling lightweight multi-tenancy at the Network's extreme edge, in *Symposium on Edge Computing (SEC), IEEE/ACM*, (2016), pp. 1–13
10. D.F. Willis, A. Dasgupta, S. Banerjee, ParaDrop: A multi-tenant platform for dynamically installed third party services on home gateways, in *Proceedings of the 2014 ACM SIGCOMM Workshop on Distributed Cloud Computing, DCC '14*, (ACM, New York, 2014), pp. 43–44
11. B. Negash, A.M. Rahmani, T. Westerlund, P. Liljeberg, H. Tenhunen, LISA 2.0: Lightweight internet of things service bus architecture using node centric networking. J. Ambient. Intell. Humaniz. Comput. **7**(3), 305–319 (2016)
12. C. Sarkar, S.N.A.U. Nambi, R.V. Prasad, A. Rahim, R. Neisse, G. Baldini, DIAT: A scalable distributed architecture for IoT. IEEE Internet Things J. **2**(3), 230–239 (2015)
13. R. Neisse, I.N. Fovino, G. Baldini, V. Stavroulaki, P. Vlacheas, R. Giaffreda, A model-based security toolkit for the internet of things, in *2014 Ninth International Conference on Availability, Reliability and Security (ARES)*, (2014), pp. 78–87
14. P. Vlacheas, R. Giaffreda, V. Stavroulaki, D. Kelaidonis, V. Foteinos, G. Poulios, P. Demestichas, A. Somov, A.R. Biswas, K. Moessner, Enabling smart cities through a cognitive management framework for the internet of things. IEEE Commun. Mag. **51**(6), 102–111 (2013)
15. P.P. Ray, An introduction to dew computing: Definition, concept and implications. IEEE Access **6**, 723–737 (2018). https://doi.org/10.1109/ACCESS.2017.2775042
16. C. Mai, R. Schöne, J. Mey, T. Kühn, U. Aßmann, Adaptive Petri Nets – A Petri Net Extension for Reconfigurable Structures. Presented at the ADAPTIVE 2018, The 10th international conference on adaptive and self-adaptive systems and applications, 18 Feb (2018)
17. D. Grzelak, J. Mey, U. Aßmann, Design and concept of an osmotic analytics platform based on R container, in *Proceedings of the 2018 International Conference on Foundations of Computer Science (FCS'18)*, (CSREA Press, Las Vegas, 2018), pp. 29–35

Chapter 10
Sensor System Optimization Under Uncertainty

Wolfgang Granig, Lisa-Marie Faller, and Hubert Zangl

10.1 Introduction

10.1.1 Motivation

Sensor systems are used in a wide range of consumer, industrial, automotive, and aerospace applications. To enable progress in novel application such as autonomous systems and robotics, e.g., autonomous automotive driving [1, 2] and unmanned autonomous vehicles [3, 4], an increased number of in-built sensors is necessary. In all those fields, it is essential to define architectures for these sensor systems, which are optimized concerning performance, operating conditions, size and production costs, etc. This chapter focuses on optimization of sensor system architectures, considering uncertainties in input parameters, design parameters, environmental parameters, and the calibration setup, to achieve given system performance requirements. When defining sensor system architectures, the goal is to achieve best possible performance together with robustness against external influences while keeping production and maintenance costs low. The challenge in this step is the lack of exact knowledge of the governing physics of each individual produced system (e.g., due to individual production variation effects, etc.) as well as the effect of external influences and deviations to the final system. A common approach is to start from a basic system model based on past experiences and building-block specifications. A generic system model for further analysis is shown in Fig. 10.1 where the real input value x is project to the estimated value $\hat{x}$ and its uncertainty

W. Granig (✉)
Infineon Technologies Austria AG, Villach, Austria
e-mail: wolfgang.granig@infineon.com

L.-M. Faller · H. Zangl
Alpen-Adria-Universität, Klagenfurt, Austria

W. D. van Driel et al. (eds.), *Sensor Systems Simulations*,
https://doi.org/10.1007/978-3-030-16577-2_10

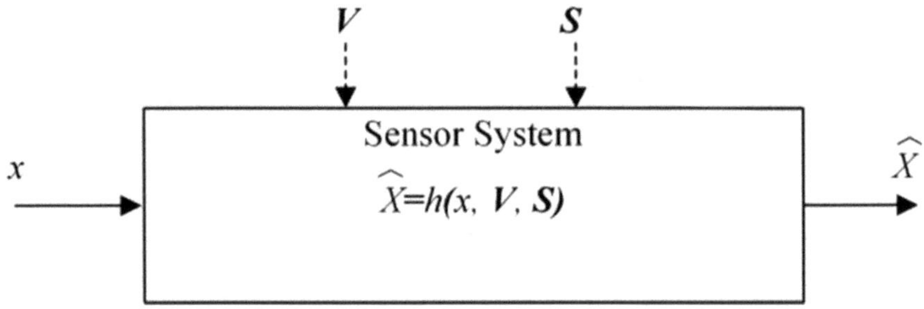

Fig. 10.1 This figure shows the sensor system model for further analysis and optimizations. The sensor projects each real input value x into estimated output $\hat{x}$. This process is influenced by design parameters $\boldsymbol{S}$ and influencing parameters $\boldsymbol{V}$ as illustrated by the dependencies of the transfer-function $h(\cdot)$. However, as the influencing parameters are random variables, also the output $\hat{X}$ becomes a random variable and the actual estimate $\hat{x}$is a realization of $\hat{X}$

u_x influenced by design parameters $\boldsymbol{S}$ and influencing parameters $\boldsymbol{V}$. Design and influencing parameters have nominal values represented in a vector (indicated here by bold letter), but in the practical realization they have random deviations and consequently they have to be described by random variables (indicated here by capital letters) generating the output random variable $\hat{X}$.

One common procedure is then to define typical operating conditions of the system and to excite the design- and influencing parameters, which are known and considered most important, up to their extreme values to assure the fulfillment of specifications [5]. Monte-Carlo simulations [6] can be performed if possible, but this is a time-consuming process, and high numbers of design parameters as well as influencing parameters increase the degrees of freedom and consequently lower the chance of finding all possible combinations which would violate the specification. In case a violation is found, and the design has to be changed to overcome the issue, the Monte-Carlo simulations have to be repeated. A reduction of iterations and a so-called pareto-front can be generated in case of multiple optimization objectives [7]. However, this process is only reasonable, if the system model adequately models the real system. Oftentimes, no such model is available during sensor system architectural exploration. The aim of this chapter is to describe a simplified and efficient way for sensor system optimization even in the architecture definition phase, based on statistical methods.

10.1.2 Statistic Considerations on Measurement Results

Real sensor systems convert the sensor output $\mathbf{y}$, now realizations for random variables and depending on influencing parameters and design parameters, to the estimated output variable. This conversion is modeled by mathematical combinations of variables and parameters, also termed estimator. A general model is shown

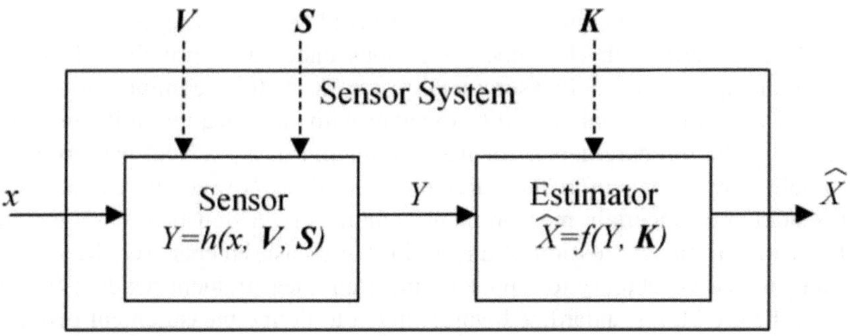

Fig. 10.2 This figure shows a block diagram of a sensor system model containing real input value x, influencing parameter $\boldsymbol{V}$, and design parameters $\boldsymbol{S}$ to generate the sensor values $\boldsymbol{y}$. In the estimator, this sensor values are processed into estimated output values $\hat{x}$ by applying estimator design parameters $\boldsymbol{K}$, e.g., calibration parameters

in Fig. 10.2. In this model, again, the real input variables x represents the true physical quantity of interest, the influencing parameters are denoted by $\boldsymbol{V}$ and the design parameters by $\boldsymbol{S}$. Additionally, the sensor system estimator parameters $\boldsymbol{K}$ (e.g., calibration parameters) need to be considered to get the output variables $\hat{x}$. The sensor functionality is modeled in terms of a mathematical description by a transfer-characteristic $h(\cdot)$ that relates the sensor output $\boldsymbol{y}$ to input values x, influencing parameters $\boldsymbol{V}$, and design parameters $\boldsymbol{S}$. The estimator which then incorporates e.g., also the calibration calculations finally provides the output $\hat{x}$.

Commonly, sensor system outputs are treated differently depending on the contributions and their characteristics, where some of them can be reduced by design or production measures. A general separation of output value contributions is shown in Eq. (10.1).

$$\hat{X}\,(x, V, S, t) = \mu_{\hat{X}}\,(x, V, S, t) + W_{\mathrm{s}}\,(x, V, S) + W_t\,(x, V, S, t) + W_{\mathrm{n}}\,(x, V, S, t) \tag{10.1}$$

The quantity of interest is represented by x, W_{s} represent deviations dependent on input x, influencing parameters $\boldsymbol{V}$, and design parameters $\boldsymbol{S}$. The (life-) time-dependent deviations are collected in W_t which also depend on x, $\boldsymbol{V}$, $\boldsymbol{S}$, and time t. Finally, in W_{n} all random deviations are modeled which depend on x, $\boldsymbol{V}$, $\boldsymbol{S}$ and even time t. All the random deviations can be treated as random variables. If the uncertainties are sufficiently small such that the system can well be approximated by the linear part of Taylor series expansion, the uncertainty of all contributors can be propagated through the sensor system in an efficient way, without time-consuming Monte-Carlo simulations. In this work, we only focus on the optimization of sensor deviations W_{s} and end of line production, using averaged sensor output values, so W_t and W_{n} are not considered.

Each time we observe a physical quantity and digitize its value for further processing, storage, or control, there is an uncertainty associated to this value. This is a well-known fact and cannot be fully resolved by improving the measurement principle as described in Heisenberg's famous uncertainty principle. In order to indicate the quality of a measurement result, it is thus common to state the measured value of physical quantities together with an indication of its uncertainty. Historically, the measurement error was considered in a practical way to describe this quality by means of a measurement deviation. Nowadays, a standardized representation of uncertain measurement quantities on a statistical basis is defined in the "Guide to the expression of uncertainty in measurements" (GUM) [8] based on standardized vocabulary to represent uncertain measurement results [9]. On the basis of the GUM, a standardized way is given to define measurement results in a universal, internally consistent and transferable way. In this context, a measurement result for a single unknown quantity has to be associated with its measurement quality expressed in measurement uncertainty, here denoted as standard uncertainty expressed as a standard-deviation. This uncertainty can be defined by a so-called Type A evaluation, using statistical analysis of series of observations, or applying a Type B evaluation other than statistical analysis. Measurement results, in this context, can be described by a set of two parameters with their estimate $\hat{x}$ and associated standard uncertainty u_x to the true value, see Eq. (10.2).

$$\text{measurement result} = \left\{\hat{x}, u_x\right\} \tag{10.2}$$

Practically, when reporting measurement results, the GUM recommends to denote uncertainty in terms of least significant digits given in brackets after the estimated value. For example, for a voltage measurement result of $U = 12.371$ V with an associated standard uncertainty of 15 mV we would write this relationship acc. Eq. (10.3).

$$U = 12.371(15)\ \text{V} \tag{10.3}$$

Such uncertain deviations u_x can result from various sources, such as thermal noise, mechanical stress, quantization-noise, general measurement errors as well as all other external influences, changing the measurement value. These random deviations are not systematically predictable, but for large numbers of measurement results, like present in mass production, they can be treated as mathematical random variable Y and are expressed in probability functions. The cumulative density function (cdf) of this random variable represents the probability of a realization of Y being smaller than y, expressed in Eq. (10.4).

$$F_Y(y) = \Pr(Y \leq y) \tag{10.4}$$

The probability density function (*pdf*) of Y is the derivative of the cumulative density function as shown in Eq. (10.5), also written as *p(y)*.

$$f_Y(y) = p(y) = \frac{F_{Y(y)}}{dy} \tag{10.5}$$

The most famous probability density function for random deviations is the Gaussian probability density function, or normal distribution function which is defined as shown in Eq.(10.6). It is also considered for measurement results in the GUM [8]. There μ_Y represents the expected value or mean value and σ_Y the standard-deviation.

$$f_{Y(y)} = p(y) = \frac{1}{\sqrt{2\pi\sigma_Y{}^2}} \cdot e^{-\frac{1}{2}\left(\frac{y-\mu_Y}{\sigma_Y}\right)^2} \quad (10.6)$$

This probability density distribution function is very common because of its regular presence in natural observations, as well as because of the central limit theorem. The central limit theorem [10] tells us that for large number of measurement results y_n, the distribution of the sums is approximately normal regardless of the form of the distribution of the individual measurement results y (see Eq. (10.7)).

$$\lim_{n\to\infty} \frac{Y_1 + \cdots + Y_n - n\mu_Y}{\sigma\sqrt{n}} \sim N\,(0;\,1) \quad (10.7)$$

10.1.3 Specification Limits of Measurement Results

Every sensor system has to transfer the parameter of interest (most of the time linearly) into a readable value, representing the value of interest. This transfer has to be performed, fulfilling certain performance requirements, defined as maximum allowed positive and negative deviations from this transfer-characteristic. For further analysis and specifications only, this deviation from the target transfer-characteristic is shown and in focus. This relationship including parameter dependence and allowed minimum and maximum deviations are drawn in Fig. 10.3.

Sensor systems for commercial use have to generate measurement results with deviations smaller than the maximum positive and larger than the maximum negative specification value with very high probability. Applying this to a large number of sensor systems, and measurement samples, using the information from the previous section, expressing deviations as statistical random variables, we get a probability of sensor systems fulfilling these specification limits. An example for two different probability density functions including deviations, stating specification limits, is shown in Fig. 10.4. Logarithmic scales are used to show the probability density regions farther from the mean values.

Such systematic deviations can occur in real sensor systems, for example when compensating a higher order polynomial behavior by a lower order approximation. In sensor system applications, such statistic descriptions are necessary to determine the probabilities of specification violations, and to specify safe operation regions for the respective systems. Additionally, for the designers it can provide valuable information to assess if their system will meet requirements in terms of resolution,

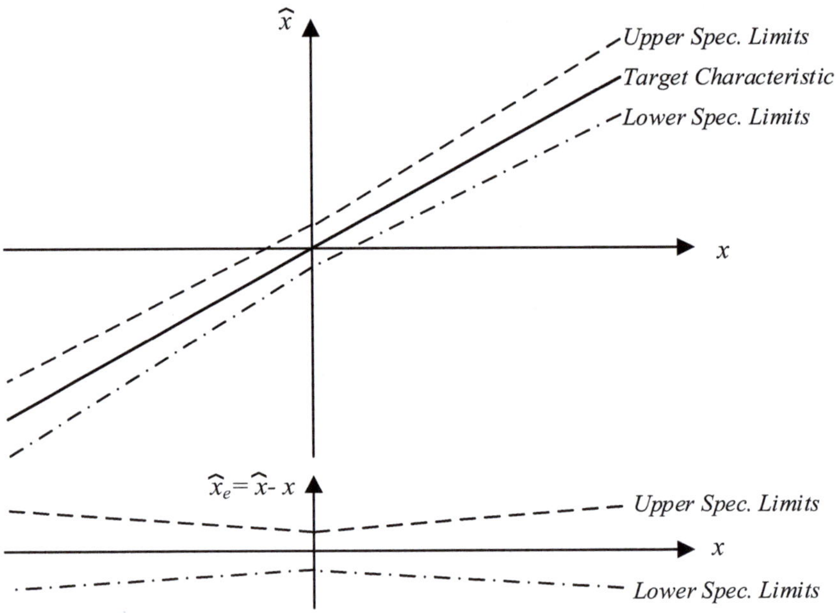

Fig. 10.3 The upper graph shows a typical sensor system transfer-characteristic mapping input parameters of interest x to output values $\hat{\boldsymbol{x}}$. The lower graph focuses only on the accepted deviations $\hat{\boldsymbol{x}}_e$defined by upper and lower specification limits for deviations

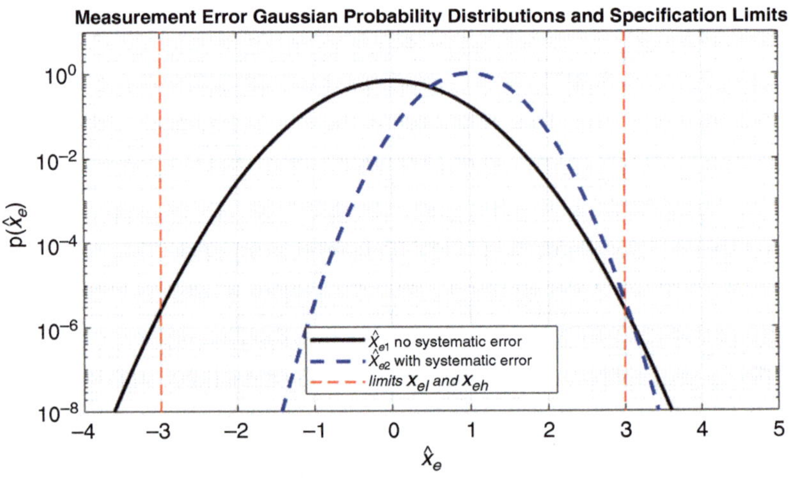

Fig. 10.4 Two possible probability density functions of deviation or error in the estimation result $\hat{X}_{e1} = \left(\hat{X}_1 - x\right)$ and $\hat{X}_{e2} = \left(\hat{X}_2 - x\right)$ approximated as Gaussian random variables with mean values $\mu_{\hat{X}e1} = 0.0$ and $\mu_{\hat{X}e2} = 1.0$ as well as standard-deviations $\sigma_{\hat{X}e1}$=0.6 and $\sigma_{\hat{X}e2}$=0.4. These graphs were drawn on a logarithmic scale to show also low probability density regions and specification limits for deviations $x_{el} = -3.0$ and $x_{eh} = +3.0$ are indicated to show the probability of specification violations

Table 10.1 Probabilities of specification violation related to Gaussian random deviations from mean values without systematic deviations $\left(\mu_{\hat{X}_e} = 0\right)$

Symmetrical limits related to the mean value	Probability of $\hat{X}_e$ to be out of specification
$\pm 1\sigma$	3.173×10^{-1}
$\pm 2\sigma$	4.550×10^{-2}
$\pm 3\sigma$	2.700×10^{-3}
$\pm 4\sigma$	6.334×10^{-5}
$\pm 5\sigma$	5.733×10^{-7}
$\pm 6\sigma$	1.973×10^{-9}

accuracy, and stability. The probabilities of deviations to remain inside a certain specification can be calculated by integration of probability densities within the specified limits according to Eq. (10.8). The probability of failing a specification can be determined according to Eq. (10.9).

$$P\left(\left(\hat{X}_e > x_{el}\right) \cup \left(\hat{X}_e < x_{eh}\right)\right) = \int_{x_{el}}^{x_{eh}} p\left(\hat{x}_e\right) d\hat{x} \tag{10.8}$$

$$P\left(\left(\hat{X}_e < x_{el}\right) \cup \left(\hat{X}_e > x_{eh}\right)\right) = \int_{-\infty}^{x_{el}} p\left(\hat{x}_e\right) d\hat{x} + \int_{x_{eh}}^{\infty} p\left(\hat{x}_e\right) d\hat{x} \tag{10.9}$$

In Table 10.1, exemplary calculations of the probabilities of $\hat{X}_e$ remaining within a symmetrical σ specification around the mean value are given together with the according probabilities of violating the specification. This can as well be considered as a measure for the reliability of a sensor system.

Random deviations are gathered in the standard-deviation σ. In any case, the overall deviations have to be within certain specification limits. In Fig. 10.4, one can also see that systematic deviations have an influence on the probability of specification limit violations. For real world production, low probabilities of specification violations are required to increase reliability and productive yield [11]. In Fig. 10.5, an overview of relations between systematic and Gaussian distributed random deviations is shown, fulfilling, e.g., symmetric deviation limits. Here we can see the need for the reduction of systematic deviations to retain enough headroom for random deviations, to guarantee a safe operation.

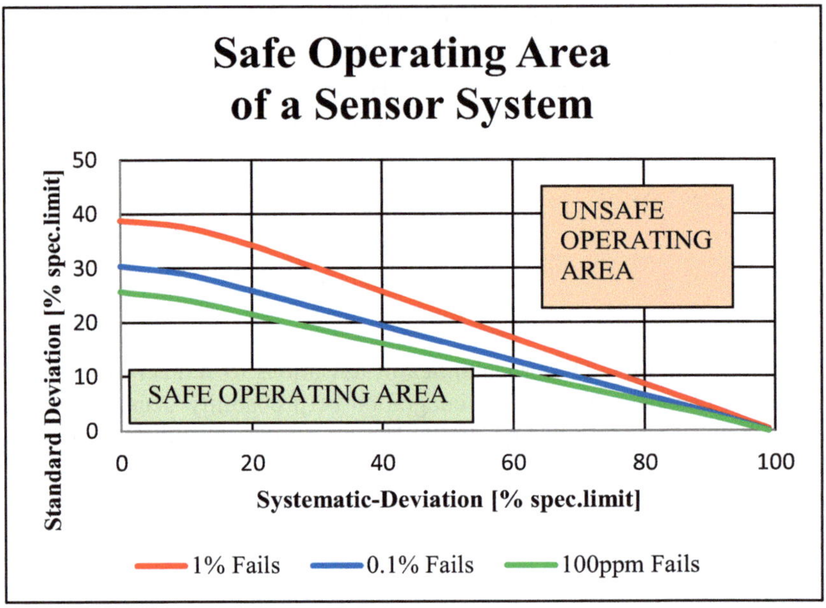

Fig. 10.5 This graph shows the boarder lines of systematic deviations versus standard-deviations needed to achieve a certain probability of specification fails, also indicated as safe- and unsafe operating areas

10.2 Uncertainty Propagation

To determine and estimate the probability of specification violations of sensor system architectures, we need to estimate the distribution of system output deviations by considering all contributing input deviations and propagate them through the system data-path. Following aspects need to be considered.

10.2.1 Correlated Parameters

Influencing parameters V, design parameters S, and noise are modeled by Gaussian probability distributions using mean values μ, standard-deviations σ, and correlations ρ between variables and parameters. When focusing on two random variables, we can analyze them jointly using joint probability density functions. A mathematical representation of two designparameters is shown in Eq. (10.10)

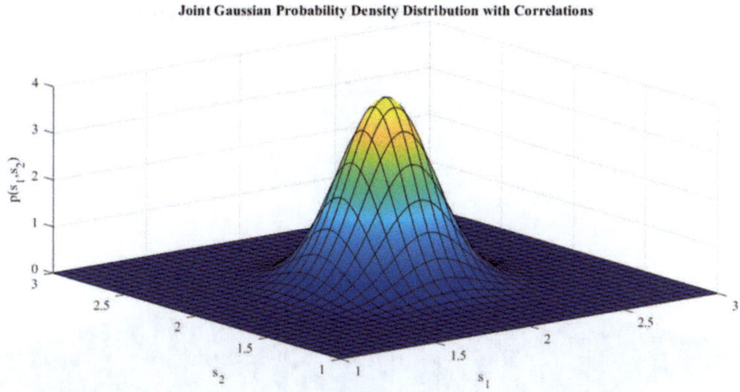

Fig. 10.6 Joint probability density function of uncorrelated random variables s_1 and s_2

with μ representing the mean values, σ representing the standard-deviations, and ρ representing the correlation coefficient between these two random variables.

$$p\left(s_1, s_2\right) = \frac{1}{2\pi\sigma_{s_1}\sigma_{s_2}\sqrt{1-\rho_{s_1 s_2}{}^2}} \cdot \exp\left(-\frac{1}{2\left(1-\rho_{s_1 s_2}{}^2\right)}\left[\frac{\left(s_1-\mu_{s_1}\right)^2}{\sigma_{s_1}} + \frac{\left(s_2-\mu_{s_2}\right)^2}{\sigma_{s_2}} - \frac{2\rho\left(s_1-\mu_{s_1}\right)\left(s_2-\mu_{s_2}\right)}{\sigma_{s_1}\sigma_{s_2}}\right]\right) \tag{10.10}$$

This joint probability density function can be represented graphically, an example of an uncorrelated joint probability density function for two variables is presented in Fig. 10.6, both means are set to 2.0 and the standard-deviations to 0.2 to better illustrate correlations. In case of a present correlation between s_1 and s_2 of $\rho_{s1s2} = 0.8$, the joint probability density function is accumulated across the diagonal as shown in Fig. 10.7.

The concept of correlated variables or parameters is summarized in Fig. 10.8 using a contour-plot. In case of no correlation, the joint probability density function of two variables is of circular shape. In case of correlations, the circular shape becomes distorted into a tilted elliptical shape, where the tilt-direction indicates a positive or negative correlation.

10.2.2 Uncertainty Propagation of Gaussian Probability Distributions

By applying Gaussian probability distributions to describe uncertainties and deviations, it is possible to propagate them through a mathematical data-path with low effort using Taylor series approximation. In Eq. (10.11), this is applied to a sensor

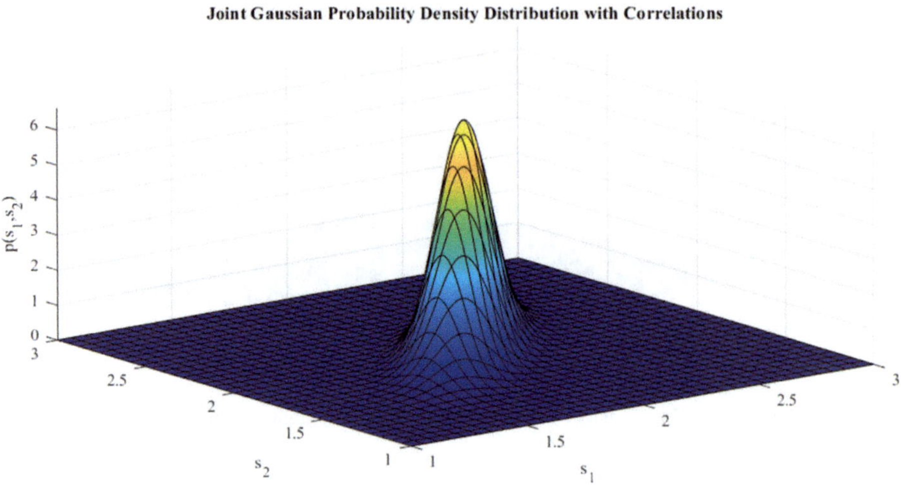

Fig. 10.7 Joint probability density function of correlated random variables s_1 and s_2 with a correlation coefficient ρ_{s1s2} of 0.8

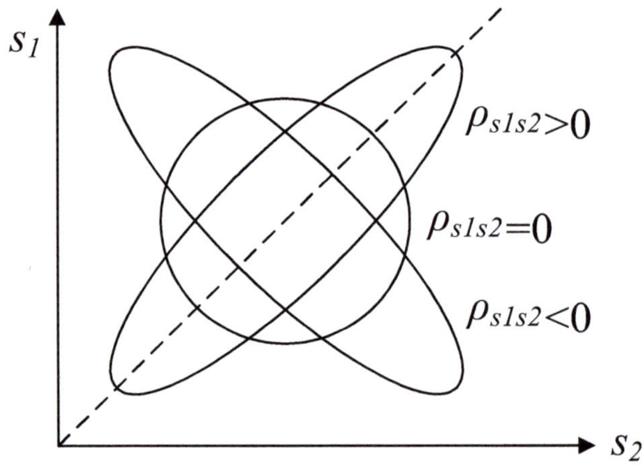

Fig. 10.8 Contour-plot of a joint probability density function of two random variables s_1 and s_2 without correlation ($\rho_{s1s2} = 0$), with positive correlation ($\rho_{s1s2} > 0$) and with negative correlations ($\rho_{s1s2} < 0$)

output signal Y depending on design parameter S, with a reduction to first order derivatives, which is considered sufficient for small deviations.

$$Y = f_{(S)} = f_{(\mu_s)} + \frac{f'_{(\mu_s)}}{1!} \cdot (S - \mu_s)^1 + \cdots + \frac{f^{(n)}{}_{(\mu_s)}}{n!} \cdot (S - \mu_s)^n \tag{10.11}$$

The basic principle used here for uncertainty propagation is adopted from error-propagation [12]. There, we separate between propagation of mean values and variances. The propagation of mean values is just the static transfer-function depending on the contributing parameters and values, which holds as we assume that higher order contributions can be neglected. Equation (10.12) shows the mean value of the sensor output signals $\boldsymbol{Y}$ depending on the mean values of influencing parameters $\boldsymbol{V}$ and design parameters $\boldsymbol{S}$ written as vectors, which is sufficiently accurate as long as nonlinearities are small enough.

$$\mu_Y = f\left(\mu_V, \mu_S\right) \tag{10.12}$$

The uncertainty of one output variable is calculated as sum of all input or parameter variances contributing to this output. The contribution of one input or parameter uncertainty is calculated via the sensitivity of the output signal related to the variable or parameter in focus. This sensitivity is calculated via the partial derivative of the output related to the variable or parameter. As an example: the sensor output variance σ_Y^2 related to a deviation of one design parameters $\boldsymbol{\sigma_S}$ including its correlations ρ shown in Eq. (13).

$$\sigma_{Y(S)}{}^2 = \sum_{i=1}^{N}\left(\frac{\partial Y}{\partial S_i}\cdot\sigma_{S_i}\right)^2 + \sum_{i=1}^{N}\sum_{j=1;i\neq j}^{M}\frac{\partial Y}{\partial S_i}\frac{\partial Y}{\partial d S_i}\sigma_{S_i}\sigma_{S_j}\rho_{S_i S_j} \tag{10.13}$$

Additionally, also all other deviations of Y (e.g., from influencing parameters V) need to be considered, even including correlations of design parameters and influencing parameters similar to that of Eq. (10.13). Design parameters and influencing parameters can be finally combined to Eq. (10.14) where no correlations between them are present and also not considered.

$$\sigma_Y^2 = \sum_{i=\{S,V\}} \sigma_{Y(i)}^2 = \sigma_{Y(V)}^2 + \sigma_{Y(S)}^2 \tag{10.14}$$

There are different methods available to de-correlate correlated probability distribution functions and treat them as weighted combination of uncorrelated random variables. Available methods are Karhunen Loève Decomposition (KLD), Principal Component Analysis (PCA), or Singular Value Decomposition (SVD). The basic principle is to define the principal orthogonal axis of the correlated probability distribution. This can be realized by a projection of covariance-values $\boldsymbol{s}$ to a new coordinate system defined by $\boldsymbol{\phi}$ as shown in Eq. (10.15).

$$\boldsymbol{s}^+ = \boldsymbol{\phi}^T \cdot \boldsymbol{s} \tag{10.15}$$

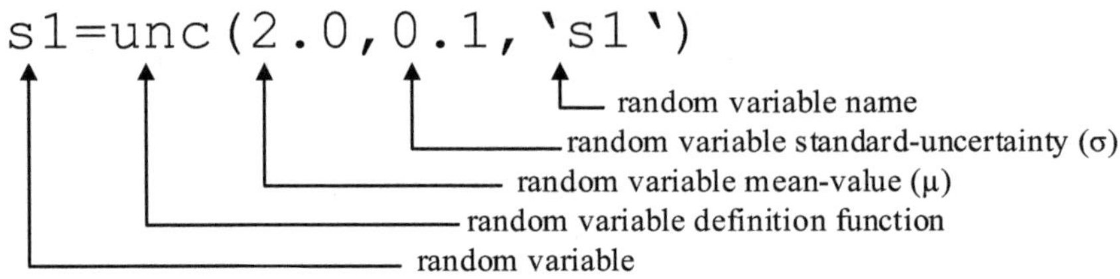

Fig. 10.9 This is an example function defining a random variable for a design parameter s_1 using the uncertainty toolbox

In our practical example, the new coordinate system $\boldsymbol{\phi}$ can be found using the eigenvectors of $\boldsymbol{s}$. When using this SVD in the data-path, the correlated variables and parameters are calculated via the transformation matrix, represented by the eigenvectors $\boldsymbol{\phi}$ and the uncorrelated values $\boldsymbol{s}^+$ as shown in Eq. (10.16).

$$\boldsymbol{d} = \phi \cdot \boldsymbol{d}^+ \tag{10.16}$$

A software tool for calculating with uncertainties and uncertain variables, especially the propagation of uncertainties through a system data-path was implemented in MATLAB [13]. A first version called “Uncertainty Toolbox” was prepared in an educational context, including some application examples [14]. The definition of an uncertain variable in MATLAB is shown in Fig. 10.9. This version was improved by adding features such as the consideration of correlated Gaussian distributions, complex number handling, and decision limits. Also, a function which compares the probability of one variable to a limit incorporating uncertain numbers was added. A practical application is given for a magneto-resistive angle sensor raw signal analysis [15]. A summary of implemented functions can be found in Appendix 1.

10.3 Sensor System Optimization Principles

10.3.1 Statistical Optimization Principles

In this section, we elaborate statistical optimization theory using the Fisher information [16, 17]. In all sensor systems, design- and input-parameter variations are present, which must be considered in optimization [18]. Optimization procedures in general need an optimization criterion. This criterion is defined by ψ and can depend on design parameters $\boldsymbol{S}$ as well as the real input value x. The optimization of design parameters can be defined as shown in Eq. (17).

$$s^* = arg \min_{S \epsilon \Sigma} \Psi\left(x, \boldsymbol{S}\right) \tag{10.17}$$

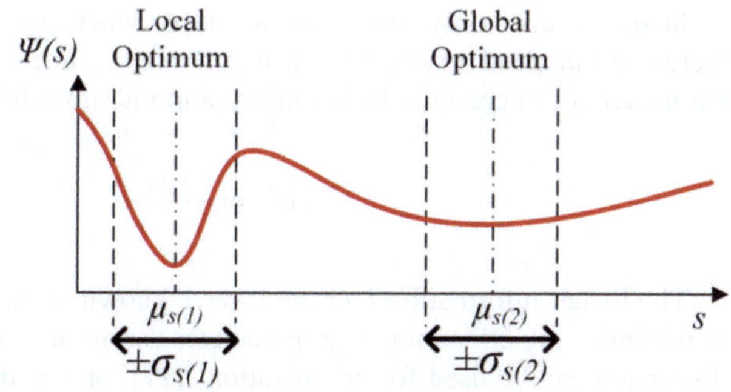

Fig. 10.10 Local and global optimum of a design considering variations of design parameters. The value of the optimization criterion achieved for location $\mu_{s(1)}$ is worse compared to that of location $\mu_{s(2)}$

The optimization shall not only optimize the sensor system in a certain local optimum, instead we are interested in a global optimum including design-parameter variations. A graphical representation of this optimization is given in Fig. 10.10.

One commonly used statistical optimization target is the Minimum Variance Unbiased Estimation (MVUE). Here, the target is to achieve the minimum variance of an estimator while it has to remain, i.e., hit the true value on average. The amount of information about the variable of interest contained in the measurement data can be described by the Fisher information I. In the linear case, the inverse of the Fisher information I^{-1} is called dispersion D as defined in Eq. (10.18) and represents a lower bound for the variance of an unbiased estimator (see Eq. (10.19)).

$$D = I^{-1} \tag{10.18}$$

$$\mathrm{var}\left(\hat{X}\right) \geq \frac{1}{I\,(x, \boldsymbol{S})} \tag{10.19}$$

A general formula for the determination of the Fisher information is given in Eq. (10.20), which is valid for additive white Gaussian deviations including considerations of their covariance C. Here x represents the real input value and $\boldsymbol{S}$ the design parameters.

$$\begin{aligned}[I\,(x, \boldsymbol{S})]_{ij}[I\,(\boldsymbol{x}, \boldsymbol{S})]_{ij} = {} & \left(\tfrac{\partial h(x,\boldsymbol{S})}{\partial x_i}\right)^T C_Y(x, \boldsymbol{S})^{-1}\left(\tfrac{\partial h(x,\boldsymbol{S})}{\partial x_j}\right)\left(\tfrac{\partial h(\boldsymbol{x},\boldsymbol{S})}{\partial s_i}\right)^T \\ & \times C_Y(\boldsymbol{x}, \boldsymbol{S})^{-1}\left(\tfrac{\partial h(\boldsymbol{x},\boldsymbol{S})}{\partial s_j}\right) \\ & + \tfrac{1}{2} tr\left(C_Y(x, \boldsymbol{S})^{-1}\tfrac{\partial C_Y(x,\boldsymbol{S})}{\partial x_i} C_Y(x, \boldsymbol{S})^{-1}\tfrac{\partial C_Y(x,\boldsymbol{S})}{\partial x_j}\right) \\ & \times\left(C(\boldsymbol{x}, \boldsymbol{S})^{-1}\tfrac{\partial C(\boldsymbol{x},\boldsymbol{S})}{\partial s_i} C_Y(\boldsymbol{x}, \boldsymbol{S})^{-1}\tfrac{\partial C(\boldsymbol{x},\boldsymbol{S})}{\partial s_j}\right)\end{aligned} \tag{10.20}$$

In the case of only linear relationships, where the derivative is constant and therefore independent on the real input value *x*, and no correlations between the parameters of interest the Fisher information is simplified to Eq. (10.21).

$$I(S) = \frac{h(S)^T h(S)}{\sigma_Y^2} \tag{10.21}$$

The Fisher information *I*, or the inverse known as dispersion *D* can now be used to minimize the effect of design-parameter variations. Several characteristics of the dispersion can be used for optimizations [17], one of the most common criteria is based on a minimization of the determinant of the dispersion, e.g., as in Eq. (10.22).

$$\Psi\,(x, \boldsymbol{S}) = \min_{s \epsilon S} |D\,(x, \boldsymbol{S})| \tag{10.22}$$

The overall formulation to find the optimum design parameters is given in Eq. (10.23).

$$s^* = arg \min_{S \epsilon \Sigma} \max_{x \epsilon X} |D\,(x, \boldsymbol{S})| \tag{10.23}$$

A practical example of a statistical optimization of design parameters for angular position sensor systems can be in [19]. The inverse Fisher information represents a lower bound for any unbiased estimator. However, such an estimator may not exist, not be found or not desirable, e.g., due to computational effort. In such cases, when a biased estimator is used, i.e., a systematic deviation (offset) occurs, the mean square error might be used as optimization criterion [20]. In this chapter we focus on Fisher information optimization only, since systematic deviations are reduced by sensor calibration.

10.3.2 Sensor Calibration

10.3.2.1 Calibration of Sensor System Output

An effective way of improving a sensor performance is to calibrate this sensor system using an accurate reference sensor system, or a reference value generator. Typically, sensor systems are calibrated during or at the end of production. This is done by applying a known physical reference value to the sensor system and compensate for the difference between the reference value and the target sensor system value. The dependence of the output and input is often estimated using a linear model. To do so, the inverse of the characteristic to be linearized has to be found. The resulting parameters and calibration values are then used to configure the sensor system. In this way, we generate compensation parameters which are correlated to the deviations of the sensor system and therefore can be used to cancel sensor system deviations in an effective manner. The necessary calibration measurements depend on the defined sensor system architecture calibration algorithms. Important aspects for the calibration of a sensor system are the respective parameters. These parameters have to be defined in advance during the architecture exploration phase, to perform a calibration procedure in the real implementation. Following deviations between reference measurement and sensor system output values can be present:

- **Offset Error:** If the measurement physical quantity is zero, the sensor system value should also indicate zero. Any deviation in this case is called an offset error.
- **Gain Error/Sensitivity Error:** When exiting the physical quantity, a certain predefined sensitivity is used to indicate it. Any deviation of the linear sensitivity-slope is known as gain error or sensitivity error.
- **Nonlinear Errors:** Since the gain error can be different at different excitations, a nonlinearity error is present. Typically, these deviations are modeled and compensated using polynomials. In practical implementations, second or third order polynomials are used.

There are more options available for calibration, such as lookup-tables and piece-wise linear interpolations which need a higher number of calibration points. Another way is using polynomial interpolations or segmented polynomial interpolations also called spline interpolations. The most practical way of calibration is a polynomial fitting because of the low number of calibration parameters and calibration points. To find the best fitting calibration parameters, one need to minimize the squared error of all calibration measurements. This can be done by polynomial regression algorithms, which are explained in detail in [21]. First one needs to define the system with linear related coefficients as shown in Eq. (10.24) using polynomials and Eq. (10.25) in matrix form.

$$\hat{x} = \sum_{i=0}^{n} y^i k_i \tag{10.24}$$

$$\hat{x} = \boldsymbol{h} \cdot \boldsymbol{k} \tag{10.25}$$

There $\boldsymbol{h}$ represents the augmented measurement vector, $\boldsymbol{k}$ represents the vector of calibration parameters, and $\hat{x}$ the value of compensated output values. In case of a polynomial calibration of order n, the information matrix is defined as shown in Eq. (10.26) where y represents the sensor output before the estimator or compensation.

$$h = \begin{bmatrix} y^0 & \cdots & y^n \end{bmatrix} \tag{10.26}$$

Here the first element represents the constant factor used for offset calculations, the second column the unmodified sensor signal, the third column the squared sensor signals (e.g., for the second order polynomial compensation), etc. The size of the parameter-vector relates to the dimension of information matrix and is shown in Eq. (10.27).

$$\boldsymbol{k} = \begin{bmatrix} k_0 \\ \vdots \\ k_n \end{bmatrix} \tag{10.27}$$

In this parameter-vector $\boldsymbol{k}$, the first parameter k_0 represents the offset value, the second parameter k_1 the linear gain value, the third parameter k_2 the quadratic polynomial compensation parameter, etc. Finding the best fitting compensation parameters $\boldsymbol{k}$ can be found by least mean square fitting according [16]. The estimation of calibration parameters is done acc. Eq. (10.28).

$$\boldsymbol{k} = \left(\boldsymbol{H}_{\text{cal}}^{T}\boldsymbol{H}_{\text{cal}}\right)^{-1}\boldsymbol{H}_{\text{cal}}^{T}\hat{\boldsymbol{x}}_{\text{cal}} \tag{10.28}$$

Here, $\boldsymbol{k}$ represents the vector of calibration parameters shown in Eq. (10.24), $\boldsymbol{H}_{\text{cal}}$ the augmented observation matrix similar to those shown in Eq. (1026), but including rows of all calibration points $\hat{\boldsymbol{x}}_{\text{cal}}$. Here, $\hat{\boldsymbol{x}}_{\text{cal}}$ represents the vector of all target output values of the sensor system, measured with a reference system or defined by known target values. In case of polynomial calibration, the number of calibration points needs to exceed the polynomial order n by at least one. The information matrix in this case becomes the matrix shown in Eq. (10.29), where $m \geq n$ in case m starts to count from 0. For increased performance of estimation, more calibration points are beneficial.

$$\boldsymbol{H}_{\text{cal}} = \begin{bmatrix} y_{\text{cal}0}^{0} & \cdots & y_{\text{cal}0}^{n} \\ \vdots & \ddots & \vdots \\ y_{\text{cal}m}^{0} & \cdots & y_{\text{cal}m}^{n} \end{bmatrix} \tag{10.29}$$

10.3.3 Optimization of Calibration Parameters

Here minimum variance unbiased estimation can be used to find the calibration points generating the lowest variation of the physical parameter of interest. This can also be based on the Fisher information or dispersion. We need to find the parameters which minimize the determinant of the dispersion matrix of the estimator with calibration parameters as values of interest (see Eq. (10.30)).

$$\boldsymbol{x}_{\text{cal}}^{*} = arg \min_{x \epsilon X} |D\left(\boldsymbol{y}, \boldsymbol{c}\right)| \tag{10.30}$$

In case of the polynomial error compensation technique, the output is linearly dependent on the values of interest (the calibration parameters), which simplifies the optimization acc. Eq. (10.31). The location of the minimum is the same with and without variances because there is no dependency on design parameters present and therefore it can be neglected.

$$\boldsymbol{x}^*_{\text{cal}} = arg \min_{x \epsilon X} \left| \left(\boldsymbol{H}^T \boldsymbol{H} \right)^{-1} \right| \tag{10.31}$$

Interestingly, in case of uncorrelated measurements, the optimization suggests to repeat measurements at the extreme points instead of distributing them uniformly over the input space. Another interesting aspect is the fact that for this calibration we only need the difference between the sensor outputs and the reference value. There is no need for an exact measurement of influencing physical quantities. Only an influencing parameter excitation is needed, the internally measured influencing parameter values are used for calibration parameter generation and later online compensation.

10.4 Sensor System Optimization Example

For educational purposes, a simple temperature sensor system application was chosen to explain the practical usage of this uncertainty toolbox and to apply statistical sensor system optimization. Target is to define a sensor system architecture using a PT1000 temperature dependent resistor of accuracy type B to set up a temperature measurement system with an accuracy of +/−3.0 °C worst case (+/−3σ standard uncertainty).

10.4.1 Sensor System Modeling Including Uncertainties

A temperature measurement system was chosen which uses a resistive Wheatstone-Bridge for generation of a differential temperature dependent voltage, which is converted via an analog-digital converter with consecutive digital calibration. The schematic of this temperature measurement system is shown in Fig. 10.11.

The whole transfer-function of this system can be described as serial connection of individual transfer-functions of the analog-part including the digital conversion with mapping to the final physical units, where resistors with indices “0” represent the resistance at 0 °C (see Eq. (10.32)).

$$\begin{aligned} \hat{T} = G \cdot VDD \cdot \Bigg(& \frac{R_{20}\,(1 + TC_2 \cdot T)}{R_{10}\,(1 + TC_1 \cdot T) + R_{20}\,(1 + TC_2 \cdot T)} \\ & - \frac{R_{40}\,(1 + TC_4 \cdot T)}{R_{30}\,(1 + TC_3 \cdot T) + R_{40}\,(1 + TC_4 \cdot T)} \Bigg) \end{aligned} \tag{10.32}$$

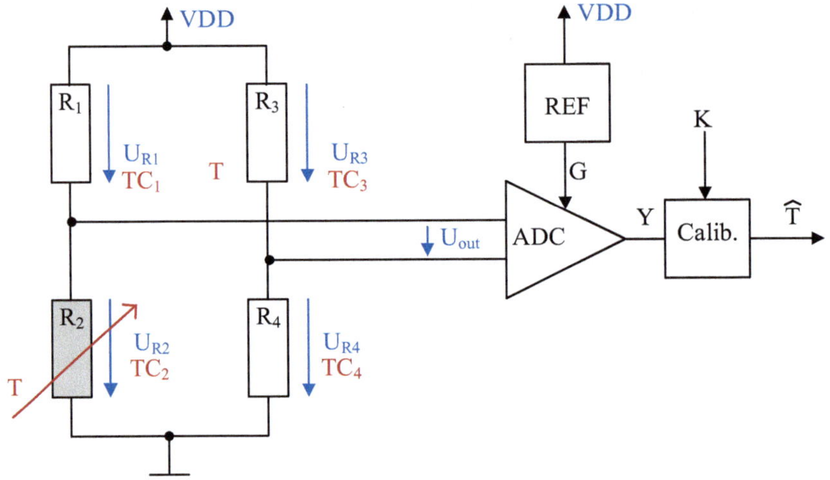

Fig. 10.11 Schematic of a resistive temperature dependent Wheatstone-Bridge with R_2 and R_3 implemented using PTC resistors and an analog-digital converter. U_{Rx} represent the voltage drops across each of the resistors, TC_x represent the temperature coefficients, U_{out} is the differential output voltage and Tout the temperature value scaled with VDD dependent ADC-Gain G. These digital data values are further used for calculating the final estimated temperature value by a calibration block using parameters K

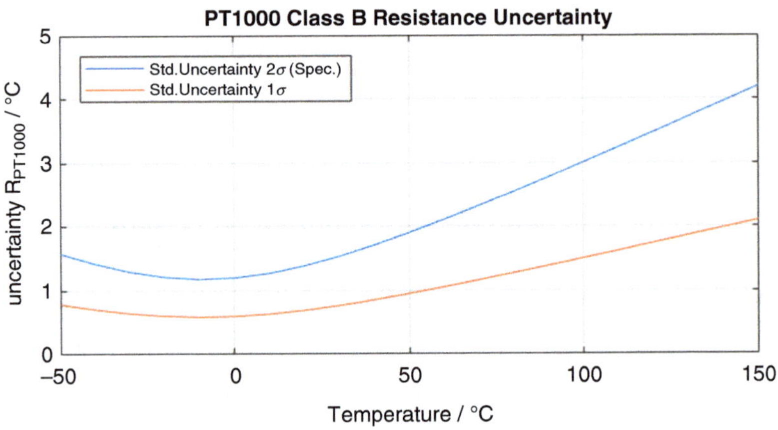

Fig. 10.12 Uncertainty of resistive sensor element PT1000, upper curve represents the specified 2 σ values of IEC60751 and the bottom curve the scaled 1 σ standard uncertainty

Uncertainties of all system components need to be known or defined to be calculated and propagated through the sensor system data-path using the "Uncertainty Toolbox" developed in MATLAB. One of the important elements in this system is the resistive temperature sensor PT1000. Its uncertainty across the whole temperature range is defined in IEC60751 [22] and shown in Fig. 10.12. Parameter uncertainties of all other system components are shown in Table 10.2. Since we have

Table 10.2 Parameter uncertainties of other components of this sensor system beside the PT1000 including possible design space

Parameter	R10 = R30	R40	ρ_{R10R30}, ρ_{R10R40}, ρ_{R30R40}	TC1, TC3, TC4	ρ_{TC1TC3}, ρ_{TC1TC4}, ρ_{TC3TC4}	VDD	G	ρ_{GVDD}
Unit	Ω	Ω	1	1/°C	1	V	°C/V	1
Typ. Range	100 ... 10,000	1000	0.0 ... 0.95	0	0.0 ... 0.95	5.0	100 ... 1000	−0.95 ... 0.0
1σ unc.Range	1 ... 5%/3	1 ... 5%/3		50 ... 200 ppm		0.5/3	5%/3	
Typ. Initial	1000	1000	0.95	0	0.95	5.0	245	0.0
1σ unc.Initial	2% /3	2%/3		100 ppm		0.5/3	5%/3	

The max deviation of standard resistors, supply voltage and gain are assumed to represent +/−3σ uncertainty the constraint that the ADC-input needs to be offset-free, R40 needs to be set to the resistance value of the PT1000 and R30 must be set to R10. Resistor values are taken according to the standard resistor series defined in IEC60063 [23]. A real system will also set the ADC-Gain depending on VDD to compensate VDD variations, here modeled as correlated uncertainties between G and VDD. The ADC-uncertainties included are collected in this gain parameter. For better visualization of consequent results from optimization and calibration, following example computes the results without uncorrelated ADC-noise.

10.4.1.1 Initial Sensor System Performance

Applying this data-path using the uncertainty toolbox is done by defining the uncertain input variables and express the equation as one is used to when using MATLAB to model the data-path. All known correlations between all input variables can also be implemented accordingly. To show the simplicity of this modeling, the example code including optional correlation definition is shown in Appendix 1. The simulation results of this MATLAB model can be seen in following figures. First, the calculation of typical values propagated via this data-path can be seen in Fig. 10.13a. There one can see that the differential temperature output voltage is

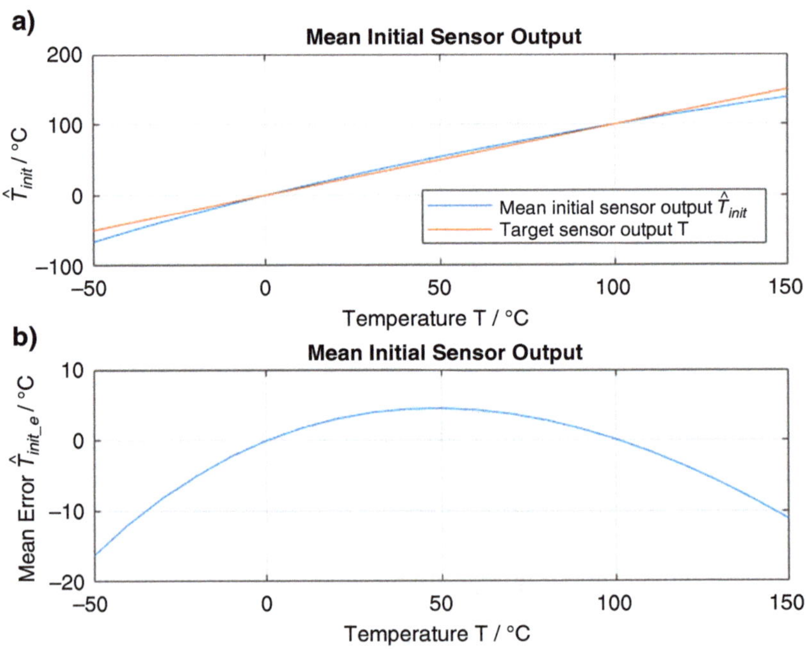

Fig. 10.13 Average transfer-characteristic (**a**) and deviation from ideal transfer-function (**b**) of ambient temperature and the measured temperature simulated using the model shown in Fig. 10.2. Initial ADC-Gain designed for fitting mean temperature at 0 and 100 °C

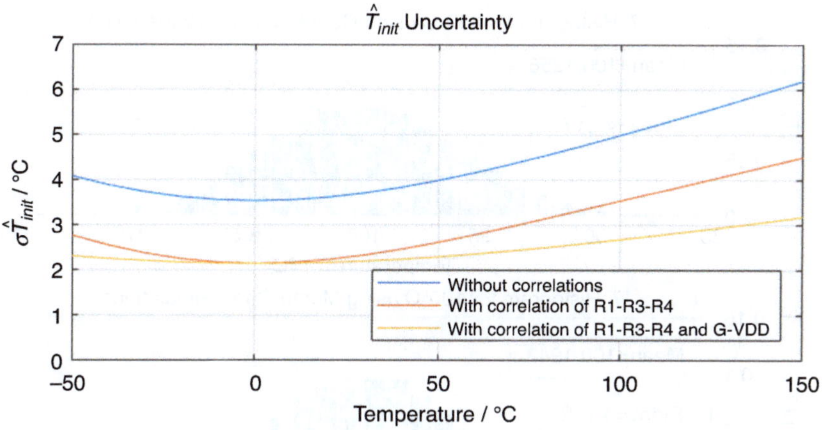

Fig. 10.14 Resulting uncertainties cumulated all uncertainties of input variables and parameters, with and without consideration of correlations of components of the same resistance type (R1, R3, R4) as well as additional correlations of ADC-Gain and VDD

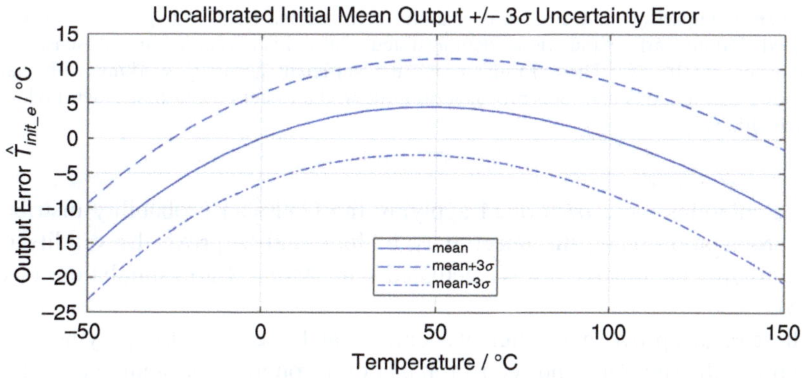

Fig. 10.15 Resulting sensor system error with systematic and random errors without any calibration and optimizations

transferred to an ideal calculated transfer-function, matched by the ideal average gain-factor of the digital transfer-gain. Without considering additional higher order effects and non-idealities, the average transfer-characteristic is close to the real transfer-characteristic, and deviations are shown in Fig. 10.13b. The ADC-Gain is designed to fit mean values at 0 and 100 °C.

Deviations are shown in Fig. 10.14. Finally, the overall sensor system error results in the performances shown in Fig. 10.15.

In practical implementations, the production uncertainty is too large to meet the defined requirements for sensor system accuracy. Further optimizations are necessary to achieve reasonable accurate results of an implemented sensor system. These

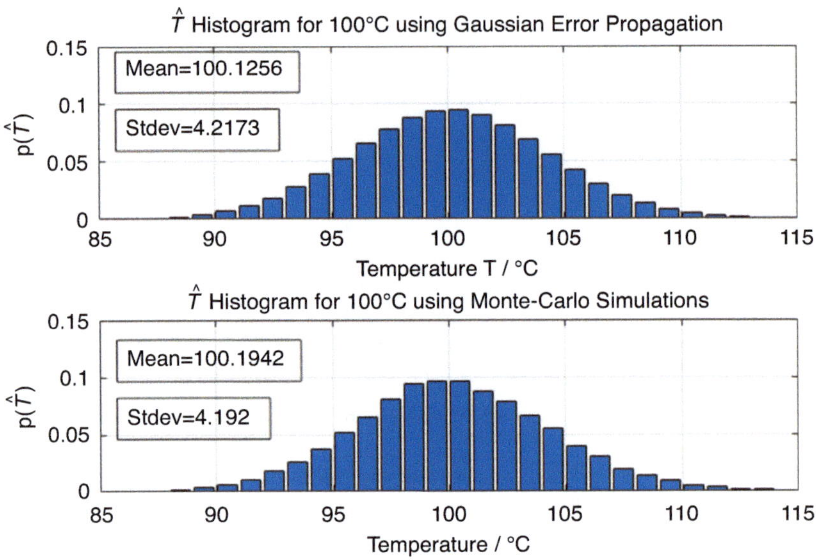

Fig. 10.16 Comparison of Gaussian probability distributions error-propagation compared with dedicated Monte-Carlo simulations demonstrated via a histogram of initial sensor system performance at 100 °C. These results prove the sufficient accuracy working with Gaussian probability distributions and their propagation through the system using first order Taylor series approximations

first calculations were performed applying the Gaussian probability distributions error-propagation using the uncertainty toolbox and to prove the validity of the output figures a comparison was calculated using Monte-Carlo simulations (see Fig. 10.16).

These results prove that sufficient accuracy can be achieved by applying Gaussian probability distributions and their propagation through the system data path using first order Taylor series approximation. Therefore, this method is used for further simulations of deviations and for system optimizations.

10.4.2 Sensor System Optimizations

In real life, all sensor systems need to be calibrated at the end of production line. This is shown in the initial sensor system calibration section. Additional to that calibration, also further statistical optimizations can be performed as shown in the following.

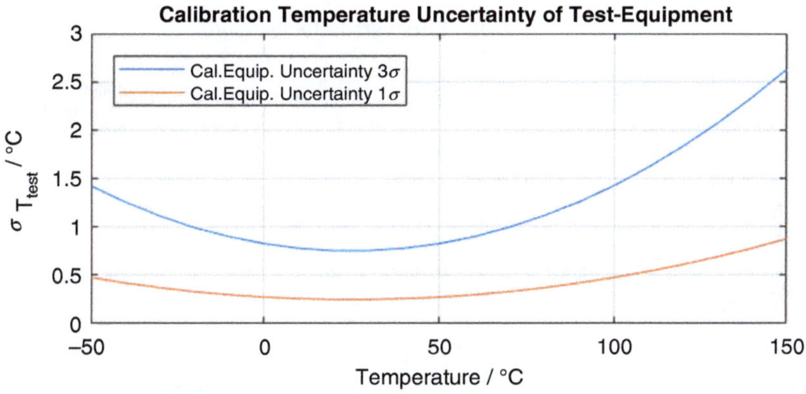

Fig. 10.17 Given measurement uncertainty of calibration temperatures with 3σ standard uncertainty and scaled down 1σ standard uncertainty. The reference temperature is most accurate around 25 °C

10.4.2.1 Initial Sensor System Calibration

This calibration is done with reference temperature performance as shown in Fig. 10.17. This is the best possible accuracy achievable by calibration, defined with 3σ uncertainties. For further statistical analysis, we need to use 1σ standard uncertainties.

This calibration can be performed on the use-case example shown in Fig. 10.9. The temperature calibration is done first at −40 °C, +20 °C, and + 100 °C. Additionally this calibration procedure of this sensor system can be performed including uncertainty propagation using the uncertainty toolbox. The example code for calibration is shown in Appendix 2 and the systematic calibration error in Fig. 10.18 as well as the measurement uncertainty after calibration can be seen in Fig. 10.19.

There one can recognize the measurement uncertainty exactly fits to the calibration uncertainty at the calibration points − 40 °C, +20 °C, and + 100 °C and dominated the remaining uncertainty. Also the remaining uncertainty without consideration of calibration uncertainty is drawn for comparison. The overall sensor system error of mean and +/−3 σ standard uncertainty is shown in Fig. 10.20. In this graph it is shown how calibration improves the performance; however, the target is not achieved in this case.

10.4.2.2 Optimum Temperature Sensor Calibration Points

By applying algorithms described by Eq. (10.29), the statistical optimum calibration temperature points are calculated. For practical optimizations, also the limits of parameters to optimize need to be defined first. In this case, the maximum

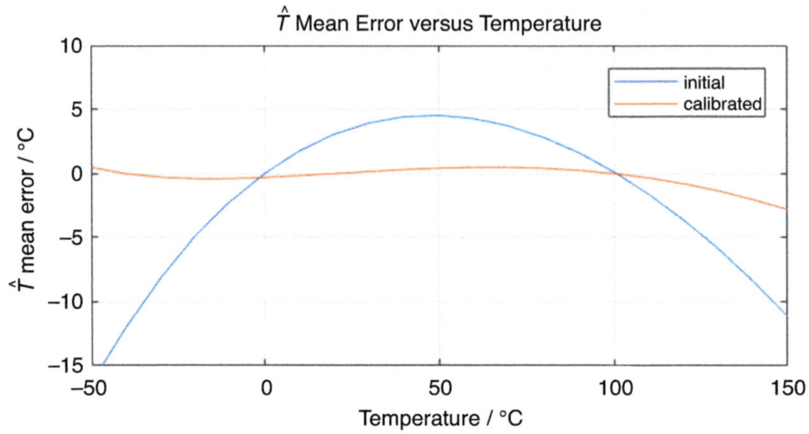

Fig. 10.18 Systematic sensor system errors after calibration. Exactly at calibration points the error is zero, at all other points a certain calibration error is present, which cannot be compensated by second order polynomial calibration functions

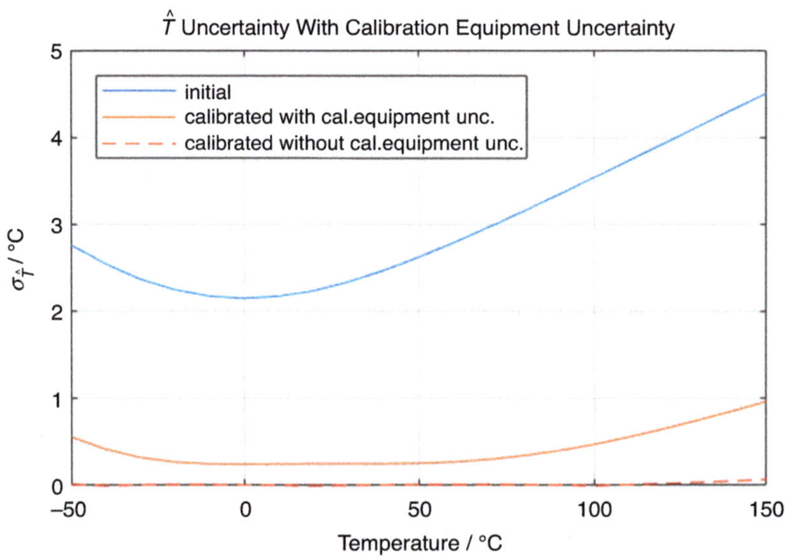

Fig. 10.19 This is the temperature measurement standard uncertainty after before and after temperature calibration at −40 °C, +20 °C, and + 100 °C including indication of uncertainty introduced by calibration equipment. Here you can see the uncertainty distribution across the entire temperature range

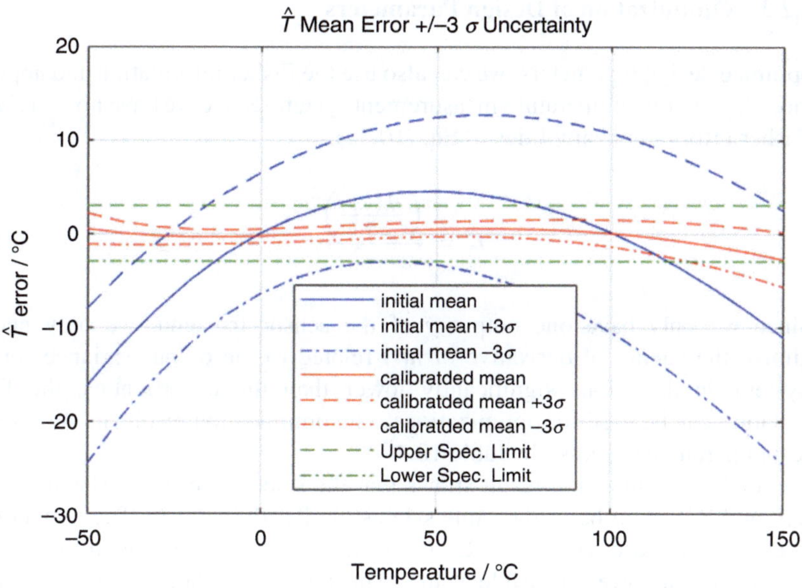

Fig. 10.20 Overall sensor system error comparison between initial no calibration and 3 point calibration at −40 °C, 20 °C, and 100 °C including +/−3σ of standard uncertainty. Calibration helps a lot but the target sensor system accuracy is not achieved

Table 10.3 Optimum temperature sensor calibration points calculated with equal and with temperature dependent uncertainties of calibration temperature

Calib. point nr.	Calibration temperature with constant uncertainty of T_{cal}	Calibration temperature with variations of T_{cal} acc. Fig. 10.16
1	−50 °C	−50 °C
2	+50 °C	+35 °C
3	+150 °C	+150 °C

temperature is +150 °C and the minimum temperature is −50 °C. Additionally, uncertainty dependencies on parameter of interests according to exemplary relationship shown in Fig. 10.14 are also considered and calculated. The results of these three-point calibrations with equal uncertainties and input-parameter dependent uncertainties can be seen in Table 10.3.

Here one can see the influence of the low uncertainty near +25 °C, which shifts the optimal center point of calibration from +50 °C to +35 °C. Also correlations can be considered between these three calibration points but they do not influence the position of optimum calibration points.

10.4.2.3 Optimization of Design Parameters

To optimize design parameters, we can also use the Fisher information and apply the methodology to our temperature measurement system. Since we have no correlation, the Fisher information simplifies to Eq. (10.33).

$$I_y = \frac{\left(\frac{\partial h(x,s)}{\partial s}\right)^2}{\sigma_y^2} \tag{10.33}$$

Since we only have one output y of the sensor frontend, we only need to maximize the square of derivative of $h(\cdot)$ related to the output variance. In case of systematic deviations significantly lower than the uncertainties, the Fisher information can be used for optimizations. The optimization results in the selection of design parameters shown in Table 10.4.

The Fisher-optimization result shows that the selection of the resistance values of R10 and R30 shall be of the same size as the PT1000 at 0°C. The temperature-coefficient standard-deviation of R3 has to be 90 ppm to match with the PT1000, others remain at lowest possible value. The resistance uncertainties need to be smallest possible. The correlation of R10 and R40 as well as the temperature coefficient correlation shall be 0.0, the other resistance correlations shall be largest possible. For best results also, the largest negative correlation between G and VDD need to be chosen to compensate VDD-variation effects. The optimization results can be also seen in the uncertainties of calibration parameters. A comparison between initial calibration and optimized parameter uncertainties can be seen in Table 10.5.

10.5 Results Including Sensor System Optimizations of Design Parameters and Calibration-Point Optimization

The results of the optimized system can be seen in the following figures where the systematic errors and uncertainty dependent on the measurement temperature are drawn in comparison to the initial defined system. Figure 10.21 shows the systematic deviations initially, after first calibration and after sensor system optimization. There can be seen that after optimization the remaining mean error is distributed more symmetrically between positive and negative deviations.

In Fig. 10.22 one can see the measurement uncertainties of the initial system, the system after calibration and after the system optimization. There in Fig. 10.22a the uncertainty without consideration of calibration equipment uncertainties is shown including the positive effect of calibration-point optimization. In Fig. 10.22b, the uncertainty with calibration equipment is drawn including presentation

Table 10.4 Optimal design parameters for the resistive temperature bridge circuit

Param.	R10=R30	$\sigma_{R10} = \sigma_{R30}$	σ_{R40}	σ_{TC1}	σ_{TC3}	σ_{TC4}	ρ R10R30	ρ R30R40	ρ R10R40	ρ TC1TC3	ρ TC3TC4	ρ TC1TC4	ρ GVDD
Unit	Ω	%	%	ppm	ppm	ppm	1	1	1	1	1	1	1
Min	100	1/3	1/3	50	50	50	0	0	0	0	0	0	−0.95
Max	10,000	5/3	5/3	200	200	200	0.95	0.95	0.95	0.95	0.95	0.95	0
Optim.	1000	1/3	1/3	50	90	50	0.95	0.95	0.81	0.95	0.95	0.81	−0.95

Table 10.5 Calibration parameters and their uncertainty before and after optimization

Parameter	P(1) Offset	P(2) Gain	P(3) 2.Order
Mean. before optim.	−0.3065	1.1274	0.0028
Mean. after optim.	−1.0024	1.1356	0.0030
Stdev. before optim.	**1.8355**	**0.0329**	**2.1538e-4**
Stdev. after optim.	**0.7097**	**0.0052**	**0.5958e-4**

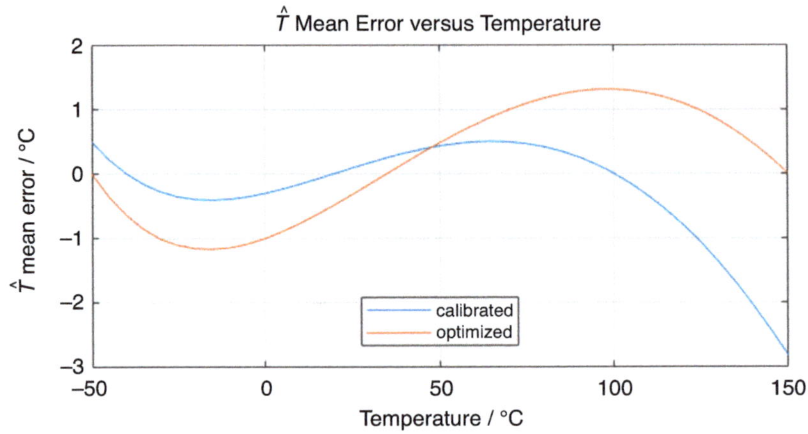

Fig. 10.21 Systematic deviations after first calibration and after optimization of the resistive bridge sensor design parameters and optimum calibration points

of uncertainty reduction after calibration-point optimization. It can be recognized that after calibration the remaining sensor system uncertainty is dominated by the uncertainty of the calibration equipment uncertainty.

In Fig. 10.23, one can see in detail the improvement of this temperature measurement system by sensor system optimization compared to the initial planned calibration. Systematic errors including +/−3σ of measurement uncertainty are plotted to see the overall remaining error which is lower than the +3.0 °C for the upper specification limit and −3.0 °C for the lower specification limit.

This example can be extended by further addition of components like leakage currents, connection resistances, amplifier, etc. and parameters like ADC resolution, thermal noise, further correlations of components and parameters. It demonstrates the applicability of first order Gaussian probability distributed uncertainty propagation and its simple execution supported by this developed uncertainty toolbox.

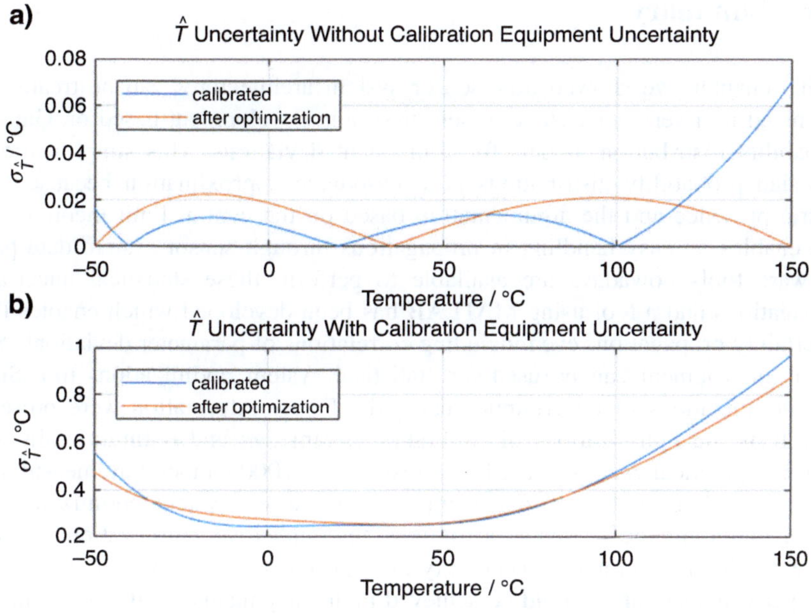

Fig. 10.22 Remaining measurement uncertainty after optimization of the resistive bridge sensor design parameters and optimum calibration points. (**a**) shows optimization results without calibration equipment uncertainty and (**b**) with equipment uncertainty versus temperature. Here it can be recognized that calibration equipment errors dominate the sensor system performance

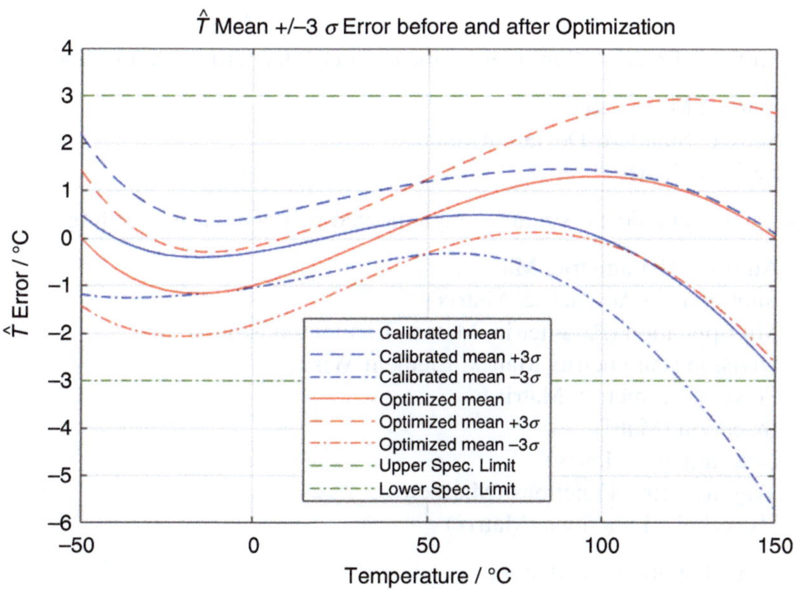

Fig. 10.23 Remaining error comparison of the initial planned calibration and after the optimization. Systematic as well as uncertainty effects are reduced best possible

10.6 Summary

In this chapter, we showed how sensor system architectures can be treated and optimized in a very early stage of sensor system development based on Gaussian probability distribution models for component deviations. This simplification to Gaussian probability distributions is a reasonable approximation because of its natural presence and the argumentation based on the central limit theorem. This also enables an easy handling in propagations through sensor system data-paths. Software tools nowadays are available to perform these statistical uncertainty propagations and a tool using MATLAB has been developed which enables these uncertainty propagations even including correlations of parameter deviations. Such a tool environment can be used for statistical system optimizations too. Statistical optimizations of uncertainties using the Fisher information were presented as well as the optimizations of calibration parameters and optimum calibration points. A practical example based on a resistive PT1000 temperature measurement bridge including calibration considerations and sensor system optimization was shown including the presentation of how this statistical optimization improves the remaining measurement uncertainty to achieve certain sensor system accuracy. This performance-target could be achieved in an easy manner using this statistical optimization procedure and according tooling.

Appendix 1: Uncertainty Toolbox in MATLAB, Implemented Functions

- Definition of random variables by mean value, standard-deviation, and name
  - Set/Get Mean value
  - Set/Get Standard-Deviation value
  - Set/Get Name
- Basic mathematical operations implemented by operator overloading
  - Addition (Parametric, Matrix)
  - Subtraction (Parametric, Matrix)
  - Multiplication (Parametric, Matrix, Element-Wise)
  - Division (Parametric, Matrix, Element-Wise)
  - Power (Parametric, Matrix)
  - Inversion (Matrix)
  - Determinant (Matrix)
  - Trigonometric Functions (Matrix)
  - Hyperbolic Functions (Matrix)
- Special Operations on Arrays
  - Determinant
  - Inverse Matrix

  - Diagonal Elements of a Matrix
  - Trace of a Matrix
  - Sum of Array Elements
  - Jacobian Matrix
  - Complex Conjugate Transpose
- Correlations
  - Set correlations
  - Get correlations
- Covariances
  - Set covariances
  - Get covariances
- Display uncertainty contribution
- Logical Comparisons (larger, smaller)
  - Between Random Variables and Parameters
  - Between Random Variables
- Complex number calculations
  - Including correlations between real and imaginary part

Appendix 2: Example Matlab Code

This is an example MATLAB code to model the resistive temperature bridge example using the uncertainty toolbox at initial architectural design state. VDD is the supply voltage, R10 ... R40 the resistance values at 0 °C, TC1 ... TC4 the temperature coefficients of each resistor. R20 represents the PT1000 which uncertainty has to be defined in Ω. "Tout" is calculated in an array from −50 to 150 °C in steps of 10 °C.

```
%Temperature Range of interest
T=-50.0:10:150.0;

VDD=unc(5.0,5.0*0.05/3,'VDD');      %+/-3sigma

R10=unc(1000,20/3,'R10');  % E48 Series Resistors 2%, +/-3sigma
R20=unc(1000,1.2/2,'R20'); % R(PT1000)=1000*(1+A*T+B*T^2),
                           % A=3.9083e-3, B=-5.775e-7
R30=unc(1000,20/3,'R30');  % E48 Series Resistors 2%, +/-3sigma
R40=unc(1000,20/3,'R40');  % E48 Series Resistors 2%, +/-3sigma

TC1=unc(0,100e-6/2,'TC1');          % E48 Series 100ppm/K,
                                      +/-2sigma
TC2=unc(0.0039083,2.5e-5/2,'TC2');% 2.8°C per 500°C,
                                      +/-2sigma
```

```
TC3=unc(0,100e-6/2,'TC3');        % E48 Series 100ppm/K, +/-2sigma
TC4=unc(0,100e-6/2,'TC4');        % E48 Series 100ppm/K, +/-2sigma

G=unc(245,245*0.05/3,'G'); %K/V, defined by Bandgap 5%, +/-3sigma

% with optional correlations R1-R3-R4
set_correl (R10,R30,0.95);
set_correl (R30,R40,0.95);
set_correl (R10,R40,0.95);
set_correl (TC1,TC3,0.95);
set_correl (TC3,TC4,0.95);
set_correl (TC1,TC4,0.95);
set_correl (G,VDD,0.0);

R1=R10*(1+TC1*T);
R2=R20*(1+TC2*T);
R3=R30*(1+TC3*T);
R4=R40*(1+TC4*T);

U12R=R2./(R1+R2);
U34R=R4./(R3+R4);
Uout=VDD*(U12R-U34R);
T_hat=Uout*G;
```

Appendix 3: Matlab Calibration Code

This is an example MATLAB code to model the calibration of the resistive temperature bridge example using the uncertainty toolbox. The calibration temperatures "Tx_calunc" are defined with −40 °C, +20 °C, and + 100 °C with given uncertainties.

```
%calibration temperatures with uncertainties
T1_calunc=unc(-40,0.25+((-40-25)^2)/25000,'T1_calunc');
T2_calunc=unc(20,0.25+((20-25)^2)/25000,'T2_calunc');
T3_calunc=unc(100,0.25+((100-25)^2)/25000,'T3_calunc');

%calibration parameters
H_cal=[1 T_hat(2)  T_hat(2)^2;...   %-40°C
       1 T_hat(8)  T_hat(8)^2;...   %20°C
       1 T_hat(16) T_hat(16)^2];   %100°C

T_cal=[T1_calunc; T2_calunc; T3_calunc];
K_cal=inv(H_cal'*H_cal)*H_cal'*T_cal;

%calibration calculation
for n=1:length(T)
  T_hat_cal(n)=([1 T_hat(n) T_hat(n)^2]*K_cal);
end
```

Glossary

$x, \boldsymbol{x}$	Measurement system real input value, input vector
$y, \boldsymbol{y}, Y, \boldsymbol{Y}$	Measurement system indication value, value-vector, RV, RV-vector
$\hat{x}, \hat{\boldsymbol{x}}, \hat{X}, \hat{\boldsymbol{X}}$	Measurement system estimated value, estimated vector, RV, RV-vector
$v, \boldsymbol{v}, V, \boldsymbol{V}$	Influencing parameter, parameter-vector, RV, RV-vector
$s, \boldsymbol{s}, S, \boldsymbol{S}$	Design parameter, parameter-vector, RV, RV-vector
$k, \boldsymbol{k}, K, \boldsymbol{K}$	Calibration parameter, parameter-vector, RV, RV-vector
$w, \boldsymbol{w}, W, \boldsymbol{W}$	Random zero mean deviations, deviation vector, RV, RV-vector
h	Augmented measurement vector
H	Augmented observation matrix
I	Fisher Information
D	Dispersion
C	Covariance Matrix

References

1. N. Beringer, Sensor-based learning—one step closer to autonomous driving, Springer Book, Fahrerassistenzsysteme 2017, May 2017
2. N. Druml, G. Fleischmann, C. Heidenreich, A. Leitner, T. Herndl, G. Hohlweg, Time-of-flight 3D imaging for mixed-critical systems, in *Proceedings of the IEEE 13th International Conference on Industrial Informatics (INDIN)*, (2015), pp. 1432–1437
3. L. Wang, A. Cavallaro, Acoustic sensing from a multi-rotor drone. IEEE Sensors J. **18**(11), 413–427 (2018)
4. M. Rossi, D. Brunelli, Autonomous gas detection and mapping with unmanned aerial vehicles. IEEE Transaction on Instrumentation and Measurement **65**(4), 765–775 (2016)
5. P.-Y. Kuo, G. Huang, D. Zhou, An efficient method for evaluating analog circuit performance bounds under process variations. IEEE Transactions on Circuits and Systems **59**(6), 351–355 (2012)
6. M. Kampl, H. Kosina, The backward Monte Carlo method for semiconductor device simulation. J. Comput. Electron. (2018)
7. A. Canelas, R. Martins, R. Povoa, N. Louenco, N. Horta, Yield optimization using K-means clustering algorithm to reduce Monte Carlo simulations, in *Design Automation & Test Conference and Exhibition (DATE)*, (2017), pp. 1201–1206
8. JCGM 100:2008, Evaluation of Measurements—Guide to the Expression of Uncertainty in Measurements (2008)
9. JCGM 200:2012, International Vocabulary of Metrology—Basical and General Concepts and Associated Terms (VIM) (2012)
10. H.P. Hsu, *Theory and Problems of Probability, Random Variables, and Random Process* (McGrawHill, New York, 1997)
11. W. Granig, L.M. Faller, H. Zangl, Sensor system optimization to meet reliability targets. Microelectron. Reliab. **87**, 113–124 (2018)
12. K.O. Arras, *"An Introduction to Error Propagation: Deviation, Meaning and Example of Equation Cy=FxCxFxT", Technical Report EPFL-AST-TR-98-01 R3* (EPFL, Lausanne, 1998)

13. Mathworks GmbH, MATLAB, version R2017b, September 14 (2017)
14. H. Zangl, K. Hoermaier, Educational aspects of uncertainty calculation with software tools. Measurement **101**, 257–264 (2017)
15. W. Granig, S. Aoudjit, L.M. Faller, H. Zangl, Statistical modeling of integrated sensors for automotive applications, in *Conference Proceedings, Automotive 2018, Milano*, (2018)
16. S. Kay, in *Fundamentals of Statistical Signal Processing: Estimation Theory*, Prentice Hall Signal Processing Series, ed. by A. V. Oppenheim, (Prentice Hall, Englewood Cliffs, NJ, 1993)
17. V. Fedorov, P. Hackl, *Model oriented Design of Experiments*, (Springer, 1997)
18. H. Zangl, G. Steiner, Optimal design of multi-parameter multi-sensor systems. IEEE Trans. Instrum. Meas. **57**, 1484–1491 (2008)
19. H. Zangl, L.M. Faller, W. Granig, Optimal design of angular position sensors. COMPEL **36**(5), 1372–1385 (2017)
20. L.M. Faller, H. Zangl, Robust design of a 3D- and inkjet-printed capacitive force/pressure sensor. IEEE Proc. EuroSimE (2016)
21. J. Fox, *Applied Regression Analysis & Generalized Linear Models*, 3rd edn. (Sage, Thousand Oaks, CA, 2016)
22. International Electrotechnical Commission, *IEC 60751: 2008 Industrial Platinum Resistance Thermometers and Platinum Temperature Sensors*, 2nd edn. (2008)
23. International Electrotechnical Commission, *IEC 60063: 2015 Preferred Number Series for Resistors and Capacitors*, Edition 3.0 (2015)

Chapter 11
Reliability of IoT Sensor Systems: New Methods for the Efficient and Comprehensive Reliability Assessment

J. Albrecht, G. Gadhiya, and S. Rzepka

Abbreviations

DA	Die attach
DA_CTE1	Coefficient of thermal expansion of the die attach material
DA_EX	Young's modulus of the die attach material
DIL	Dual in-line package
DMA	Dynamic mechanical analysis
DoE	Design of experiment
FOWLP	Fan-out wafer-level package
IoT	Internet of things
LF	Leadframe
MC_CTE1	Coefficient of thermal expansion of the encapsulation (mold compound) material
MC_EX	Young's modulus of the encapsulation (mold compound) material
MC_TG	Glass transition temperature of the encapsulation (mold compound) material
obj_ST_DP_MC	Interface stresses at the interface between mold-die pad as an objective for the optimization
PCB	Printed circuit board
PoP	Package on package
RDL	Redistribution layer
SiP	System-in-package
ST_DP_DA	Interface stresses at the interface between die pad-die attach

J. Albrecht (✉) · G. Gadhiya · S. Rzepka
Micro Materials Center at Fraunhofer ENAS, Chemnitz, Germany
e-mail: jan.albrecht@enas.fraunhofer.de

W. D. van Driel et al. (eds.), *Sensor Systems Simulations*,
https://doi.org/10.1007/978-3-030-16577-2_11

ST_LEAD	Interface stresses at the interface between mold-lead pins
ST_DP_MC	Interface stresses at the interface between mold-die pad
TMA	Thermo-mechanical analysis
Z_DA	Thickness of die attach

11.1 Introduction

The internet of things (IoT) with its many applications in the fields of automotive, industrial, and personal electronics require innovative electronic systems offering various new functionalities within shortest time-to-market yet with high reliability, functional safety, and at costs low enough to be acceptable for the mass market. Highly integrated circuits as well as packaging solutions are necessary to fulfill those demands. System-in-package (SiP) based on fan-out wafer-level package (FOWLP) [1], package on package (PoP), or interposer solutions as well as 3D technologies (die stacks, wafer stacks, package stacks) are able to meet these technical requirements and performance expectations. Typically, those new miniaturized and highly integrated systems comprise different active and passive components made of different materials, which all have their specific and very diverse properties regarding thermal expansion and stiffness. Hence, they behave very differently under thermal and mechanical loading conditions resulting in thermo-mechanical stresses inside the package, which finally may lead to fatigue and failures. These conditions occur during fabrication as well as during the service life so that they need to be captured by the qualification tests. Therefore, the thermo-mechanical risks shall be assessed during the development of the package in order to select the right materials and choose the most suited process steps for maximum production yield but also to fulfill the reliability demands. Unfortunately, comprehensive thermo-mechanical assessments and reliability tests are very costly and time consuming. Here, virtual prototyping based on numerical models offers an attractive alternative [2, 3].

The utilization of virtual techniques like finite element (FE) simulation can provide deep insights into the thermo-mechanical stress state of those systems during fabrication and service life. Furthermore, they can be used for lifetime prediction. By this, it is possible to reduce the time for design and testing. While simulations are done within few hours, the physical tests would take several months. However, the most important prerequisite is to have accurate input parameters, which means geometry parameters, well-suited material models with all the required material data as well as the consideration of all relevant process steps and service or test conditions, respectively. The collection of the input parameters is frequently a complicated, long and costly process. However, once these parameters are available, virtual prototyping becomes a very efficient method not just for the specific design but for many other similar developments as well. Of course, the models need to be validated before being used for the assessment of new products. This is done by comparing the simulation results to those of experimental tests in order to ensure the correctness of the predictions based on the simulations.

This chapter describes the methodology of acquiring the input data and of performing the validation for the virtual prototyping of DIL packages as an instructive example.

11.2 Models for Virtual Prototyping of IoT Sensor Systems

For virtual prototyping based on numerical models, fully parametric FE models are an essential requirement. There are different ways to define parametric FE models. Mostly, parametrization includes the variation of material properties or geometrical dimensions of a component as a parameter, e.g., the thickness of the layer of materials. It does not change the basic geometry such as the number of keypoints, lines, areas, or volumes. Therefore, the FE mesh can be the same during these variations. A more complex way would also include changes in the basic geometry of the components or changes in the number of components within same model. This leads to a change in the number of keypoints, lines, areas, volumes and requires a different meshing. So far, most of the attempts to create parametric FE models follow the first way [3–5] while less work has been reported for also including basic geometry variations or topological changes. One example shall briefly be outlined here. The parametric FE modeling is capable of covering wide variations in SiP products based on the FOWLP technology. It is based upon a clear hierarchical architecture approach defining groups of similar SiP structures (Fig. 11.1) divided into several levels of building blocks, for which a library of parametric models has been established using ANSYS™ [7] and comprehensively pre-calibrated. All level of packaging components: multiple dies, molding compound, redistribution layers (RDLs), solder balls, vias, passives, and PCB board are integrated in this comprehensive set of models.

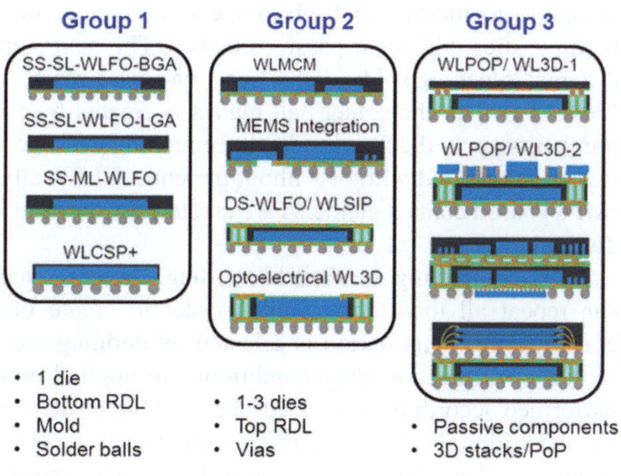

Fig. 11.1 Groups of similarly complex SiP components ([6], with permission)

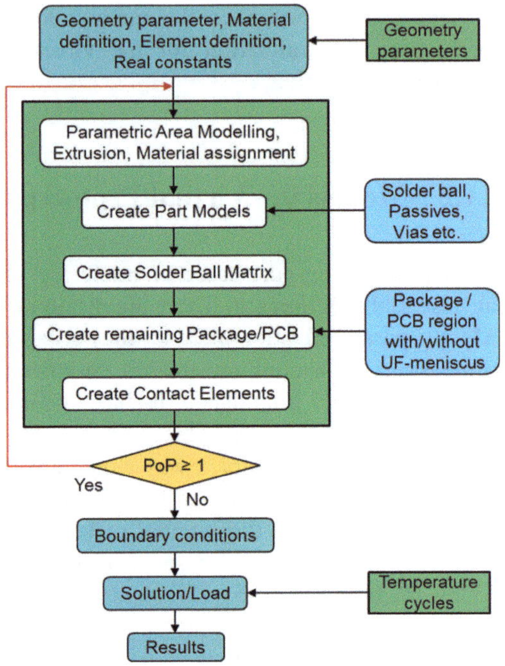

Fig. 11.2 Flowchart for parametric FE modeling for a system of models ([8], with permission)

The ANSYS APDL macros (Ansys parametric design language) for creating the FE models have been represented in flowchart form in Fig. 11.2. The process starts by defining the geometry parameters, material definitions, element definitions, and real constants followed by five main modeling steps. The first step creates parametric areas, performs the complex areas subtraction retaining the given area numbers, creates areas mesh, extrudes areas to create volumetric mesh and assigns material properties to volume elements. This step gives the flexibility to create complex geometries and FE models for any combination of components like multiple dies, vias, and passive devices. The next step creates FE part models of different types of solder balls, vias, and passives by calling different sub-macro files. After that, the solder ball matrix is created by copying the part models from previous step. In the fourth step, the remaining package region outside the solder ball matrix is modeled with or without the underfill. Finally, the Multi-Point Constraint (MPC) based contact elements are created between the FE models which have mesh dissimilarity.

After completing the model for a single layer of electronic packages, the routine can repeat all the steps again in order to create the second layer for the PoP configuration. This option is selected by defining the switch parameter PoP = 1. If PoP = 0 the boundary conditions are applied and finally, the FE analysis is performed according to the load steps and the results are extracted. The part models created within the modeling process can be assembled very flexibly—controlled by another dedicated parametric script. In this way, the system of fully parametric FE models is very well capable of replicating many different SiP products covering all three groups as shown in Fig. 11.1.

The sets of parametric FE models can also be used to perform optimization studies. Combining the FE solver code (e.g., ANSYS) with a DoE software tool such as optiSLang™ [9], the best geometrical layout and the best possible material parameters can be identified for a freely selected target function such as the minimum risk of interfacial and bulk damage in the packages during temperature cyclic loads. By use of these commercially available software tools and scripting languages like Python, the optimization process can even be created for automated execution. In these studies, a challenging first part is to define the parameters, which really have to be optimized. These key parameters need to be identified from the larger set of impacting parameters. Typically, a sensitivity study is performed to identify the most relevant parameters and their behavior (linear/non-linear). Therefore, all parameters are changed in a small range and the resulting impact to the target parameters is measured.

11.3 Example: Design Optimization of Eight-Pin DIL Package

In this section, the previously mentioned methods are explained in more detail using a classical DIL eight-pin package. For the sake of simplicity, the example focuses on studies that allow modification of parameters without changing the basic structure. The explanation starts with the presentation of the parametric FE simulation model followed by the description of the material characterization and material modeling process. Afterwards, the typical failure modes observed in those packages are shown. Also, the model validation has been illustrated and the sensitivity analysis with optimization of design and material parameters has been explained.

11.3.1 Simulation Model

Figure 11.3 shows an eight-pin package design modeled as a parametric finite element model using double symmetry by APDL language (ANSYS Parametric

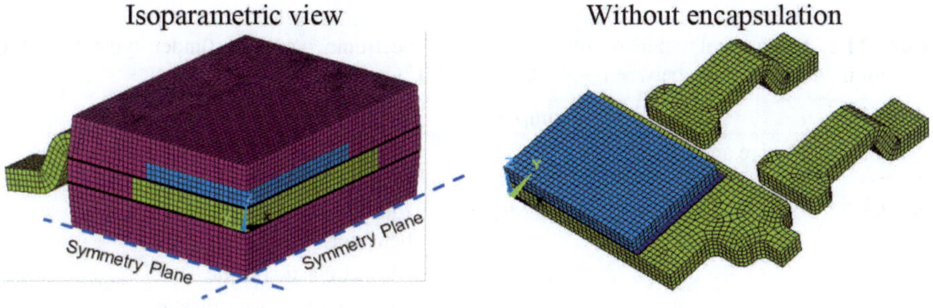

Fig. 11.3 Parametric finite element model of the DIL package

Design Language) within the commercial FE software ANSYS. The base of the model is the silicon die in the middle of the package. It is attached on the leadframe (copper) by a die attach. The housing of the chip is an encapsulation material made of an epoxy filled with silica particles. The electrical contacts between the chip and the lead pads are realized by bond wires, which do not have to be considered in the model as they have no significant mechanical effect. The FE model consists of nearly 55,000 nodes and elements. For the validation of the model, the initial input parameters like material properties still may need to be changed depending on experimental measurements.

11.3.2 Material Characterization

Table 11.1 summarizes the typical materials used for DIL packages and the considered thermo-mechanical behavior. Intensive investigations in the field of polymers (adhesives, encapsulation) have shown that the properties of these materials are strongly time and temperature dependent. They show relaxation (decreasing stress under constant deformation state) and creep behavior (increasing deformation under constant mechanical loading). A linear visco-elastic material model can describe this time and temperature dependency. The material characterization can be done by dynamic mechanical analysis (DMA), where a frequency sweep is used to extract the elastic response of a material under variation of the loading frequency and temperature. This results in a characteristic curve as shown in Fig. 11.4. Further materials of a classical DIL package are silicon, which is modeled as linear elastic structure, and copper with elastic-plastic behavior. The mechanical behavior of these materials is dependent upon the field of application (thermal exposure).

Materials used in a DIP package are typically copper (leadframe), silicon (chip), the encapsulation material (mold compound), and the die attach (polymeric adhesives), see summary in Table 11.1. The objects of the optimization routine for the DIL package are the interfaces between the materials. Therefore, critical values for the interface toughness are necessary. Table 11.3 summarizes the interfaces considered here and an assortment of typical test methods to determine the interface toughness parameters.

Table 11.1 Mechanical behavior of materials for electronic packages (under typical loading condition—thermal cycle between −50 °C and +150 °C)

Material	Example	Mechanical behavior
Silicon (with active and passive layers)	Die	Linear elastic
Solder	Solder balls	Elastic, elastic plastic, creep, cyclic behavior
Metals	Leadframe	Elastic, elastic plastic
Adhesives	Die attach	Visco-elastic, visco-plastic
Filled polymers	Encapsulation	Visco-elastic

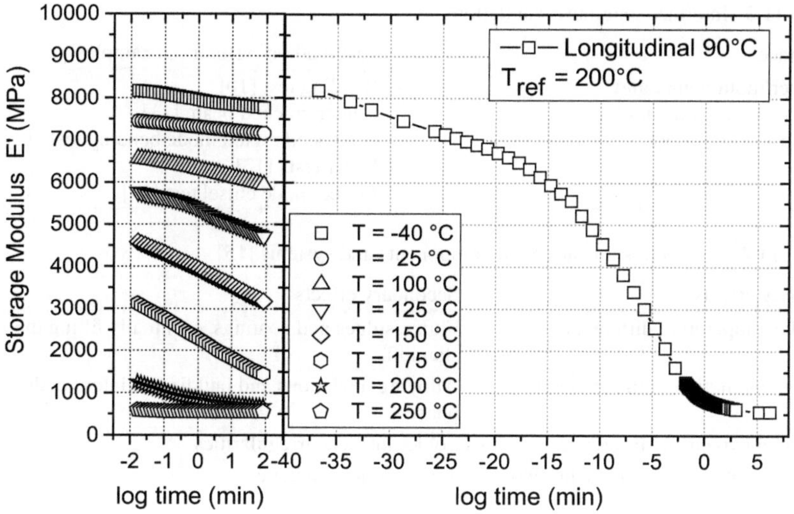

Fig. 11.4 Typical dependency of the storage modulus on the temperature for an epoxy material determined by DMA

Table 11.2 Measurement methods

Material	Material behavior	Young's modulus	Poisson ratio	CTE
Copper	Elastic, elastic plastic	Tensile test, bending test, nanoindentation	Tensile tests	TMA
Die attach	Visco-elastic	DMA	Tensile tests	TMA
Mold compound	Visco-elastic	DMA	Tensile tests	TMA

Table 11.2 The thermal as well as the mechanical behavior of these materials is essential for the thermo-mechanical reliability of the packages. In detail, this means that the time-temperature dependency of the coefficient of thermal expansion (CTE) and modulus of elasticity (Young's modulus) for the polymeric materials has to be considered.

The objects of the optimization routine for the DIL package are the interfaces between the materials. Therefore, critical values for the interface toughness are necessary. Table 11.3 summarizes the interfaces considered here and an assortment of typical test methods to determine the interface toughness parameters.

11.3.3 Thermo-mechanical Failure Modes

Thermo-mechanical failures, which may occur in packages and systems, are widespread. Typical failures are bond wire lift off (not considered in the example),

Table 11.3 Interface toughness validation

Interface	Test method (reference examples)
Copper/mold compound Silicon/mold compound Leadframe/die attach	Bending test [10] Center cracked beam [11] Advanced mixed mode bending test [12] Shear tests [13] Mixed mode chisel test [14]

Table 11.4 Parameters for fatigue of components and systems [15]

Primary effects	Secondary effects
Cyclic temperature difference	Flux residues and coatings, eventually filling the standoff
Size/construction of the component	Alloying deviations and building of intermetallic phases
Thermal expansion difference	Solder intrinsic irregularities
Coupling effect of components and board/systems	Temperature gradients
	Mounting related interactions
	Voids, microstructure, single grain joints

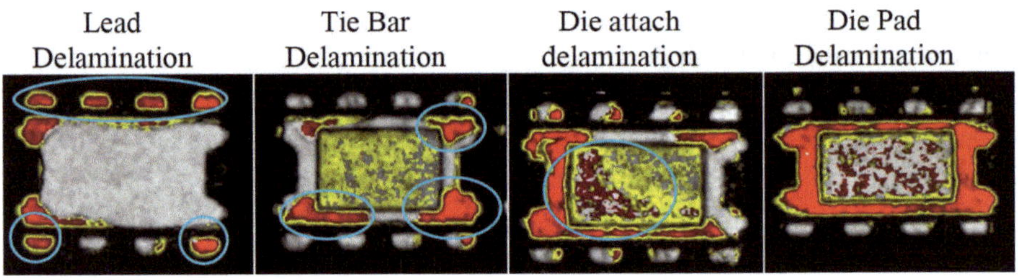

Fig. 11.5 Failure modes within a DIL eight-pin package analyzed by SAM (Scanning Acoustic Microscopy) [provided by Infineon]

solder fatigue, and interfacial delamination. The impacting parameters for the fatigue of the components are listed in Table 11.4.

The delamination at various interfaces inside a real DIL package has been detected by scanning acoustic microscopy. Figure 11.5 shows the status after thermal cycle tests (150 °C to −55 °C).

11.3.4 Model Validation

A very important issue for models used in virtual prototyping is their validation based on experimental tests. This includes the determination of material parameters under consideration of environmental conditions addressed to the package or

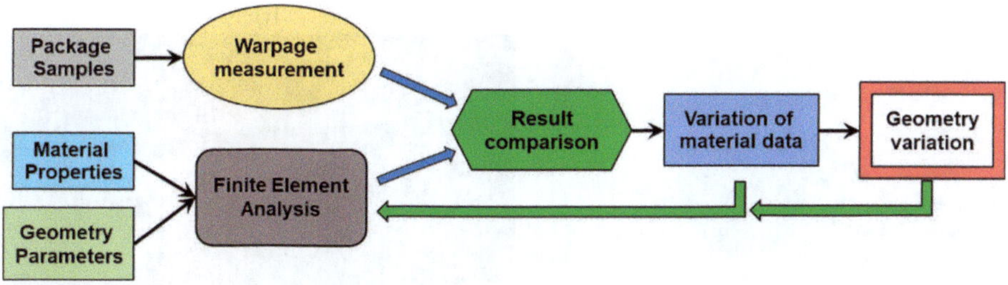

Fig. 11.6 Schematic of a typical validation procedure

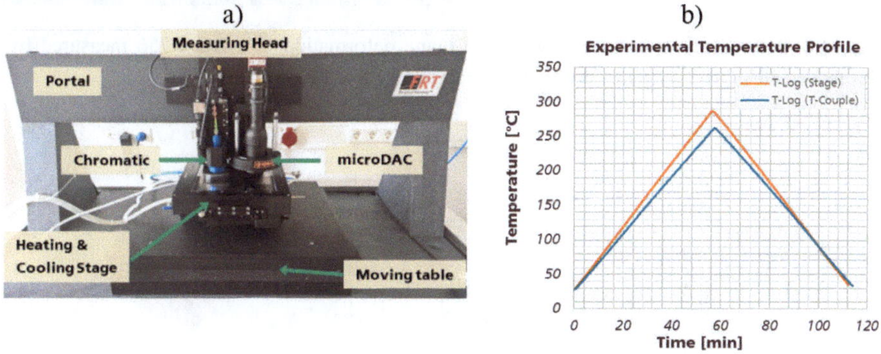

Fig. 11.7 (**a**) MicroProf® Test System for thermal out-of and in-plane measurements, (**b**) temperature profile

system. In detail, this means the temperature range, humidity, and mechanical loads. A typical validation procedure is shown in Fig. 11.6.

Beside other methods, one way to directly get the information of the thermal-mechanical response of a package or system is to use the deformation behavior under thermal impact. This deformation can be measured in-plane or out-of-plane (warpage). Figures 11.7 and 11.8 show the MicroProf® measurement system for both kinds of setups. The utilization of a chromatic sensor allows precisely measuring the out-of-plane deformation, whereas an image correlation analysis by means of the VEDDAC system enables indentifying the in-plane deformations with high resolution. Both deformation states can then be compared to simulated findings.

A comparison between the experimental and simulated warpage of the DIL package is shown in Fig. 11.9a. The curvature bandwidth calculated from the measurements for five samples is plotted for single die package. The result of a simulation, in which no delamination has been considered, is found out of measured curvature bandwidth. Hence, simulations have also been performed that consider a delamination at the interface between die attach and die pad (DA delam) as well as between leadframe and mold compound (LF delam), respectively (Fig. 11.9b). The

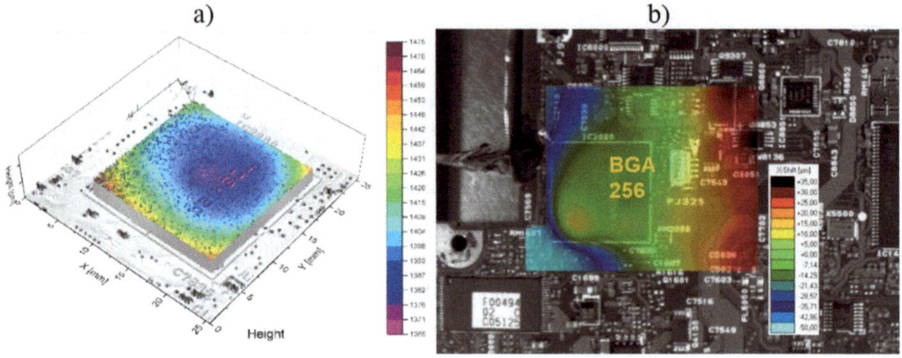

Fig. 11.8 MicroProf® test results: (**a**) out-of-plane deformation of a BGA 256 measured by a chromatic sensor, (**b**) in-plane deformation analyzed by microDAC® [16]

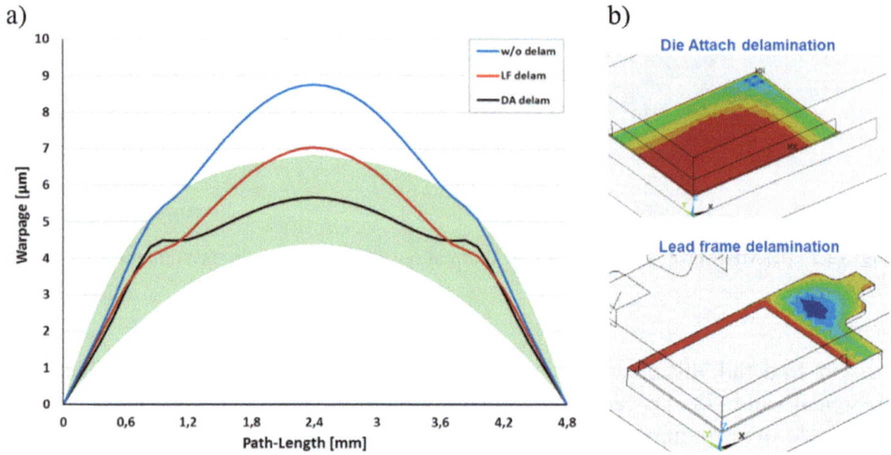

Fig. 11.9 (**a**) Comparison between the experimental and simulated warpage of the DIL package (experimental results are shown as a grey band) and (**b**) FE simulations considering the die attach and lead frame delamination

results of those simulations show a significant change in the warpage magnitude and fall nicely into the measured interval. Hence, the DIL package under test suffered from delamination as already seen in Fig. 11.5.

11.4 Sensitivity Analysis and Optimization

Sensitivity study and optimization based on finite element modeling were done using optiSLang 5.0 embedded in ANSYS workbench. In this example, the aim of the sensitivity analysis is to identify most relevant geometrical and material parameters, which have the maximum impact on the desired objective parameter. So, it is very important and also challenging to define correct objective parameters which aim to reduce the risk of interfacial and bulk damage in the package.

For that, the reference case is simulated first with the reference values of geometrical and material parameters. The results are analyzed to finalize the objectives. Typical objective criteria used in this kind of reliability analysis refer to warpage, stress, strain, or mechanical energy values, respectively. The aim of the optimization is to minimize these values, i.e., the objectives establish the optimization criteria.

In this example, the interface stresses which are extracted at the interfaces using contact elements are defined as the optimization criteria, whose minimization would lower the delamination risk. Therefore, the FE model was modified and contact elements were included at the different interfaces. The interface stresses are extracted at the interface between mold-lead pins (ST_LEAD), mold-die pad (ST_DP_MC), and die pad-die attach (ST_DP_DA) as shown in Fig. 11.10. The total interface stress includes the tensile normal stress and two shear stresses. The reduction of the numbers of objectives is done by combining the interfacial stress to an effective stress parameter:

$$\sigma_{\mathrm{eff}} = \sqrt{\langle {\sigma_z}^2 \rangle + \tau_{xz}^2 + \tau_{yz}^2}$$

with σ_z, τ_{xy}, and τ_{yz} are the three stress components, one in normal and two in shear direction. $\langle i \rangle$ is with the MaCaulay's brackets which takes negative values for i as zero.

There are eight geometry parameters and eight material parameters to be optimized, which lead to many design legs in the sensitivity analysis. So, the first sensitivity analysis has been performed only with eight geometry parameters. Extra variables and constraints are defined in optiSLang to allow the variations of geometry dimensions in an acceptable range, which results in successful meshing and FE model creation for each design. The analysis results showed the chip dimensions and the die attach thickness (Z_DA) as the most important out of the eight geometry parameters. Hence, the final sensitivity analysis considered only

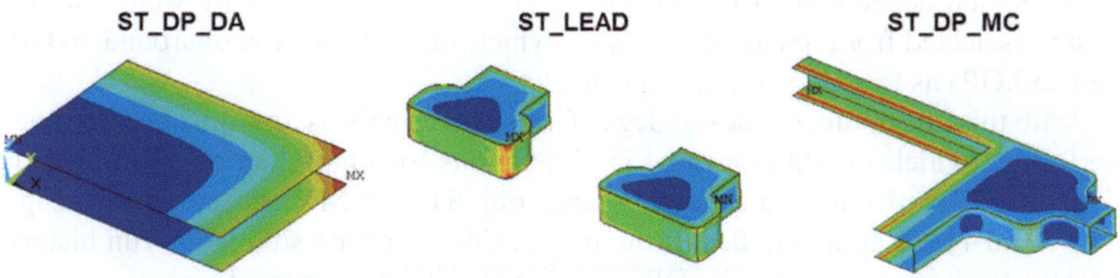

Fig. 11.10 The total interface stresses (including compressive stress) of the contact elements at different interfaces showing critical regions in red

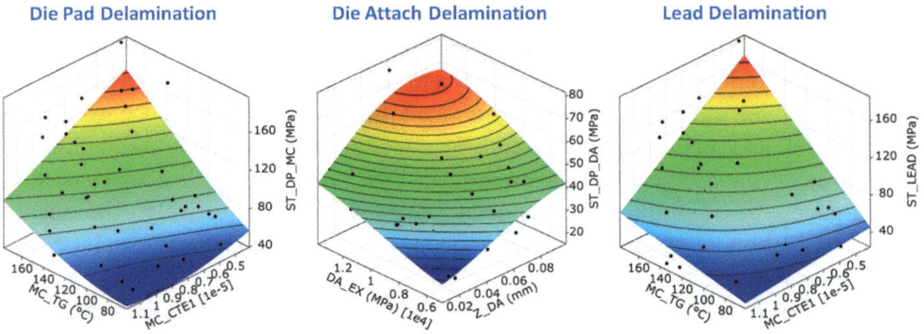

Fig. 11.11 The results of the sensitivity analysis from 55 design legs

those geometry parameters in addition to eight material parameters (e.g., CTE1, CTE2, Young's modulus E and glass transition temperature of mold T_g, and die attach). For each of the material parameters, a 50% variation around their reference was considered while the prony series parameters have just been adopted to the respective T_g, CTE, and E magnitude but not changed in their value. The cooling down from stress free temperature 260 °C to −50 °C was simulated to occur in 1000 s. The DoE plan has been generated using Advanced Latin Hypercube Sampling.

The results of the sensitivity analysis are depicted in Fig. 11.11. It is assumed that the total interface stress criteria represent the delamination mechanism for that respective interface. For example, ST_DP_MC value represents the interface delamination or damage risk at mold–die pad interface. If this stress value is high, the delamination risk is high. This assumption is valid for geometry variations but it might not be straightforward but need adjustments in case of varying the material properties since the stiffness and the strengths at the interfaces may vary with the differences in the material properties.

According to this simplified assumption, the die pad delamination (LF delam) is influenced mostly by MC_TG and MC_CTE1 followed by MC_EX and die attach thickness, which also create some effect. Die attach delamination (DA delam) is most susceptible towards die attach thickness, DA_EX and DA_CTE1. Lead delamination depends on MC_TG, MC_CTE1, and MC_EX. There were few best designs selected from sensitivity analysis, which suggest MC_EX of around 70 GPa and 130 GPa as favorite values, respectively.

With this pre-optimized design legs of sensitivity analysis, optimization has been performed which in addition run 155 simulations for finding an optimum design. The evolution of one objective parameter obj_ST_DP_MC can be seen in Figs. 11.11 and 11.12. It also indicates the best designs over the simulation run history. Finally, the best designs for the DIL package have been suggested.

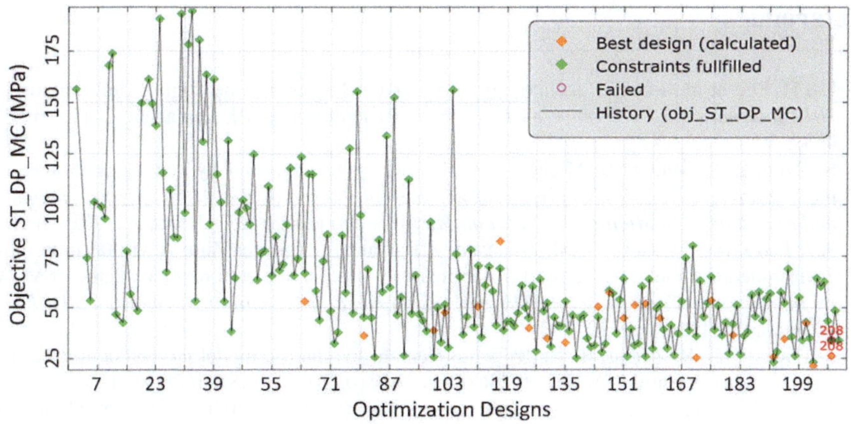

Fig. 11.12 Objective history during the optimization process

11.5 Summary and Conclusion

The focus of this chapter was to give a brief overview of a modular system of parametric FE models. Such kind of modeling methods are highly needed for automated virtual prototyping of current and future IoT solutions.

1. A fast model generation approach was shown based on parametric modeling for FOWLP based SiP products. This was reached by use of a library structure. The approach is very flexible for different kinds of packages.
2. Based on an example of a classical dual-in-line package, this approach was used to generate a parametris FE model. The model was validated based on a novel thermo-mechanical defomation analysis setup (MicroProf®).
3. Based on the validated parametric FE model, an optimization strategy was followed to reduce the risk of thermo-mechanical driven delamination inside the package.

Summarizing, it was shown that physically meaningful FE model can be generated as long as appropriate material parameters are available. With the help of these models, comprehensive experimental thermo-mechanical assessments and reliability tests which are very costly and time consuming can be reduced.

References

1. L.S.H. Lee et al., FOWLP technology as wafer level system in packaging (SiP) solution, in *2017 International Conference on Electronics Packaging (ICEP)*, Yamagata, 2017, pp. 491–493
2. S. Rzepka, A. Müller, B. Michel, Virtual prototyping advanced by statistic and stochastic methodologies, in *2010 11th International Thermal, Mechanical & Multi-Physics Simulation, and Experiments in Microelectronics and Microsystems (EuroSimE)*, Bordeaux, 2010, pp. 1–9
3. A.J. Przekwas, Y. Jiang, M.M. Athavale, Z.J. Wang, J. van der Zijp, A virtual prototyping environment for multi-scale, multi-disciplinary simulation of electronics packaging of MCMs, in *InterSociety Conference on Thermal Phenomena in Electronic Systems, I-THERM V*, Orlando, FL, 1996, pp. 352–358
4. J. Tang, P. Xi, B. Zhang, B. Hu, A finite element parametric modeling technique of aircraft wing structures. Chin. J. Aeronaut. **26**(5), 1202–1210 (2013)., ISSN 1000-9361
5. J. Fan, Z. Yang, J. Wang, S. Ding, C. Liu, Z. Shi, Parametric finite element modelling and nonlinear analysis of vehicle brake, in *2009 International Conference on Mechatronics and Automation*, Changchun, 2009, pp. 1762–1766
6. A. Cardoso et al., Development of novel high density system integration solutions in FOWLP—complex and thin wafer-level SiP and wafer-level 3D packages, in *IEEE 67th Electronic Components and Technology Conference*, 2017, pp. 14–21
7. ANSYSTM, *Multiphysics, Version 18, User's Manual* (Ansys Inc., 2018)
8. G. Gadhiya, B. Brämer, S. Rzepka, Automated virtual prototyping for fastest time-to-market of new system in package solutions, in *2018 7th Electronic System-Integration Technology Conference (ESTC)*, Dresden, 2018, pp. 1–7
9. OptiSlang™, *User Manual* (Dynardo GmbH, Weimar, 2018)
10. V. Sundararaman, B.D. Davidson, An unsymmetric double cantilever beam test for interfacial fracture toughness determination. Int. J. Solids Struct. **34**(7), 799–817 (1997)
11. P.G. Charalambides, J. Lund, A.G. Evans, R.M. McMeeking, A test specimen for determining the fracture resistance of bimaterial interfaces. J. Appl. Mech. **56**(1), 77–82 (1989)
12. J.R. Reeder, J.R. Crews Jr., Mixed-mode bending method for delamination testing. AIAA J. **28**(7), 1270–1276 (1990)
13. R. Dudek, B. Braemer, R. Pufall, H. Walter, B. Seiler, B. Wunderle, Determination of interface fracture parameters by shear testing using different theoretical approaches, in *Proc. 13th EuroSimE Conf.*, 2012
14. G. Schlottig, Reliability at the Chip interfaces: delaminating the silicon die from molding compound, Dissertation, 2012, ISBN: 978-94-6186-072-9
15. R. Dudek, Mounting effects on board level reliability, in *EuWoRel-Workshop*, Berlin, 2015
16. B. Seiler, R. Dudek, L. Scheiter, K. Kreyßig, S. Haase, *Characterization of the Thermo-Mechanical Deformation Behaviour of Composite Materials for Power Electronics Using Digital Image Correlation Method microDAC®, Micromaterials and Nanomaterials*, 15, ISSN 1619-2486, MicroCar, 2013, pp. 204–207

Chapter 12
Sensor Testing for Smart Mobility Scenarios: From Parking Assistance to Automated Parking

J. Murgoitio Larrauri, E. D. Martí Muñoz, M. E. Vaca Recalde, B. Hillbrand, A. Tengg, Ch. Pilz, and N. Druml

12.1 Introduction

At present, road vehicle automation is one of the major challenges of nowadays' transport system and its goals are to achieve the ideal energy efficiency, the minimum environment impact, and the highest safety rate as well as to improve the drivers' comfort. Due to this fact, currently, it is not only the trend to develop systems that help the driver in controlling the own vehicle in the driving process such as ADAS, but also Cooperative ITS for automating road transport as well.

These cooperative systems are based on communications in real time among IoT devices. However, the connectivity between these devices and the vehicle, together with the data obtained by the own vehicle, with the purpose to manage an automated and adequate decision making, represents a new step in the evolution of this concept.

So, the concept of Internet of Vehicle (IoV) is the next step for future smart transportation and mobility applications and requires creating a new IoT ecosystem based on trust, interoperability and security. IoT infrastructure (based on sensors, ubiquitous connectivity, information processing) will transform the environment

J. M. Larrauri (✉) · E. D. M. Muñoz · M. E. V. Recalde
Tecnalia Research & Innovation, Parque Científico y Tecnológico de Bizkaia, Derio, Bizkaia, Spain
e-mail: jesus.murgoitio@tecnalia.com; enrique.marti@tecnalia.com; myriam.vaca@tecnalia.com

B. Hillbrand · A. Tengg · C. Pilz
Virtual Vehicle Research Center, Graz, Austria
e-mail: Bernhard.Hillbrand@v2c2.at; Allan.Tengg@v2c2.at; Christoph.Pilz@v2c2.at

N. Druml
Infineon Technologies Austria AG, Graz, Austria
e-mail: Norbert.Druml@infineon.com

W. D. van Driel et al. (eds.), *Sensor Systems Simulations*,
https://doi.org/10.1007/978-3-030-16577-2_12

and economics of mobility scenarios in the future: first through the connections and secondly through the emergence of an ecosystem of services around smarter vehicles.

In this context, IoSense is the ECSEL project which will deploy new capabilities (Sensors, Components and Systems) through several demonstrators, one of them using the SmaBility use cases: The connection of vehicles to their surrounding environment and to Internet will offer a great variety of new possibilities that will bring new functionality to citizen and organizations and will make the transport more efficient and safer. Intelligent perception and decision making for safer and autonomous driving are the main objectives of the SmaBility demonstrators, which clearly search for safer driving and new application ecosystem associated to mobility: Automated parking is one of the Smability use cases.

The called "Automated parking" IoSense SmaBility use case impacts the mobility application domain by enabling and advancing concepts for Internet of Vehicle (IoV) as the next step for future smart transportation and mobility applications creating a new IoT ecosystem based on trust, interoperability and security. This demonstrator of the IoT use case enables a significant progress beyond the state-of-the-art on high performance information processing through data fusion algorithms that combine V2V and V2I information with on-board sensor information.

The results of this demonstrator will accomplish the MASP defined in the [1]. It will be built on advanced systems for driver assistance, **cooperative systems**, and driver status monitoring as well as **environment perception**. Systems will be validated under virtual, semi-virtual and **real-world conditions** (i.e. weather conditions). This requires dependable solutions for advanced sensors and actuators, **data fusion**, efficient use of **connectivity** human interaction technologies, CPS, and (real-time) simulation concepts.

This chapter is structured by the following sections:

- Introduction
- Selection of adequate simulation methodologies
- TOF-Concept and design of sensors system
- Simulation model development: Simulation elements, Algorithms, and Testing
- Summary and conclusions

12.2 Selection of Adequate Simulation Methodologies

This chapter lists the capabilities in the design, modelling and simulation area of each partner (TECNALIA, IFAT and VIF) involved on the title "From Parking Assistance to Automated Parking" within the Smart Mobility scenarios (Smability).

12.2.1 Tecnalia

Tecnalia, as a research company with around 1500 workers, most of them engineers, uses different simulation tools. Anyway, the department involved on this project is focused on intelligent transport systems, and more specifically we are interested on smart mobility scenarios using ADAS systems and automated ground vehicles (UGV).

From this point of view, vehicle simulators are very valuable for validation and testing of new functionalities for automated driving. In order to increase security and safety in new algorithms, realistic simulators are necessary in the automotive sector.

From partially automated to fully automated systems (SAE J3016 levels), some of the key components are communicated between the different on-board elements (perception, human interaction and control algorithms). In this sense, a simulator that combined all these features to reduce the real test effort is very convenient. In the last decades, the relationship between control algorithm development and system dynamics implementation is increasing. In order to develop better decision and control features for automated vehicles, a precise multi-body model to describe the vehicle dynamics in different scenarios is still missing.

Robust dynamic models, based on a correct environment modelling, are going to lead a reduction of development cost of the new ADAS. In this section, a description of the current simulator available on market used by us is presented.

12.2.1.1 Civitec Pro-Sivic

This tool has been designed as a sensor simulator by the company CIVITEC. It gives the opportunity to simulate a variety of complex scenarios like intersections, roundabouts, multiple vehicles and pedestrians on the road, and changes in the weather conditions rain, fog, snow, brightness and others.

Moreover, it allows to adapt the parameters and behaviours of sensors and weather, and it has the potential of being connected with RTMaps platform to test new algorithms, from the data fusion and control point of view.

There are some technologies that have been tested on this platform on last years. Real-time road lane detection and tracking is one most interesting. Another application is the use of the platform Simulink/Matlab to communicate it with the Pro-SIVIC simulator to test cooperative algorithms, like CACC (Cooperative Adaptive Cruise Control), on automated vehicles. Pro-Sivic is mainly oriented to perception and ADAS developments.

12.2.1.2 CarSim and TruckSim

CarSim and TruckSim are products of the Mechanical Simulation Company, those are dynamics vehicle simulators, for passenger cars and light trucks, respectively.

Those offer some capabilities like software-in-the-loop (SiL), model-in-the-loop (MiL), hardware-in-the-loop (HiL), driver-in-the loop (DiL), supporting vehicle sensors and vehicle-to-vehicle (V2V) communication. Additionally, they have a standard interface to Matlab/Simulink. This simulator is mainly oriented to validation and ADAS, i.e.: obstacle avoidance algorithm using model predictive control, robust control for in-wheel motor vehicles and dynamic trajectory generation.

12.2.1.3 IPG Automotive CarMaker and TruckMaker

CarMaker and TruckMaker are other dynamic simulators, with open integration and tested with other platforms like Matlab/Simulink and capable to test HiL and multi-HiL on real time.

Some of the works made on this platform on the last years are test of a vision-based lane keeping assistance control triggered ECU.

12.2.1.4 Tass International PreScan

PreScan of Tass international is a simulator that is specialized on sensor modelling like GPS, vision, laser, radar, accelerometer and odometry. It has the capability of managing n-vehicles and n-sensors in each one. It is also based on Matlab/Simulink.

Some of the most important works made on this platform are a full spectrum camera simulation for reliable virtual development and validation of automated driving applications, among others.

12.2.1.5 SCANeR

SCANeR Studio is a software suite for Human-In-the-Loop driving simulations; it is developed by OKTAL and is based on works of the Vehicles Simulation and Perception research group of Renault and works of SERACD.

Several European projects have been used as background for the development of the software, examples are Prometheus, TRaCS (TRuck and Coach Simulator) and CARDS (Comprehensive Automobile R&D Simulator).

12.2.1.6 Tecnalia Dynacar

Dynacar (see Fig. 12.1) is a simulation tool developed by Tecnalia which provides a real-time vehicle model covering multiple domains. It focuses on two main domains. The first is vehicle dynamics. It provides a high-fidelity vehicle physics simulation basing a multi-body vehicle dynamics model. This is combined with a Pacejka tire model, and submodels for elements like the engine, transmission, steering system, braking system, aerodynamics, etc. The second domain is the Electric-Electronic architecture of the vehicle.

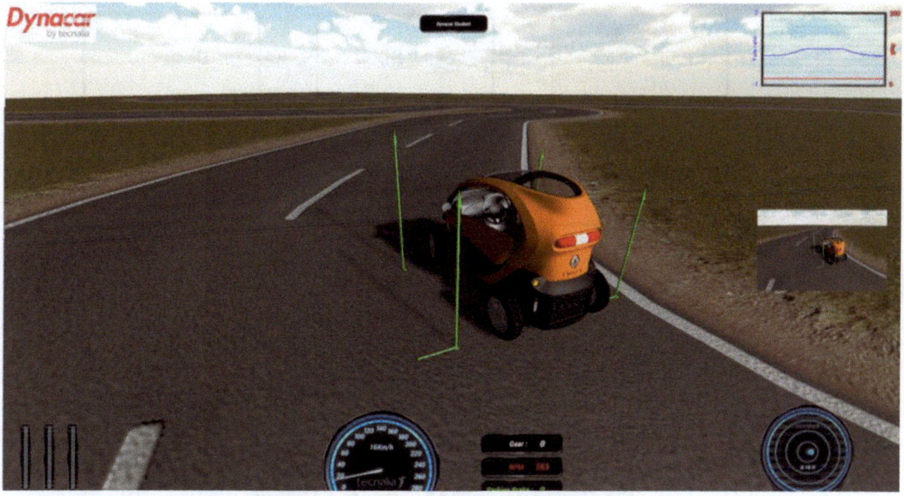

Fig. 12.1 Dynacar by Tecnalia

Here subsystem models (from ECUs to components) can be connected for development of control functions.

Dynacar permits real-time simulations, besides accelerated time simulations. The real-time capability is very valuable, as, combined with its notable modularity and interfacing options, it permits to execute tests with driver-in-the-loop (DiL) and hardware-in-the-loop (HiL) setups, for instance for ECU (Electronic Control Unit) development or also motor test-bench testing. The flexibility is further improved because it can run as standalone or be integrated into a Simulink model.

Many works in Tecnalia are based on Dynacar capabilities, using them to implement automated driving manoeuvres. A testing methodology for the validation of control algorithms for future automated vehicles has been designed. The remarkable modularity and adaptability have been exploited in the designs, which enables a good trajectory definition, cooperative manoeuvres and virtual validation. Different kind of vehicles and scenarios can be tested (Fig. 12.2).

12.2.2 IFAT

Infineon employs a major internal tool flow that copes with all important aspects of chip development. For the development of the next generation Time-of-Flight 3D sensor, the following approach is employed (listed in a simplified way):

- Analogue design
  - Cadence's Spectre, fast SPICE tools (e.g. UltraSim), and internal tools are used for simulation purposes.

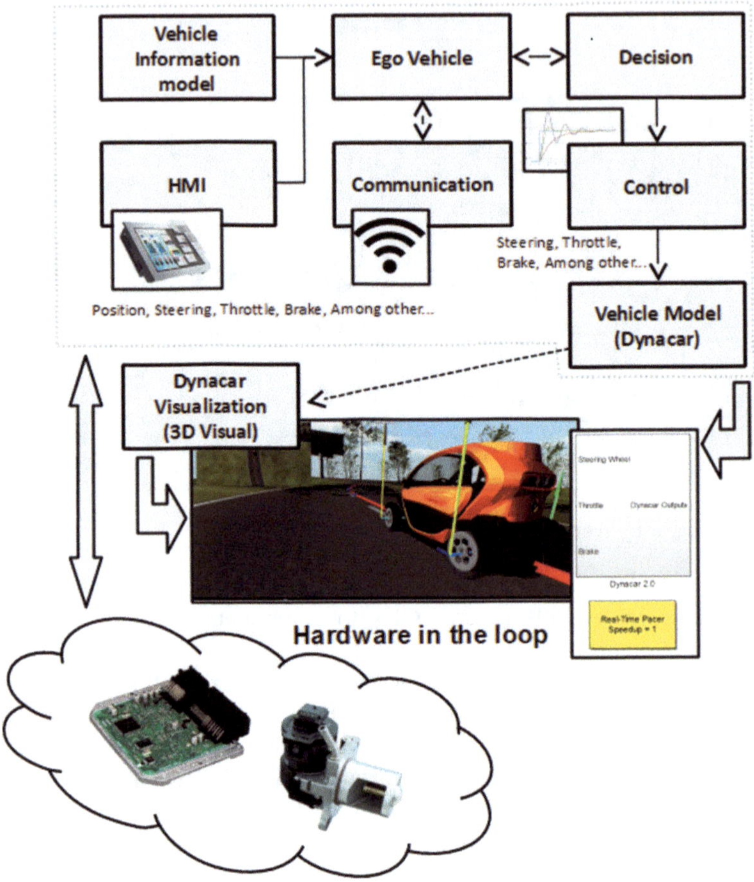

Fig. 12.2 Architecture framework for automated driving

  - Designs are developed with the Cadence Design Suite.
  - Models of the analogue components are written in VHDL or Verilog.
- Digital design
  - Designs are described either in the languages VHDL or Verilog.
  - Synopsis or Mentor Graphics (e.g. ModelSim) tools are used for simulation purposes.
- Top level
  - SystemVerilog and a constrained random verification principle are used as functional verification solution, which includes digital designs, models of the analogue components, and firmware.

- System level
  - Matlab is used as the most important tool to model and simulate Time-of-Flight camera system concepts (sensor + illumination unit).

12.2.3 Virtual Vehicle (VIF)

In the past few decades, numerous specific simulation tools have been established in the automotive industry. However, these tools typically specialize on individual areas of expertise. There is very limited support for a heterogeneous simulation environment. The goal of "co-simulation" is to overcome this limitation and to merge the challenges from different areas. Common co-simulation platforms are often limited to a single area of expertise (e.g. the design of a thermal management system with a heterogeneous tool landscape). Such platforms typically address problems from a very specific, restricted dynamic range, which means that the different models exhibit similar time behaviour.

However, the development of modern, mechatronic systems requires a much broader approach. The interactions between sub-systems from different areas have to be taken into account through a suitable interconnection of the parts. The coupling of existing (specific) simulation programmes (and the models implemented therein) from different areas of expertise represents a promising approach for the simulation of the complete system.

With the introduction of co-simulation in the development process, the task of developing complex mechatronic systems can be solved in a very efficient way. For example, the integration of Finite Element Methods (FEM) is supported for applications in the field of "Integral Safety". Another example is the integration of electrical and thermal components into existing drivetrain concepts in the field of "alternative drives", such as hybrid or electric vehicles. In all of these considerations the vehicle itself is not the exclusive focus, but rather the interactions with the vehicle's environment and the influence of the real driver are taken into account.

The task of a co-simulation platform is to take the complex interactions of the various simulation models in a suitable and correct way into account. The platform has to enable the precise co-working of different simulation tools.

The co-simulation platform ICOS (Independent Co-Simulation; see [2]) developed at VIF fulfils all these requirements (Fig. 12.3). Over the past years, ICOS has proven both its flexibility and its reliability to succeed in a vast amount of different automotive use cases. ICOS is included in the software AVL Model.CONNECT™[1]. Apart from a user-friendly graphical interface, Model.CONNECT™ offers powerful model integration capabilities as well as an efficient post-processing and visualization engine.

[1]The Model.CONNECT™ software suite is distributed by AVL. For research projects—like IoSense—VIF can provide licenses free of charge but limited to the project duration and project content.

Fig. 12.3 ICOS allowing for interdisciplinary co-simulation (source: [2], with permission)

A key feature is the support of a vast amount of simulation tools, such as Adams, AMESim, AVL CRUISE, CarMaker4Simulink, CarSim, Dymola, Flowmaster, Java, KULI, LabVIEW, MATLAB, Microsoft Excel, OpenModelica, SIMPACK, or dSpace VEOS. Additionally, models provided in the standardized FMU interface are supported. Supplementary software can be added by using a generic Dll-based interface. The list of supported simulation tools is continuously increasing.

Furthermore, Model.CONNECT™ supports advanced model variation capabilities. Values of model parameters can be prescribed, and the co-simulation is performed automatically for all variants. The results can then be visualized and compared in a user-friendly and intuitive way.

In addition to the aforementioned simulation tools, Model.CONNECT™ can also be connected to real-time systems. Currently, UDP and CAN are supported as communication protocols.

12.2.3.1 Proposal for "Rationalization of Simulation Methodologies"

A fundamental objective within IoSense is the appropriate multi-domain modelling of the demonstrator. In general, it is not feasible to combine all of the required domain-specific models to one monolithic simulation because of the completely different dynamics. Consequently, it is state-of-the-art to use co-simulation. This means that each of the subsystem models (e.g. vehicle components, sensors, control units, driver, environment . . .) is modelled using the approach and simulation tool suited best for its domain. Subsequently, the simulations are executed and coupled by making use of a specialized co-simulation tool (see Fig. 12.4). A fundamental advantage of this method is that the functionalities of the single subsystem models do not need to be changed. As a prerequisite, it is only necessary to select and specify the interface specifications (e.g. data type, dimension, range . . .) of the variables to be coupled.

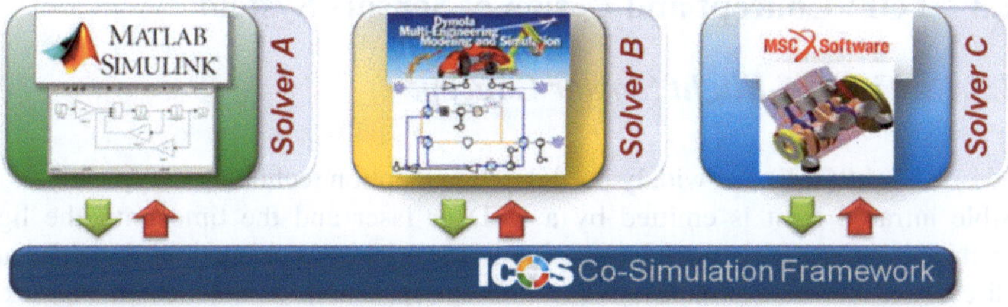

Fig. 12.4 Simulator coupling via co-simulation (source: [3])

Table 12.1 Overview of all software used related to the Smability scenarios and automotive applications by which partner

Software name	Used in	Partner ID
Ansys [8]	Smability(LiDAR), simulation of chip and coils, RFID transponders; high-speed hardware designs	IFAT-Graz
Cadence Design Suite	Analogue and mixed-signal chip design, Smability (LiDAR), automotive applications	IFAT-Graz, TECNALIA
CarMaker (IPG)	Smability	VIF
CarSim and TruckSim	Automotive applications	TECNALIA
Civitec Pro-Sivic	Automotive applications	TECNALIA
Comsol Multiphysics	Smability (LiDAR)	IFAT
Matlab	System level, Smability (LiDAR)	IFAT, IFAT-Graz, VIF
Mentor Graphics ModelSim	Circuit simulations	IFAT-Graz
Python	Smability (LiDAR)	IFAT
SCANeR	Automotive applications	TECNALIA, VIF
Sumo (DLR)	Smability	VIF
Tass international PreScan	Automotive applications	TECNALIA
Tecnalia Dynacar	Automotive applications	TECNALIA
ViSim (PTV)	Smability	VIF
VTD (Vires)	Smability	VIF

The co-simulation software AVL Model.CONNECT™ (implementing the ICOS framework) already supports many simulation tools. It is continuously being improved and extended by support of further specialized tools.

12.2.3.2 Software Selection

The table below summarizes the software, where they are used and by which partners. Co-operation between partners is connected to using the same kind of software (Table 12.1).

12.3 TOF: Concept and Design of Sensors System

12.3.1 *Time-of-Flight Sensor Concept*

Time-of-Flight (ToF) is a widely used depth perception technology. Typically, non-visible infrared light is emitted by a LED or laser and the time until the light is reflected from the scenery is measured. There are direct (such as LiDAR) and indirect measurement principles. The indirect approach, as detailed in Fig. 12.5, evaluates the distance by means of the phase shift between the emitted and received continuous wave signals with the help of photonic mixing devices (PMD). If integrated together with a strong illumination unit, such Time-of-Flight cameras can perform range measurements of up to 50 m. The raw data, which is gathered by the Time-of-Flight sensor, is then post-processed by a processing pipeline that is typically implemented in software and that exploits GPU hardware accelerators. Recently, these PMD-based depth perception solutions became quite famous thanks to miniaturization improvements and their low computational performance requirements, which enabled the integration into small embedded devices such as smart phones (see also Google Tango).

12.3.1.1 Time-of-Flight Sensor Design

During the IoSense project, IFAT developed in close collaboration with CTR the next generation of its Time-of-Flight sensor which implements the indirect photonic mixing device principle. Figure 12.5 depicts the high-level sensor design.

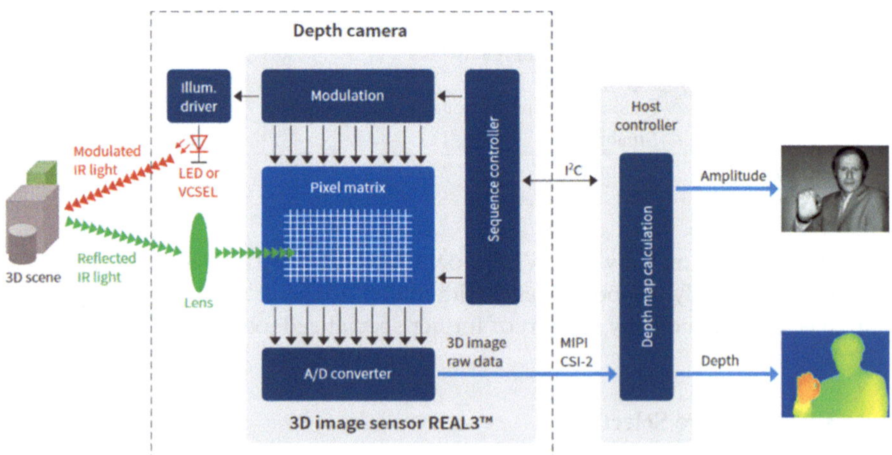

Fig. 12.5 Basic sensor design and working principle of the Time-of-Flight 3D imaging technology

- An I^2C interface is used to configure and control the sensor. Lots of registers, within a dedicated register container, are implemented in order to enable high flexibility and to support various types of applications (such as 3D scanning, gesture recognition, driver monitoring).
- A small sized and power-efficient internal controller is used to control the sensor SoC and its components. This proprietary controller developed by Infineon features not only modulation control, illumination control, pixel actuation, but also supports autonomous imaging phase sequences (such as combination of four-phase Time-of-Flight measurements used for gesture recognition with eight- or nine-phase Time-of-Flight measurements used e.g. by Google Tango) or coded modulation.
- The modulation and phase shifter component is one of the central elements of the sensor. Typically, one Time-of-Flight measurement requires shifting the phase of the emitted light several times (e.g. four times). Therefore, the modulation and phase shifter component implement a lot of complex logic. Furthermore, this block will support the innovative approach of coded modulation, which will drastically boost the performance of the next generation sensor.
- An illumination driver is implemented in order to emit modulated infrared light with the help of an external LED or VCSEL (laser).
- As shown in Fig. 12.5, after emitting modulated infrared light, the reflected light is focused by several lenses onto the PMD pixel matrix. Since PMD pixels are larger than RGB pixels, the light sensitive area of each pixel is consequently smaller. This smaller fill-factor results in reduced efficiency in converting light to electrons. This drawback is addressed by implementing micro-lenses on top of each pixel in order to focus the incoming light to the pixel's light sensitive area.
- High-speed A/D converters read out the pixels and transform their raw analogue values into the digital domain.
- Finally, a high-speed CSI-2 serial interface is used to transmit the sensor data to the application processor. Thanks to this standardized interface, the sensor can be easily integrated into small embedded systems such as smart phones.

IFAT's and CTR's activities not only fully addressed IoSense's objectives of reducing sensor manufacturing costs and sensor manufacturing time, the partners fully strived for improving sensor performance (e.g. in terms of power consumption) and measurement accuracy. These objectives lead to the following exemplified conceptual and design improvements developed during IoSense:

- The sensor's chip area is shrunk by more than 30% which consequently reduces manufacturing costs and improves the yield. Furthermore, an improved yield also improves margins accordingly.
- Given the fact that the Time-of-Flight sensor implements a rather complex optical measurement principle, also wafer-level testing is more complex than the testing of standard CMOS RGB sensors. Because test time directly influences the costs, it is of high importance to integrate sufficient test capabilities and thus to reduce test times as much as possible. Therefore, the next generation Time-

of-Flight sensor integrates several new test functions which can be executed through the integrated controller and firmware. The next generation sensor chip reduces the wafer-level test time by a lot more than 30%, thus reducing the total manufacturing time and manufacturing costs.

- The core pixel technology is further developed and advanced aiming for high measurement performance and improved sensing accuracy.
- From a system/camera perspective, guaranteeing eye-safety is absolutely crucial. In current camera/module generations, eye-safety is guaranteed with the help of external circuitries made of discrete components. These external components require space and increase the Time-of-Flight sensor module's bill of materials. Therefore, a major objective is to integrate the eye-safety circuits into the sensor and thus to reduce module costs significantly.
- Several new and innovative concepts are researched, developed and integrated that improve the sensor's measurement performance and accuracy drastically. One of these concepts aims for improving the absolute measurement accuracy. This improvement will be achieved by the integration of certain reference pixels that measure internal and external delays caused by wires, drivers, etc. By knowing these delays, the post-processing software can then compensate the acquired sensor data. Thus, higher absolute measurement accuracy will be achieved with the next generation Time-of-Flight sensor.

12.3.1.2 Link to Smability Demonstrator and KCIs

IFAT's current Time-of-Flight 3D imaging sensor generation is already available within reference cameras developed and built by the partner company PMDTech in Germany. These reference cameras will be used within the Smability demonstrator. Together with the partner ViF and Tecnalia, such a reference Time-of-Flight camera will be integrated into a partly automated vehicle and will be evaluated for short-range ADAS use cases (such as parking assistance).

On the demonstrator level, defining figures for the top ten measurable Key Competitiveness Indicators (mKCIs), the weight of those parameters depends on use case and market conditions: An overview of the mKCIs for the ToF sensor/SmaBility demonstrator is presented in Table 12.2.

12.4 Simulation Model Development

12.4.1 Time-of-Flight Simulation Based on ROS

The camera that is used in the IoSense project is a Pico Monstar camera (Fig. 12.6) provided by the IoSense partner Infineon Technologies Austria AG.

It has a fish eye lens, which grants a Field of View (FoV) of $100° \times 85°$ (H $\times$ V) while still providing very accurate depth measurements (Table 12.3).

Table 12.2 Overview of mKCIs for the ToF sensor/SmaBility demonstrator

KCI name	KCI description	Measurable target for IoSense
Functionality/feature set	Meet full functionality	100% of functionality met
Performance	Range measurement error	1%
Form factor/size; weight	Chip area reduction compared to previous generations	30%
Power consumption	Use-case dependent	–
Safety	Meet eye-safety requirements	100% of requirements met
Security	–	–
Reliability	Maintain reliability goals of previous generations	100% of reliability of previous generations met
Manufacturability	Reduction of test time	30%
Ease of integration	Reduction of BoM and external components	10% BoM reduction
Cost	Reduction of manufacturing costs	30%

Fig. 12.6 Pico Monstar ToF camera

Table 12.3 Parameter of the camera

Parameter	
Dimensions	62 mm × 66 mm × 29 mm
Measurement range	0.5–6 m
Framerate	5, 10, 25, 35, 45, 60 fps (3D frames)
Acquisition time per frame	5 ms @ 60 fps
Resolution	352 × 287 (100 k) px
Viewing angle (H × V)	100° × 85°
Depth resolution	≤1% of distance (1–6 m @ 5 fps)
	≤1% of distance (0.5–2 m @ 60 fps)

12.4.1.1 Characterization of the Camera

To generate a simulation model for a given camera, it was necessary to find out the characteristics by conducting several measurements in different scenarios. For evaluation and validation of the characteristics of the camera, a specific range of test objects resembling real-life features had been selected. These objects were placed in front of the camera under various lightning conditions (see Fig. 12.7). All test scenarios were performed with the longest illumination time available to obtain the best image quality while reducing the frame rate to 5 fps.

As the camera has a fish eye lens and a frontal infrared flash, the derivation in accuracy as well as the visibility of objects at the borders and edges of the field of view had to be checked. Therefore, a measurement was taken from a centre point compared to a point (1) at a vertical edge-pixel, (2) at a horizontal edge-pixel and (3) at the outermost corner-pixel of the video stream.

Summing up the data collected during these measurements, it turned out that as long as the object is visible to the camera, the distance can be measured with an accuracy of <1 cm. The software provided with the camera automatically corrects the fish eye effect of the lens and provides a 3D representation of the scene (see Fig. 12.8).

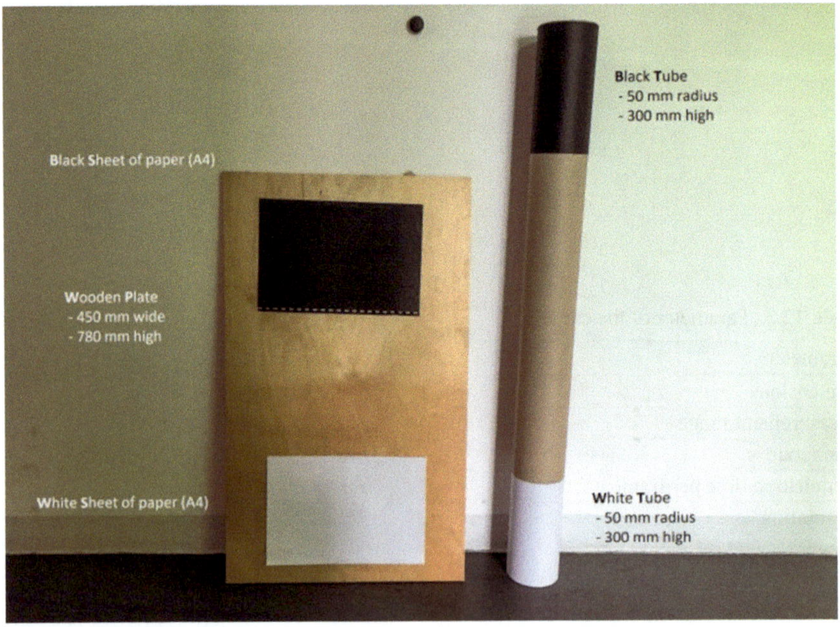

Fig. 12.7 Test objects used for characterizing the TOF camera provided by IFAT: black sheet of paper, white sheet of paper, wooden board, black tube, white tube and white wall

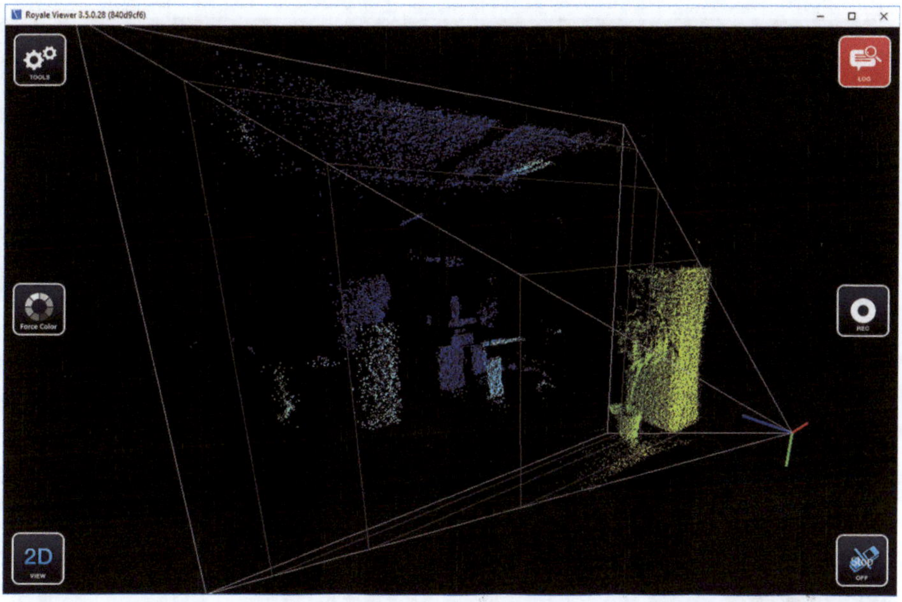

Fig. 12.8 A depth image taken in an office: A cupboard and a plant are visible (yellow/green). Additional cupboards and a desk can be identified (cyan/blue) too. Even the ceiling with its lights can be seen

While testing, a few issues have been noticed:

Based on the principle functionality of a ToF camera it is obvious that *background light* might pose a problem. The conducted measurements show that natural light sources with limited spectra like torches do not influence the performance significantly. The same applies to artificial light sources such as neon tube illumination in underground garages. However, sunlight in direct alignment with the camera angle and even indirect sunlight degrades its functionality noticeably and narrows its field of view.

Reflective materials like water-repellent paint in parking garages or mirrors or reflective number plates create artefacts or increase the noise by overexposing the image sensor. Another problem with highly reflective objects is the phase wrap that occurs when objects move beyond the wavelength of the modulation frequency (in this case approx. 7.5 m). They appear a lot nearer than they really are (see Fig. 12.9).

Light-absorbing materials like black paper or dark cars can create the so-called *black hole effect*. This occurs when an object reflects too little light and therefore vanishes from the image leaving a black spot. The problem is that black usually means "infinite distance". An example of this effect is depicted in Fig. 12.10.

Another reason for black holes is objects rotated at a *point angle* towards the ToF camera so that only very little of the emitted light is returned to the image sensor.

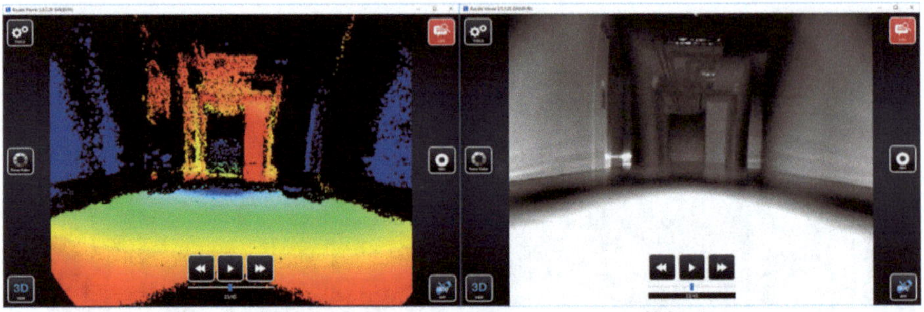

a) Depth Image b) Gray Image

Fig. 12.9 The ground in front and the highly reflective door far away appear to have the same distance. (**a**) Depth image. (**b**) Grey image

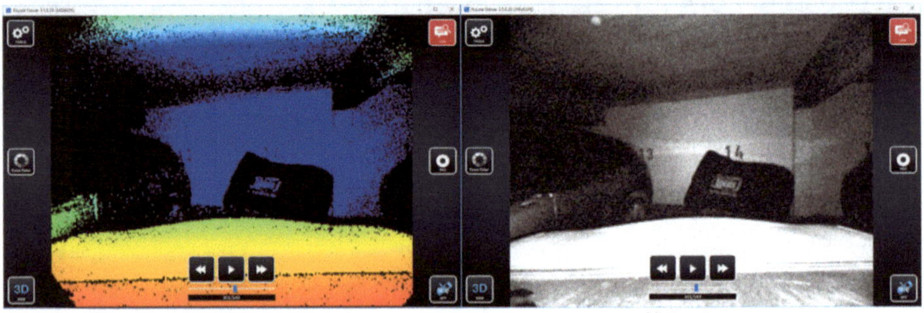

a) Depth Image b) Gray Image

Fig. 12.10 Black hole effect in a parking garage. A dark covered motorbike is reported "infinitive away". (**a**) Depth image. (**b**) Grey image

12.4.1.2 Selecting a Simulation Tool for the ToF Camera

After characterizing the ToF camera provided in various scenarios and conditions, several existing simulation software packages have been investigated to find a software package that can be extended by a realistic model of the present time-of-flight camera. The goal was to simulate autonomous parking manoeuvres. Therefore, the software is required to provide an image stream just like a real ToF camera would provide for a given scenario. Furthermore, the simulator should grant access to the camera intrinsic and parameters to match them with those of the real ToF camera.

Several tools were under consideration to build up a simulation model. After evaluating the tools SCANeR Studio, CarMaker, Vires VTD and ROS, it was decided that ROS would be most suitable for this use case.

The Robot Operating System (ROS) administered by the Open Source Robotics Foundation (OSRF) is gaining more and more attention in the automotive industry. Originally it is a software framework and middleware targeting research robots. On

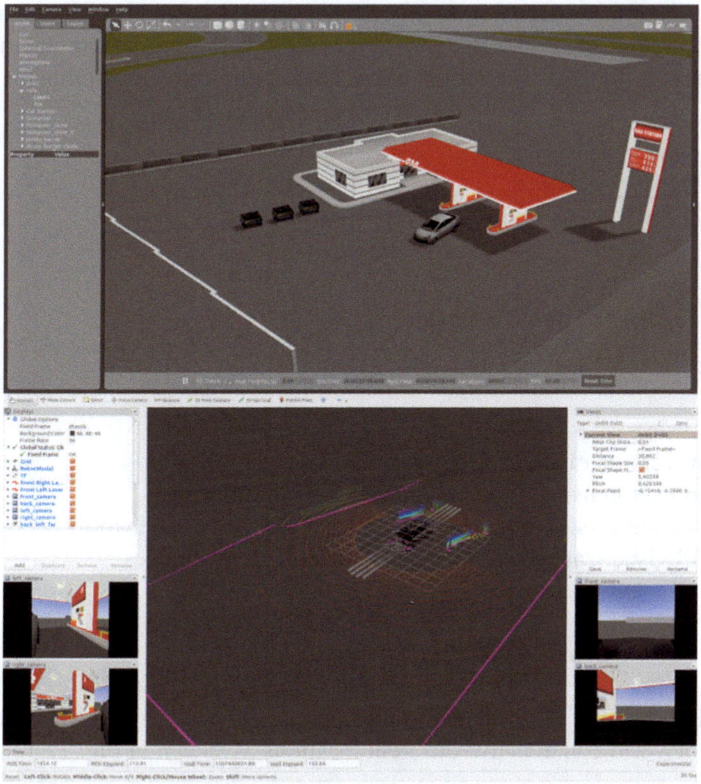

Fig. 12.11 Gazebo example "CarDemo3"

top of ROS it is possible to run a 3D robot simulation environment, called Gazebo. This simulation environment can also be used to model automotive scenarios.

A sample scenario is available in Gazebo that includes a small scene and a car equipped with camera, LIDAR and radar sensors (Fig. 12.11). The ROS community provides several additional sensor implementations—for the real world as well as for the simulation world—which include 3D and Time-of-Flight cameras. Because of the open source policy, these sensors are available as source code and can be parameterized and extended to approximate the characteristics of a real camera.

Because of its open source policy, the combination ROS with Gazebo seemed to be the best choice for setting up a simulation environment that could handle TOF cameras for driving- and parking scenarios.

12.4.1.3 Hardware Requirements

The need for powerful computation equipment depends primarily on the planned simulation. For non-complex worlds and lightweight sensors like cameras a state-of-the-art upper mid-range PC would be sufficient. Using an NVIDIA graphics card is beneficial since CUDA-features can be used to speed up the simulation. For the development of the ToF simulation environment, the following computer had been selected:

- CPU: Intel i7-7700
- GPU: NVIDIA GT 730
- RAM: 8 GB

12.4.1.4 Software Requirements

For the ToF simulation based on ROS, the following software environment is needed:

- a recent version of Linux Ubuntu (16.04 LTS)
- a ROS-desktop installation to simulate the camera
- a compatible version of Gazebo to create and simulate the world
- the ROS visualization tool *"rviz"* to show the depth images

12.4.1.5 Test Environment

For the simulation of realistic parking situations, a Gazebo-world was created. This ParkingLotWorld contains five different locations as shown in Fig. 12.12.

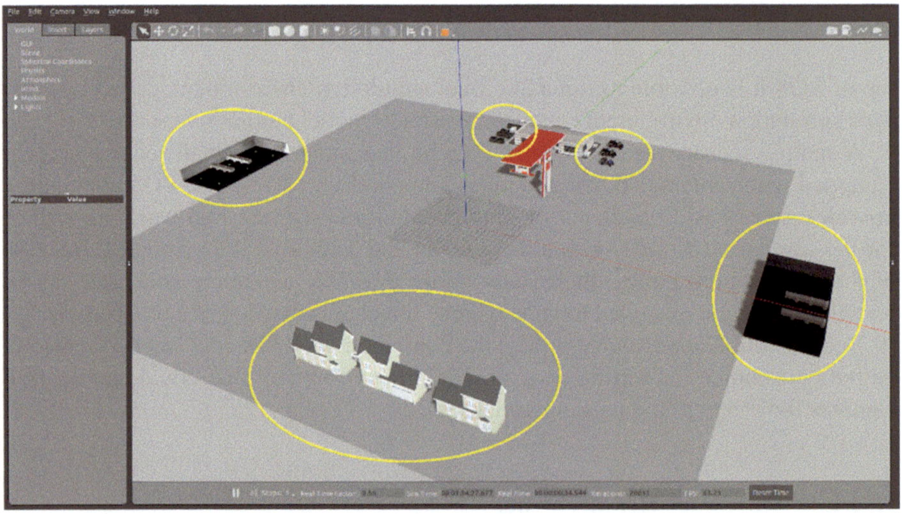

Fig. 12.12 Gazebo Simulation: ParkingLotWorld

Scenarios

The five different locations model very diverse situations. There are two parking garage scenarios (dark/light) and three standard outdoor scenarios.

The DepthCameraPlugin in ROS depends on RGB values. That means, the dark parking garage scenario represents less reflective, dark painted walls and objects in the parking garage. The light parking garage scenario on the other hand acts as a model for highly reflective objects.

The locations with the dumpster and the houses at the bottom represent standard parking situations between dark and light objects. The parking scenario with the buggys is more interesting as it provides a test environment with two different materials (black body, green hood).

Car Model

Since Virtual Vehicle has a Ford Mondeo as physical demonstrator, a simulated Ford Mondeo model provided by Dataspeed was adapted. Figure 12.13 shows the RVIZ-view of three different models. The left model has Monstar ToF cameras integrated into the front grill and the trunk lid. The right model has four additional cameras, which are mounted in the headlights and taillights. These additional cameras are z-rotated to increase the FoV. The model in the middle is slightly different, as the cameras are mounted behind the windshield and the back window. The position of the cameras behind the windows already shows a restricted FoV as the bottom part is blocked by the car's chassis.

12.4.1.6 Implementation of the ToF Camera Within a ROS Node

There are two plugins within ROS that are suitable for depth camera simulations with Gazebo. The *ROS depth camera* simulates a generic depth camera and the *ROS Openni Kinect* simulates an Xbox Kinect sensor. As both plugins have overlapping

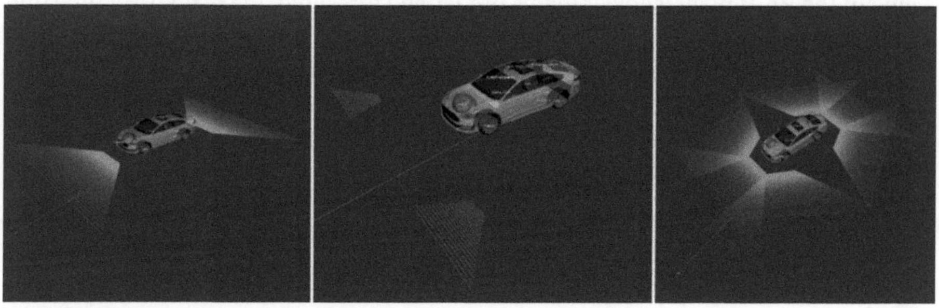

Fig. 12.13 Ford Mondeo with three different camera locations

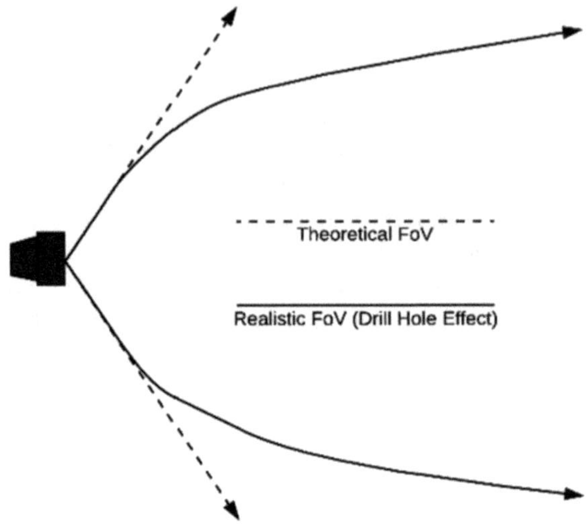

Fig. 12.14 Ideal- and realistic view of a ToF camera

functionality, they might be merged in the future. Anyway, for the simulation of the ToF provided by IFAT and its specific characteristics the ROS depth camera was chosen, as the implementation is less focused on a specific type of camera. The existing code had to be partly rewritten to match the characteristics.

The original *ROS depth camera* already supports a cut-off for the minimal distance to the camera. For the maximal range of the camera only the far clipping plane is defined. The real camera however has a more restricted field of view as shown in Fig. 12.14. Therefore, an additional cut-off region for the maximal distance had to be implemented.

The black hole effect needed a compromise for the simulation. The *ROS depth camera* provides an RGB image of the current simulated world. In other words, it does not emit a flash and take the exposed values, but it presents a rendered image as if it were a normal camera. Therefore, a post rendering filter decides if the RGB value is bright enough to be detected by the ToF camera. When the averaged grey value of the RGB pixel is lower than a predefined threshold, the pixel is set invisible.

The range decreasing effect in bright sunlight could be implemented by conducting a separate simulation with decreased detection range. This workaround was necessary because the implementation of the inherited *ROS depth camera* plugin in combination with the compromise for the black hole effect excludes the possibility to use existing lighting as reference.

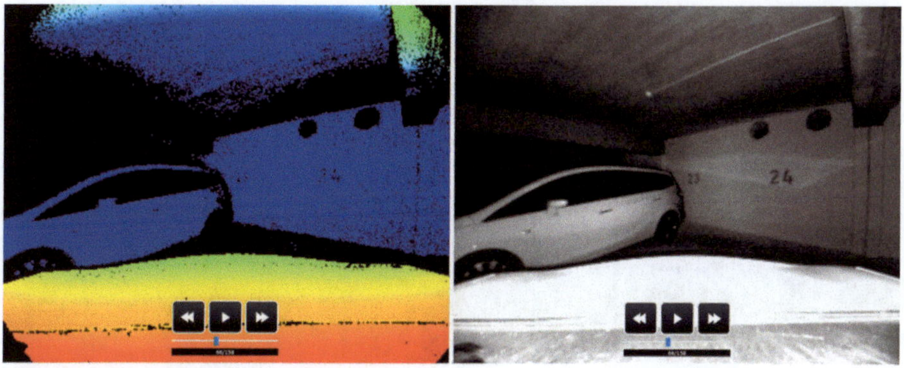

a) b)

Fig. 12.15 Parking manoeuvre in a bright garage using the ToF camera from IFAT. (**a**) Depth image. (**b**) Grey image

a) b)

Fig. 12.16 Parking manoeuvre in a bright garage using Gazebo and ROS simulation. (**a**) Depth image. (**b**) RGB image

12.4.1.7 Results

The optimal environment for the ToF camera provided by IFAT is a garage with white walls and a bright car. Figure 12.15 shows the real-life camera frame from a parking manoeuvre. The car, as well as the walls, is clearly visible.

The simulation of the same scene is pictured in Fig. 12.16. Both, the car and the walls are perfectly traced. There is only one detail missing: the random salt and pepper noise. But this effect can either be included by some random noise in the simulation or easily be removed within the original camera frame in a post-processing step.

A second test with the ToF camera and its associated simulation model has been conducted outside in ambient sunlight. Figure 12.17 shows the real-life camera

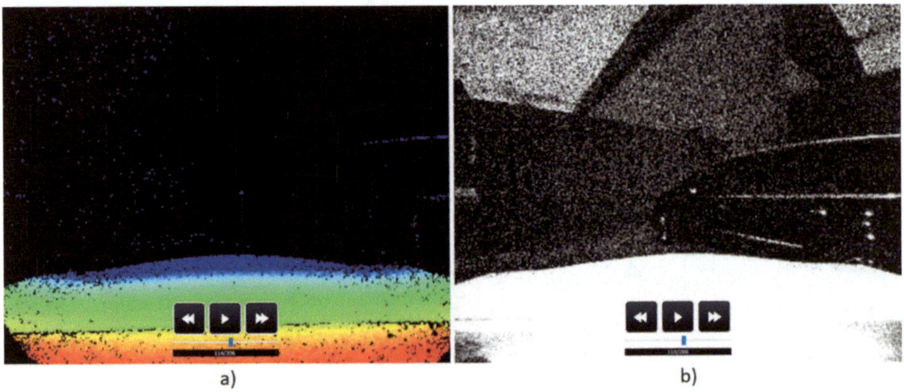

Fig. 12.17 Parking manoeuvre outside using the ToF camera from IFAT. (**a**) Depth image. (**b**) Grey image

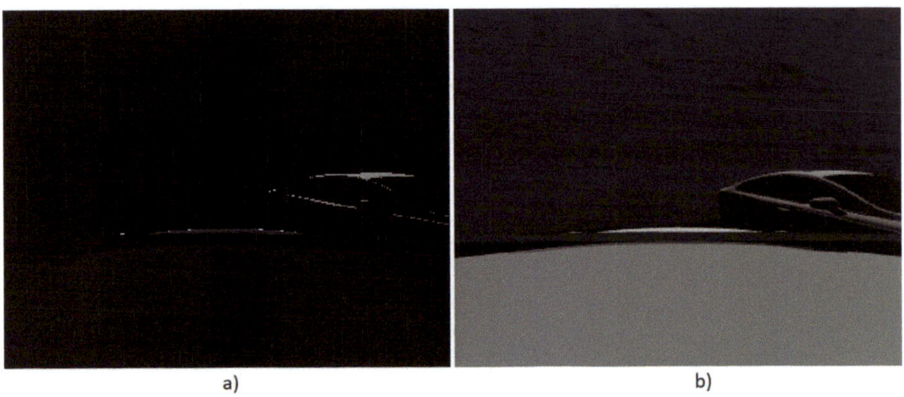

Fig. 12.18 Parking manoeuvre outside using Gazebo and ROS simulation. (**a**) Depth image. (**b**) RGB image

frame from a parking manoeuvre with a black car and a container. The container on the left is virtually invisible due to its dark colour. The car is hardly detectable, only the bright chrome bars of the car reflect the emitted light back to the camera.

The simulation pictured in Fig. 12.18 comes close to the reality again. The walls are invisible as they are too dark. Except for the light tip on top of the trailer and some brighter edges on the side, the car is invisible. The main difference is again the missing salt and pepper noise.

12.4.2 Parking Assistance Scenario with Time-of-Flight Camera

In parking assistance scenario, a ToF camera, similar to the previous one, is considered as substitute for ultrasonic range sensors. The expected advantages of using such camera include faster response, better resolution and object recognition capabilities. Combining depth information with a vehicle geometry model and ego-information (position, speed, steering angle), it is possible to estimate distance to collision point and time to collision (TTC) with great accuracy. Thus, parking assistance can be pushed up to a manoeuvre-aware level as shown in Figs. 12.19 and 12.20. Furthermore, advanced identification algorithms can be used to discern between regular obstacles and pedestrians for increased safety.

The simulation part of this scenario has been decomposed in two subgroups: Simulation elements and Algorithms. Next sections describe tasks in detail.

12.4.2.1 Simulation Elements

Simulation elements are independent components needed to run tests in the simulator. We discern three tasks:

- Development of a model of demonstrator vehicle, including appearance, behaviour, sensors and actuators.
- Development of ToF sensor model, to be included in the vehicle.

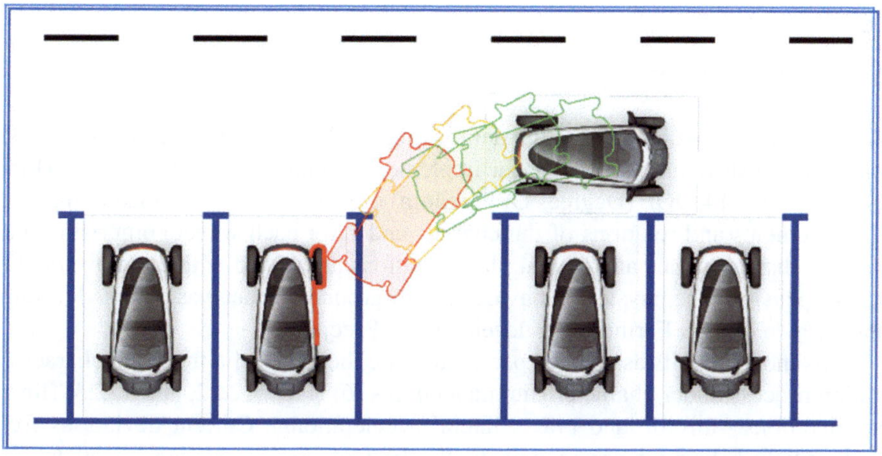

Fig. 12.19 Parking assistance in a collision trajectory case, with manoeuvre-aware detection of obstacles

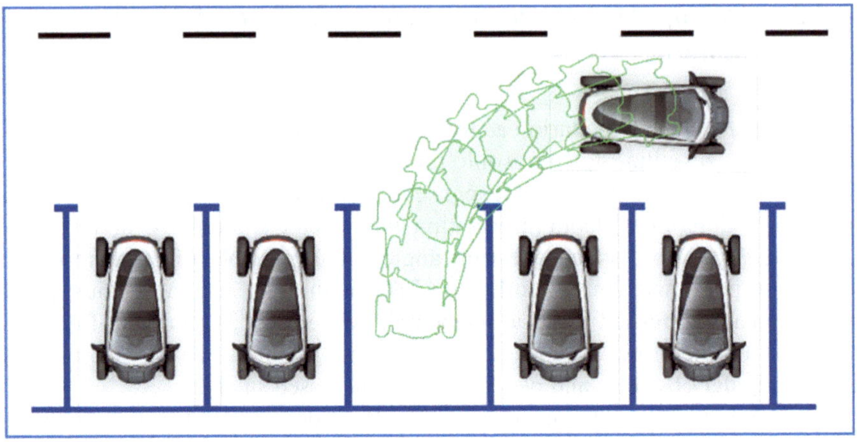

Fig. 12.20 Parking assistance with manoeuvre-aware detection of free path

- Creation of a suitable base scenario for the tests. This requires configuring the environment (road, parking places, other elements as buildings or trees), other vehicles (either circulating or parked), and pedestrians.

The two first tasks are completely independent and can be performed in parallel. Current development is being supported by Dynacar, the internal simulation environment of TECNALIA.

The scenarios proposed in *Smart Mobility (SmaBility)* need to be defined in the simulator, to carry out the first tests.

Vehicle/Demonstrator

A physically realistic model of the demonstrator (electric Renault Twizy Urban 45) has been finished and used. Dynacar implements a multi-body formulation [4] that is applied to a 14-DOF (Degrees Of Freedom) model (Fig. 12.21), considering 6 for displacements and rotations of the chassis, and 2 for each wheel comprising their vertical displacements and frontal plane rotations associated with the movement of the suspension. The physical behaviour of tire-ground contact was obtained through the *Magic Formula* Formulation developed by Pacejka [5].

This model and its associated parameters are the result of a thorough characterization process under characteristic manoeuvres [6] (Figs. 12.22, and 12.23) (linear accelerations—throttle and brake, lateral accelerations). Current deviations from real vehicle behaviour are related, to the best of our knowledge, to other non-modelled effects such as the mechanical tolerances of throttle/brake pedals and steering wheel, or the delays of control PLC and other electric/electronic devices.

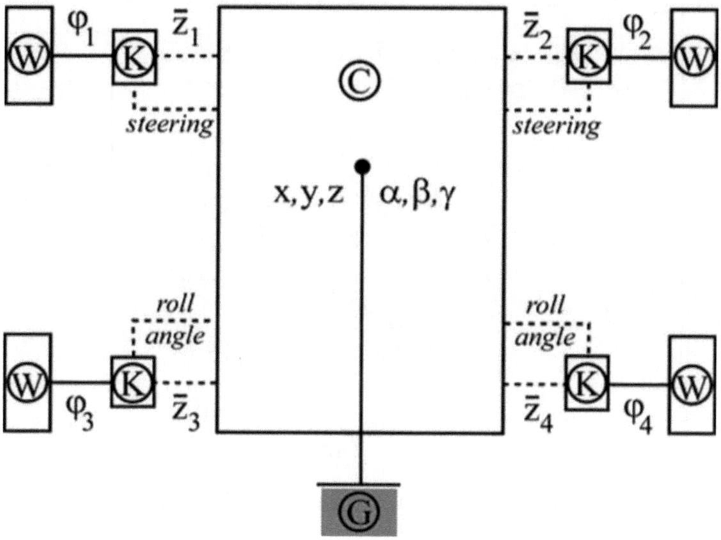

Fig. 12.21 14-DOF multi-body formulation implemented in Dynacar

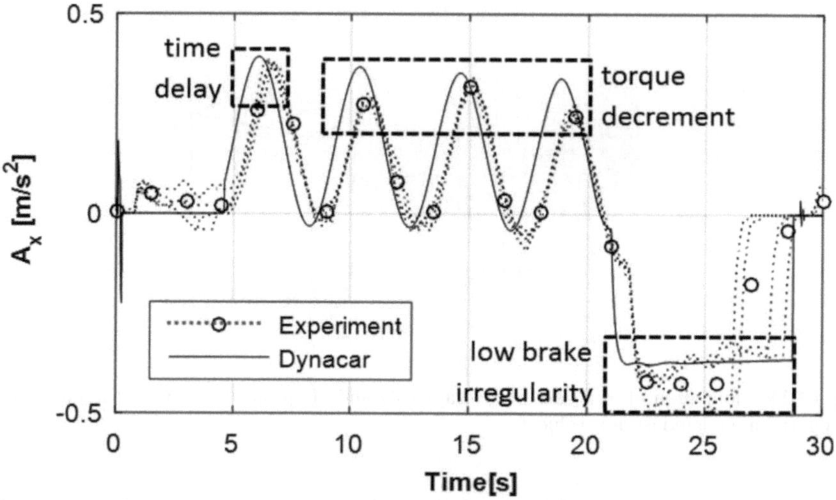

Fig. 12.22 Samples taken from the analysis of simulation model accuracy: Longitudinal acceleration

ToF Sensor

Sensor simulation, in this task, is interpreted from the point of view of usage in the scenario. It does not consider any electric, electronic, thermal nor any other physical or internal device simulation. A pure geometric approach is applied to 3D mesh

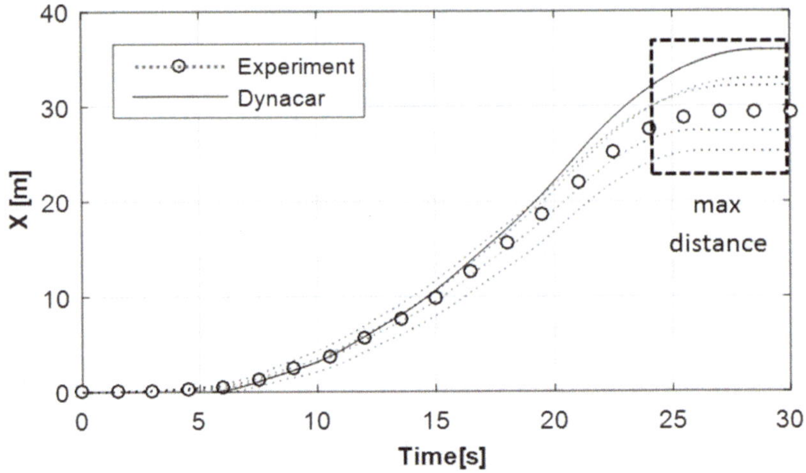

Fig. 12.23 Samples taken from the analysis of simulation model accuracy: longitudinal position

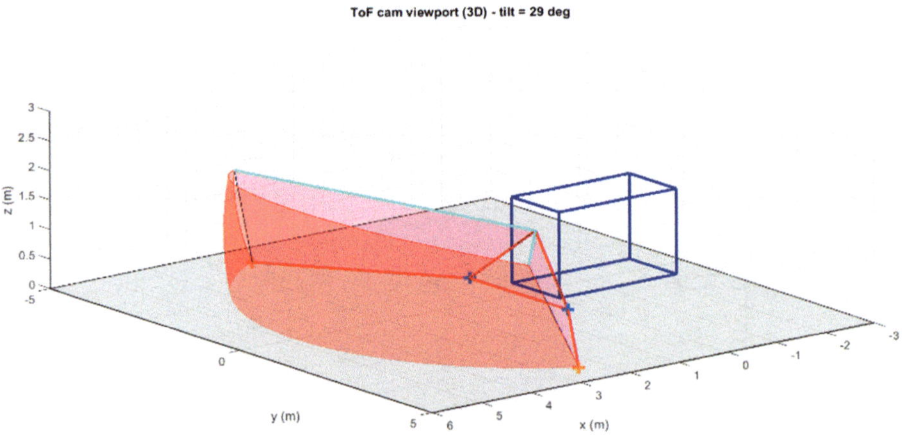

Fig. 12.24 Simulation of ToF sensor coverage (3D and 2D projection) for a specific mounting configuration in demonstrator vehicle

representing the scenario, although full 3D data has been used solely for deciding the location of the sensor in the demonstrator platform, as shown in Figs. 12.24 and 12.25.

Simulation results were used to evaluate several options for sensor integration in the demonstration platform and decide the final location. This saved testing time in the vehicle and prototyping efforts. Figure 12.26 shows the final integration in the demonstrator.

The algorithms designed for parking assistance are two-dimensional. All the algorithms involved in parking assistance (trajectory prediction, occupancy map

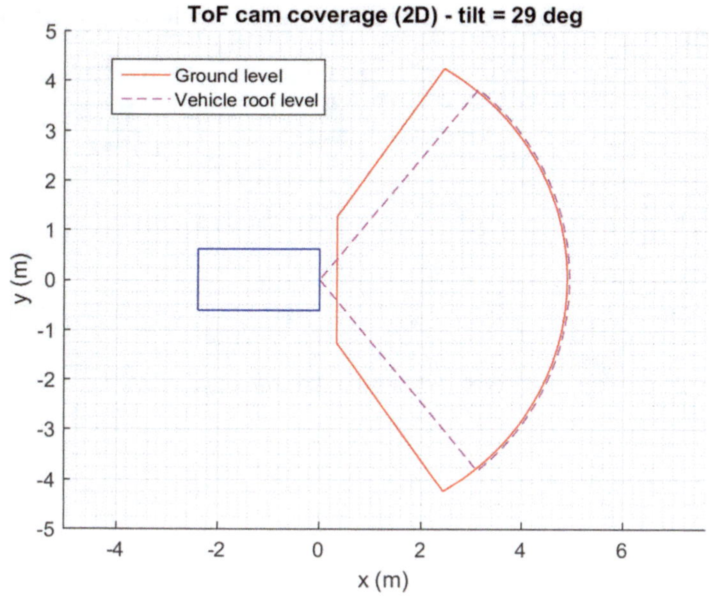

Fig. 12.25 Simulation of ToF sensor coverage (3D and 2D projection) for a specific mounting configuration in demonstrator vehicle

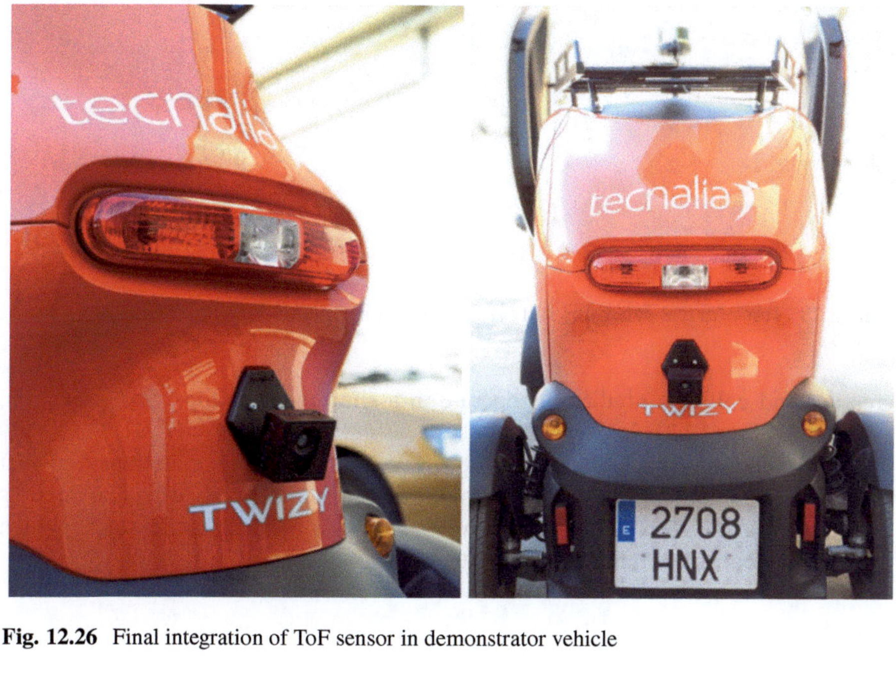

Fig. 12.26 Final integration of ToF sensor in demonstrator vehicle

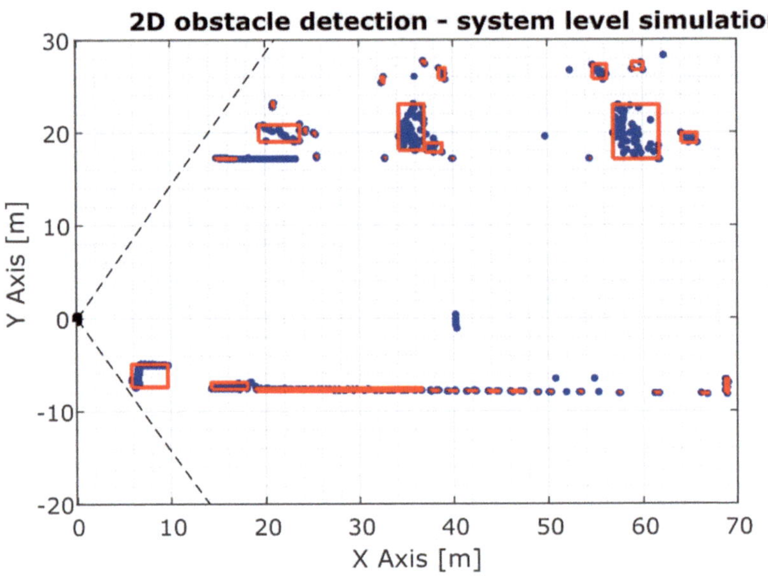

Fig. 12.27 ToF camera simulation as 2D projection, including moving obstacles

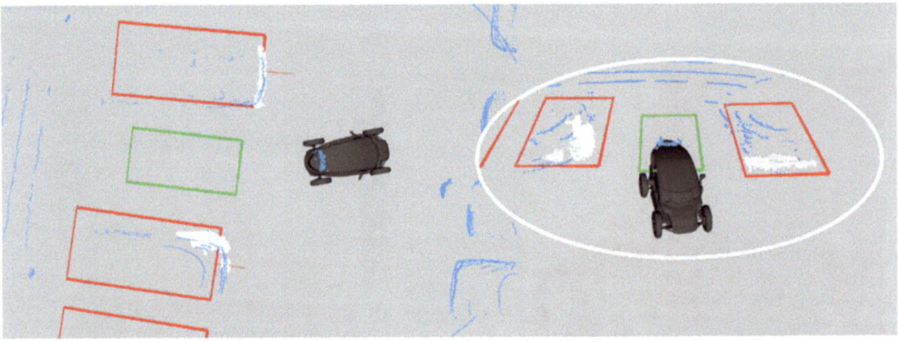

Fig. 12.28 Test with real data using software developed in simulation environment

creation and collision prediction) operate in a simplified 2D top-view projection, as shown in Fig. 12.27.

The generated obstacle data has been successfully used in simulations to detect collisions and plan parking manoeuvres. Furthermore, later tests with real data (as shown in Fig. 12.28) required only minor tweaks of the software stack, which resulted in a considerable reduction of the implementation effort.

Scenarios Configuration

A first test with the real sensor revealed several limiting factors that need to be considered in the design of the system. A preliminary set of scenarios was defined, covering combinations of four different variables:

- Parking type: parallel, perpendicular
- External lighting conditions: high (outdoors, sunny), medium (outdoors, cloudy/dusk/dawn), low (indoors or outdoors by night)
- Optical characteristics of external materials: vehicles paint colour, shiny surfaces, etc.
- Moving obstacles: present, not present

Two of those factors (lighting conditions and optical characteristics of external materials) have been translated into a simpler threshold for detection range: far (5 m) and close (2 m).

Moving obstacles (pedestrians) have been successfully added to the simulated scenarios. Due to technical limitations, these elements are not visible in the graphical interface of the simulation environment. However, they are simulated at sensor system and taken into account into the collisions detection algorithms.

The final set of simulation scenarios is composed by a total of eight samples, according to the table below (Table 12.4).

The particularities of the ToF sensor and its dependency of external materials required a modelling detail that surpasses the capabilities of the selected simulation tools. This, the presented set of scenarios is intended to assist the integration of assisted parking system components, rather than the implementation of the low-level algorithms for processing ToF camera data (Figs. 12.29 and 12.30).

12.4.2.2 Algorithms

The designed parking solution involves the development of four algorithmic modules.

- Vehicle egomotion estimation
- Vehicle trajectory prediction
- ToF camera obstacle detection + identification
- Time-to-collision reasoning

Vehicle Egomotion Estimation

Sensor acquisition and data fusion for estimation of vehicle position, speed, angles and other egomotion variables. It is based in a standard implementation of the Cartographer algorithm based on other vehicle sensors: GPS, IMU and Lidar.

Table 12.4 The final set of simulation scenarios

#	Moving obstacles	Parking	Detection range	Expected outcome
1	Not present	Parallel	Far (5 m)	• **Difficulty:** easy • **Free-space reconstruction:** good • **Good test for trajectory** and collision prediction (confidence on odometry)
2			Close (2 m)	• **Difficulty:** medium • **Free-space reconstruction:** difficult • **Good test for trajectory** and collision prediction (confidence on odometry)
3		Perpend	Far (5 m)	• **Difficulty:** easy • **Free-space reconstruction:** good
4			Close (2 m)	• **Difficulty:** easy-medium • **Free-space reconstruction:** medium
5	Present	Parallel	Far (5 m)	• **Difficulty:** medium • **Free-space reconstruction:** good • **Good test for trajectory** and collision prediction (confidence on odometry)
6			Close (2 m)	• **Difficulty:** hard • **Free-space reconstruction:** difficult • **Good test for trajectory** and collision prediction (confidence on odometry)
7		Perpend	Far (5 m)	• **Difficulty:** medium • **Free-space reconstruction:** good
8			Close (2 m)	• **Difficulty:** medium-hard • **Free-space reconstruction:** medium

Fig. 12.29 Simulation of parallel parking case. Manoeuvring

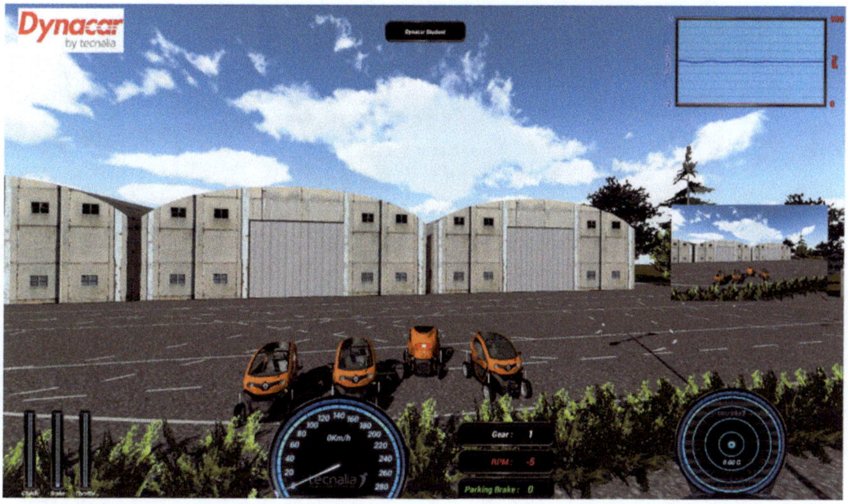

Fig. 12.30 Simulation of perpendicular parking case. Manoeuvring (front view)

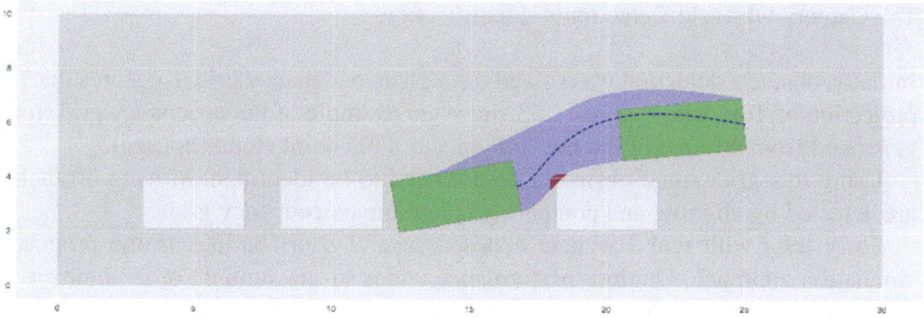

Fig. 12.31 Vehicle trajectory prediction in parallel parking scenario

For real implementation, a mix of vehicle odometry, steering wheel angle and video-based odometry can provide similar results.

Vehicle Trajectory Prediction

In order to estimate the TTC, parking assist system needs to estimate future positions of the vehicle. This problem is solved by a dynamic model of the vehicle that combines current egomotion with vehicle actuators (steering wheel position and throttle/brake pedals).

The selected dynamic model is a kinematic bicycle model. At low speeds it is accurate enough for the reduced dimensions of demonstrator vehicle (Figs. 12.31 and 12.32).

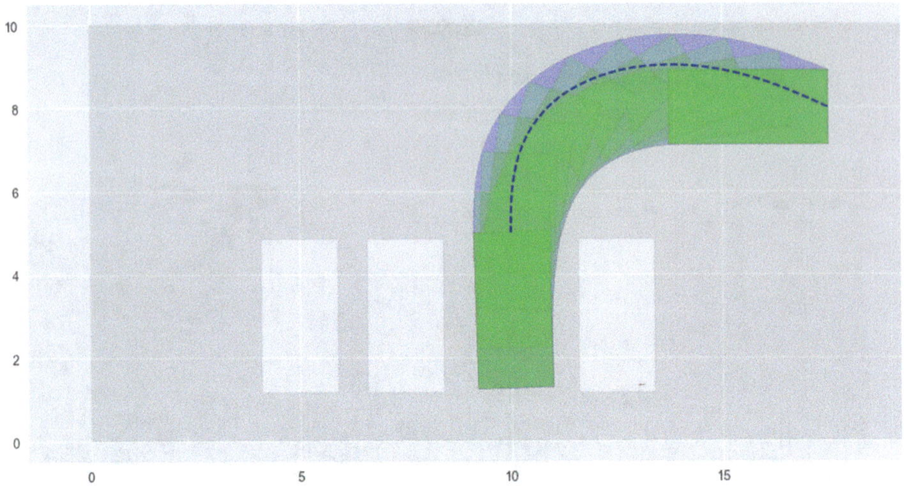

Fig. 12.32 Vehicle trajectory prediction in perpendicular parking scenario

ToF Camera Obstacle Detection + Identification

Initially, obstacle detection was solved through an occupancy grid that stores the 2D projection of ToF data. Figure 12.33 shows an example of the occupancy grid (top) generated from a frame of the raw ToF data as a 3D point cloud (bottom).

Using this grid, static obstacles do not need to be identified. Moving obstacles are detected by aligning and comparing consecutive occupancy grids.

Early tests with real ToF data exposed several shortcomings in the proposed simulation approach. Outdoor performance under strong sunlight or in some cases (e.g. vehicles painted in white or light colours) is exceptionally poor over 1–2 m range. This issue was solved by developing specific algorithms exclusively over real data and adapting its output to the 2D approach expected by the rest of the framework.

An example is the paper [7], that describes a specific data fusion algorithm that combines video processing over ToF intensity image with 3D processing over depth data to detect vehicle plates, allowing to detect vehicles in a parking lot up to 10 m away from sensor (Fig. 12.34). Realistic simulation involves not only working over the simulation tools (sun position, light intensity, atmospheric conditions), but also over the scenarios to account for the different materials. The estimated effort was considered too high for this project.

Time-to-Collision Reasoning

This model has been finally implemented to calculate distance to collision, since it is more useful for parking assistance purposes. The model has been implemented

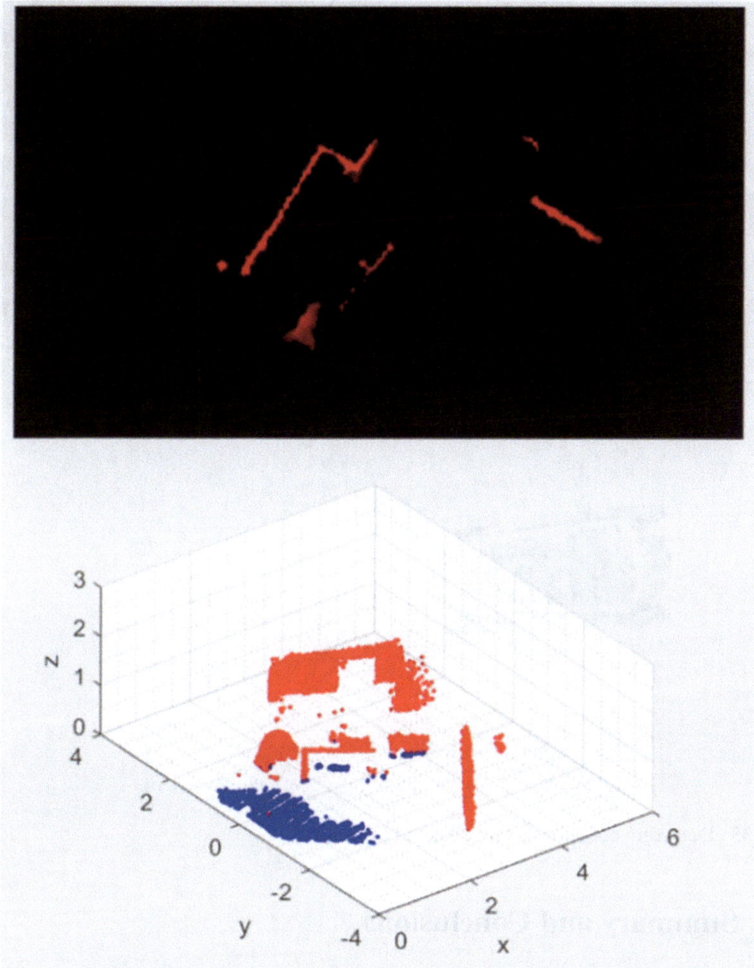

Fig. 12.33 Occupancy grid for obstacle detection (top) from 3D ToF data (bottom)

as the distance travelled by the mass centre of the vehicle along manoeuvre curve when the collision first occurs.

The algorithm is split in two phases. First, points are determined to be in collision trajectory by applying a ray-casting algorithm to taking the polygon that describes the area covered by the vehicle in its predicted trajectory (Fig. 12.35). Then, collision distance is calculated as the lateral projection of the output of collision detection system over the mid axis of expected manoeuvre.

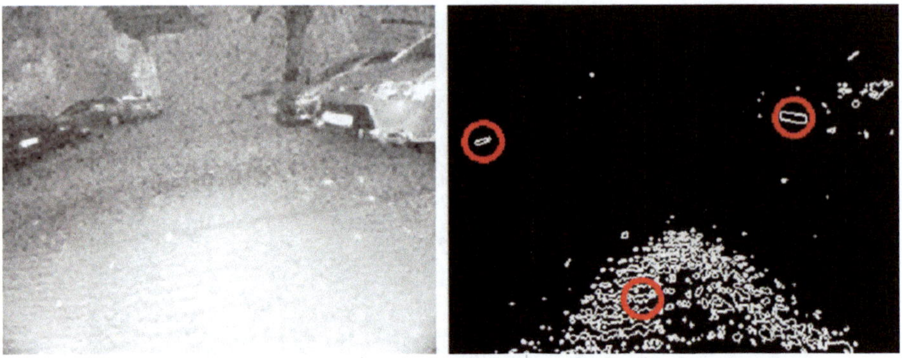

Fig. 12.34 Sample of real data showing the importance of advanced material simulation

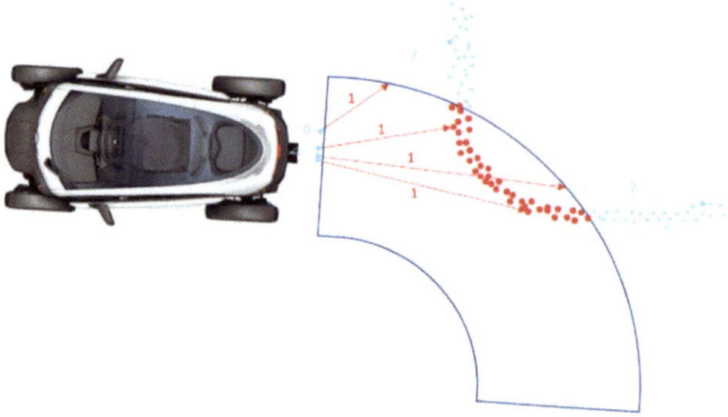

Fig. 12.35 Detecting collisions over predicted trajectory

12.5 Summary and Conclusions

Developing a full simulation model has been a challenging task that in some cases required looking for alternative solutions:

- Dynacar (TECNALIA internal simulator) has been selected as final simulation environment due to issues with the planned tool. This choice had a minor impact, such as not being able to visualize moving obstacles (e.g. pedestrians) in the graphical interface, but it has no effect in the development of the system.
- System-level simulation of ToF sensor has been simplified to a 2D approach. This has no impact in the development of the system since the first stage of sensor data processing consisted in calculating that projection through a trivial mathematical process.

The simulation environment has proved to be a useful tool through all the phases of the project. It has been used for:

- Developing algorithms and computational elements.
- Integrating computational elements in a parking assistance solution.
- Optimizing sensor location in demonstrator vehicle.
- Running unitary tests for system components.
- Running functional assessment tests for the system.

Some features have not been implemented in the system-level simulation tool, because the effort required was estimated to surpass the potential benefits. An example is the inclusion of advanced materials (with specific optical characteristics) in the simulation tool, which is required to enable the accurate sensor models implemented by other partners. Consequently, some algorithms had to be built and tuned using real data—which did not arrive until late stages of the project, increasing the risk of missing deadlines.

In spite that the total effort has been kept low in this example, the implementation of a full-scale simulation model would have been beneficial for later projects using the same sensor and could have helped detecting issues with raw data in early stages.

References

1. B. Williams, The CALM handbook, continuous air-interface long and medium range, ISO TC204: ETSI ERM TG37, 2004
2. ICOS 3.3, Independent Co-Simulation, Virtual Vehicle Research Center, Graz, Austria, http://www.v2c2.at/icos
3. M. Benedikt, J. Zehetner, D. Watzenig, J. Bernasch, Modern coupling strategies—is co-simulation controllable? in *NAFEMS Seminar*, 2011
4. J. Cuadrado, D. Vilela, I. Iglesias, A. Martín, A. Peña, A multibody model to assess the effect of automotive motor in-wheel configuration on vehicle stability and comfort, in *Proc. ECCOMAS Multibody Dynamics*, 2013, pp. 1083–1092
5. H.B. Pacejka, I. Besselink, *Tire and Vehicle Dynamics* (Butterworth-Heinemann, Oxford, 2012)
6. J.A. Matute, M. Marcano, A. Zubizarreta, J. Perez, Longitudinal model predictive control with comfortable speed planner, in *2018 IEEE International Conference on Autonomous Robot Systems and Competitions (ICARSC)*, Torres Vedras, 2018, pp. 60–64. https://doi.org/10.1109/ICARSC.2018.8374161
7. L. Paarup, M.E. Vaca, E. Martí, J. Murgoitio, J. Pérez, (Submitted) Car parking assistance based on Time-of-Flight camera, in *2019 IEEE Intelligent Vehicles Symposium (IV'19)*
8. Ansys, http://www.ansys.com/

Chapter 13
Health Monitoring for Lighting Applications

W. D. van Driel, L. M. Middelburg, B. El Mansouri, and B. J. C. Jacobs

13.1 Introduction

The history of reliability as we know it now goes back to the 1950s, when electronics played a major role for the first time [1, 2]. Now, 70 decades later, the electronic industry is facing a continuous increase of early and wear-out failures with accompanying consequences. Figure 13.1 depicts the struggle for the different high-tech industries, ranging from harsh environment suitability to long lifetime and warranty coverage. Nowadays, products with high failure rates are logged on the web leading to bad reputation for a company. In many ways, reliability is part of everyday life and part of consumer expectations. The word reliability is extensively used by the general public and the technical community, as illustrated by the following: there are over 3000 published books whose title or keywords contain the word reliability; the web of science lists some 10,000 technical papers with "reliability" as a keyword (since 1973); and the popular search engine Google lists over 12 million occurrences of "reliability" on the world wide web.

Solid state lighting applications are now at the doorstep of massive market entry [3, 4]. The penetration will grow most rapidly in the street and roadway and general service submarkets in terms of the percentage of total lumen-hour sales [5–7]. Scenarios estimate the expected future adoption of LEDs based on historical data

W. D. van Driel (✉)
Delft University of Technology, EEMCS Faculty, Delft, The Netherlands

Signify, HTC48, Eindhoven, The Netherlands
e-mail: willem.van.driel@signify.com

L. M. Middelburg · B. El Mansouri
Delft University of Technology, EEMCS Faculty, Delft, The Netherlands

B. J. C. Jacobs
Signify, HTC48, Eindhoven, The Netherlands

W. D. van Driel et al. (eds.), *Sensor Systems Simulations*,
https://doi.org/10.1007/978-3-030-16577-2_13

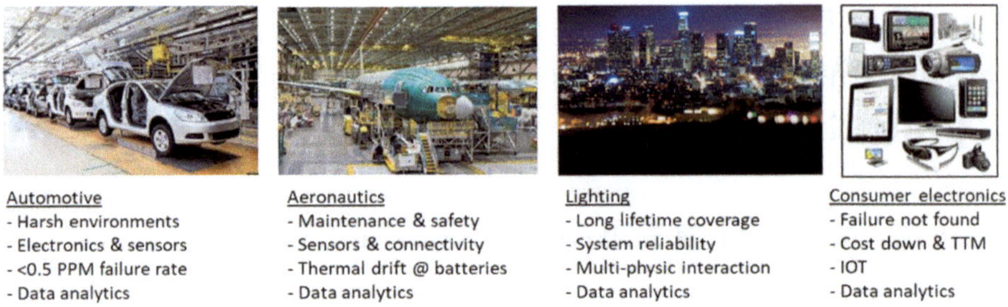

Fig. 13.1 The major concern of high-tech industries: direct finance loss; delayed product release; liability and reduced consumer confidence. Each industry has its own key reliability focus areas, as listed here

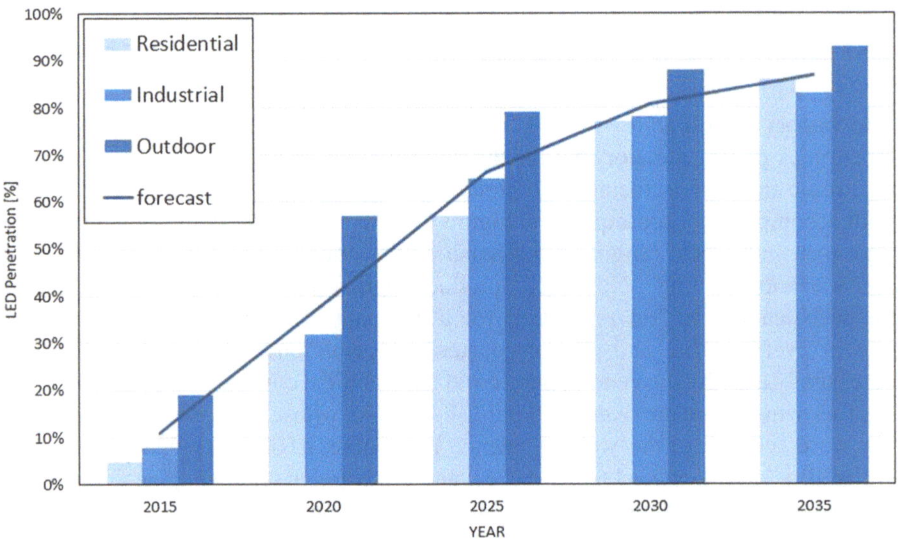

Fig. 13.2 LED penetration levels in three applications, solid line indicates overall trend (Data from [3–5]; with permission)

and the current trajectory for the technology. LEDs are predicted to comprise over 90% installed penetration by 2025 and nearly 100% by 2035. Figure 13.2 depicts the projections indicating the fact that LED penetration is growing rapidly.

Accompanied with the LED penetration, the lighting industry also experiences an exponential increasing impact of digitization and connectivity of its lighting systems [8]. The impact of this digitization is far beyond the impact on single products and extends to an ever-larger amount of connected systems. Continuously, more intelligent interfacing with the technical environment and with different kind of users is being built-in by using more and different kind of sensors, (wireless) communication, and different kind of interacting or interfacing devices. Figure 13.3 gives two examples towards these controlled and connected systems, just to highlight the possible scale of it.

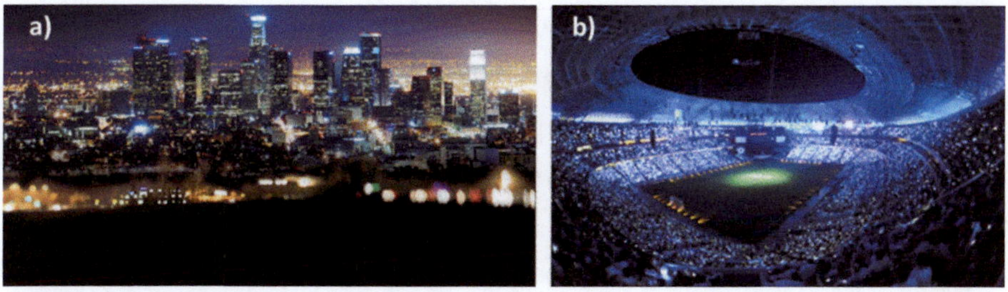

Fig. 13.3 Examples of controlled and connected systems: (**a**) connected luminaires in one city and (**b**) connected luminaires in a stadium

13.2 From Traditional Test-to-Pass to Health Monitoring

Traditional lighting is shifting towards connected lighting and as a result companies are also enabled to shift more towards an information-based environment [9]. The use of information from connected sources can be described as a revolution named big data. With big data, data analytics from live connections of "intelligent" systems can be used to determine the system prognostics. Due to these changes in technology, the next generation of product data will be much richer in information [10, 11]. Reliability and availability will become enablers for product designs. Big data will bring detailed understanding of failure mechanisms, usage scenarios, technology, and optimal designs. For example, products can be outfitted with sensors that can be used to capture information about how and when and under what environmental and operating conditions products are being used. But the data can also be used for pure reliability analysis. Examples are signal-detection algorithms to detect unsafe operating conditions or precursors to system failure that can be used to protect a system by shutting it down or by reducing load to safe levels. And on top of this there can be a need to predict the remaining life of the system (or the remaining life of its most important life limiting components). This topic is named as prognostics and health monitoring (PHM). PHM refers to the process of predicting the future reliability or determining the remaining useful lifetime of a product by assessing the extent of deviation or degradation of a product from its expected normal operating conditions [12]. Today, we predict failure rates on system level following classical reliability approaches, where standardized testing and experimental failure analysis are used in order to obtain conservative bounds from the failure models. However, except in the case of reliability "incidents," there is only limited feedback with which we can judge the effectiveness of our reliability approach. Prognostics and monitoring are not just about creating a more reliable product: it is about creating a more predictable product based on real-world usage conditions. Data analytics is a necessary part of this, but it is not enough. In order to add value, product insights need to be leveraged into the technologies that are

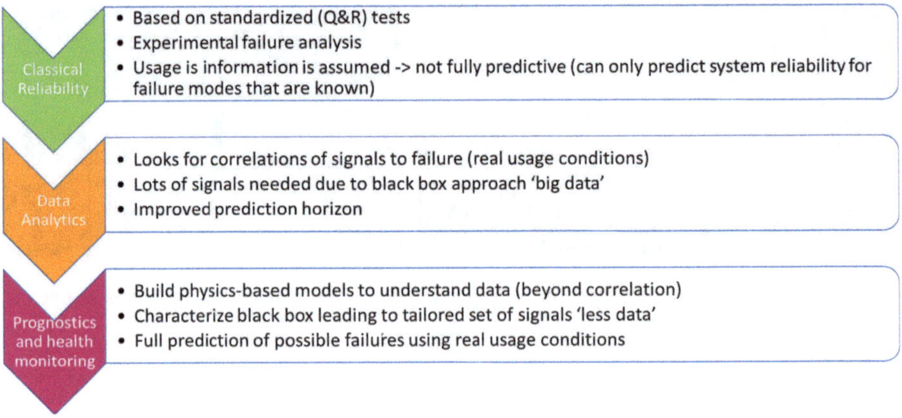

Fig. 13.4 Changing from test-to-pass to prognostics and health monitoring

used in order to differentiate from others. Prognostics and monitoring are not about troubleshooting reliability issues; rather, it is a new control point enabled by the transition to a lighting services business. It is the combination of data and deep physical (and technological) insight that will give a unique "right to win" in the lighting industry. The future possibilities for using big connected data in reliability applications are unbounded. Lifetime models that are based on this data have the potential to explain much more variability in field data than has been possible before. Figure 13.4 schematically displays the change from a classical reliability test-to-pass approach to prognostics and health monitoring.

13.3 Simulations for Lighting Applications

Predictive (reliability) modeling capabilities may enable the performance over time for a product. Figure 13.5 depicts the currently available toolset for lighting applications ranging from LED chip, package to complete and connected systems. Clearly, tools do exist and are used by academia and industry but direct coupling of them and inclusion of time needs further attention. On the lowest level, GaN chip is already manufactured to feature size down to several micrometer and nanometer and also modeling the light output of GaN requires knowledge from across different field such as quantum efficiency and their interaction between different loading conditions. For packaging and module level modeling, FEM is a well-established technique for predicting thermomechanical behavior of the LED packages. However, since the performance of LEDs is highly dependent on the light quality, there is a need to develop techniques that could predict the light

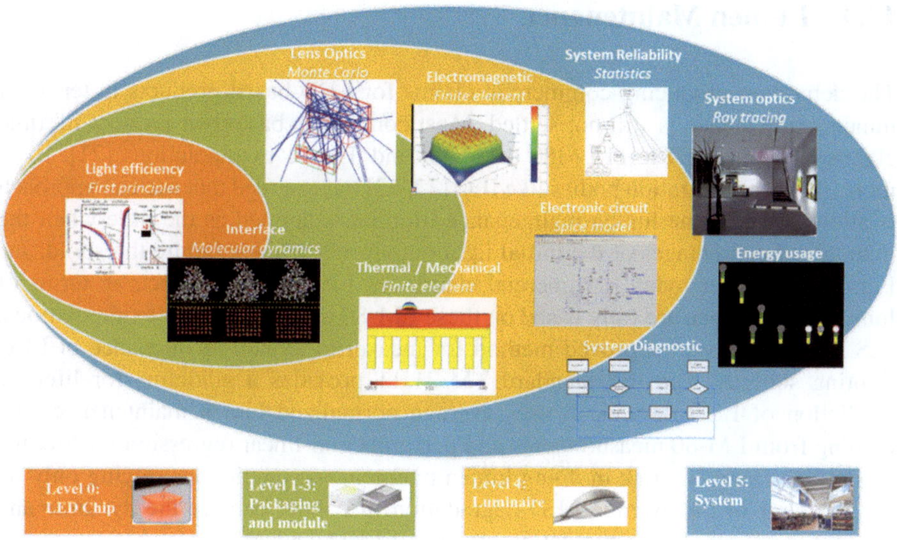

Fig. 13.5 The prediction landscape for SSL products. Coupling and inclusion of time is needed to cover reliability predictions (courtesy by C.A. Yuan, [19])

behavior in the optical system as the package degrades. Tapaninen et al. [13] presented a test case for coupling two physical aspects of an LED, optical and thermal, using specific simulation models coupled through an open source platform for distributed multi-physics modeling. They showed how to connect a Mie theory based scattering calculator with ray tracing. Alexeev et al. [14] followed this approach by connecting a ray-tracing model for the light conversion in LightTools® to a thermal model in ANSYS®. Tarashioon et al. [15] introduced a multi-physics reliability simulation approach for solid state lighting (SSL) electronic drivers. This work explored the system-level degradation of SSL drivers by means of applying its components reliability information into a system-level simulation. An automatic coupling between electrical simulations in SPICE® with thermal simulations was established in order to perform thermal-electrical reliability predictions. Sun et al. [16–18] continued this work by improving the thermal part through automatic coupling with ANSYS®. These multi-physics modeling attempts are needed to cover the grand challenge for reliability modeling in lighting applications. Simply because all failure modes and mechanisms such as electronic drift, browning, coating degradation, color shift, lumen decay, water ingress, corrosion, etc. are results of strong multidisciplinary interactions.

13.4 Lumen Maintenance

The debate on producing commercial claims for LED-based products in terms of lumen maintenance is still not settled. Most companies base their product lifetime claims fully on LED-level LM80 data [20] and TM-21 extrapolations [21]. Even more, the standardization bodies like IEC [22, 23] have agreed that such an approach is allowed. Here, the lumen maintenance lifetime is defined as the time when the maintained percentages of the initial light output fall below a failure threshold, see Fig. 13.6. So, per today, commercial claims for LED-based products in terms of lumen maintenance are fully based on these LM-80 data and TM-21 extrapolations. IES LM-80-08 is an approved method for measuring lumen maintenance of LED lighting sources. The IES standard TM-21-11 provides a guideline for lifetime prediction of LED devices. It uses average normalized lumen maintenance data coming from LM-80 measurements and performs non-linear regression for lifetime modeling. There is a risk in doing lumen maintenance predictions relying only on the behavior of the average LED degradation. It cannot capture the dynamic and random variation of the degradation process of LED devices.

Alternative approaches are rare as only few other publications build upon the TM-21 method. Fan et al. from the CALCE institute of technology [24] have used the degradation-data-driven method (DDDM) which is based on the general

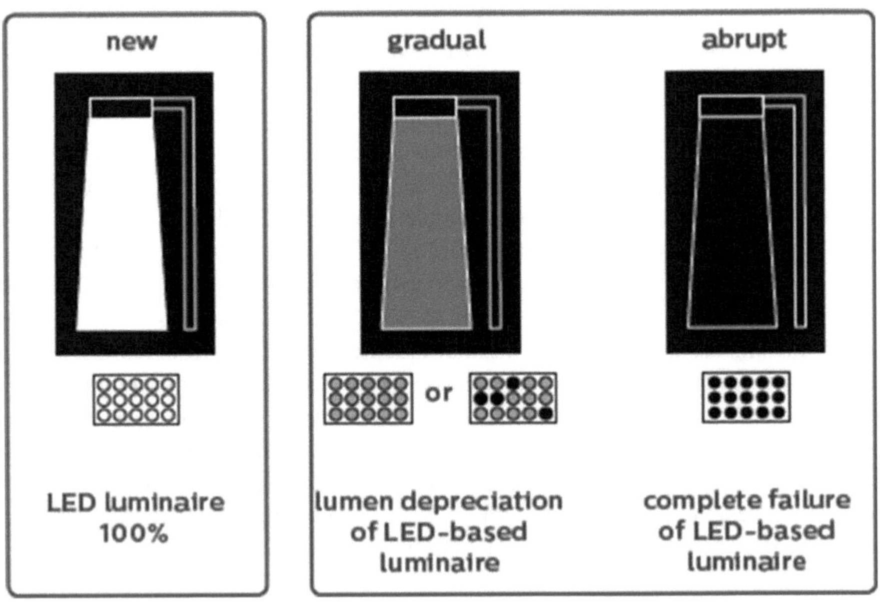

Fig. 13.6 Over time performance of an LED-based system

degradation path model. They use it to predict the reliability of HP LEDs through analyzing the lumen maintenance data collected from the IES LM-80-08 lumen maintenance test standard. Their method can get much more reliability information out of the data (e.g., mean time to failure, confidence interval, reliability function). In an accompanying paper, Fan et al. [25] describe a particle filter-based (PF-based) prognostic approach based on both Sequential Monte Carlo (SMC) and Bayesian techniques. These techniques are used to predict the lumen maintenance life of LED light sources. Also, here the alternative approach achieves better prediction performance, with an error of less than 5% in predicting the long-term lumen maintenance life of LED light sources. Lall et al. [26] follow up on this approach by using Bayesian Probabilistic Models for the assessment of the onset of degradation in solid state luminaires. The failure threshold decay rate has been calculated using an Arrhenius model, neglecting the effects of current density and humidity. The statistical approach is quite valid but also seen as complicated. Quan et al. [27] describe an in-situ method to monitor the lumen degradation of LED packages. They conclude that the luminous flux of the LEDs shows a steady and slow depreciation, but no proper statistical analysis was performed on their measured data. Huang et al. [28–30] investigated the degradation mechanisms of mid-power white-light LEDs. In their studies, a modified Wiener process was employed for the modeling of the LED devices' degradation, following the earlier work of Tsai et al. [31]. Using this method, the dynamic, random variation, as well as the non-linear degradation behaviors of the LED devices were described. They applied the Hallberg-Peck's model to describe the effects of temperature and humidity on LED degradation, thereby ignoring the crucial effects of the current density on this degradation.

Another alternative approach, used in this work, is to study the "degradation" data of each LED individually [32]. It means that for each individual LED a model as stated in Eq. (13.1) is fitted. Then, we can predict L70 values for each LED and turn degradation values into failure times. The mixed effects model is one of the most popular approaches in degradation analysis [33–35]. In order to describe the unit-to-unit variations of the test units, the unknown parameters of the mean degradation path are described in terms of the mixed (or random) effects. Often the mixed effects formulations do not take the time-dependent error structure into consideration. Therefore, the stochastic process formulation or Gauss-Markov method can be an alternative approach to model the product's degradation path. Dealing with those more complex models, to find the maximum likelihood estimates (MLEs) of the unknown parameters, the mixed effects model is computationally intensive. STATA or *R* can be used [36]. Here, the method by Meeker and Weaver [33] to analyze a large set of LM80 data is chosen. In our approach, we have gathered LM80 sets of High Power (HP) LEDs, see table below. In total, we analyzed approximately 27,000 data points (Table 13.1).

Table 13.1 LED dataset details

LED	If [A]	Tcase [degC]	Read points
1	0.7; 1.0	55; 85; 105	1680
2	0.35; 0.5; 0.7; 1.0; 1.5	55; 85; 105; 125	3880
3	0.5; 0.7; 1.05; 1.2	55; 85; 105	4200
4	1.5; 2.1; 3.0	55; 85; 105	1950
5	0.35; 0.7	55; 85; 105	2500
6	1.5; 2.1; 3.0	85; 105	1300
7	0.5; 0.7	85; 105	1550
8	0.5; 0.7; 1.0;	55; 85; 105; 120	3850
9	0.7; 1.2; 1.5	55; 85; 105; 120	2880
10	0.5; 0.7; 1.0; 1.5	55; 85; 105; 125	2277

The degradation of lumen for an LED at time t [h] and accelerating factors temperature T [°C], and current I [A] is given by:

$$\Phi(t) = \exp\left(-\alpha t^{\beta}\right) \tag{13.1}$$

With

$$\alpha = CI^{n} \exp\left(B/\left(T + 273.15\right)\right) \tag{13.2}$$

where $C > 0$, $n > 0$, and $B < 0$

We can use the linear mixed-effects models available in Stata. These models are also known as multilevel models or hierarchical linear models. The overall error distribution of the linear mixed-effects model is assumed to be Gaussian, and heteroskedasticity and correlations within lowest-level groups also may be modeled. The key to fitting mixed models lies in estimating the variance components, and for that there exist many methods. Most of the early literature in mixed models are dealt with estimating variance components in ANOVA models. For simple models with balanced data, estimating variance components amounts to solving a system of equations obtained by setting expected mean-squares expressions equal to their observed counterparts.

The transformed observed lumen degradation Y at time t is:

$$Y = \ln\left(-\ln\left(\Phi(t)\right)\right) = \beta \cdot \ln(t) + B/\left(T + 273.15\right) + n \cdot \ln(I) + \ln(C) + \varepsilon \tag{13.3}$$

With:

$$\varepsilon \sim^{iid} N\left(0, \sigma^{2}\right)$$

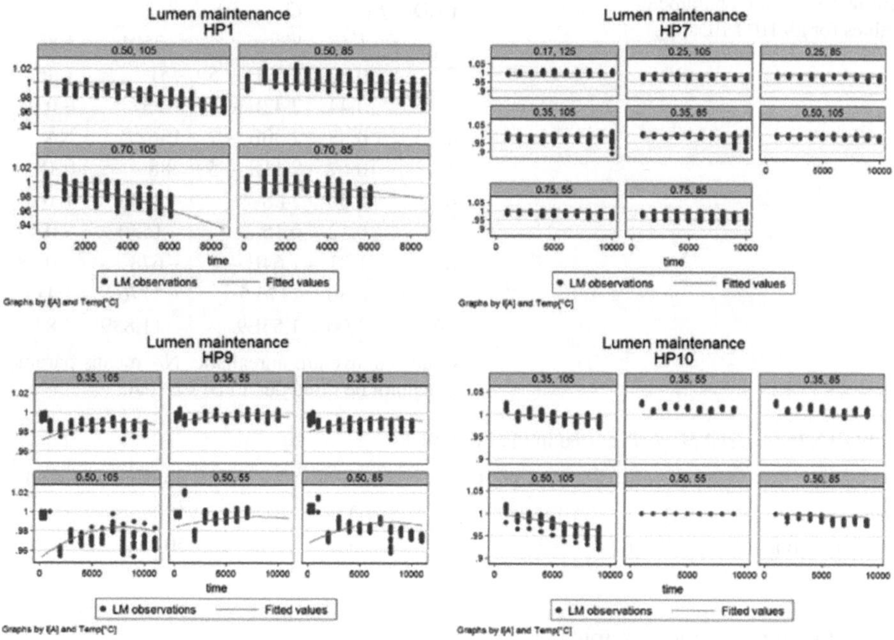

Fig. 13.7 Four typical degradation curves for the HP LEDs analyzed

We assume that the variability in the regression parameters can be described by a bivariate normal distribution. This assumption reflects the LED-to-LED variability in the degradation intercepts and slopes:

$$(\ln(C), \beta)' \sim N_2\left(\theta, \begin{pmatrix} \sigma^2_{\ln(C)} & \rho\sigma_{\ln(C)}\sigma_\beta \\ \rho\sigma_{\ln(C)}\sigma_\beta & \sigma^2_\beta \end{pmatrix}\right) \tag{13.4}$$

Figure 13.7 depicts four typical degradation curves of the LED LM80 data, including the fitted behavior (following Eq. (13.2)). The different graphs represent different setting of current and temperature. For each LED a model can be found, having all conditions in it. Looking at the figures, a wide variety of degradation can be found, e.g.:

- Remain stable at the low-stress conditions
- Increase then decrease
- Gradually increase
- Gradually decrease

Also, it is not given that higher stress conditions lead to higher lumen decrease. There can be multiple reasons for such non-theoretical behavior, e.g.:

- Insufficient data integrity
- Large noise over signal values

Table 13.2 Fitted parameter values for all HP LEDs

LED	β	C	B	n
1	1.47	88.0	−7501	1.83
2	0.87	1.32E − 5	451	1.86
3	0.44	2.42E − 3	−537	−0.08
4	0.83	2.30E − 7	1562	NA
5	0.67	2.11E − 5	388	0.18
6	0.30	74.8	−3729	1.16
7	0.12	0.15	−1530	−0.45
8	1.31	1.64E − 8	−678	−0.48
9	−0.30	1175.5	−2776	1.41
10	1.06	1.55E9	−11,889	3.83

Values cursive are unrealistic. NA means parameter cannot be fitted due to lack of data

Table 13.3 Material properties used in the FE model

Material	FR4 PCB	Silicon	Solder 60Sn-40Pb
Heat capacity [J/(kg*K)]	1369	703	150
Thermal conductivity [W/mK]	0.3	163	50

- Not using reference samples
- Corrections during the measurements (for instance at 6000 h)
- Differences between test houses
- Exposure to chemical incompatible substances from air pollutants or from outgassing of neighboring materials

Table 13.2 lists all the fitted parameters for the HP LED dataset. The ranges for the parameters underline the differences in degradation behavior, as mentioned above. Looking at the parameters one can state the following:

- β: TM-21 assumes that this parameter should be 1.0. Table 13.3 clearly identifies that this is a strong approximation as the data set finds realistic values in the range of 0.1–1.5 with an average of 0.8.
- C: this is a scaling factor and all values can be the found. An average value makes no sense.
- B: this value is the temperature acceleration, the average value of 4091 reflects an activation energy of 0.35 eV, which is quite reasonable.
- n: reflects the influence of current, negative values can be discarded. In the current data set, we find realistic values in the range of 0.2–3.8 with an average value of 1.7. It is known that current acceleration for HP LEDs can be quite substantial.

Based on the average data, the overall lumen maintenance behavior of the LEDs is depicted in Fig. 13.8 (L80 values). These are gradual degrading function that can be used to forecast the remaining life at any moment of time, with a given temperature and current.

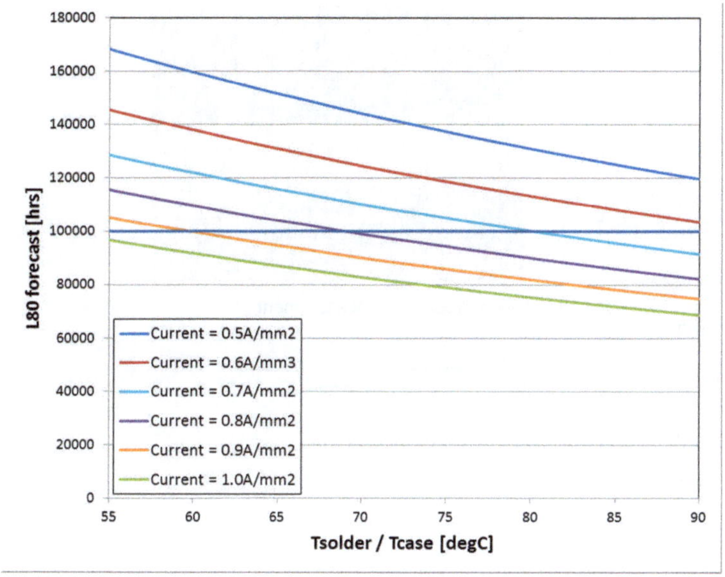

Fig. 13.8 Lumen maintenance L80 curves as function of the temperature for given current densities

13.5 Model Verification

The above-mentioned models and algorithms need to be verified. On a single LED level, LM80 tests are used to verify these models. On a system level, the verification is more complex. Here, a series of lifetime tests are being executed.

Twenty pieces of a retrofit LED lamp are put on test, see Fig. 13.9 on the left the "10 mm free air" sleeve test and on the right the prediction versus measurement.

A second verification concerns an LED module, tested now for 10,000 h. The test chamber is shown in the left picture, the verification in the right picture but only up to 6000 burning hours (Fig. 13.10).

Then a final verification is done on luminaire level. Total burning hours are 12,400 h. LED configuration is 16 HP Cree LEDs. The comparison of the prediction model to the measurement results is shown below (Fig. 13.11).

In conclusion, this verification shows that:

- The prediction models are pessimistic, it tends to give an early warning.
- B50 (50% of the population) is within 4%.

The implement hardware algorithm gives a total burning hour of (based on model, B50L94.7) 14,900 h.

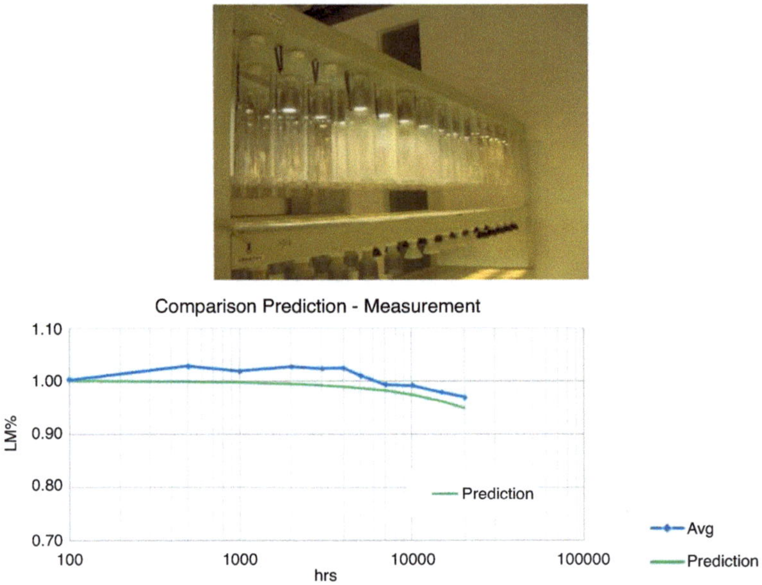

Fig. 13.9 Model verification using retrofit LED lamps

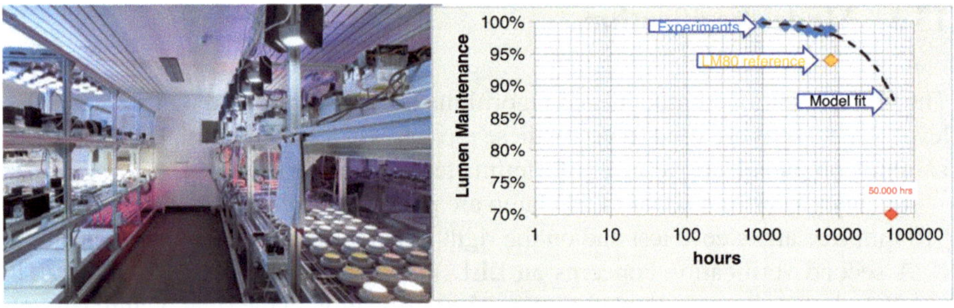

Fig. 13.10 Model verification using an LED module

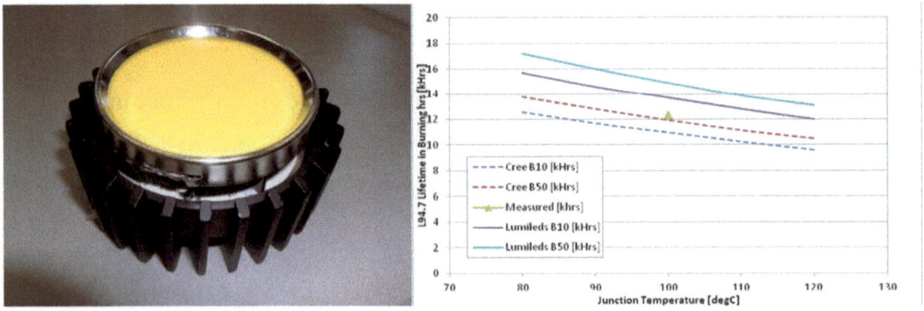

Fig. 13.11 Comparison of the prediction model to the measurement results

13.6 Uncertainty

Lighting systems are composed of many components (electronics, LEDs, custom components) assembled in a specified architecture and subjected to a certain user profile. With the smart lighting demonstrator in mind, it is important to be able to predict the failure of the components with a certain acceptable level of uncertainty. Here the starting point are the failure time distributions of the separate components, and how these distributions depend on physical circumstances such as current, temperature, relative humidity, burning hours, and switches per day. The output consists of system failure percentiles over time for several failure criteria, lumen output over time, and Paretos describing what failures are most important. The distributions of the components are assumed to be true, and all output is conditional on this assumption. For instance, the failure time of an electronics may be $T\sim$ Weibull (α, β) where we take the given values of α, β for granted. However, in practice, these may be estimated from a life time test with limited sample size, so that the sampling uncertainty is actually larger: if you would do the same test with new systems, you would get different estimates for α, β. Note that in many cases there is an additional layer: α may depend on temperature and current via additional coefficients. The question that needs to be answered is: *what are the impacts of sampling uncertainty of component parameters on the spread of system failure probabilities?*

Currently the system failure probability is considered at a given time, e.g., L70 at 50 kh. These are in fact properties derived from probability distributions of the system. Here one is to add a statement to the value of B50L70 of how certain that value is in view of the uncertainty of the components' lifetime, in the form of say a 95% confidence interval that reflects that uncertainty. There are various random quantities: for each component, and then usually a few physical failure mechanisms (e.g., for LEDs: solder failure, wirebond failure, speed of gradual decline of flux). If L70 is the random variable for the point in time at which the system crosses the 70% of initial output level, we can view L70 as a function of those many random variables (per component, per failure type):

$$\begin{aligned} L70 &= f\left(X_1, \ldots, X_p\right) \\ X_1 &\sim \text{Distribution } (\alpha_1, \beta_1), \ldots \end{aligned} \tag{13.5}$$

The uncertainty on the many parameters of the input such as α_1, β_1 may be modeled as (α_1, β_1) following some multivariate distribution, close to the sampling distribution. In terms of Bayesian statistics, we specify some prior distribution on parameters such as (α_1, β_1). Between components and failure types these will be independent. Under this prior distribution of the components, L70 has a prior distribution as well. Bayesian inference usually goes on by combining new data with the prior beliefs, but that step is not going on in our case: just getting to know the prior distribution of quantities like L70 will be enough.

As demonstrated above, lumen maintenance follows an exponential decay with L70 as the point in time where lumen output drops below 70% of the initial value. L70 follows a lognormal distribution as L70(location, scale), with:

- location = log(C) – n * log(current) + B/Temperature
- scale = sigma_lm from measurements

This results into the fact that the uncertainty depends on two parameters only, being location and scale. Typically, life time tests vary current and temperature and are used to estimate the regression coefficients.

The covariance between sigma and (C, n, B) will be zero, a property of the normal distribution. So, it is a 3 × 3 covariance matrix.

$$\log(-\log(L)) = \beta\log(t) + B\cdot\frac{1}{T} + n\cdot\log(I) + \log(C) + \epsilon \qquad (13.6)$$

As fitted in statistical software Stata [36]:

$$\log(-\log(L)) = \underline{\beta_k}\log(t) + B_k\cdot\frac{1}{T} + n_k\cdot\log(I) + \underline{c_k} + \epsilon \qquad (13.7)$$

where (c, β) follows a bivariate normal distribution per LED and k is type. Let the mean be (c_k, β_k) and write $\left(\underline{c_k}, \underline{\beta_k}\right) = (c_k, \beta_k) + \left(\epsilon_c, \epsilon_\beta\right)$. Then the model is

$$\log(-\log(L)) = \left(\beta_k + \epsilon_\beta\right)\log(t) + B_k\cdot\frac{1}{T} + n_k\cdot\log(I) + c_k + \epsilon + \epsilon_c \qquad (13.8)$$

Taking $L = 0.7$ and solve for t where $\epsilon = 0$, first set $d = \log(-\log(0.7)) = -1.03$:

$$d - B_k\cdot\frac{1}{T} - n_k\cdot\log(I) - c_k - \epsilon_c = \left(\beta_k + \epsilon_\beta\right)\log(t) \qquad (13.9)$$

The stochastics will be complex, assuming random components being zero, it results:

$$\begin{aligned}\text{location} = \log(L70) &= \tfrac{1}{\beta_k}\left(d - B_k\cdot\tfrac{1}{T} - n_k\cdot\log(I) - c_k\right)\\ &= \log\left(\exp\left(\tfrac{d-c_k}{\beta_k}\right)\right) - \left(\tfrac{n_k}{\beta_k}\right)\cdot\log(I) + \left(-\tfrac{B_k}{\beta_k}\right)\cdot\tfrac{1}{T}\\ &= \log(C_{lm}) - n_{lm}\cdot\log(I) + B_{lm}\cdot\tfrac{1}{T}\end{aligned} \qquad (13.10)$$

The variance-covariance equals $\frac{1}{n}\cdot V$ with $n = nr$ of LEDs and V as follows:

symmetric avg$V$[3,3]

_nl_1 _nl_2 _nl_3

_nl_1 92.889581
_nl_2 −13.417044 5.8164606
_nl_3 −29731.637 2119.4056 11051779

Now the variance of log($L70$). The stochastics come from $+(\epsilon_c, \epsilon_\beta)$. We have

$$\log(L70) = \frac{d - B_k \cdot \frac{1}{T} - n_k \cdot \log(I) - c_k - \epsilon_c}{\beta_k + \epsilon_\beta} = \frac{R}{S} \tag{13.11}$$

And can be approximated as follows:

$$\begin{aligned}
\mathrm{Var}\,(R/S) &\approx \tfrac{1}{(\mu_S)^2}\mathrm{Var}(R) + 2\tfrac{-\mu_R}{(\mu_S)^3}\mathrm{Cov}\,(R, S) + \tfrac{(\mu_R)^2}{(\mu_S)^4}\mathrm{Var}(S) \\
&= \tfrac{(\mu_R)^2}{(\mu_S)^2}\left[\tfrac{\mathrm{Var}(R)}{(\mu_R)^2} - 2\tfrac{\mathrm{Cov}(R,S)}{\mu_R\mu_S} + \tfrac{\mathrm{Var}(S)}{(\mu_S)^2}\right] \\
&= \tfrac{(\mu_R)^2}{(\mu_S)^2}\left[\tfrac{\sigma_R^2}{(\mu_R)^2} - 2\tfrac{\mathrm{Cov}(R,S)}{\mu_R\mu_S} + \tfrac{\sigma_S^2}{(\mu_S)^2}\right] \\
\mu_R &= d - B_k \cdot \tfrac{1}{T} - n_k \cdot \log(I) - c_k \\
\mu_S &= \beta_k \\
\mathrm{Var}(R) &= \mathrm{Var}\,(\epsilon_c) \\
\mathrm{Var}(S) &= \mathrm{Var}\,(\epsilon_\beta) \\
\mathrm{Cov}\,(R, S) &= \mathrm{cov}\,(\epsilon_c, \epsilon_\beta)
\end{aligned} \tag{13.12}$$

An example calculation with uncertainty in the lumen maintenance performance of the LED system is depicted in Fig. 13.12. It shows the calculated results in the

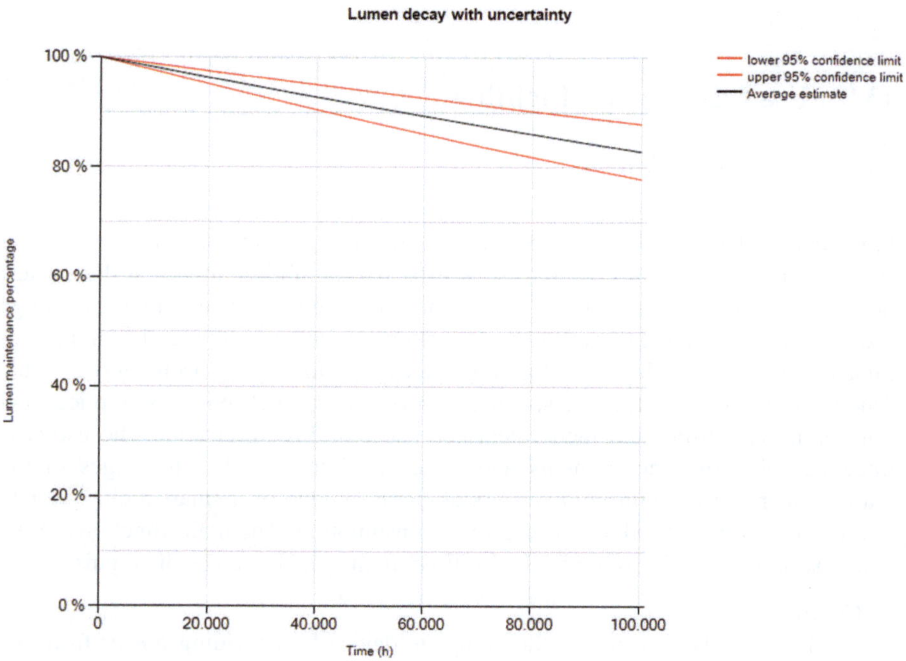

Fig. 13.12 Calculated uncertainty in the lumen decay

uncertainty of the lumen decay. It indicates that L80 at 100 kh can be met in most of the cases.

13.7 Towards a Digital Twin

Combining the algorithms with sensor data, the so-called digital twin comes into sight, which is no more than just a mathematical model of a physical object [37, 38]. Digital twin refers to a digital replica of physical assets, e.g., a luminaire, that can be used for various purposes. The digital representation provides both the elements and the dynamics of how the device operates throughout its life cycle. In the lighting case it can be the lumen maintenance over time. Definitions of digital twin technology used in prior research emphasize two important characteristics. Firstly, each definition emphasizes the connection between the physical model and the corresponding virtual model. Secondly, this connection is established by generating real-time data using sensors. Given the lumen maintenance model parameters listed above, the influence of temperature and current on the lifetime of the LED is established. These models can be diverted such that the temperature increase (coming from sensor data) directly relates to the time until failure or the remaining useful life. Here, a look-up table approach will be used, programmed into an eight-bit processor. The overall flow of the algorithm is depicted in Fig. 13.13.

13.8 Use Case: Smart Lighting

13.8.1 Introduction

Predictive and preventive maintenance are the key development targets for the smart lighting use case. Preventing and/or discovering failure modes at the earliest possible integration level will enable smart maintenance and, obviously, huge cost savings. An increase of the temperature is believed to be the signal for lumen maintenance. The smart lighting use case includes the specifications of the functionality and performance specification of the thermal, photonics/optical and light emitting sensors and their integrated packaging requirements. The use case integrates electronic components and systems addressing the challenges of the massive increase of connected sensors as the backbone of a smart digital society that needs significant reduction of energy consumption. The main functions in the demonstrator are depicted in Fig. 13.14. Four main systems can be distinguished:

1. Controls

   The controls will master the complete device by providing power from the mains. It will also serve to gate to the external world.

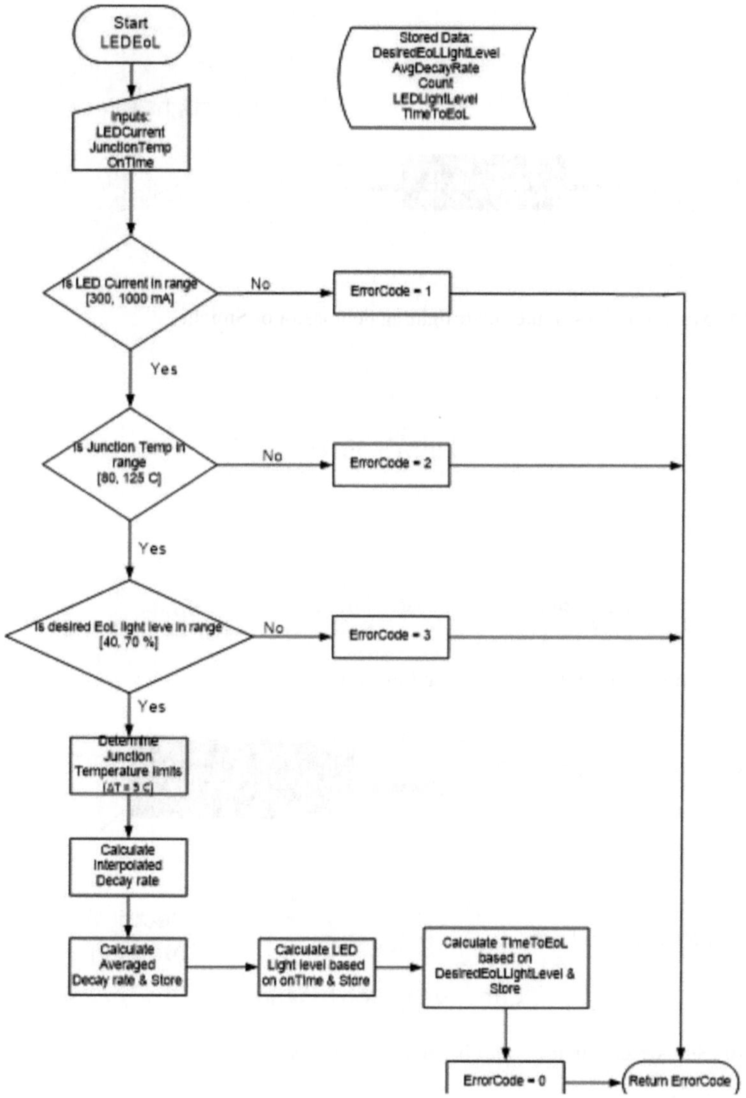

Fig. 13.13 Flowchart of getting the current status of the LED-source calculated

2. Power supply
   The power supply provides the correct power to the light source and the health monitoring device by a stable current and/or voltage to it. It will also serve to gather important signals from the light source and/or the health monitoring device.
3. Light source
   The light source produces the light that is needed in the application.

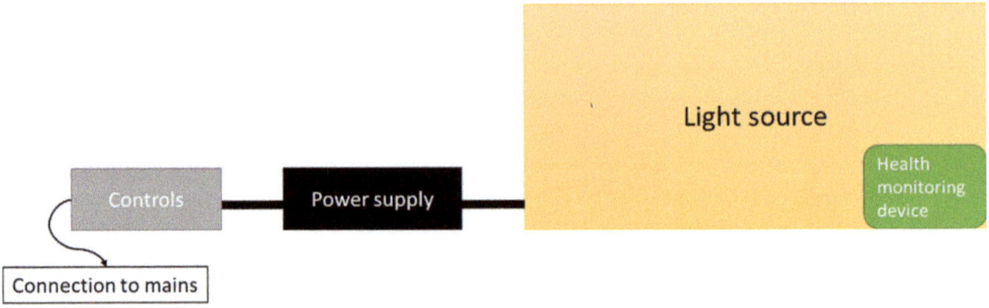

Fig. 13.14 Main functions in the smart lighting demonstrator Smight

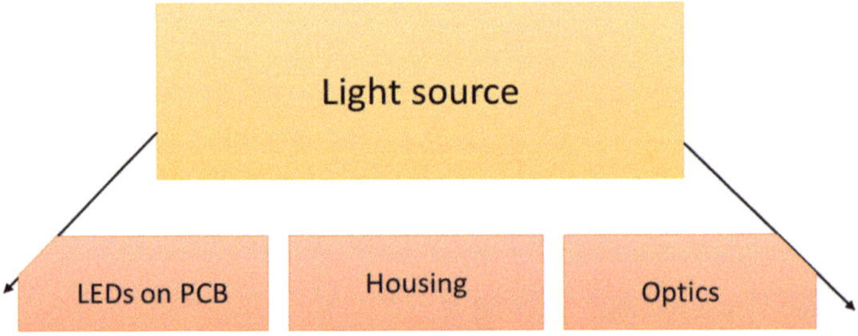

Fig. 13.15 Sub-systems in the light source function

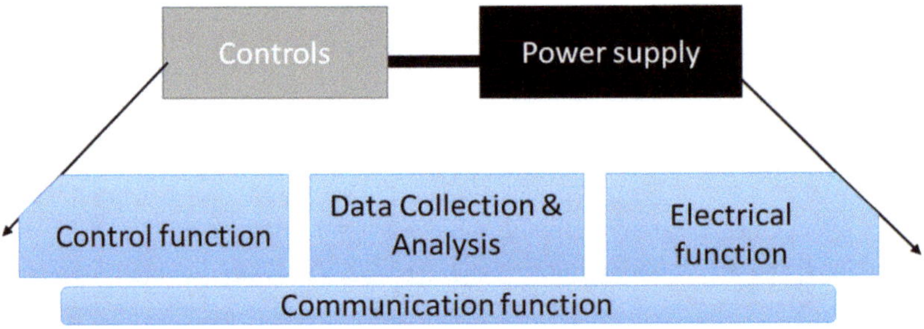

Fig. 13.16 Sub-systems in the controls and power supply

4. Health monitoring device
   The health monitoring device monitors the performance of the product.

Further breakdowns of the above main systems are depicted in Figs. 13.15, 13.16, and 13.17.

1. Light source
   We will use a typical light source consisting of several LEDs on a printed circuit board, placed in a housing and surrounded by optical elements. Optical elements are reflective materials and an exit window.

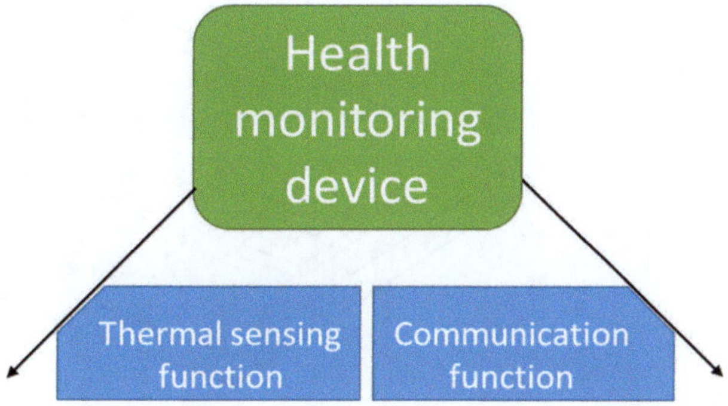

Fig. 13.17 Sub-systems in the health monitoring device

2. Controls and power supply
   We can specify the control and power supply to provide the control and electrical function. On top of this either one of the two should include the algorithms in order to collect and analyze the data coming from the health monitoring device. Finally analyzed data should be provided to the external world for maintenance purposes.
3. Health monitoring device
   The add-on health monitoring device should cover the maintenance and thus provide the thermal sensing function. It should also communicate the data to the power supply.

13.8.2 Simulations: LED on PCB Level

A finite element (FE) simulation model of the LED light source is created to identify the hotspots and the optimum position of the thermal sensor. The junction temperature and the thermal conductivity of the surrounding substrate material are of great influence on the thermal behavior. When this behavior is identified extensively, this information can be used to optimize the system design and to use for lifetime prediction by the application of temperature sensors. The expected results are an improved design for reliability due to optimized chip-layout, minimization of thermal interfaces, material resistance, and stresses.

Doing thermal simulations on LED systems deliver important information on where the thermal hotspots do arise, and at which locations the temperature sensors should be placed for adequate thermal monitoring. Information about temperature at different locations in the system can be related to the LED junction temperature, but also the temperature of the surrounding environment.

With the aid of a COMSOL FEA model consisting of an LED die and an LED package put together on a PCB board consisting of a predefined material using solder paste. Such a scenario mimics a practical situation, where usually multiple

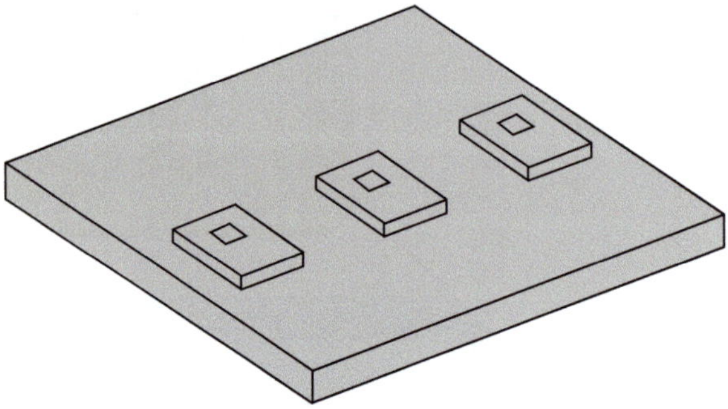

Fig. 13.18 Details of the model

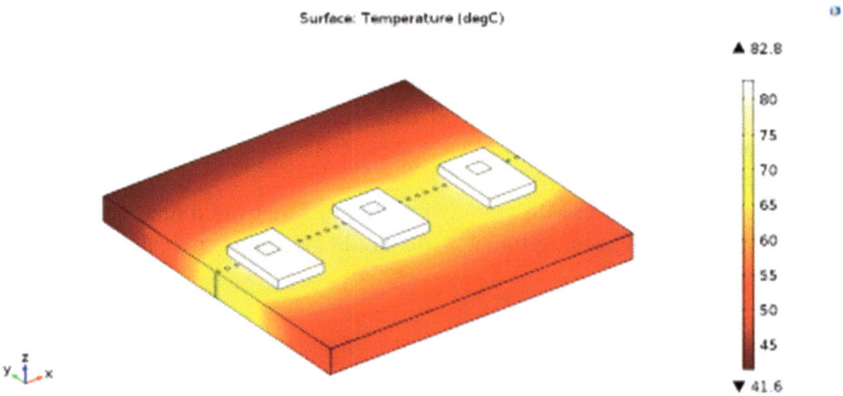

Fig. 13.19 Calculated thermal behavior on PCB level

LED packages are placed on a single PCB board to increase light output of 1 luminaire.

In this thermal problem, the two main figures of the used materials which determine the steady state and dynamic thermal behavior are the heat capacity and the thermal conductance. In this simulation, the LED is modeled as heat source with a fixed amount of power. The values in the below table are used in the simulation.

For finding the optimal position of the other LED dies and temperature sensors to be used for monitoring, an FE simulation is done. The model contains LED dies mounted on top of a PCB substrate using solder joints, see figure below (Fig. 13.18).

The calculated thermal behavior of the LEDs at a power of 0.3 W is given in the figure below (Fig. 13.19).

As can be seen the LEDs generate a significant amount of heat which also raises the temperature of the PCB underneath it, requiring a proper heatsink. The LEDs should be placed far enough from one another to reduce further heating. Figure 13.20 shows the temperature for the LEDs as a function of distance.

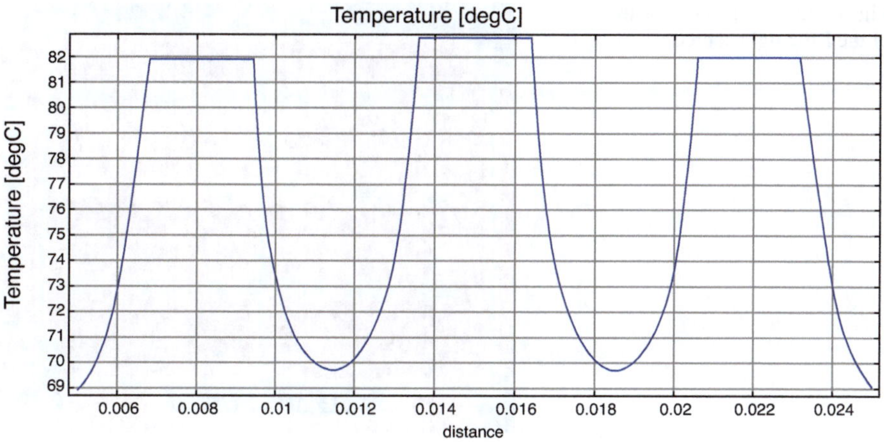

Fig. 13.20 Temperature as function of the distance, PCB level

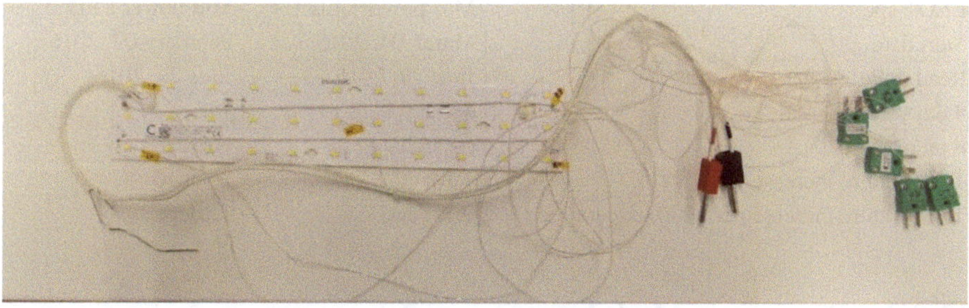

Fig. 13.21 Light source with thermo-couples as currently being tested

As can be seen from the above figure, the temperature drops outside the LEDs which shows that a minimum distance between the LEDs needs to be considered. The same can be done for placing temperature sensors near the LED as well as at a certain distance near the edge of the PCB to show the difference in the temperature distribution.

Experimental verification was done with a commercially available LED board. Figure 13.21 shows the light source with the thermo -couples placed on top. Figure 13.22 shows a temperature scan of the PCB to indicate the heat distribution from a top view. Notice the uniform distribution of temperature, which is vital to pick the location of the sensors. The resulting temperature increase is listed in Table 13.4. These results are input for the design of the thermal sensing function.

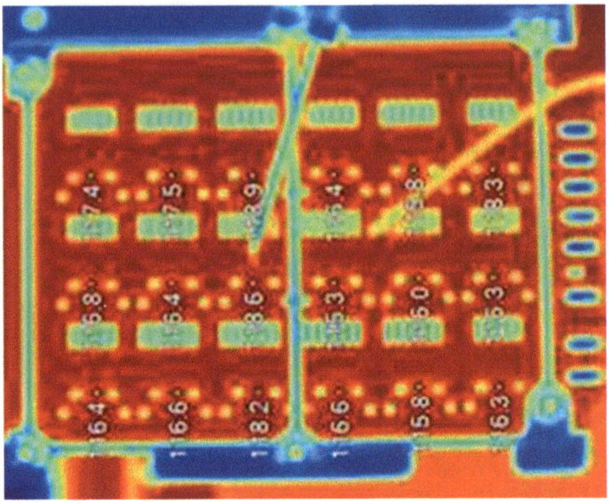

Fig. 13.22 Temperature scan of the LED light source

Table 13.4 Lifetime test results, readings for two products

Item	Product A	Product B
LED Type	Mid-power LED I	Mid-power LED II
Start date	March 04, 2015	February 02, 2016
Read point	June 29, 2016	June 29, 2016
# days testing	483	148
# on/off	69,552	102,225
Temperature increase	1.3degC	2.3degC
Temperature increase per 1000 h operation	0.15degC	1.1degC

13.8.3 Diagnostic Module

Remote diagnostics via sensing solutions and data analytics lowers maintenance costs by sending inspection and repair crews only when and where needed, improving operational efficiency. Based on the luminaire health status, its operation can be adapted to preserve the right light and schedule preventive maintenance. Also, service offerings can be tailored based on usage patterns and geographic conditions of the lighting network enabling more accurate lifetime predictions of luminaires.

The diagnostic module adds the sensors to luminaires (LED-based product). It is used for incident detection and failure analysis by monitoring environmental conditions that influence the luminaire life time. The module consists of (see also Figs. 13.23 and 13.24):

- An Onboard MCU for failure detection and sensor data aggregation
- Sensor data available to a DALI master (e.g., the IoT Client)
- Compatible with a DALI enabled LED driver

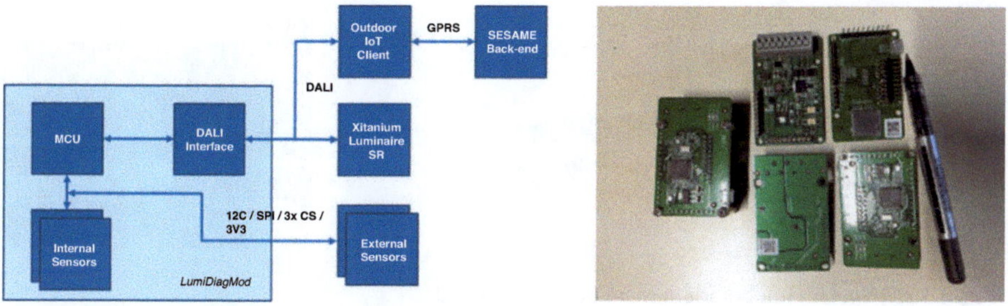

Fig. 13.23 Prototype of the diagnostic module

Fig. 13.24 View on the prototype (light source is not visible)

13.8.4 Integration into Test Beds

System integration in order to demonstrate the concept is done in an outdoor test bed, depicted in Fig. 13.25.

The details of this test bed are:

- Temperature as function of days is monitored
- Almost 1 year testing performed
- Under worst case conditions, where the thermal performance is worsened as to provoke failures

The results are shown in Fig. 13.26. As the figure shows, indeed the temperature increases prior to failure demonstrating the concept of health monitoring.

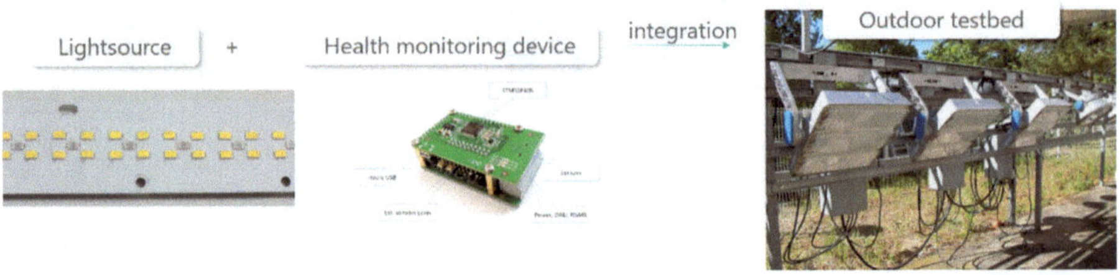

Fig. 13.25 System integration into an outdoor test bed

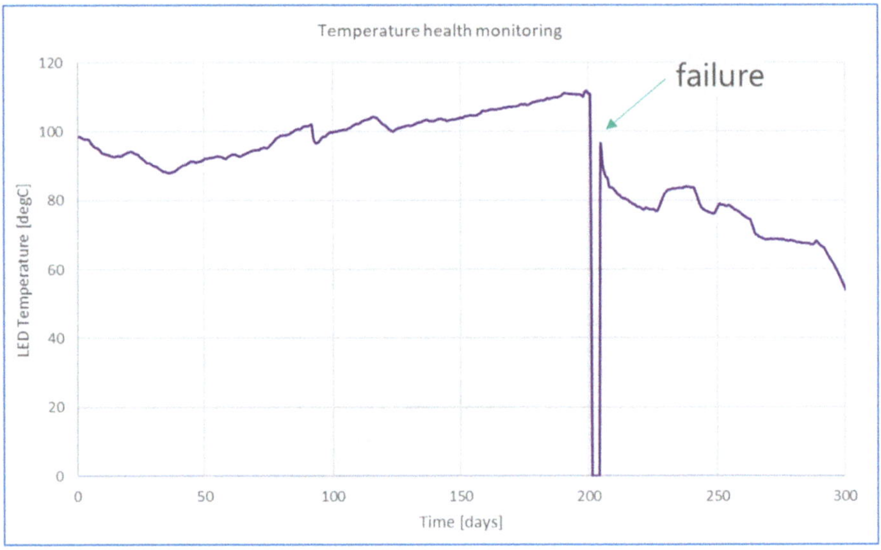

Fig. 13.26 Temperature as function of days for the test bed

A second test bed was installed at the premises of the Technical University of Graz in Austria, see Fig. 13.27. The data is generated on-line. The integrated health monitoring device is placed within the cabinet of the luminaire. The test bed has the following features.

- An environmental sensor monitors temperature, humidity, barometric pressure as well as acceleration in three axes and as vector.
- It operates autonomously and sends measured data daily to a functional email account.
- The measurement modules are small in size and thus can be integrated easily in a luminaire.
- The autonomous data measurements and transmission is independent from AC mains voltage for about 1 day thanks to an integrated energy storage battery.

Fig. 13.27 Test bed at the Technical University of Graz

13.8.5 *Life Time Prediction*

With the incoming data from the test bed and the physical description of it, it is possible to predict the lifetime of these products. There are three main paradigms for solving classification and detection problems in sensor data: data-driven approach, model-driven approach, and fusion-driven approach which combines the first two approaches. Data-driven is a new way of thinking, enabled by machine learning which is learning patterns from historical data. Results are dependent on the availability of both good quality data. As more and more data has become available, developing good performing classifiers using machine learning has become more and more feasible. Data-driven approaches can be very effective for electronic systems, considering that the capability of realizing complex physical models for the system is reduced. However, in most of the cases the parameters monitored have no connections to the real failure. So there is need for a method to link the actual failure with the monitored parameters, which is the fusion-driven approach. Here, a fusion-driven method is proposed to predict the catastrophic failure of luminaires only based on historical measured time series. In this case, it is the (rising) temperature for which the failure occurrence is known by the models described in the previous chapter.

For predictive maintenance modeling, the goal is to predict whether the luminaire experiences its failure in the next few days. The prediction of catastrophic failure is defined as "Given time-series of features (such as current, energy, burning hours, etc.), for n number of days, predict if a catastrophic failure will happen within the next *m* days." This is indicated in Fig. 13.28.

Figure 13.29 shows the mean value of five test accuracy from five machine learning algorithms over prediction days. Here the observation days are fixed to 10. With the increase of the prediction days, the mean accuracy value decreases and then back to fluctuating values. The observation infers that the effect of observation days on the accuracy is limited. Further observations are:

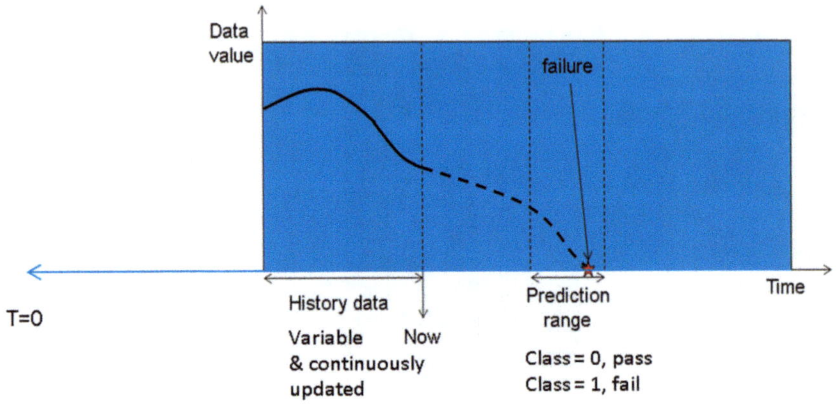

Fig. 13.28 Predicting the time frame in which the failure will occur

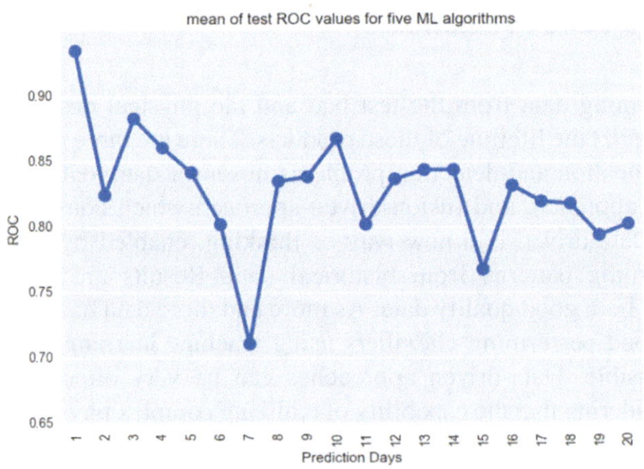

Fig. 13.29 Mean value of five test accuracy from five machine learning algorithms over prediction days

- Models show balanced accuracy of 77–83%, sensitivity (predicting failures) of 55–85%, and specificity (predicting non-failures) of 81–99% using prediction windows of 2 days. This suggests failure prediction using temperature data is feasible.
- Prediction accuracy decreases by 5–10% when the prediction window increases to 10 days.

The next step is to validate, test, and further develop algorithms using more incoming data.

13.8.6 Final Remarks

In the past 4 years, we have witnessed a substantial change in the lighting industry. Traditional companies have changed their strategy and upcoming competition has pushed down prices for LED-based products considerably. LED penetration levels increased so as the diversity of commercially available replacement products. New processes and materials were introduced, and consequently new failure modes appeared. This trend will continue as the lighting industry is getting connected and large amount of user data is being analyzed. New components are needed to deliver this functionality (sensors, actuator IoT modules) and, as such, the diversity from an architectural point of view will also increase. Gradually but slowly the term reliability will be replaced by availability and "smart" maintenance will distinguish good from bad products. In this chapter, we have presented a smart lighting use case in which the temperature rise served as indicating parameter for lumen maintenance. A fusion-driven approach combining sensor test data with a physical system description was developed. In fact, these are our first steps towards a digital twin for a connected luminaire.

Acknowledgments This work was supported by the European project "Iosense: Flexible FE/BE Sensor Pilot Line for the Internet of Everything." This project has received funding from the Electronic Component Systems for European Leadership Joint Undertaking under grant agreement No 692480. This Joint Undertaking receives support from the European Union's Horizon 2020 research and innovation programme and Germany, Netherlands, Spain, Austria, Belgium, Slovakia.

Thanks to the support from Harry Broers for providing the health monitoring device. Thanks to the support from Ulrich Boeke for designing and organizing the test bed in Graz and thanks to the Technical University of Graz for hosting this test bed.

References

1. J. McLinn, A short history of reliability, J. Reliab. Inf. **30**, 8–15 (2011)
2. W. Denso, The history of reliability prediction. IEEE Trans. Reliab. **47**(3-SP), 321–328 (1998)
3. W. D. Van Driel, X. J. Fan (eds.), *Solid State Lighting Reliability: Components to System* (Springer, New York, 2013)., ISBN 978-1-4614-3067-4
4. W. D. Van Driel, X. J. Fan, G. Q. Zhang (eds.), *Solid State Lighting Reliability: Components to System Part II* (Springer, New York, 2017)., ISBN 978-3-319-58174-3
5. Navigant Consulting, Inc., Energy Savings Forecast of Solid-State Lighting in General Illumination Applications, report prepared for the U.S. Department of Energy, August 2014
6. Navigant Consulting, Inc., Energy Savings Forecast of Solid-State Lighting in General Illumination Applications, report prepared for the U.S. Department of Energy, September 2016
7. Estimated LED penetration of the global lighting market from 2010 to 2020, http://www.statista.com/statistics/246030/estimated-led-penetration-of-the-global-lighting-market/ (last visited on 8/25/2016)
8. D. Schenkelaars, W.D. van Driel, M. Klompenhouwer, I. Flinsenberg, R. Duijve, Towards prognostics & health management in lighting applications, in *European Conference of the Prognostics and Health Management Society 2016*, open access journal, vol. 7, p. 7, 2016, http://www.phmsociety.org/node/2090/

9. S. Ismail, *Exponential Organizations: Why New Organizations Are Ten Times Better, Faster, and Cheaper Than Yours (and What To Do About It)* (Diversion Books, 2014)
10. W.Q. Meeker, Y. Hong, Reliability Meets Big Data: Opportunities and Challenges (2013). Statistics Preprints. Paper 82. http://lib.dr.iastate.edu/stat_las_preprints/82
11. W.Q. Meeker, L.A. Escobar, *Statistical Methods for Reliability Data* (Wiley, New York, 1998)
12. M.G. Pecht, *Prognostics and Health Management of Electronics* (Wiley, Hoboken, 2008)
13. O. Tapaninen, P. Myöhänen, A. Sitomaniemi, Optical and thermal simulation chain for LED package, EuroSimE (http://www.eurosime.org/), 17th–20th of April 2015, Montpellier (France)
14. A. Alexeev, W. Cassarly, V.D. Hildenbrand, O. Tapaninen, A. Sitomaniemi, Simulating light conversion in mid-power LEDs, EuroSimE (http://www.eurosime.org/), 17nd–20th of April 2016, Montpellier (France)
15. S. Tarashioon, W.D. van Driel, G.Q. Zhang, Multi-physics reliability simulation for solid state lighting drivers. Microelectron. Reliab. (2014). https://doi.org/10.1016/j.microrel.2014.02.019
16. B. Sun, X.J. Fan, C. Qian, G.Q. Zhang, PoF-simulation-assisted reliability prediction for electrolytic capacitor in LED drivers. IEEE Trans. Ind. Electron. **63**(11), 6726–6735 (2016)
17. B. Sun, X.J. Fan, W.D. van Driel, H.Y. Ye, J.J. Fan, C. Qian, G.Q. Zhang, A novel lifetime prediction for integrated LED lamps by electronic-thermal simulation. Reliab. Eng. Syst. Saf. **163**, 14–21 (2017)
18. B. Sun, X.J. Fan, H.Y Ye, G.Q. Zhang, A reliability prediction for integrated LED lamp with electrolytic capacitor-free driver, in *IEEE Transactions on Components, Packaging and Manufacturing Technology*, 2017, p. 99
19. Communication with Dr. C.A. Yuan, owner of Ichijouriki LS R&D Co., Ltd, http://www.ichijouriki.com/, 26-09-2016
20. IES LM-80-08: Approved method for measuring maintenance of Led light sources
21. IES TM-21-11: Projecting Long Term Lumen Maintenance of LED Light Sources
22. IEC 62722-2-1, Luminaire performance—part 2-1: particular requirements for LED luminaires, IEC standard, Edition 1.0 2014-11
23. IEC 62717, LED modules for general lighting—performance requirements, IEC standard, Edition 1.1 2015-09
24. J. Fan, K.-C. Yung, M. Pecht, Lifetime estimation of high-power white LED using degradation-data-driven method. IEEE Trans. Device Mater. Reliab. **12**(2), 470–477 (2012)
25. J. Fan, K.-C. Yung, M. Pecht, Predicting long-term lumen maintenance life of LED light sources using a particle filter-based prognostic approach. Expert Syst. Appl. **42**(5), 2411–2420 (2015)
26. P. Lall, J. Wei, P. Sakalaukus, Bayesian models for life prediction and fault-mode classification in solid state lamps, in *16th International Conference on Thermal, Mechanical and Multi-Physics Simulation and Experiments in Microelectronics and Microsystems*, 2015
27. C. Quan, L. Xiaobing, C. Qi, W. Kai, L. Sheng, L. Jingyana, Research on lumen depreciation related to LED packages by in-situ measurement method. Microelectron. Reliab. **55**, 2269–2275 (2015)
28. J.L. Huang, D.S. Golubović, S. Koh, D.G. Yang, X.P. Li, X.J. Fan, G.Q. Zhang, Degradation mechanisms of mid-power white-light LEDs under high temperature-humidity conditions. IEEE Trans. Device Mater. Reliab. **15**(2), 220–228 (2015)
29. J.L. Huang, D.S. Golubović, S. Koh, D.G. Yang, X.P. Li, X.J. Fan, G.Q. Zhang, Degradation modeling of mid-power white-light LEDs by using Wiener process. Opt. Express **23**(15) (2015). https://doi.org/10.1364/OE.23.00A966
30. J.L. Huang, D.S. Golubović, S. Koh, D.G. Yang, X.P. Li, X.J. Fan, G.Q. Zhang, Optical degradation mechanisms of mid-power white-light LEDs in LM-80-08 tests. Microelectron. Reliab. **55**(12, Part B), 2654–2662 (2015)
31. T.-R. Tsai, C.-W. Lin, Y.-L. Sung, P.-T. Chou, C.-L. Chen, Y. Lio, Inference from lumen degradation data under Wiener diffusion process. IEEE Trans. Reliab. **61**, 710–718 (2012)
32. W.D. van Driel, M. Schuld, B. Jacobs, F. Commissaris, J. van der Eyden, B. Hamon, Lumen maintenance predictions for LED packages. Microelectron. Reliab. **62**, 39–44 (2016)

33. W.Q. Meeker, L.A. Escobar, A review of accelerated test models. Stat. Sci. **21**(4), 552–577 (2006)
34. Y. Hong, W.Q. Meeker, Field-failure predictions based on failure-time data with dynamic covariate information. Technometrics **55**(2), 135–149 (2013). https://doi.org/10.1080/00401706.2013.765324
35. B.P. Weaver, W.Q. Meeker, Methods for planning repeated measures accelerated degradation tests. Stat. Qual. Prod. **30**(6), 658–671 (2014)
36. STATA multilevel mixed effects reference manual, release 15, Stata Corp LLC, College Station, TX
37. Minds + Machines: Meet A Digital Twin, YouTube, GE Digital. Retrieved 26 July 2017
38. Introduction to Digital Twin: Simple, but detailed, YouTube, IBM Watson Internet of Things. Retrieved 27 June 2017

Chapter 14
Ultraviolet Sensing in WBG: SiC

B. El Mansouri, W. D. van Driel, and G. Q. Zhang

14.1 Introduction

The main application for the ultraviolet (UV) sensor is the detection of partial discharge associated with power grid equipment, but not limited to this. The UV electromagnetic radiation occupies the spectral range of 10–400 nm. The UV radiation is used in many applications ranging from chemical processes such as lithography to biology. The UV spectral range is typically divided into four wavelength (λ) regions as illustrated in Fig. 14.1.

The photodetection is then also classified according to the UV range of interest [1–4]. The photodetectors are referred to as solar-visible for the spectral range above UVA [1]. Another special range is the deep UV (DUV) range of 220–280 nm [1, 3]. Here the DUV radiation from the sun is mostly absorbed by the ozone layer leaving longer wavelengths than 280 nm (UVC) to reach earth. This gives the opportunity to use photodetection in the lower wavelength region without being affected by the sun, hence the classification as solar blind photodetection. As for wavelengths shorter than 200 nm, these will be absorbed by the oxygen in the atmosphere. The photodiodes can further be classified depending on the used materials, structure of the sensor, insulation, filtering and so on.

Most common photodetectors are semiconductor based, i.e. photoconductors, photodiodes such as PN, pin and MS and MSM diodes, and phototransistors.

B. El Mansouri · G. Q. Zhang
Delft University of Technology, EEMCS Faculty, Delft, The Netherlands
e-mail: B.ElMansouri@tudelft.nl; G.Q.Zhang@tudelft.nl

W. D. van Driel (✉)
Delft University of Technology, EEMCS Faculty, Delft, The Netherlands

Signify, HTC48, Eindhoven, The Netherlands
e-mail: willem.van.driel@signify.com; willem.van.driel@philips.com

W. D. van Driel et al. (eds.), *Sensor Systems Simulations*,
https://doi.org/10.1007/978-3-030-16577-2_14

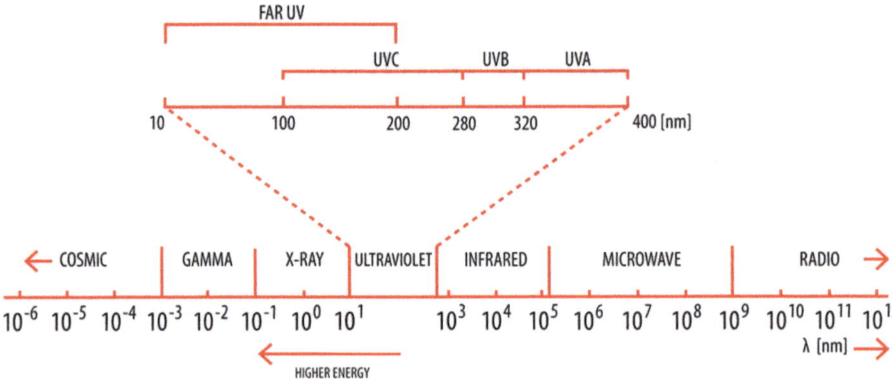

Fig. 14.1 The electromagnetic spectrum with the highlighted UV range

Semiconductor devices are attractive thanks to their small size, lightweight and insensitivity to magnetic fields. Moreover, they can have good sensitivity, linearity and high-speed operation [5]. The devices, each have strengths and weaknesses enabling the detection of various optical signals in different fields ranging from simple applications such as sunshine detection, to industrial or military applications such as corona discharge, flame detection or missile plume detection.

The aim here is to use SiC as a starting material. SiC is a WBG semiconductor and has the advantage of high responsivity, high thermal stability, robust radiation hardness, high response speed and high power [1]. Next to this, the photodetector should be sensitive to low light level in the range of fW–aW/cm^2 while having a clear distinction from the noise level and maintaining a reliable signal.

The photodetectors can be selected based on the detection signal. Each optical source may emit photons at a different spectral range and have a certain signal strength. This translates in various requirements which also motivates the device type selection. The requirements may include:

- Spectral range of the optical signal.
- The optical signal strength.
- Linearity of the output signal.
- Response time (speed).

Each of these requirements has additional restraints on the selection of the device. Take for example the minimum detectable signal which sets the noise floor and dark current requirements. Or the upper limit of the signal detection which is determined by the maximum current that the detector can handle without becoming saturated. All these restraints are then translated in the minimum sensitivity or responsivity, the detectivity and NEP requirements.

14.1.1 Applications

UV photodetection has potential for many applications in both civil and military fields which may include flame detection, missile threat detection, water treatment, astronomical observations, telecommunications, environmental monitoring and many more. One of the most interesting applications is the corona discharge at the electrical power networks. This is key because it manages the energy needs of societies in the form of electricity. It is then important for critical applications such as missile threat or flame detection, requiring good temperature stability of the photodetectors, to have a reduction in errors for a reliable signal [1, 2, 5–7]. In space applications, for optical communications such as inter-satellite communications, $\lambda < 280$ nm can be useful and for the industry related applications such as instrumentation and UV lithography emitter calibration can be used [5]. Next to the application, the device type is selected based on the signal type, signal strength for detectability, environmental conditions and so on. Table 14.1 lists some examples of the applications.

14.1.2 State of the Art

The UV detection in the state grid is currently applied using UV cameras such as "UV-260 Corona Discharge Camera" and "DayCor Superb UV imager" which superimposes the detected UV light on a normal image to show the location [8]. The "DayCor Superb UV imager" has a UV sensitivity of 2.2×10^{-18}W/cm^2 and a minimum visible light detection of 0.1Lux. The minimum discharge detection is 1 pC at a distance of 10 m. The spectral range is kept within 240–280 nm as this corresponds to the solar blind range allowing for daytime detection. The discharge severity can then be estimated based on the photon number [8]. However, this method is only used at close range while needing someone to operate it. This

Table 14.1 Examples for UV detection applications

Applications	Harsh environment
UV lithography (193 nm)	–
UV curing processing (365 nm)	–
Disinfection of water and air (240–290 nm)	–
Detection of corona discharges (<280 nm)	YES
Early missile (plume) threat detection	YES
Chemical and biological threat detection	–
Flame detection/combustion monitoring	YES
Environmental monitoring (ozone/pollutants)	YES
Optical communications, e.g. space (<280 nm)	YES
Astronomical studies	–

shows that developing sensors that can be employed over the entire grid offer better monitoring of the power grid.

Next to the camera, an imaging system was also developed to detect ultraviolet detection [9, 10]. The imaging sensor is able not only to detect hydrogen flames but also corona discharge at a proximity. The UV light intensity detected for the latter is 37 nW/cm^2 with a 400 lux background light intensity. The imaging sensor is based on CMOS technology while the photodetection is obtained using an organic photoconductive film (OPF). The film is not only sensitive in the UV range but also in the visible range and therefore a smart imaging scheme was used to detect both the visible and UV radiation at the same time. This enables a normal image with an imposed UV image.

14.2 Device Types

There are various technologies used for photodetection application ranging from vacuum tube devices used as photomultipliers (PMTs) to semiconductor devices such as photoconductor, Schottky diodes and pn-junctions including their adaptations [1, 5]. Here, two structure types are used, the MSM and device types. Therefore, the discussion in this chapter will limit itself to the discussion and development of these devices.

14.2.1 The Photoconductor

The photoconductor, also referred to as a photoresistor, is an ohmic photodetector. The device has a simple design comprised of a pair of ohmic electrodes with on the top surface a highly doped semiconductor absorbing layer and an electrode on each side, as shown in Fig. 14.2.

The MSM photoconductor allows for simple fabrication not only reducing cost but is an advantage as no additional epitaxial layer is needed and hard materials such as SiC are less of an issue regarding implantation, diffusion and etching. Further the device can realize a large area of detection. The conductivity of the semiconductor

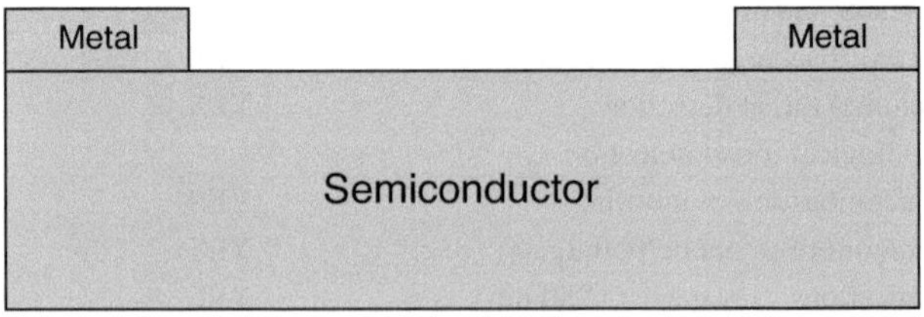

Fig. 14.2 Metal-semiconductor-metal photoconductor

is dependent on the amount of incident photons which in turn excite electrons to the conduction band enabling a higher conductivity of the material. When the electrodes are biased, the electrical field will drive the photogenerated e-h pairs to separate and move towards the electrodes resulting in a photocurrent in the external circuit proportional to the photon lux [1]. In the absence of incident photons, the dark current of the device is given by

$$I_{\mathrm{d}} = \frac{V}{R_{\mathrm{d}}} \tag{14.1}$$

where R_{d} is the resistance of the material in the dark. Since the conductivity is dependent on the incident photons with sufficient energy, the resistance of the semiconductor will drop resulting in a photocurrent given by

$$I_{\mathrm{p}} = V\left(\frac{1}{R_{\mathrm{i}}} - \frac{1}{R_{\mathrm{d}}}\right) \tag{14.2}$$

where R_{i} is the resistance under illumination. Figure 14.3 illustrates the *I-V* characteristics of the photoconductor in the dark and under illumination [2].

The full expression for the photocurrent in a photoconductive device is given by [11]

$$I_{\mathrm{p}} = \left(\frac{P_{\mathrm{opt}}}{hv}\right) q\eta G = \left(\frac{P_{\mathrm{opt}}}{hc}\right) \lambda q\eta G \tag{14.3}$$

with

$$G = \frac{\mu\tau V}{s^2} \tag{14.4}$$

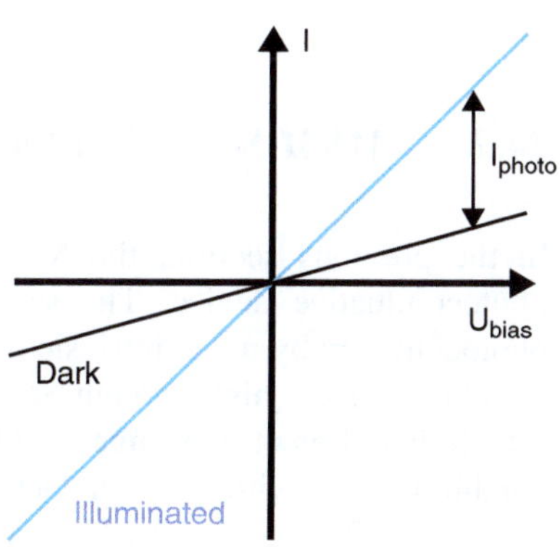

Fig. 14.3 *I-V* characteristics of a photoconductor

where P_{opt} is the incident optical power, V is the applied voltage, q is the electron charge, η is the quantum efficiency (QE), hv is the photon energy, λ is the wavelength, c is the speed of light, G is the photoconductive gain, μ is the electron mobility, τ is the carrier lifetime and s is the inter-electrodes spacing.

The performance of the photoconductor depends on a number of physical parameters. For instance, a trade-off between a high gain G and a fast response has to be found for longer recombination lifetime of the holes making the device slower. The photoconductive nature of the device is that it requires a bias which inherently adds dark current to the noise contributions which reduces the minimum detectable optical signal. Moreover, the electric field effects such as space charge limited current, avalanche and dielectric breakdown limit the photocurrent and therefore become an issue for scaling down of photoconductor devices for higher end applications. The scaling down is further limited by the area required for optical sensing to obtain any useful optical signal especially for devices with metallic electrodes which reflect light. This drawback can be reduced through transparent electrodes such as Indium Tin Oxide (ITO) or 2D materials such as graphene which is the case for this work. The device can further be enhanced by using nanomaterials on the top surface which results in a high photo conductive gain G and high responsivity [12].

Some of the drawbacks such as large area requirement render the device only useful for applications with slower speeds. Further the output signal of the device will also depend on previously detected signals due to a photo memory effect. The photoconductive nature of the device also gives rise to a nonlinear relation between the resistance and illumination E_i as given by equation

$$R = AE_i^{-\alpha} \tag{14.5}$$

where A and α are constants depending on the semiconducting material and the processing used to manufacture the device. Since the device changes resistance, it is also highly temperature dependent adding to uncertainty of the optical signal in the form of thermal noise.

14.2.2 MSM Schottky Photodetector

In the previous section, the MSM structure was shown to be applicable as a photoconductive device. The same structure can be utilized for Schottky type photodetectors by using intrinsic or low doped semiconductors with an appropriate metal to create a high enough Schottky barrier [13]. The contacts can be made as interdigitated electrode fingers (IDT) which result in back-to-back diodes and by combining this with semiconductors having a wide bandgap, UV photodetection is enabled [2, 13, 14]. The biasing of the two diodes will always result in reverse operation of one diode while the other is in forward bias enabling the conduction of the current. As was discussed in the previous text for the photoconductor, the device is simple to fabricate, but has additional advantages such as low dark current, low junction capacitance, high bandwidth (BW) and is CMOS compatible [13, 15, 16].

To analyse the behaviour of the MSM, first consider the metal-semiconductor (MS) Schottky diode shown in Fig. 14.4.

As was mentioned in earlier text, when a metal is brought into contact with a semiconductor the electrical behaviour at the junction either takes the ohmic or Schottky characteristics. However, this device type is asymmetrical in nature in the sense that the Schottky junction is only one-sided. Figure 14.5 shows the band diagram of a metal and an n-type semiconductor before and after equilibrium.

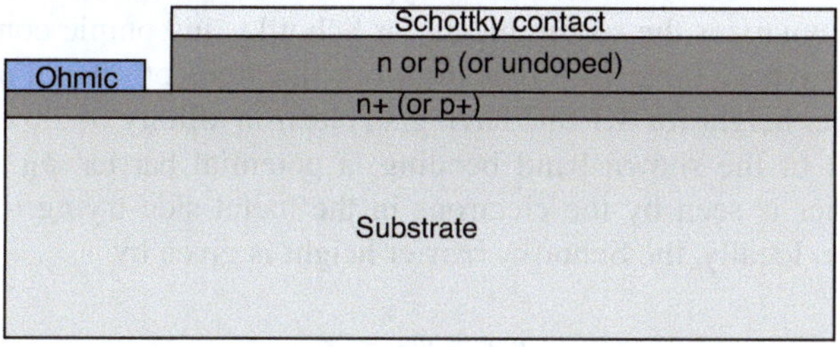

Fig. 14.4 Metal-semiconductor Schottky contact

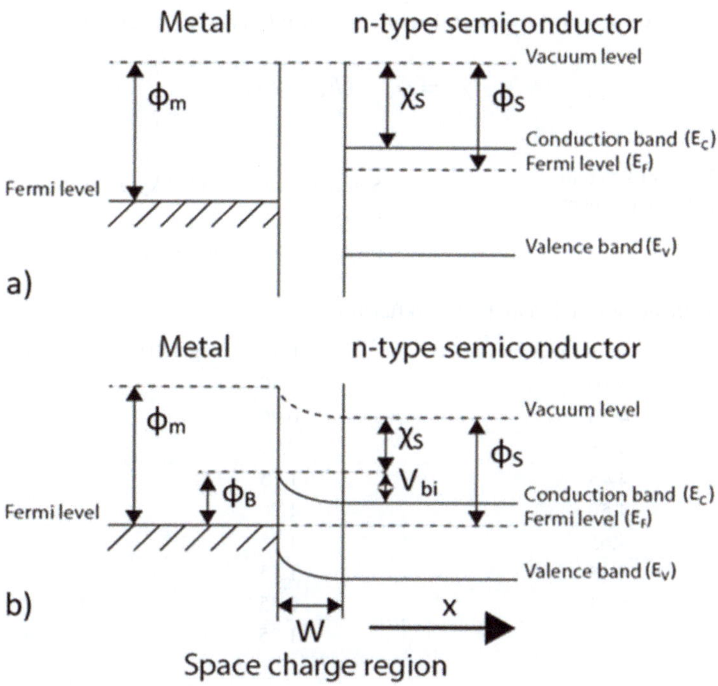

Fig. 14.5 Band diagram of a metal and n-type semiconductor: (**a**) before equilibrium and (**b**) after equilibrium

In the above figure, χ_s is the semiconductor electron affinity, ϕ_s is the semiconductor workfunction, ϕ_m is the metal workfunction, ϕ_B is the Schottky barrier height, V_{bi} is the built-in voltage and W is the space charge region (SCR) width along the distance x. Figure 14.5a shows the Fermi levels of the metal and semiconductor before equilibrium is reached. Upon contact between the two materials, the Fermi energy level becomes constant throughout the structure reaching an equilibrium. This is thanks to a flow of electrons from the semiconductor into the lower energy states of the metal leaving positively charged ions behind, thus creating a depletion region (SCR). It should be noted that Fig. 14.5b assumes the metal workfunction to be larger compared to that of the semiconductor else an ohmic contact will be the result. For p-type semiconductors this is the opposite. Table 14.2 summarizes the requirements for Schottky and ohmic contacts for both semiconductor types. Table 14.3 lists the work functions of several metals and the resulting barrier height for 4H-SiC having an electron affinity of 3.3 eV.

As a result of the shown band bending, a potential barrier ϕ_B known as the Schottky barrier is seen by the electrons in the metal side trying to move to the semiconductor. Ideally, the Schottky barrier height is given by

$$\phi_B = \phi_m - \chi_s \tag{14.6}$$

The equation shows the Schottky contact dependence on the semiconductor electron affinity (χ_s) and the metal workfunction (ϕ_m) $\cdot$ χ_s is given by

$$\chi_s = \phi_s - (E_c - E_f) \tag{14.7}$$

Table 14.2 Schottky contact workfunctions requirement

Semiconductor	Schottky (rectifying)	Ohmic
n-type	$\phi_m > \phi_s$	$\phi_m < \phi_s$
p-type	$\phi_m < \phi_s$	$\phi_m > \phi_s$

Table 14.3 Relevant metals and their workfunctions

Metal	Workfunction (ϕ_m) eV	Barrier height 4H-SiC (ϕ_m) eV
Au	5.10	1.80
Ag	4.26	0.96
Al	4.28	0.98
B	4.45	1.15
Cr	4.50	1.20
Cu	4.65	1.35
Ni	5.15	1.85
Pt	5.65	2.35
W	4.55	1.25

On the other hand, after the band bending the electrons from the semiconductor side will also see a built-in potential barrier known as the built-in voltage V_{bi} given by

$$V_{bi} = \phi_m - \phi_s = \phi_B - (E_c - E_f) \tag{14.8}$$

As discussed before, the semiconductor should be low doped to achieve Schottky barrier behaviour. The reason behind this is the dependence of the built-in voltage on doping similar to the pn-junction. This dependence is shown below equations.

$$E_c - E_f = \frac{k_B T}{q} \ln\left(\frac{N_c}{N_d}\right) \tag{14.9}$$

Yielding

$$V_{bi} = \phi_m - \chi_s - \frac{k_B T}{q} \ln\left(\frac{N_c}{N_d}\right) = \phi_B - \frac{k_B T}{q} \ln\left(\frac{N_c}{N_d}\right) \tag{14.10}$$

where k_B is the Boltzmann's constant, T is the temperature and N_d is the donor concentration. The built-in voltage is the barrier seen by electrons in the semiconductor which prevents further flow of electrons into the metal. N_c is the effective density of states in the conduction band and is given by

$$N_c = 4\sqrt{2}\frac{\sqrt[3]{\pi m^* k_B T}}{h^3} \tag{14.11}$$

where m^* is the effective electron mass and h is the Planck's constant ($h = 6.63 \cdot 10^{-34}\ \mathrm{m^2 \cdot kg \cdot s^{-1}}$). Using the above equation for the built-in voltage, a graph can be constructed for the dependence on the doping concentration of n-type 4H-SiC, see Fig. 14.6.

The above figure assumes Nickel is used as a Schottky metal contact at a temperature of 300 K. To calculate the effective density, the anisotropic nature of the 4H-SiC needs to be considered since the effective mass varies for different crystal orientations.

When a Schottky contact is biased, either reverse or forward, the barrier height will change as shown in Fig. 14.7a, b.

The electrons in the semiconductor will easily diffuse across the SCR into metal for a sufficiently reduced barrier height due to forward bias ($V_{bi} - V_F$). On the other hand, for $V_{bi} + V_R$ the barrier will be increased blocking the electrons. Furthermore, the above diagrams are very similar to those of a pn-junction resulting in similar *I*-*V* characteristics with exponential behaviour. However, the current mechanism in the MS junction is due to the flow of majority carrier electrons as opposed to the pn-junction, allowing for high-speed operation. The high speed is also thanks to the one-sided depletion region of the Schottky diode. Similar to a pn-junction the

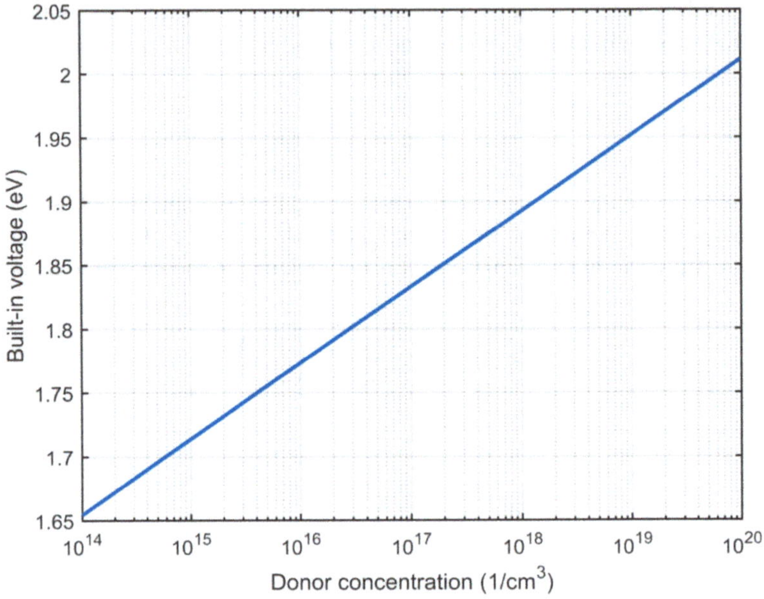

Fig. 14.6 Built-in voltage as a function of the donor concentration in the n-type semiconductor

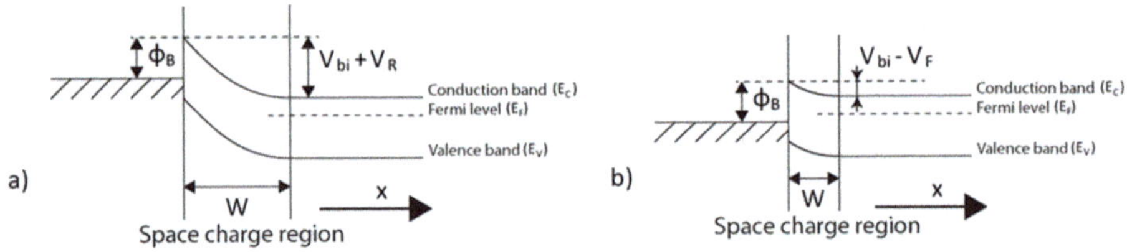

Fig. 14.7 Ideal band diagram of a metal and n-type semiconductor: (**a**) under reverse bias and (**b**) under forward bias

electrostatic properties, the electric field and SCR width, can be determined starting with Poisson's equation. Consider

$$\frac{dE}{dx} = \frac{\rho(x)}{\epsilon_0 \epsilon_r} \tag{14.12}$$

where $\rho(x)$ is the SCR volume density and $\epsilon_0 \epsilon_r$ is the permittivity of the semiconductor. Integrating the above equation yields

$$E = \int \frac{qN_\mathrm{d}}{\epsilon_0 \epsilon_r} dx = \frac{qN_\mathrm{d}x}{\epsilon_0 \epsilon_r} + C_1 \tag{14.13}$$

To solve for the integration constant, assume electric field at the SCR edge of the semiconductor to be zero.

$$C_1 = \frac{qN_d x}{\epsilon_0 \epsilon_r} \tag{14.14}$$

Assuming the SCR starts at $x = 0$ and extends to $x = W$, the electric field at any point y in the x-direction can be written as

$$E = -\frac{qN_d (W - x)}{\epsilon_0 \epsilon_r} \tag{14.15}$$

The SCR width can be determined by assuming a highly doped p layer to represent the metal in a p^+n junction. The SCR in the n-type semiconductor under a reverse bias then becomes

$$W = \sqrt{\frac{2\epsilon_0 \epsilon_r (V_{bi} + V_R)}{qN_d}} \tag{14.16}$$

It should be noted that the photogenerated e-h pairs outside the SCR will recombine before reaching the electrodes. Charge carriers generated in the SCR will be collected at the electrodes before recombination takes place. Now that the single junction contact is introduced, this can be extended with a second contact. Consider Fig. 14.8a, b showing the MSM band diagram after equilibrium and the back-to-back diodes.

As can be seen from the above figure, the two contacts to the semiconductor are symmetrical of nature. However, when a bias is applied to the device this symmetry is broken where one barrier height is increased, and the other is lowered. This results in an operation where one diode is always reverse biased and the other forward biased allowing the flow of charge carriers to form a current in an external circuit. As a consequence, the MSM Schottky device will exhibit nonlinear I-V characteristics. This method allows for a low dark current, hence the minimum detectable signal

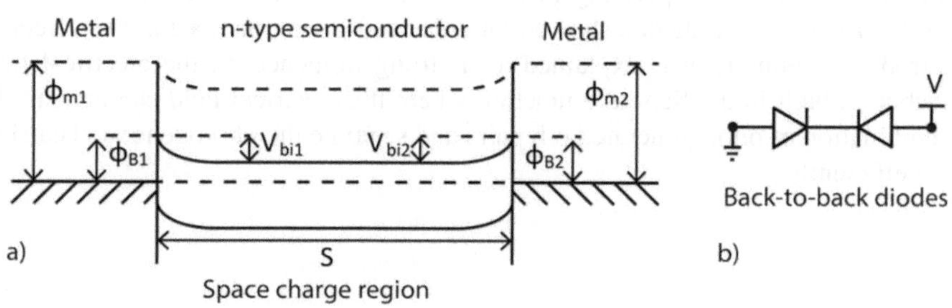

Fig. 14.8 (**a**) MSM band diagram after equilibrium and (**b**) MSM back-to-back diodes

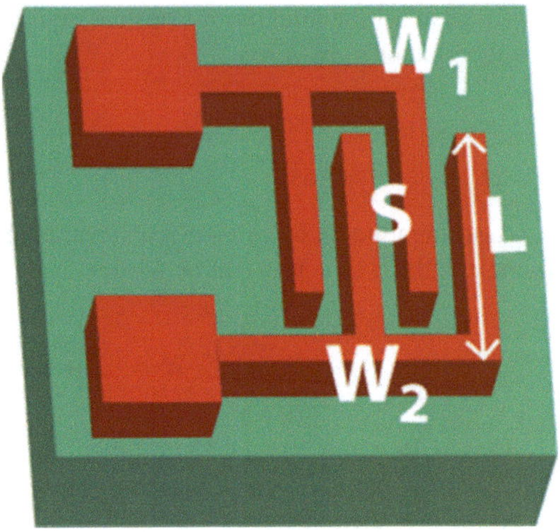

Fig. 14.9 MSM structure

capability is enhanced. Of course, to keep the dark current as low as possible, the crystal quality of the semiconductor should be as high as possible while the metal selected should result in an as high as possible barrier height [17]. This can also result in higher responsivities. Similar to the single Schottky contact, high speeds can be achieved thanks to the low junction capacitance and high BW is available. Further the simple structure is CMOS compatible while allowing also for less conventional processes such as the addition of nanomaterials for enhancing the device responsivity usually found in photoconductive devices. Consider Fig. 14.9 illustrating an MSM device where L is the length of a single finger electrode, S is the separation between the opposing electrodes, and W_1 and W_2 are the widths of the top and bottom finger electrodes, respectively.

The reason for having two electrode widths is enabling asymmetry by geometrical design to enhance the performance of the device. Changing both electrode width can also have a significant influence on the performance similar to increasing the bias voltage [18]. An Au/ZnO MSM device was designed with two electrode widths where one electrode was fixed and the other decreased [19]. The device showed at a ratio of 20:1 an enhancement in the responsivity from 0.3 to 20 mA/W at the same conditions while operating in photovoltaic mode at 0 V as a self-powered device. Moreover, the device showed an enhanced operation speed. The effect of electrodes asymmetry was explained as a strong influence on the electrical field distribution built in the Schottky junction where the electrical field can prevent the recombination of photogenerated e-h pairs and separate the photogenerated carriers more efficiently.

14.2.3 MS Schottky Contact Current Mechanism

The current mechanisms in Schottky contact devices is due to three main mechanisms: thermionic emission (TE), thermionic field emission (TFE) and field emission (FE) [17]. Figure 14.10 illustrates these effects.

Thermionic emission describes the electron flow over the potential barrier when the charge carriers gain enough thermal energy while assuming the barrier height to be larger than the thermal voltage ($k_B T/q$). This is dominant in moderately doped semiconductors operating at room temperature and is used to describe the *I-V* characteristics of the MS Schottky contact.

The current density given by TE is according to Eq. (14.17).

$$J = J_s \left(\exp\left(\frac{qV}{k_B T} \right) - 1 \right) \tag{14.17}$$

where V is the applied voltage. The saturation current J_s is given by

$$J_s = A^{**} T^2 \exp\left(-\frac{q\phi_B}{k_B T} \right) \tag{14.18}$$

where A^{**} is the effective Richardson constant. This equation allows for the calculation of the barrier height using the *I-V* characteristic of a MS Schottky device. J_s is found experimentally.

Thermionic field emission results from high electric field application to the semiconductor under temperatures relatively higher than room temperature. As a result, the current density due to this mechanism will be higher as compared to TE and FE mechanisms.

Field emission, also known as quantum tunnelling, is due to the narrowing of the potential barrier as a result of high electric field giving the electrons a higher probability of tunnelling through the barrier. This mechanism is more dominant

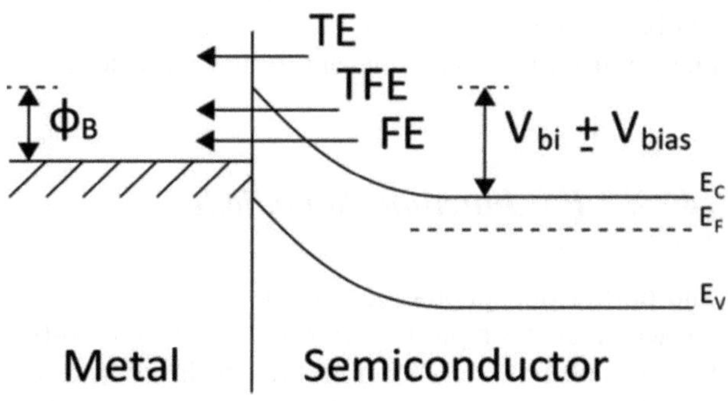

Fig. 14.10 Schottky contact current mechanisms

for high doping concentrations in the semiconductor or at low temperatures. The tunnelling current is given by

$$J_{\mathrm{t}} = J_{\mathrm{ts}} \exp\left(-\frac{q\phi_{\mathrm{B}}}{E_{00}}\right) \quad (14.19)$$

where J_{ts} is the tunnelling saturation current and E_{00} is the characteristic tunnelling energy related to the tunnelling effect transmission probability. E_{00} is given by

$$E_{00} = \frac{qh}{4\pi}\sqrt{\frac{N_d}{\epsilon_{\mathrm{s}} m^*}} \quad (14.20)$$

This parameter helps predict which mechanism will be dominating depending on three conditions given by the ratio $E_{00}/k_B T$ as shown by

$$\text{Condition 1}: \frac{E_{00}}{qk_{\mathrm{B}}T} \leq 0.2 \quad (14.21)$$

$$\text{Condition 2}: 0.2 < \frac{E_{00}}{qk_{\mathrm{B}}T} \leq 5 \quad (14.22)$$

$$\text{Condition 3}: \frac{E_{00}}{qk_{\mathrm{B}}T} > 5 \quad (14.23)$$

From these equations, it should be possible to calculate at room temperature and see that TE dominates at $N \leq 3 \times 10^{17}\mathrm{cm}^{-3}$ and TFE at $3 \cdot 10^{17}\mathrm{cm}^{-3} \leq N \leq 2 \cdot 10^{20}\mathrm{cm}^{-3}$. At higher doping levels, FE becomes dominant. Recombination in the SCR and recombination in neutral region under forward bias also contribute to the current. In this work TE is considered since low doped semiconductors are used for the fabrication of the MSM Schottky devices. There are other effects in the device associated with the metal-semiconductor contact. Such can be the image-force, interface layers, traps and other defects, etc. Such effects may be discussed when relevant for the characterization of the device.

14.2.4 PN-Junction Photodiode

The pn-junction photodiode is one of the first junction semiconducting devices created. It is a simple structure consisting of a p-type layer on top of an n-type layer or substrate. By bringing the two-layer types a potential barrier exists under equilibrium and no net current flow through the diode exists. Figure 14.11 shows an illustration of the device along with its electrical symbol.

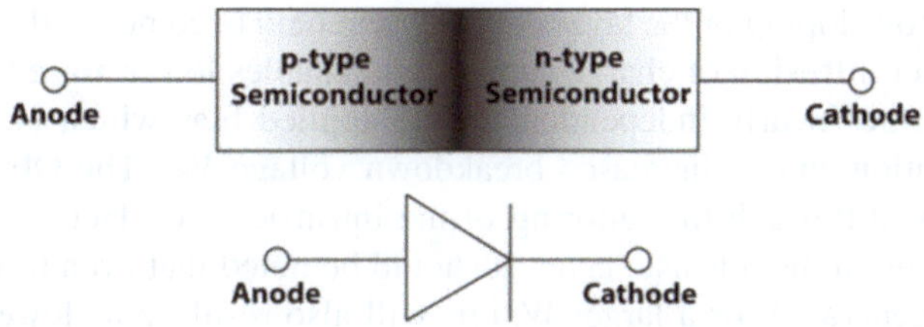

Fig. 14.11 PN-junction diode illustration and its electrical symbol

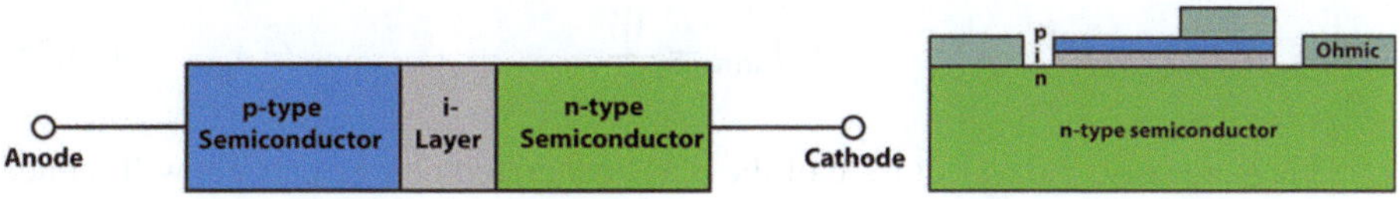

Fig. 14.12 Pin-junction diode illustrations

The device has a response to incident photons in the form of free charge carriers generation. Each incident photon generates an e-h pair which are separated by an electrical field generated by reverse biasing the device. Here the electrons at the p-side move through the depletion region towards the n-type side while the holes at the n-type side move towards the p-type side for collection resulting in a small electric current.

The pn-junction has drawbacks. The first drawback is a large dark current restricting low light level detection. The depletion region of the device is essentially a capacitor which is too large due to the small width of the SCR. This results in limiting fast response detection and high frequency modulation. Another disadvantage is a low QE at long wavelengths, although this is not an issue in this work for the application of UV sensing.

14.2.4.1 Pin-Junction Photodiode

To overcome some of the limitations associated with the pn-junction, an intrinsic layer is added between the p- and n-type layers as shown in Fig. 14.12.

The depletion region is enhanced by the width of the intrinsic layer. First this can be tailored to reduce the capacitance of the junction for the same optical sensing area enabling faster response and higher frequency modulation with bandwidths typically in the range of 10 MHz. The capacitance is then typically in the order of pF. This is because the capacitance of the device becomes bias independent as shown in Eq. (14.24).

$$C = \frac{\varepsilon_0 \varepsilon_r A}{W} \tag{14.24}$$

Thanks to low doping of the layer the electrical field becomes uniform which also presents further diffusion of charge carriers and enables lower noise [2]. Moreover, the SCR becomes nearly independent of the applied bias which is beneficial for a stable operation and an increased breakdown voltage V_{br}. The QE of the device is also enhanced through the tailoring of the intrinsic layer. Here the photons are mostly absorbed in the intrinsic layer. It should be noted that even though more e-h pairs can be generated for a larger *W*, this will also result in a slower response as shown in Eq. (14.25).

$$T_{\mathrm{drift}} = \frac{W}{V_{\mathrm{d}}} \tag{14.25}$$

This is because the transit time of the photogenerated charge carriers will reduce the response time of the photodiode. This results in a trade-off between speed and responsivity of the device due to reduced photon absorption at smaller *W*.

For the ohmic contacting to the layers additional n^+ and p^+ layers can be used as a connection to the metals for low contact resistances [2]. This property also results in a usually higher detectivity compared to Schottky photodiodes mainly due to their lower dark current (related to noise) [13]. Next to this depending on the specific design, responsivity can be significantly increased at reverse voltages in Schottky and p-n photodetectors [13]. However, the increased bias also reduces the UV-to-visible ratio.

This type of device has been used in various research and showed good performance. A 6H-SiC UV pin-junction photodetector is already commercially available [20]. Here the n^+ layer is always doped with large dose typically $>10^{19}$ cm^{-3} where the devices showed an extremely low reverse current, and a peak responsivity of 150–175 mA/W range at 270 nm, corresponding to a 70–85% quantum efficiency. Another work showed a 4H-SiC based pin PD was designed to have a responsivity of 0.13 A/W at a wavelength of 270 nm, an EQE of ~61% and a UV-to-visible ratio exceeding 10^3 for the wavelength of 270 and 380 nm. The PDs usually require passivation which is done by coating with SiO_2 layer for passivation and antireflection where the SiO_2 layer has an absorption at $\lambda < 200$ nm [5, 21]. However, surface recombination plays an important role especially for far UV photodiodes [5]. Thus, the control of the semiconductor to SiO_2 interface is important for the design [5]. The thickness of the SiO_2 layer should also be as thin as possible to improve the penetration depth of the photons with higher energy which can be the case for VUV photodiodes [5]. In another research, the performance of a p-n device was improved without the need of this passivation layer. Here a 4H-SiC pn-junction UV photodetector was designed and achieved a responsivity of 0.03 A/W at 280 nm. The photocurrent was found to be four orders of magnitude larger than the dark current [22].

Since the pin photodiode is a multi-layer device, some care should be taken into the design. The absorption losses in the field-free layers should be avoided as much as possible. This can be accomplished by using very thin p- and n-layers. Furthermore, when selecting a wide bandgap for the doped top layer, it should have an absorption in a spectral range which allows for light absorption in the intrinsic layer.

14.2.5 The Avalanche Photodiode

An important disadvantage of the pn and pin-junctions is the lack of gain, i.e. a successfully absorbed photon only generates a single e-h pair. Amplifying the output signal using an external gain will add a significant amount of noise which will drown weak signals of very low optical input signal. In the past other type of devices were used for this, e.g. a photomultiplier (PMT), i.e. a vacuum tube which has the high sensitivity requirement. However, the PMT has a bulky design, low QE, limited linearity, a narrow spectral operation region and generates heat. The limitations of the PMT and pin photodiodes can be overcome using a pin photodiode designed to have internal gain, also known as avalanche gain, hence the name avalanche photodiode (APD).

To achieve the internal gain, a large reverse bias voltage is applied to the pin diode near the breakdown voltage of the junction. The pin device is however modified to achieve this result by adding a thin layer of p-type doped material between the intrinsic and n-type layers as illustrated in Fig. 14.13. Here the electrical field is shown to be maximum at the pn^+-junction.

The increased reverse bias allows for a higher electric field close to the junction breakdown enabling impact ionization [2]. Each time a photon is absorbed it will generate an e-h pair. The electric field then accelerates the electron which gains enough kinetic energy to cause impact ionization resulting in an avalanche multiplication of electrons, e.g. an internal gain is obtained. The APD knows some adaptations which include heterostructured devices and multi-quantum well. The APD presents many advantages such as high speed, high sensitivity, high

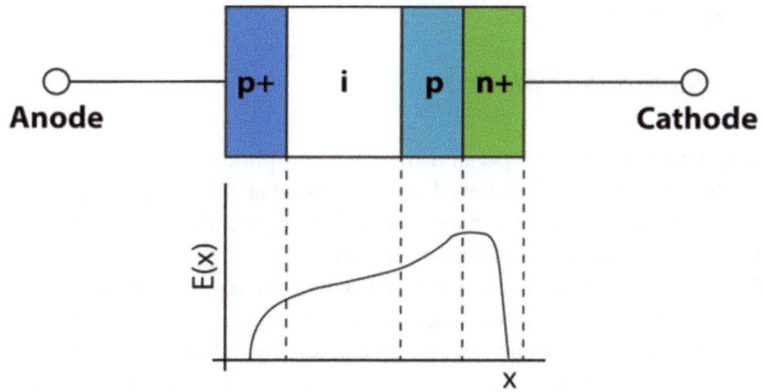

Fig. 14.13 Avalanche pin-junction illustration and the electrical field across the layers

responsivity, high internal gain and even low noise. For example, 4H-SiC APDs exhibit low noise thanks to the large ionization coefficient ratio of $\sim$10 between holes and electrons [5, 23]. The APD can also be based on other types of junctions such as the MS junction [2] but is by far based on the pin-junction. The latter shows trade-offs between the maximum achievable responsivity, speed and operating voltage [2, 23].

Of course, the APD also has disadvantages. The multiplication effect not only results in a higher photocurrent, but also in a higher dark current and lower noise performance while also adding noise from the multiplication process itself. The bandwidth of the device becomes smaller while the response time increases due to the avalanche build-up time and holes transit time through the absorption region.

The devices can be improved using various techniques. A pin-junction photodetector was fabricated and showed a dark current density of 63 nA/cm^2 at 90 pA, a QE of 40% and a gain of 10^3. The same device was optimized using a recessed-window structure and by improving the antireflection (AR) coating layer [24]. The work showed a responsivity of 136 mA/W at 262 nm, with an EQE of 60%, avalanche gains of over 10^6, an excess noise factor characterized by k value of $\sim$0.1, and a spatially uniform response [25]. The pin-junction device performance can be further improved by using a separate absorption and multiplication (SAM) structure [23]. A SAM-APD has separate high-field multiplication and absorption regions and by optimizing the thicknesses and doping concentrations of these layers, some of the issues associated with the pin-structure APD can be solved [23]. For instance, the SAM-APD structure injects only a single type of carrier into the multiplication region which reduces the multiplication noise that arises from the stochastic nature of the multiplication process [23]. The fabricated device achieved an optical gain higher than $1.8 \cdot 10^4$ at 90% of the breakdown voltage of about 55 V. At 42 V reverse bias and 270 nm wavelength, the peak responsivity increased to 0.203 A/W, corresponding to a maximum EQE of $\sim$93% [23].

14.3 Operation Modes

14.3.1 Photovoltaic Mode

The photodetectors can be operated in either the photovoltaic (PV) or photoconductive (PC) modes. Devices based on Schottky barrier and p(i)n junctions such as diodes and transistors have been applied using the PV mode [11]. In the PV mode the diode is operated in open circuit similar to a photocell, without an external bias. When the optical active area is illuminated, the equilibrium at the junction is disturbed causing e-h pairs to be generated at the depletion region. The electric field across the junction then separates charge carriers by pulling the electrons towards the n-side and holes towards the p-side. Then a current flows through the diode from the n- to the p-side. As the barrier further decreases more charge carriers can cross this and an increase in the forward current. The forward current is then balanced out by the photocurrent so that no net current can flow. The barrier lowering is then seen as a measurable voltage across the open circuited device.

The governing equations of the photovoltaic mode are then started with equating the forward- and photocurrents as shown below.

$$I_{\mathrm{f}} = I_{\mathrm{ph}} \tag{14.26}$$

with

$$I_{\mathrm{ph}} = I_0 \left(\exp\left(\frac{qV}{k_{\mathrm{B}}T}\right) - 1 \right) \approx I_0 \exp\left(\frac{qV}{k_{\mathrm{B}}T}\right) \tag{14.27}$$

The above equation assumes the exponential term to be much greater than unity. The open circuit voltage is then given by

$$V = \frac{k_{\mathrm{B}}T}{q} \ln\left(\frac{I_{\mathrm{ph}}}{I_0}\right) \tag{14.28}$$

Notice that the voltage is a nonlinear function of the current, hence also the illumination. This readout method has a slow response which depends on the absorbing layers thickness. However, a major advantage is the lack of dark current, hence lower noise, thanks to the absence of biasing.

14.3.2 Photoconductive Mode

The photodiodes can be operated in PC mode by applying a reverse voltage bias. Here the n-side is connected to the positive terminal while the p-side is connected to the negative terminal. The electric field separates the electrons and holes in the depletion region widening it further and increasing the energy barrier height. Thanks to this the flow of charge carriers is almost completely stopped and only a small reverse current still flows, also known as a dark current, I_{d}. This current is the result of thermally generated charge carriers which is always present for temperatures $T > 0$ K.

As the device absorbs incident photons, extra charge carriers are generated and separated by the electric field in the depletion region. This results in a photocurrent I_{ph} in the same direction as the dark current.

Every method knows advantages and disadvantages. The output signal to be measured is the photocurrent which is a linear function of the illumination as compared to the open circuit voltage for the PV mode which is nonlinear. The PC mode also results in a higher response as compared to the PV mode. Since the reverse biasing results in a higher electric field across the depletion region widening the depletion region, the response speed is also enhanced as the transit time for the charge carriers is reduced. Of course, the electrical biasing results in an increased noise due to the dark current.

14.4 Photodetector Characteristics

Standard metrics of photodetectors have been defined to describe the performance of photodetectors. This enables the selection of devices for the appropriate applications. This section will describe the relevant parameters used for characterizing the photodetectors.

14.4.1 Photoconductive Mode

The incoming photons incident on the optical active area of the photodetector are not all collected or converted to e-h pairs. To describe this, the quantum efficiency (QE) is defined as the ratio of the number of generated electrons to the number of incident photons [2, 5]. This is given by

$$\eta = \frac{\text{\#of generated electrons}}{\text{\#of incident photons}} \times 100\% \tag{14.29}$$

The QE can then be related to the responsivity (*R*), defined as the ratio of the photocurrent and the incoming optical power with the unit A/W [2], given by

$$\eta = \frac{I_{\text{ph}}/q}{P_{\text{opt}}/h\nu} = \frac{R \cdot h\nu}{q} = R \cdot \frac{hc}{\lambda q} \rightarrow R = \frac{I_{\text{ph}}}{P_{\text{opt}}} = \frac{\eta\lambda q}{hc} \tag{14.30}$$

where ν is the frequency, c is the speed of light and λ is the wavelength of interest. The above equation also shows that the responsivity is a linear function of the wavelength and that even though photons at shorter wavelengths have more energy, the responsivity will be lower [2, 5, 11].

$$R = \frac{\eta\lambda q G}{hc} = \frac{\eta\lambda G}{1240} \quad [\text{nm}] \tag{14.31}$$

The responsivity further also depends on photoconductive gain G and the frequency modulation. Here G is equal to unity for diodes without multiplication or defect enhanced to trap carriers and generate gain by $\tau_{\text{lifetime}}/\tau_{\text{transit}}$.

$$R = \frac{\eta\lambda G}{1240} \cdot \frac{1}{\sqrt{1 + (2\pi f \tau)^2}} \tag{14.32}$$

where the 3 dB frequency is at $1/\sqrt{2}$ of the responsivity value under illumination. It should be noted that a gain increase will reduce this value. As a result of reflection at the IDT contacts of the device, the QE is reduced and is characterized by the fill factor given by

$$F = \frac{W}{S + W} \tag{14.33}$$

where W is the width of an individual finger electrode and S is the separation between opposing finger electrodes. For asymmetrical design, this factor changes. This figure can be enhanced when semitransparent electrodes are used, e.g. Indium Tin Oxide (ITO) or graphene electrodes.

14.4.2 UV-to-Visible Discrimination Ratio

The UV-to-visible is also a figure of merit which is defined as the ratio between the peak responsivity and that below the bandgap, thus the UV-to-visible ratio [13]. This ratio is an indication of the quality of the layer and can be defined as the number of nm needed to reduce the responsivity by a decade as is indicative for electrical filters [13].

14.4.3 Spectral Response/Operation Region

The spectral response is associated with the cut-off wavelength which is often given as an indication of the operating point or region of a photodetector [2]. The spectral cut-off wavelength can be defined as the wavelength for which the responsivity drops by either 10 or 50%. This also classifies the photodetector as visible-blind, solar-blind, etc.

14.4.4 Signal-to-Noise Ratio

One of the most important figures of merits is the signal-to-noise ratio (SNR) which describes how good the signal can be distinguished from the noise. There are several noise contributions imposed on the output signal of the photodetector which will be discussed in the coming text. The SNR is given by Eq. (14.34).

$$\mathrm{SNR} = \frac{I_{\mathrm{ph}}^2}{I_{n,\mathrm{total}}^2} \tag{14.34}$$

In the case of a high photocurrent, the above equation can be approximated by:

$$\mathrm{SNR} \approx \frac{I_{\mathrm{ph}}^2}{I_{n,\mathrm{s}}^2} \tag{14.35}$$

where only the shot noise is taken into account. Here it shows that the lowest detectable power is SNR = 1.

14.4.5 *Signal-to-Noise Ratio*

Next to the SNR the noise equivalent power (NEP) is also specified for photodetectors as a figure of merit. The NEP is defined as the ratio of the noise current to the responsivity R. The NEP, expressed in Watt, is usually calculated at the wavelength of interest at a BW of 1 Hz and room temperature. Further the parameter can be normalized for the BW [5] by

$$\mathrm{NEP}^* = \frac{\mathrm{NEP}}{\sqrt{\mathrm{BW}}} \tag{14.36}$$

However, the NEP is not adequate for comparing different types of photodetectors as this is dependent on parameters such as area, temperature, biasing voltage and modulation frequency. To enable comparison between different device types, the devices should be independent of the area A and BW [2, 5, 13]. This value is used to show the minimum detectable optical power of a photodetector which is limited by various noise sources and can be from the detector properties, readout setup such as the noise in an amplifying transistor, or from the statistical fluctuations of the optical signal and any significant background radiation during the detection process [2]. The detectivity D is given by

$$D = \frac{1}{\mathrm{NEP}} \tag{14.37}$$

and is independent of area A and BW by

$$D^* = \frac{\sqrt{\mathrm{ABW}}}{\mathrm{NEP}} \tag{14.38}$$

expressed in $\mathrm{cm}\sqrt{\mathrm{Hz}}/\mathrm{W}$ or Jones with typical values around $10^{11} - 10^{13}$ Jones or even higher for high end photodetectors such as single photon detectors (SPAD). This shows that it's important to consider all the aspect of the measuring system including the photodetector type and design but also the readout components [2].

Thus any noise contributions from the readout circuit should be kept to a minimum when designing for high speed and precision applications.

14.4.6 Minimum Detectable Optical Power

The signal of interest which can be seen by the photodetector has two limits: (1) an upper limit at which the signal detection is determined by the maximum current that the detector can handle without becoming saturated and (2) a lower limit set by the noise floor. The minimum output current signal will then be

$$\text{Noise floor} = \frac{\text{Total noise } (A)}{\text{Responsivity } (A/W)} \tag{14.39}$$

To gain an idea of what the noise sources are, consider below figure. The noise sources associated with photodetectors are thermal, shot, 1/*f*, and photon noises.

Thermal noise, also known as Johnson noise, is encountered in resistive materials for temperatures higher than 0 K. The contributions are from the shunt resistor of the photodetector and any other resistances encountered by the output current signal such as the series resistances and load resistance to convert the current to a voltage. The noise is the result of thermal energy generating charge carriers in conducting materials. Here the electrons are constantly randomly moving and colliding with the atoms of the material. Each collision then contributes to the noise current. However, the sum of all these contributions is zero and thus no net current can flow. This is given by

$$I_{n,t} = \sqrt{\frac{4K_{\mathrm{B}}T\mathrm{BW}}{R}} \tag{14.40}$$

where K_{B} is the Boltzmann constant, T is the temperature, BW is the operating bandwidth and R is the resistance. Notice that this noise can be reduced by cooling the system, operating at a lower BW or reducing the resistances. Operation at a lower BW is not straightforward as the $1/f$ noise dominates. Reducing the load resistance will reduce also the strength of the converted signal.

The shot noise arises from fluctuations in the stream of electrons towards the electrodes. This was first derived from vacuum tube technology. The noise is also referred to as generation-recombination (g-r) noise. The shot noise is given by

$$I_{n,\mathrm{s}} = I_{\mathrm{sat}}\left(\exp\left(\frac{qV}{K_{\mathrm{B}}t}\right) - 1\right) = -I_{\mathrm{d}} \tag{14.41}$$

with I_{sat} being the reverse saturation current. For dark current I_d the noise becomes

$$i_d^2 = 2q\ \mathrm{BW} I_d \tag{14.42}$$

As for the case when there is illumination (quantum noise), the equation becomes

$$i_d^2 = 2q\ \mathrm{BW} I_{ph} \tag{14.43}$$

The shot noise can be minimized by minimizing the DC component of the current (dark current). This can be further minimized by keeping the application BW as low as possible, at 1 Hz as is commonly done. Later in this text the signal-to-noise ratio (SNR) will be shown where the shot noise can be used as an estimate of the lower limit of the noise when the photocurrent is sufficiently large.

The 1/*f* noise, also referred to as excess noise, is due to a number of effects such as contacts and surface trap states but is not well understood. Thus, by good contact fabrication and reduction of surface traps this noise may be reduced. The 1/*f* noise is inversely proportional to the frequency and is given by

$$I_f^2 \sim \frac{I^2 \mathrm{BW}}{f} \tag{14.44}$$

With f being the frequency, as shown in Fig. 14.14 this noise is large at low frequencies and can be reduced if f is sufficiently large, usually >1 kHz for photodetectors.

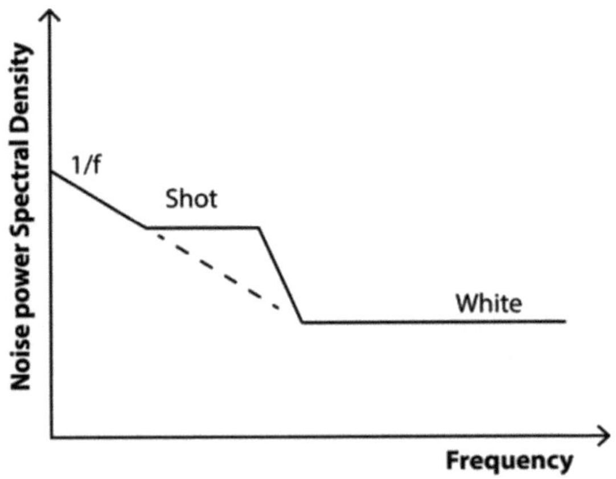

Fig. 14.14 Noise sources

14.4.7 Optical Absorption Coefficient

The absorption coefficient (α) indicates the capability of a material to absorb photons with the enough energy. A material through which electromagnetic waves can travel without loss of energy is transparent. On the other hand, when electromagnetic waves are completely reflected (or absorbed), they are called opaque. Semiconductors are semitransparent which means that electromagnetic waves can possibly travel through the material while an attenuation of the intensity occurs. Electromagnetic waves penetrating into a semiconductor will travel into the material for a length of $1/\alpha$, also referred to as penetration depth. This means that the absorption material should have a thickness at least larger than the penetration depth. The coefficient allows for determining the number of photogenerated free charges (e-h pairs) which can be generated at a spectral range and what the attenuation is of the optical power P for different depths [2]. This is defined by Eq. (14.45) as

$$\alpha_{\text{opt},i}(\lambda) = \frac{4\pi \kappa_i(\lambda)}{\lambda} \tag{14.45}$$

Figure 14.15 shows the spectra of absorption coefficients at room temperature for various direct and indirect semiconductors [2].

As can be seen 4H-SiC has an absorption in the UV range while it has a strong attenuation at longer wavelengths (smaller bandgap).

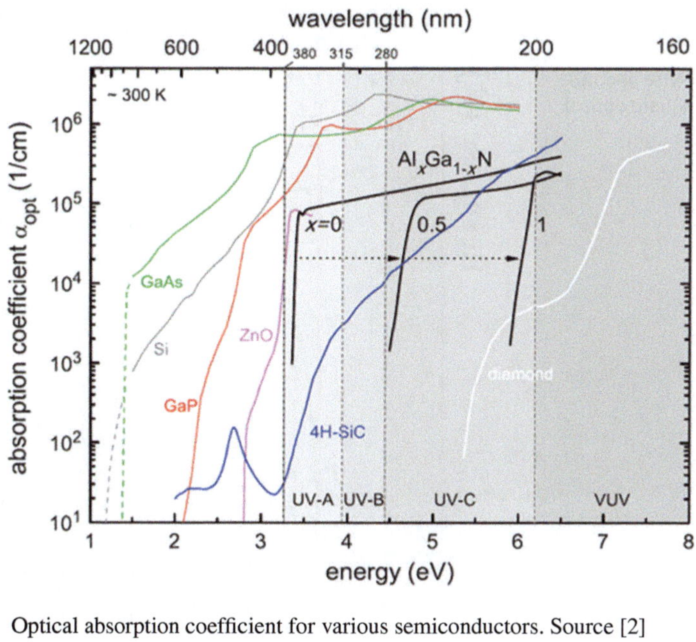

Fig. 14.15 Optical absorption coefficient for various semiconductors. Source [2]

14.4.8 Response Time and Persistent Photoconductivity

Response time gives the operation speed of a photodetector in terms of rise and fall times. Here both the rise and fall times are used and are defined as the time it takes for the detector to detect a rectangular light pulse excitation and have its response go from 10 to 90% of its value, and vice versa as is shown in Fig. 14.16 [2, 5].

The response time is determined by a number of factors which limits the bandwidth of the photodiode operation. The contributions can be from:

- The series resistance and the capacitance of a junction which give the RC time.
- The drift of photogenerated carriers in high-field regions such as the intrinsic layer in a pin photodiode which gives the drift time.
- The diffusion of photogenerated minority carriers in field-free regions such as the p- and n-regions in a pin photodiode which gives the diffusion time.

The times from each contribution are then added as a square and taken square root of the total value. Some photoconductors may show large rise and fall times which may be even in excess of several 1000 s [2]. Such phenomena are known as persistent photoconductivity (PPC) and may render photodetectors non-useful for many applications such as corona discharge where the frequency of the plasma is two times the net-frequency at 100 Hz or 120 Hz. These phenomena are often found to let itself be described by a stretched exponential of the form [2]

$$I(t) \sim \exp\left(-\frac{t}{\tau}\right)^{\beta} \tag{14.46}$$

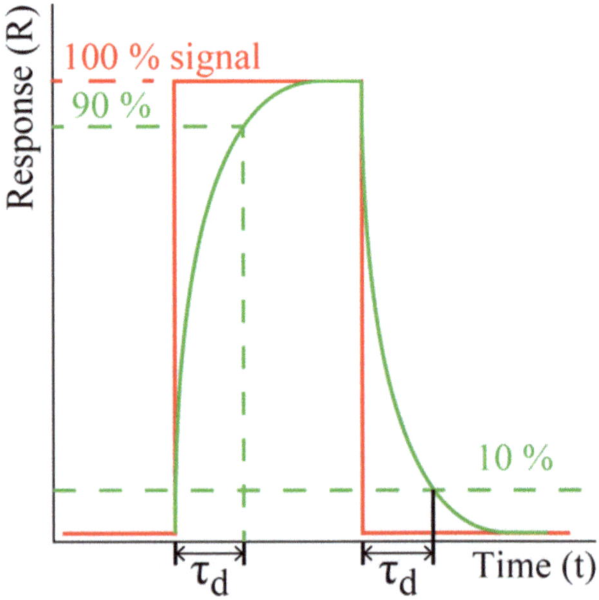

Fig. 14.16 Rise and fall times for a square optical signal

where τ is a time constant and $0 < \beta \leq 1$ is the Kohlrausch stretching parameter. The latter accounts for the microscopic nature of the electronic and atomic relaxation processes underlying the non-exponential change in measured conductivity [2].

14.4.9 Bandwidth

Bandwidth (BW) is defined as the frequency at which the photocurrent is (3 dB) lower than the low-frequency value [5]. This can be related to the decay time (τ) for the special case of exponential transient response by

$$\mathrm{BW} = \frac{1}{2\pi\tau} = \frac{2.20}{2\pi\tau_{\mathrm{d}}} = \frac{2.20}{2\pi\tau_{\mathrm{r}}} \tag{14.47}$$

As was shown in Fig. 14.16, to get a reliable result for the bandwidth, the excitation pulse needs to be rectangular with a rise time much shorter and pulse width much larger than the rise time of the response, respectively. This is required in order for the photocurrent to reach a steady state value [5].

14.4.10 Linearity

The linearity between the photocurrent and the optical power incident on the active area of the device is an important property which allows for a power-independent responsivity (R_{sp}) as given by Eq. (14.31) and a reliable device operation [2]. The linearity is affected by a number of processes such as the carrier recombination kinetics in the absorber material, the equivalent electrical circuit of the photodetector and the readout circuit [2].

14.4.11 Dynamic Range

The dynamic range gives the range over which the detector response is linear with respect to the incident light intensity [11].

14.5 Final Remarks

Optical detections are required in many applications, ranging from civil to military fields. Depending on the optical source, each sensing element needs to have distinct properties with the spectral range at the top. Choices such as sensitivity and environment play an equally important role, if not more important. The properties of the sensors can be tailored by selecting a proper material for a proper photodetector device type. In this chapter, photodetection principles and requirements are discussed, including design considerations and relevant parameters.

References

1. L. Sang, M. Liao, M. Sumiya, A comprehensive review of semiconductor ultraviolet photodetectors: from thin film to one-dimensional nanostructures. Sensors **13**, 10482 (2013)
2. M. Kneissl, J. Rass, III-nitride ultraviolet emitters, in *Springer Series in Materials Science* (Springer, 2016)
3. H. Chen, K. Liu, L. Hu, A.A. Al-Ghamdi, X. Fang, New concept ultraviolet photodetectors. Mater. Today **18**, 493 (2015)
4. A. Aldalbahi, E. Li, M. Rivera, R. Velazquez, T. Altalhi, X. Peng, P.X. Feng, A new approach for fabrications of SiC based photodetectors. Sci. Rep. **6**, 23457 (2016)
5. E. Monroy, F. Omnès, F. Calle, Wide-bandgap semiconductor ultraviolet photodetectors. Semicond. Sci. Technol. **18**, R33 (2003)
6. L. Luo, Y. Zhang, S.S. Mao, L. Lin, Fabrication and characterization of ZnO nanowires based UV photodiodes. Sensors Actuators A Phys. **127**, 201 (2006)
7. G. Li, J. Zhang, X. Hou, Temperature dependence of performance of ZnO-based metal-semiconductor-metal ultraviolet photodetectors. Sensors Actuators A Phys. **209**, 149 (2014)
8. Z. Zhang, W. Zhang, D. Zhang, Y. Xiao, J. Deng, G. Xia, Comparison of different characteristic parameters acquired by UV imager in detecting corona discharge. IEEE Trans. Dielectr. Electr. Insul. **23**, 1597 (2016)
9. T. Okino, S. Yamahira, S. Yamada, Y. Hirose, A. Odagawa, Y. Kato, T. Tanaka, Ultraviolet and visible spectral imaging of hydrogen flames using an organic photoconductive film CMOS imager, in *Proceedings of the International Image SensorWorkshop*, Hiroshima, vol. 30 (2017), pp. 188–191
10. T. Okino, S. Yamahira, S. Yamada, Y. Hirose, A. Odagawa, Y. Kato, T. Tanaka, A real-time ultraviolet radiation imaging system using an organic photoconductive image sensor. Sensors **18**, 314 (2018)
11. T.D. Moustakas, R. Paiella, Optoelectronic device physics and technology of nitride semiconductors from the UV to the terahertz. Rep. Prog. Phys. **80**, 106501 (2017)
12. S. Shinde, C. Bhosale, K. Rajpure, N-doped ZnO based fast response ultraviolet photoconductive detector. Solid State Electron. **68**, 22 (2012)
13. E. Munoz, E. Monroy, J. Pau, F. Calle, F. Omnes, P. Gibart, III nitrides and UV detection. J. Phys. Condens. Matter **13**, 7115 (2001)
14. Y. Duan, M. Cong, D. Jiang, Z. Guo, X. Zhou, N. Hu, K. Yu, ZnO ultraviolet photodetector-based metal-semiconductor-metal structure, *in Young Scientists Forum 2017*, vol. 10710 (International Society for Optics and Photonics, 2018), p. 107100X
15. H.-P. Lin, X.-J. Lin, D.-C. Perng, Electrodeposited CuSCN metal-semiconductor metal high performance deep-ultraviolet photodetector. Appl. Phys. Lett. **112**, 021107 (2018)
16. H. Ferhati, F. Djeffal, New high performance ultraviolet (MSM) TiO2/glass photodetector based on diffraction grating for optoelectronic applications. Optik Int. J. Light Electron Opt. **127**, 7202 (2016)
17. S. Mohammadnejad, S.E. Maklavani, E. Rahimi, Dark current reduction in ZnO based MSM photodetectors with interfacial thin oxide layer, in *International Symposium on High Capacity Optical Networks and Enabling Technologies, 2008 (HONET 2008)* (IEEE, 2008), pp. 259–264

18. X. Yang, D. Jiang, Z. Guo, W. Zhang, N. Hu, Y. Duan, S. Gao, Q. Liang, T. Zheng, J. Lv, Improving performance of the MgZnO ultraviolet photodetectors by changing the interdigital electrode width. Mater. Res. Bull. **98**, 275 (2018)
19. H.-Y. Chen, K.-W. Liu, X. Chen, Z.-Z. Zhang, M.-M. Fan, M.-M. Jiang, X.-H. Xie, H.-F. Zhao, D.-Z. Shen, Realization of a self-powered ZnO MSM UV photodetector with high responsivity using an asymmetric pair of au electrodes. J. Mater. Chem. C **2**, 9689 (2014)
20. J. Edmond, H. Kong, A. Suvorov, D. Waltz, C. Carter Jr., 6H-silicon carbide light emitting diodes and UV photodiodes. Phys. Status Solidi (a) **162**, 481 (1997)
21. X. Chen, H. Zhu, J. Cai, Z. Wu, High-performance 4H-SiC-based ultraviolet p-i-n photodetector. J. Appl. Phys. **102**, 024505 (2007)
22. S. Biondo, M. Lazar, L. Ottaviani, W. Vervisch, V. Le Borgne, M.A. El Khakani, J. Duchaine, F. Milesi, O. Palais, D. Planson, 4H-silicon carbide thin junction based ultraviolet photodetectors. Thin Solid Films **522**, 17 (2012)
23. H. Zhu, X. Chen, J. Cai, Z. Wu, 4H–SiC ultraviolet avalanche photodetectors with low breakdown voltage and high gain. Solid State Electron. **53**, 7 (2009)
24. X. Bai, X. Guo, D.C. Mcintosh, H.-D. Liu, J.C. Campbell, High detection sensitivity of ultraviolet 4H-SiC avalanche photodiodes. IEEE J. Quantum Electron. **43**, 1159 (2007)
25. H. Liu, D. Mcintosh, X. Bai, H. Pan, M. Liu, J.C. Campbell, H.Y. Cha, 4H-SiC pin recessed-window avalanche photodiode with high quantum efficiency. IEEE Photon. Technol. Lett. **20**, 1551 (2008)

Chapter 15
Integrated Photonic Microsystems

J. Kjellman, N. Hosseini, Jeong Hwan Song, T. Tongnyuy, S. Dwivedi, B. Troia, B. Figeys, S. Kerman, A. Stassen, P. Neutens, S. Severi, R. Jansen, P. Soussan, S. S. Saseendran, A. Marinins, and X. Rottenberg

15.1 Introduction

Photonics, the science of generating and/or processing light at the wavelength-scale, e.g., micro/nanometer-scale, is enabling evermore applications, including LED-lighting, fiber-to-the-home internet, solar panels, displays, and image sensors. Recent progress in nano-fabrication now also allows to produce large-scale photonic circuits on wafer-scale with CMOS-compatible processes, providing cost-effective high-quality optical systems.

The development of these wafer-scale photonics platforms was first led by the ever-increasing requirements of bandwidth for high-speed communication. The increase in complexity of the optical interconnections forced indeed to consider moving away from fiber-based macrosystems to microchip-based ones, reproducing in compact form factor and at low costs the functions of bulky, failure-prone and expensive optical macrosystems. The key platform technology for this was and still is Si-photonics [1].

Recently, the maturing integrated photonics platforms emerge as key enablers for a novel family of more diverse applications. In particular, the hybridization of Si-photonics platforms with novel material systems, e.g., SiN-core waveguides, opens broader wavelength reach than previously addressed. SiN can indeed be used to guide light on-chip across the visible and through the NIR, extending till $\sim$8 μm. SiN is instrumental to enable for example industrial and biomedical sensing [2] (Fig. 15.1).

J. Kjellman · N. Hosseini · J. H. Song · T. Tongnyuy · S. Dwivedi · B. Troia · B. Figeys
S. Kerman · A. Stassen · P. Neutens · S. Severi · R. Jansen · P. Soussan · S. S. Saseendran
A. Marinins · X. Rottenberg (✉)
IMEC, Leuven, Belgium
e-mail: Xavier.Rottenberg@imec.be

W. D. van Driel et al. (eds.), *Sensor Systems Simulations*,
https://doi.org/10.1007/978-3-030-16577-2_15

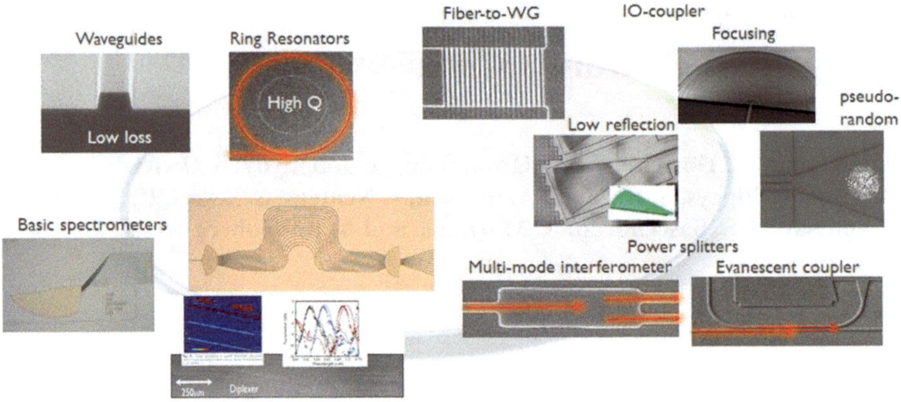

Fig. 15.1 Examples of integrated photonic components and sub-systems produced in SiN-based platform at IMEC [2]

15.2 Case Study: Optical Beamformer

Large amounts of applications emerge, or re-emerge in new form factors, at novel price points, in new use case scenarios, that rely on the controlled emission and/or reception of free-space electromagnetic energy. LiDAR, LiFi, wireless powering, presence detection, light treatment, V2V or V2I interaction, 3D displays, ... are a few of those. Key sub-system at their heart are beam throwers and catchers or, in general, optical beamformers.

Optical beamformer sub-systems are composed of a set of relatively standard photonic building blocks organized in a photonic integrated circuit, the principle of which is described in Fig. 15.2. This schematic will be the basis for the discussion in the following sections of the key building blocks of a beam scanner.

15.3 Typical Device Building Blocks

In this section, we will introduce the basic components used in an optical beamformer and describe their function, simulation approach, and key relevant performance.

15.3.1 In- and Out-Couplers

An in-coupler is a photonic device that receives light from a source (e.g., fiber or laser) and couples the light into photonic integrated circuits (PICs). The silicon (or

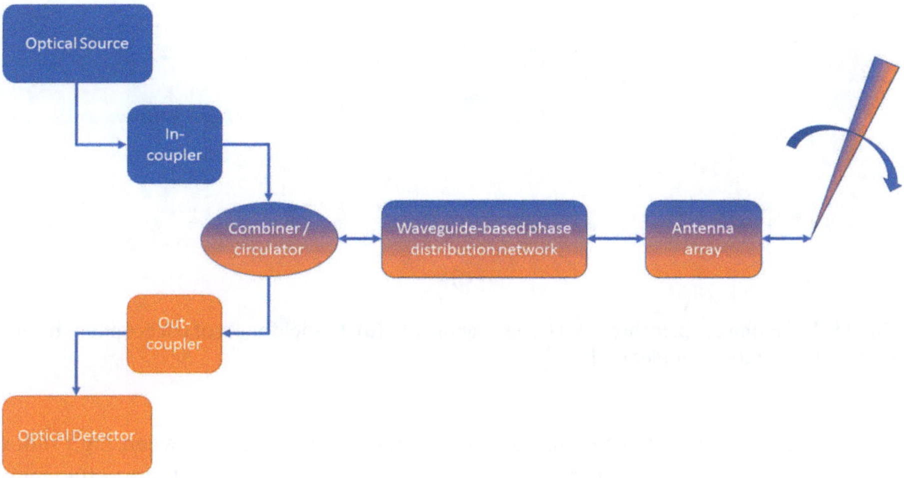

Fig. 15.2 Principle of optical beamformer function and building blocks

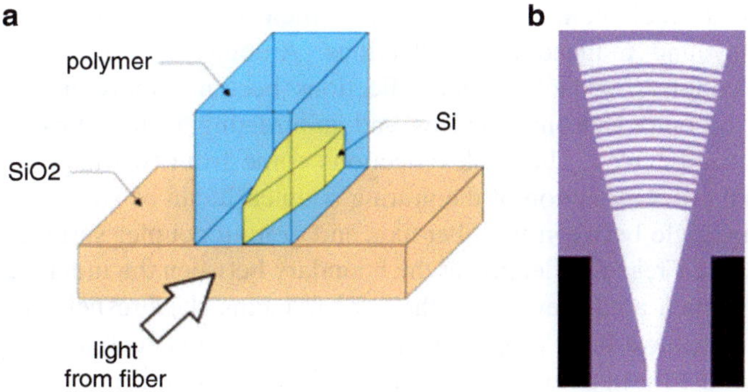

Fig. 15.3 (**a**) Typical edge mode size expander [3] and (**b**) grating coupler

silicon nitride) based PIC enables the realization of high-density circuits due to unique properties of the high refractive index contrast with silicon oxide. However, the high index contrast can result in a few hundred nanometers fundamental mode sized waveguides. It occurs a large mismatch loss and tight tolerance between the fundamental mode of the waveguide and that of the single mode fiber (SMF). Therefore, the in-coupler consists of a mode expansion to match the waveguide mode size to the optical fiber mode to allow for high coupling efficiency to the optical fiber.

There are generally two types of mode size expanders. One is the edge coupler and the other is the grating coupler. An inverse tapered waveguide covered by a polymer waveguide is a common edge coupler type mode size expander [3] as shown in Fig. 15.3a. A grating coupler (Fig. 15.3b) is the vertical coupling type mode size expander.

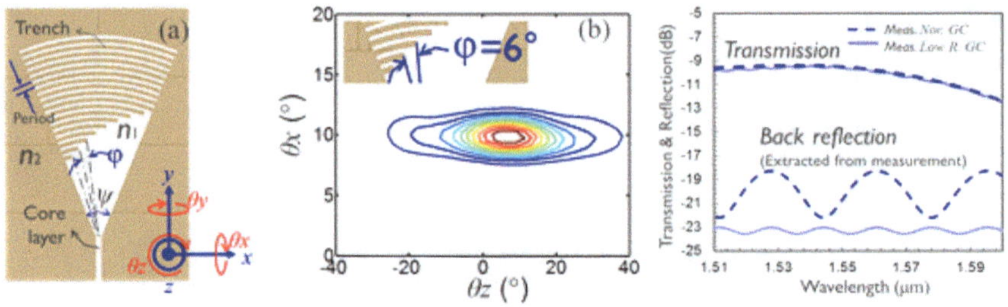

(a) (b) (c)

Fig. 15.4 Grating coupler shape and design parameters (**a**), far-field angle of out coupling (**b**), and results of the grating coupler (**c**) [9]

The grating couplers provide outstanding interconnectivity between high index contrast waveguides and fibers. The grating coupler has high coupling efficiency, large alignment tolerance, and large bandwidth up to several tens of nanometers. Therefore, grating couplers are intensely investigated to optimize performance characteristics which includes high coupling efficiency, polarization diversity or/and independency and reduced back-reflections. Among the various characteristics of grating couplers, reducing back-reflections becomes more important for the integration of optoelectronic devices and minimizing instabilities and errors of interferometric devices. The back-reflections come from two major sources. One is the second order reflection of the grating coupler. It can be suppressed by setting the coupling angle between the fiber axis and grating coupler surface to a 10° tilt. The other is the Fresnel reflection at the boundary between the input waveguide and the grating trench area. Several methods of reducing the Fresnel reflection in the grating coupler have been proposed by many researchers such as a rib waveguide mode [4], a tilted elliptical grating coupler [5], an apodized grating coupler [6], a subwavelength structured grating [7], and asymmetric grating trenches. Asymmetric grating trenches for low back-reflections have been reported by IMEC [8, 9]. Here, the low back-reflection grating couplers will be focused as an in-coupler in this chapter.

The shape of the grating coupler is shown in Fig. 15.4a. The principal of the low back-reflection grating coupler is that Fresnel reflection will be reduced by the asymmetrically curved boundary from the trenches which alters the reflected light path. The output coupling will be tilted by the asymmetric trench as shown in Fig. 15.4b.

The result of the low back-reflection is as shown in Fig. 15.4c. The coupler is fabricated on a 200-mm silicon wafer with a CMOS-compatible process. A 2.3 μm oxide under-cladding layer was deposited using a high-density plasma chemical vapor deposition (HDPCVD) process. Subsequently, the core layer (300 nm of silicon nitride) was added using plasma-enhanced chemical vapor deposition (PECVD). The grating couplers and core waveguides were patterned using 193 nm optical lithography and a full etch (300 nm) was performed using an inductively coupled plasma reactive ion etch (ICP-RIE) process based on fluorine etch chemistry. A top cladding layer of 2 μm oxide was then deposited.

The fabricated grating couplers in C-band are less efficient than reported grating couplers on silicon nitride waveguide platforms [10]. The grating coupler measured was fabricated on a silicon nitride platform used for multi-project wafers (MPWs), which had a lower index contrast and thinner thickness than a conventional silicon nitride platform (silicon nitride index = 2.0, thickness = 400 nm). Additionally, the waveguide thickness of the platform used was 300 nm, which was a limiting factor of the directionality that could be achieved from the grating design.

There is another example of the low back-reflection grating coupler for visible lights [11]. The mode field diameter (MFD) of visible light at 630 nm from the fiber is calculated to be ~4.5 μm using a numerical aperture of ~0.12 at 630 nm, which is from the specification of the fiber we used. The simulated coupling efficiency of low back-reflection grating coupler is −6.6 dB at the peak corresponding to a wavelength of 640 nm. The measurement result is −7.5 dB at 644 nm. For comparison, the normal grating coupler was measured, and the efficiency was −6.8 dB at a wavelength of 644 nm. Moreover, the coupling penalty was 0.7 dB as shown in Fig. 15.5a. This penalty is likely to derive from the shape and size mismatch between the fiber mode and the mode from the grating coupler. Simulations and measurements of back-reflection characteristics are in good agreement. Furthermore, the reflection of the low back-reflection grating coupler is 8 dB less than that of normal grating coupler as shown in Fig. 15.5b.

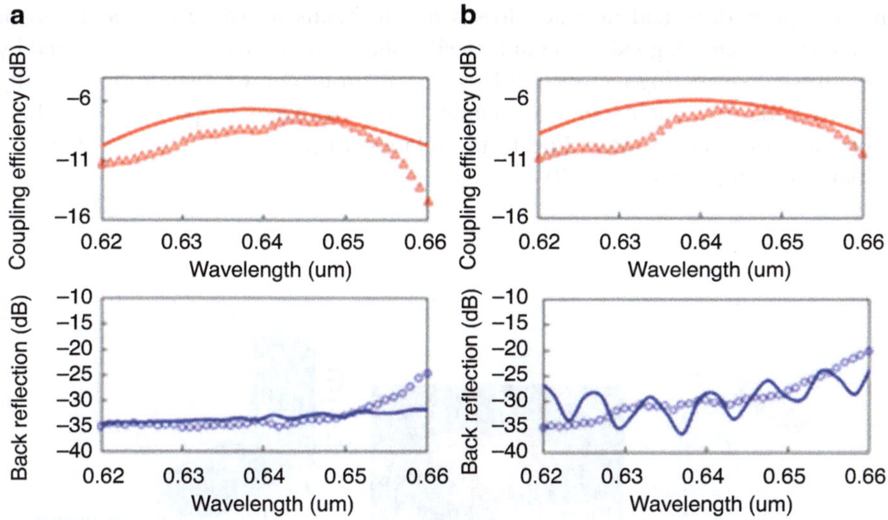

Fig. 15.5 Coupling efficiency and back-reflection (triangles and circles are measurements, and solid lines are simulations) of (**a**) low back-reflection grating coupler and (**b**) conventional designed grating coupler [11]

The back-reflections of the normal and designed grating couplers were −23 dB and −31 dB at full width half maximum (FWHM), respectively.

This grating coupler will be a viable solution for an in-coupler which requires low back-reflections such as PIC using interferometric sensing or PIC with active devices.

15.3.2 Waveguide

Waveguides are at the core of all photonic integrated circuits (PICs). Among other functionalities, waveguides enable guiding, coupling, switching, splitting, multiplexing, and demultiplexing of optical signals. Waveguides of various geometries can be realized in different material systems. The strip waveguide implementing silicon or silicon nitride cores and silicon oxide as cladding, as showed in Fig. 15.6, are most commonly used. Each material system has their own advantages and drawbacks.

Key waveguides metrics for optical beam forming are their losses, coherence length, and power handling.

Waveguide losses impact the overall efficiency of the beamformer, playing an important role in defining the requirements for their surrounding components and sub-systems, e.g., light sources and detectors, to realize an application-specific power link budget. These losses are typically caused by three major factors, i.e., scattering loss mainly due to surface roughness, absorption losses due to intrinsic material properties, and radiation losses due to bends in waveguides and loss of light confinement. A good waveguide design should minimize the losses caused by these three contributing factors. All loss contributions in a waveguide are typically lumped in the loss parameter α accounting for the exponential power decay along propagation distance z according to the relation $P(z) = P_0 e^{-\alpha z}$, where P_0 is the initial power at $z = 0$ (Fig. 15.7)

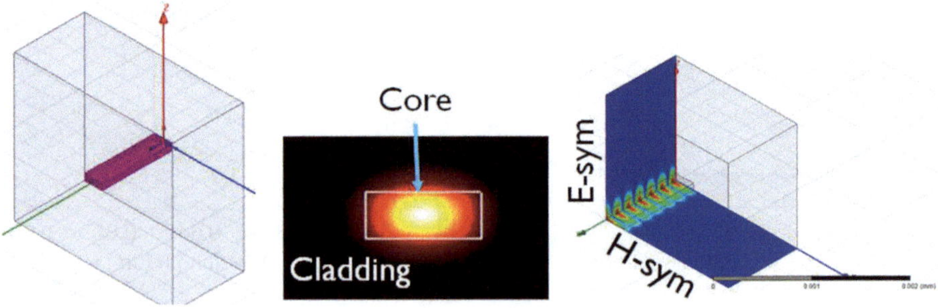

Fig. 15.6 (Left) 3D view of strip waveguide in ANSYS/HFSS [12]; (center) field intensity in the cross section; (right) instantaneous E-field distribution in a quarter waveguide simulated using internal symmetries

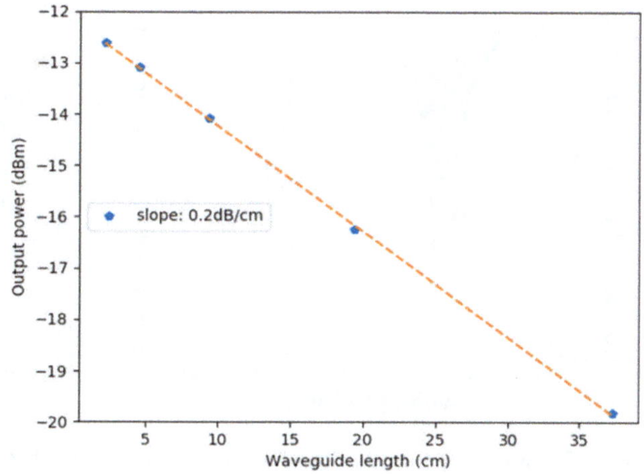

Fig. 15.7 Measured output power exiting waveguides of increasing length and extracted power loss per unit waveguide length; measurement performed at 905 nm on strip waveguides with PECVD SiN core and SiO cladding

Waveguide coherence length is a metric that is used to characterize the introduction of random phase errors in long waveguides. These phase errors are crucial in beamformers that rely on the proper definition of phase relations between distant radiating elements, e.g., ~1 cm or ~10,000 wavelengths apart. Fabrication non-uniformity and non-idealities introduce minute changes in the effective refractive index of a waveguide non-traceable through the waveguide losses themselves. The phase jitter introduced is most noticeable in long interferometers, e.g., Mach Zehnder interferometers (MZIs), that compare the optical length of two long paths.

To measure the coherence length of a waveguide platform, e.g., Si-photonics, a set of nominally identical MZIs is mapped over a representative number of wafers. The jitter observed in the measured interferograms of Fig. 15.8, obtained for MZIs of increasing lengths, is used to determine the coherence length [13] of the platform. For comparison, the same can be performed for SiN-based platform, showing much lower jitter, thus better coherence length.

The **waveguide power handling** is defined by the emergence of dominant non-linear loss in the waveguide platform at high input power. It is key for beam throwers, e.g., LiDAR illuminators, telecom transmitters, and LiFI source, that are to deliver large power to their environment. While various non-linear effects can lead to non-linear absorption, e.g., Kerr effect [14], the two-photon absorption is known to dramatically limit the power handling of silicon-on-insulator waveguide platforms around their typical target wavelength, i.e., 1.55 μm. Indeed, as power levels increase, photon densities increase in the waveguide cores. Thus, the probability of two photons interacting with the same Si atom increases, resulting in the generation of free-carriers modulating the transparency of the bulk Si material. To mitigate this, one approach consists of broadening the waveguides to lower their effective photon density at given power. However, this quickly leads to multimode waveguide geometries where the risk of unwanted mode generation and the ensuing uncontrolled power and phase distributions are non-practical for optical beamformers.

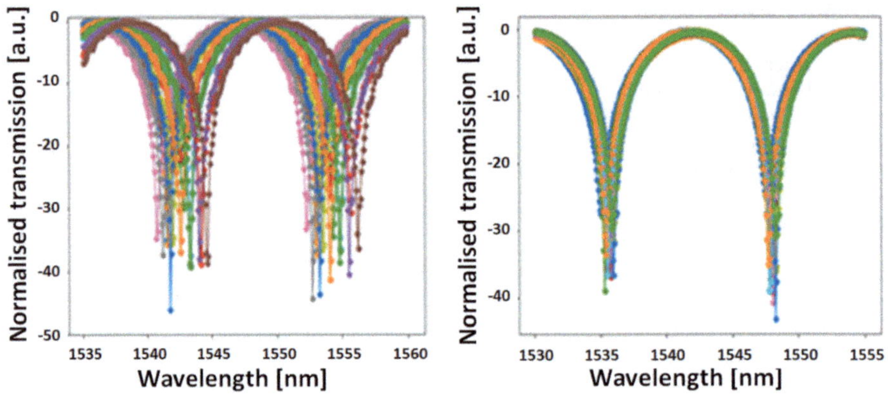

Fig. 15.8 Measured transmission of nominally identical Mach Zehnder interferometers produced using left and right respectively Si and SiN cores, depicting respectively large and small jitter, depicting large and short said coherence length

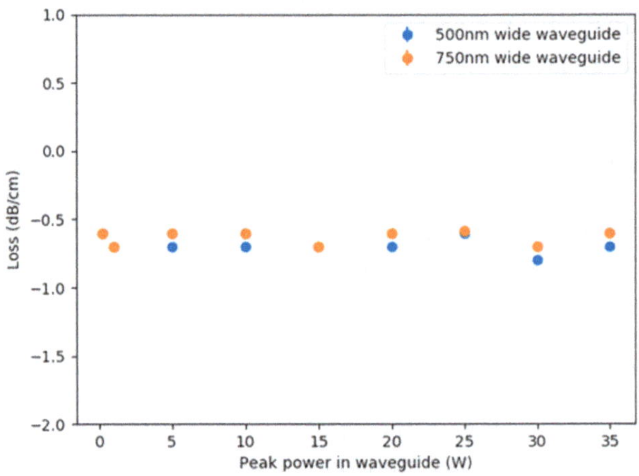

Fig. 15.9 Power loss per unit length measured at 905 nm wavelength for a LPCVD SiN core waveguide at increasing peak power, demonstrating power handling above 35 W

In comparison, SiN does not suffer, or suffers much less, from these non-linear effects. To study this, we designed, fabricated, and measured single mode SiN waveguides. While the power handling of Si-waveguides was limited to a few tens of mW, SiN waveguides were shown to handle up to 35 W peak power. This power handling measurement, reported in Fig. 15.9, was only limited by the capabilities of our measurement system, in particular the used picosecond laser source.

In conclusion, we have presented waveguide platforms and their three most important metrics in view of optical beamformer applications. While more work is ongoing to further lower insertion loss and increase power handling and coherence

Table 15.1 Indicative measured waveguide platform performance

Wavelength [nm]	905	1550
Waveguide core	SiN	Si
Waveguide loss [dB/cm]	0.2	1.4
Power handling [W]	35	10–20 e − 3
Waveguide coherence [cm]	1	0.1

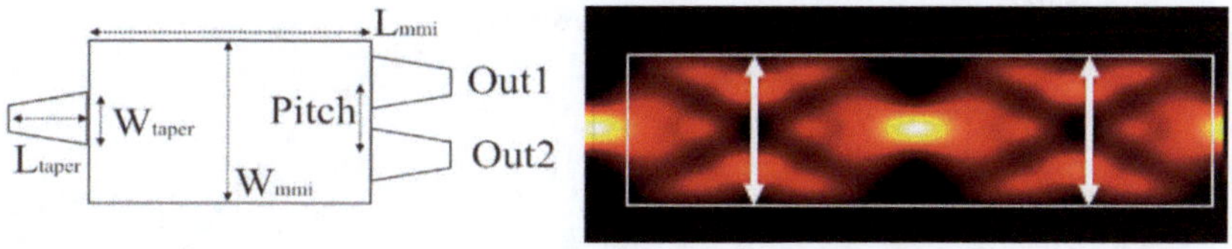

Fig. 15.10 Schematic overview of 1 × 2 MMI power splitter (left) and simulated intensity distribution in the structure at its target work wavelength (right)

length, Table 15.1 reports already outstanding indicative measured performance for IMEC platforms.

15.3.3 Power Splitters

Power splitters are key components in PICs and in particular optical beamformers. These can be realized in different ways such as Y-splitters, multimode interference (MMI) couplers, and evanescent directional couplers. From those embodiments, MMI couplers are often the components of choice. Indeed, they combine compactness and low insertion loss with a broadband optical response that is quite tolerant to fabrication variations. In applications that require many subsequent levels of splitters, e.g., optical beamformer, it is critical that the power split is nicely balanced both in amplitude and phase, as even small imbalances on the individual splitter level will result in dramatic variations at the final stage of the splitter tree. Due to their symmetry, 1 × 2 3 dB MMI power splitters as showed in Fig. 15.10 split optical power evenly over wide bandwidth.

At the input of the MMI, a single mode waveguide is tapered out to widen its mode for insertion in the multimode slab section. In this section, multiple symmetric modes are excited. All excited modes interfere while propagating as they present different effective refractive index, thus, different phase velocities. The interferogram (interference pattern) produced over the length of slab section can be seen in Fig. 15.10 where self-imaging for example can be seen. To design a 1 × 2 splitter, one can locate the positions along the MMI in the interferogram where two pockets of even power are produced, e.g., white arrows. Placing there two symmetric tapered waveguides allows defining the two output ports of the 1 × 2 3 dB MMI splitter. Further, the simulation tools can be used to optimize split ratio, fine sweeping slab length, slab width, input taper waveguide width and pitch.

Table 15.2 Parameters of 1 × 2 MMI split input power evenly

W_{wg} [μm]	T_{wg} [μm]	W_{taper} [μm]	L_{taper} [μm]	W_{MMI} [μm]	L_{MMI} [μm]	Pitch [μm]
0.45	0.15	1.15	30	3	12.9	1.57

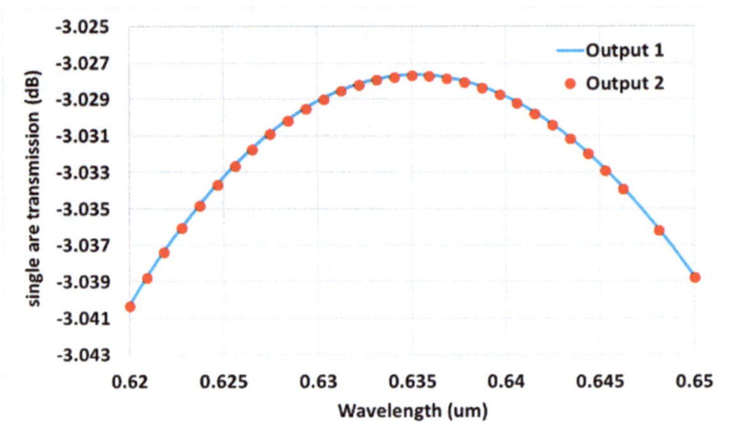

Fig. 15.11 Simulated spectral response of 1 × 2 MMI power splitter with parameters reported in Table 15.2

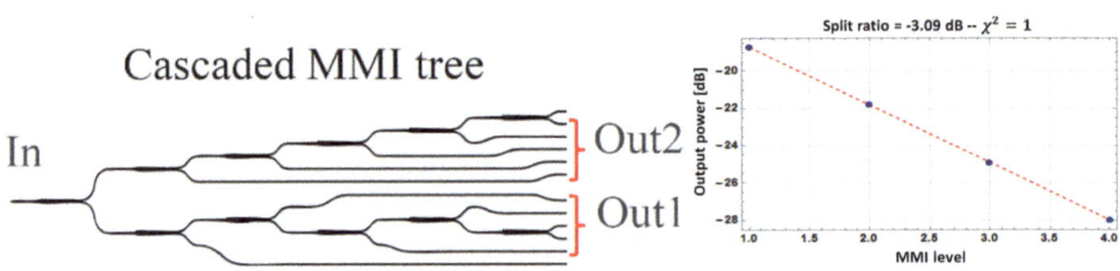

Fig. 15.12 Schematic of MMI tree test structures to measure their power splitting ratio (left) and typical measured output power at tree outputs depicting a linear relation (right)

In order to design and optimize the MMI power splitter at the desired wavelength, commercial software are used, e.g., photon design and mode-solution of Lumerical relying on eigen-mode expansion (EME) method for simulations or ANSYS-HFSS and COMSOL relying on Finite Element Method (FEM). For example, we designed a 1 × 2 3 dB MMI power splitter operating at the wavelength of 638 nm. The device parameters are presented in Table 15.2. Its simulated spectral response, Fig. 15.11, shows only ∼0.01 dB variation over a wavelength range of 30 nm.

To accurately measure the split ratio and imbalance of MMI power splitters, cascaded tree structures shown in Fig. 15.12 are typically implemented. The power of all outputs is measured at the wavelength of 638 nm. As shown in Fig. 15.12, a very good linear fit is obtained to extract MMI split ratio in one point of a given wafer. The MMI tree is repeated over different positions on wafer to map the wafer-level variation of MMI split ratio, i.e., measured to be -3.12 ± 0.04 dB in this case.

15.3.4 Phase Shifters and Modulators

Phase modulators are crucial for dynamic photonics. They control and change the phase of guided light waves, and hence enable on-chip (spatial) switching of light, as well as beam shaping and steering. Applications such as optical beamformers, optical phased arrays, quantum optics, and telecommunications require energy efficient, compact on-chip optical phase shifters and modulators.

Si and SiN integrated photonics are by nature respectively semi-passive and passive platforms. In Si-platform, state-of-the-art modulators typically implement free carrier plasma dispersion in p-n junction or heterogeneous integrations. However, many of these solutions cannot be applied for silicon nitride platforms due to the insulating nature of the platform. Other tuning methods such as thermal, stress-optical, Pockels-effect, and MEMS based modulators have been proposed for the both platforms [11, 15, 16]. Among these, thermal phase shifters have the advantage of small footprint, low fabrication complexity, and ease of integration in CMOS-compatible wafer production line. They however suffer from low energy efficiency and slow modulation speed.

Figure 15.13a illustrates an asymmetric Mach Zehnder interferometer (MZI) with heater system on top of one of the arms. By applying electrical voltage to the heater layer, current flows across the heater. The structure heats up, producing a thermal gradient around the waveguide core as shown in Fig. 15.13b. The thermal gradient induces refractive index change both in core and clad sections. In order to estimate the waveguide effective index, the configurations are simulated in the COMSOL Multiphysics finite-element solver. Figure 15.13b shows the thermo-optical simulation results for a configuration where the heater width is 2 μm. As can be seen, temperature distribution has its maximum at heater layer and

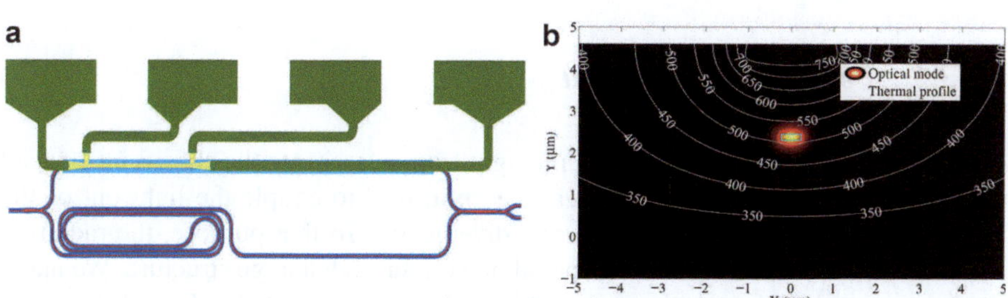

Fig. 15.13 (**a**) Schematic of MZI with heater system to measure thermal phase shift efficiency. (**b**) Induced temperature distribution around waveguide

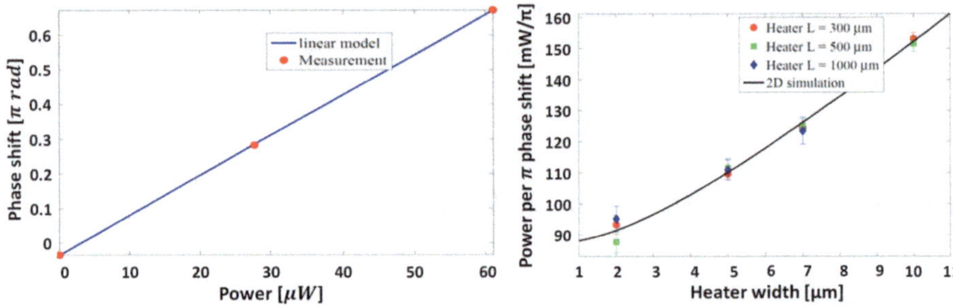

Fig. 15.14 (**a**) Measured phase shift by applying electrical power to heaters. (**b**) Measured P_π of different heater configuration with the variation of heater width and length

decreases to the room temperature where it comes close to the silicon substrate layer. The waveguide phase modulation as function of temperature can be expressed as $\Delta\varphi = 2\pi L/\lambda_n \Delta n_{\text{eff}}$, where L is the phase shifter length, λ_0 is the free-space wavelength, and Δn_{eff} is thermally tuned effective refractive index and depends on the induced temperature.

The straight arm of MZI is heated up by applying voltage difference between two heater pads. The spectral response of MZI is measured for each applied voltage and the resonance shift of the MZI is extracted by modeling the MZI relation to the measured data. Figure 15.14a indicates measured phase shifts for two applied power. As can be seen, phase shift linearly depends on the applied power. By fitting a linear relation, we can simply extract P_π, power for one π-phase shift, for each device. We used a wafer-level measurement setup to get statistic measurements of devices over all areas of wafer. Wafer-level measurements of P_π are illustrated at Fig. 15.14b. Circular, squared, and diamond markers indicate the measured P_π of configurations with respectively heater length of 300 μm, 500 μm, and 1000 μm vs the heater width. As can be seen, the smaller the heater width, the more efficient heater modulator, while heater length has negligible effect on the performance of heaters. The measurement variation stays below 4% over different positions in wafer.

15.3.5 *Leaky Wave Antenna*

A grating coupler can be used to couple light from a laser into the chip as introduced in Sect. 15.3.1. A grating coupler can be optimized to couple the light out of the chip at a desired elevation with narrow divergence. To that purpose, the radiating structure needs to emit light, leak optical power, through a large structure. We name such a grating coupler a leaky wave antenna (LWA) as depicted in Fig. 15.15.

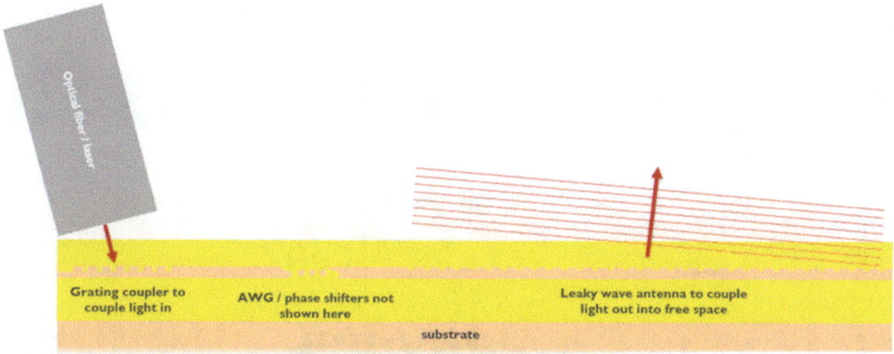

Fig. 15.15 Principle cross section of a typical optical phase array for beam steering: Light couples into a chip via vertical grating coupler or laser and then it goes through the photonic integrated components for example an array waveguide grating (AWGs) or phase shifters and comes out through a long and weak grating and we can name them a leaky wave antenna

A small periodic perturbation is required to couple the light gradually out of the chip. The emission angle can be calculated through a simple Bragg equation:

$$n_{\text{eff}} - \frac{\lambda_0}{P} = \sin \theta_0$$

where n_{eff} is effective mode index, P is a pitch or period of an antenna, λ_0 is a wavelength in the vacuum, and θ_0 is the out-coupling angle. The formula is valid for small P values, i.e., a grating without grating lobes.

Different designs of LWAs can be realized on a silicon or silicon nitride platform depending on the different perturbations that are used. Some proposed design of antennas is given below and shown in Fig. 15.15:

(a) Ridge waveguide: Partial etch of 70 nm on 220 nm silicon-on-insulator platform with waveguide width of 600 nm
(b) Grating elements on the top of the waveguide
(c) Grating elements on side of the waveguide

These above perturbations cause a portion of the guided light within the waveguide to be scattered. The scattered light-out or an antenna radiation pattern in the far-field can be designed based on these perturbations/gratings. So, it is extremely important to have the right design parameters mentioned above to get a desired radiation profile. The other important matrix is a strength of grating which mainly depends on how the field decay over the length of the antenna. In order to design LWA, the designer should keep in mind about the radiation profile, the strength of the grating, and of course the fabrication limitations for the platform (Fig. 15.16).

For example, the partial etch ridge waveguide design allows to implement a weak grating strength, ideal for 100–500 μm long antennas, in a standard silicon

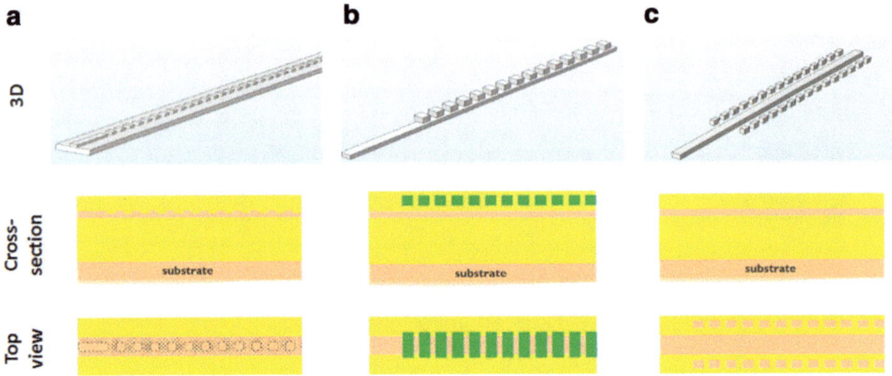

Fig. 15.16 Different types of proposed LWAs: (**a**) partial etch Si ridge waveguide, (**b**) SiN top gratings on Si waveguide, and (**c**) grating elements on side of the waveguide

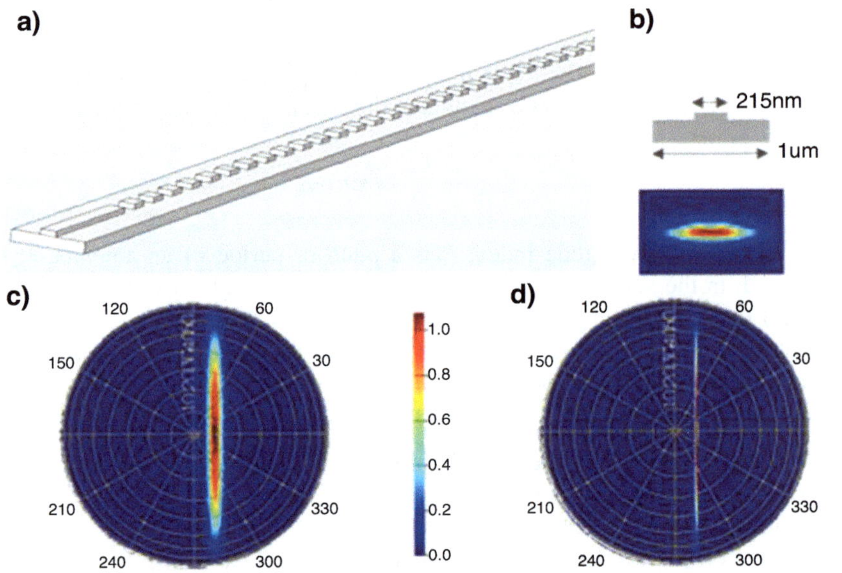

Fig. 15.17 (**a**) Ridge waveguide based LWA, (**b**) LWA cross section and the optical mode, (**c**) simulated far-field radiation pattern of 10 μm long antenna, and (**d**) simulated far-field radiation pattern of 100 μm long antenna

photonics flow. The longer the antenna, the smaller the divergence in the polar direction. The simulation of the ridge waveguide based LWA for two different lengths are shown in Fig. 15.17.

In the uniform perturbation as shown above, in the different type of antennas, there is a fixed delay rate along the antenna, so corresponding near field profile is exponential. It is a physical length because a certain portion of antenna emits more

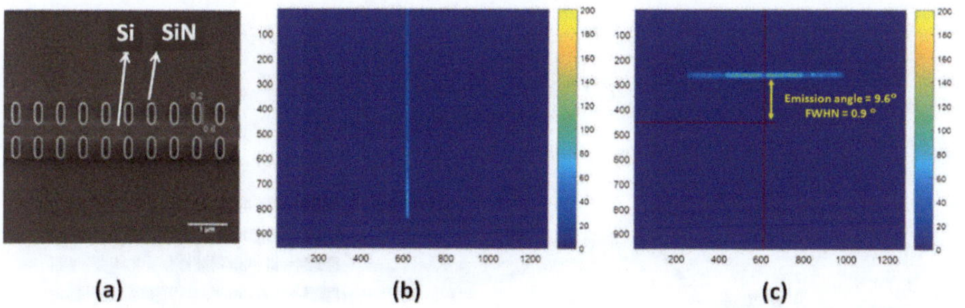

Fig. 15.18 (**a**) SEM image of LWA, with SiN perturbations, (**b**) measured near field profile showing the exponential decay, and (**c**) measured far-field radiation pattern showing the emission angle and the divergence

than others. There are different LWA antenna design challenges in order to achieve the desired far-field radiation pattern, for example, Gaussian (near field is Gaussian), Sinc (near field is rectangular or uniform), etc. For example, in order to design the uniform near field pattern the gratings positions need to be non-uniform or apodized.

Another important phenomenon in LWA design is that these antennas leak not only upwards but also downwards, towards the substrate they are defined on, with parasitic back-reflections as a result. One way to solve this issue is to design directional leaky wave antennas by using Si-SiN dual layer structures that can increase the amount of power for beamforming and prevent the undesired back-reflections. The way it works is the upward light is constructively interfering and the light which goes into the substrate interferes destructively. This is a part of ongoing research.

A fabricated Si LWA with SiN perturbations is shown in Fig. 15.18a. The measured LWA near field profile and the far-field are shown in Fig. 15.18b, c respectively.

15.4 System-Level Design and Performance

Taking the components described in the previous sections, an optical beamformer can be constructed. A typical architecture is shown in Fig. 15.19. Light is coupled on left hand side, power is split by a binary tree of 1 × 2 MMIs, and phase shifters are used to control the phase of individual leaky wave antennas. Controlling the phase of each antenna steers the beam(s). For beam steering with an array of constant pitch and uniform distribution of power between the antennas, a phase shift of

$$\phi_n = n \times \Delta\phi \text{ modulo } 2\pi$$

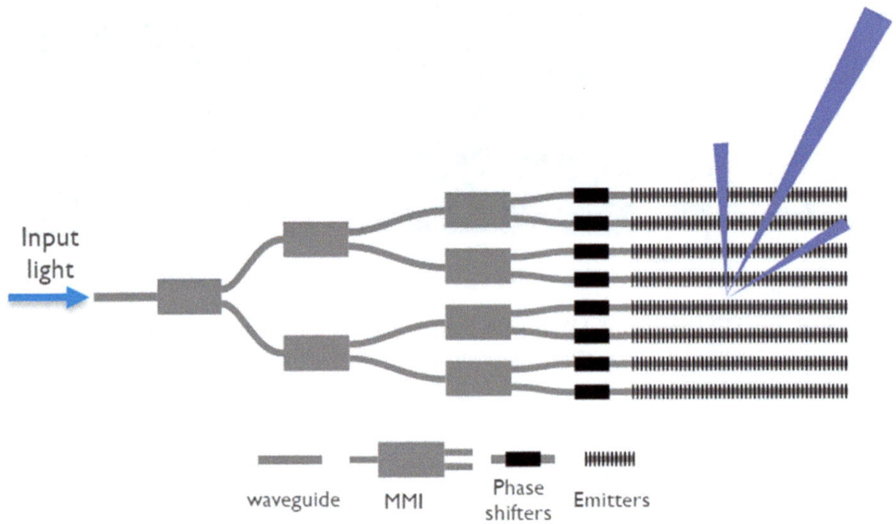

Fig. 15.19 Binary tree architecture for optical beam steering

applied to the *n*-th phase shifter will steer the beam to an angle of

$$\theta = \sin^{-1}\frac{\Delta\phi}{2\pi}\frac{\lambda}{d},$$

where λ is the wavelength and *d* is the antenna pitch.

When designing an optical beamformer, various considerations have to be made. For simplicity, we will illustrate this on an array with a constant pitch and uniform distribution of power, also known as a uniform linear array (ULA). For such an array it is known that when the pitch $d > \lambda/2$ so-called grating lobes will appear. These are generally undesired beams in addition to the central main beam. The appearance of grating lobes reduces the power in the main beam and is a key driver to minimize the pitch. However, this must be balanced against considerations of power coupling between waveguides and antennas which may prevent proper functioning of the device. For an ULA the first grating lobe can be found at an angle of $\theta_1 = \sin^{-1}\lambda/d$ for $d > \lambda$ (otherwise $\theta_1 = 90°$). This grating lobe separation is also the limit of the beam steering range if unambiguous steering of a single beam is desired.

Another consideration is the beam divergence which is often application specific. At zero beam deflection (i.e., $\Delta\phi = 0$), divergence is given by $\Delta\theta = 2\sin^{-1}\lambda/(Nd)$, where *N* is the number of antennas. As can be seen, to make a low divergence beam it is necessary to have a large aperture *Nd*. Combined with the fact that it is desirable to minimize *d*, we realize that the number of antennas can quickly become large, which again implies a large number of phase shifters, which again implies a large number of electrical connections and driver circuit complexity. Thus, beam divergence, optical loss, beam steering range, and electrical interface are all connected and must be carefully balanced by the designer.

15.4.1 Array Factor

More generally than an ULA, an optical beamformer as described is an array of electromagnetic emitters, and for such systems it is well known that the far-field pattern is given by $\mathrm{FF}(\theta, \phi) = \mathrm{EF} \times \mathrm{AF}$, where EF and AF are the so-called element factor and array factor, respectively. FF, EF, and AF are functions of spherical coordinates, where θ and ϕ are the polar angle and azimuthal angle, respectively. Being in the so-called far-field implies that this far-field illumination pattern is independent of the distance to the optical beamformer as long as it is "far away." What constitutes "far away" is a matter of definition and beyond the scope of this work and we simply state that the illumination pattern will approach this limit asymptotically as the distance from the optical beamformer increases.

The element factor is extracted by simulating a single antenna using for instance FDTD and then performing a near-to-far-field calculation. Such calculations can often be performed by the FDTD software. The array factor, on the other hand, can be calculated numerically as a complex sum of contributions from each antenna.

To illustrate this, Fig. 15.20a–c shows, in two dimensions, the element factor of a single, exponentially decaying leaky wave antenna, the array factor of a uniform, linear and the array factor. Since the element factor is in many cases approximately a straight line near the zenith, it is practical to only consider the array factor in one dimension along this line as we shall see later. Lastly, we observe that this

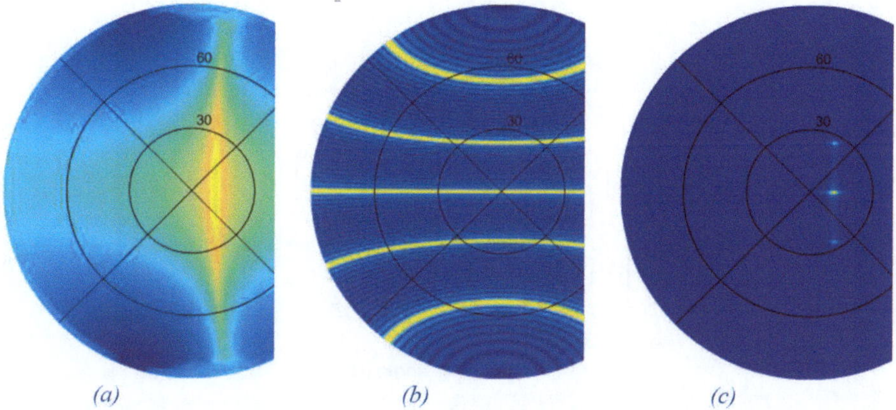

Fig. 15.20 (**a**) The element factor of an exponential decaying antenna; (**b**) the array factor of a linear uniform array with 16 antennas on a pitch of 2.5 wavelengths; and (**c**) the product of the element factor and array factor representing the far-field illumination pattern

illustration shows the grating lobes in the array factor and how they transfer into the full far-field illumination pattern.

15.4.2 Phase Error

One important aspect of the design is to consider the effect of the phase error introduced by the waveguides. We find that the best approach is to model this with numerical software such as Matlab, Python, or Octave. The 1D array factor can be easily calculated for an array of leaky wave antennas where each antenna n is given a phase ϕ_n. To emulate a system with phase errors $\phi_n = \phi_n^{\text{ideal}} + \delta_n$, where δ_n is randomly drawn from a normal distribution with a certain standard deviation σ. With this model, multiple, independent calculations of the 1D array factor can be performed to accumulate statistics on the optical beamformer performance under the effect of phase errors. In Fig. 15.21, we see the array factor in one dimension with and without phase errors. For these calculations $\sigma = 60^{\circ}$ and 250 unique, random sets of phase errors have been generated. Each set can be considered analogous to a single optical beamformer with phase errors stemming from fabrication variations.

Qualitatively Fig. 15.21 illustrates how severely the side lobes suppression ratio can be degraded and how the peak power is reduced. In this particular case, the

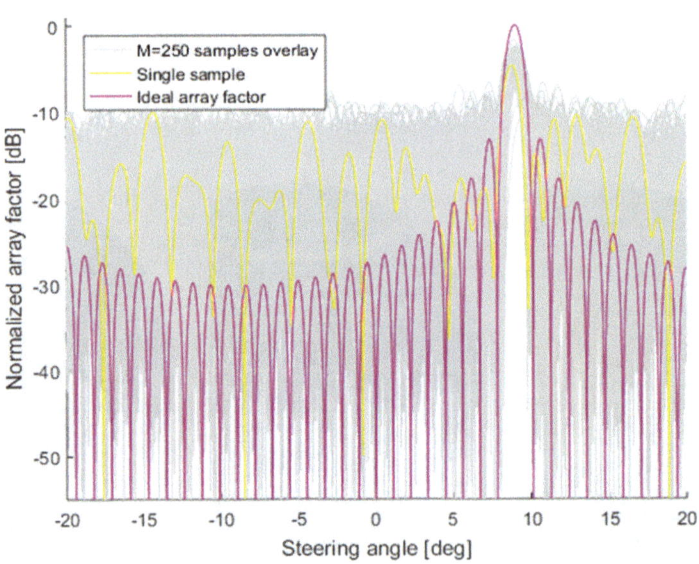

Fig. 15.21 Effect of phase error on the array factor of a 32 element ULA with 1.6 wavelength pitch ($\sigma = 60^{\circ}$). Yellow and gray curves represent unique sets of phase errors analogous to unique devices with fabrication variations

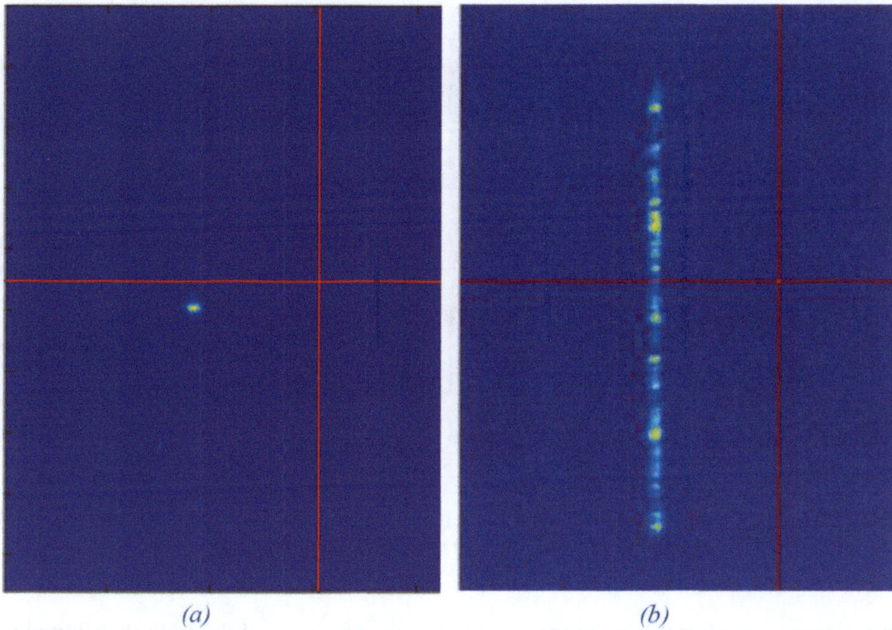

Fig. 15.22 Measurement of far-field illumination from optical beamformer shows (**a**) a well-defined spot for a device realized in silicon nitride and (**b**) a poorly defined spot due to phase errors in a device realized in a combination of silicon and silicon nitride

devices have the peak power reduced by as much as 7 dB compared to the ideal, phase error free case.

The effect can be shown in practice too. Figure 15.22 shows images of the far-field of two optical beamformers. One of the optical beamformers is realized entirely in silicon nitride (Fig. 15.22a) while the other is realized in silicon nitride with silicon phase shifters distributed throughout the power splitting tree (Fig. 15.22b). This clearly illustrates the detrimental effect of phase errors and the difference in phase error performance between silicon and silicon nitride waveguides.

To compensate for the phase errors in the optical beamformers, several approaches can be considered. The first is to minimize phase errors by carefully selecting waveguide material, geometry, and processing. The previously discussed Fig. 15.22 clearly shows the impact of Si on phase errors. However, it is not always practical to choose a low phase error material such as silicon nitride. In such cases, calibration becomes necessary. One way to achieve this is through an off-chip approach where the beam is imaged with a camera and the beam intensity in the desired direction is maximized by iterative optimization of the phase shifter set-points. Figure 15.23 shows how the spot of an eight-antenna array in silicon was experimentally optimized using this method.

An on-chip approach is also possible by monitoring the phase of each antenna. This can be done using interferometry either between the waveguides just before

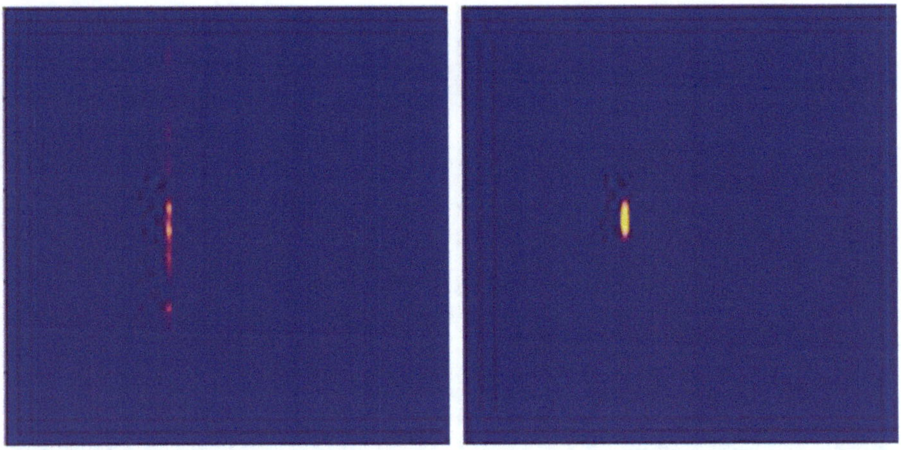

Fig. 15.23 (**a**) Un-optimized far-field from eight element, silicon optical beamformer; and (**b**) far-field from the same beamformer after iterative optimization of phase shifter set-points

the antennas or after the antennas. In either case, a fraction of the light distributed to each antenna is tapped off and then combined using an MMI, waveguide *y*-branch or similar structures. The combined, interfered light will have a power that depends on the phase between the two inputs. With an on-chip power monitor it is possible to analyze this optical signal and determine if there is phase error and what bias to add to the associated phase shifter in order to minimize the phase error.

15.5 Conclusions and Perspectives

This chapter introduced the basics of integrated photonic (sub-)system design using optical beamformers/scanners as case study. Two major families of technological platforms were described and used throughout the chapter, namely Si- and SiN-based waveguide platforms. Devices were described in both platforms, and key performance metrics were mentioned and put in perspective. For example, while Si-photonics is an established and mature platform with compact high-speed low power phase shifters, the emergent SiN-photonics platform presents exquisite power handling and coherence length combined with the capability to manipulate broadband optical signals (VIS/NIR) in a CMOS-compatible monolithic integration.

This chapter only scratched the surface of actual beamformer design and is by no means an exhaustive description of the field. The device optimization, from waveguides in novel material systems to the development of advanced phase shifters based on MEMS or phase transition materials, but also the architecture optimization, moving away from cascaded tree power and phase distribution networks to alternatives like the switched Rotman lens approach are ongoing vibrant activities in the integrated photonics community that cannot be summarized in these short pages. The enthused reader is encouraged to join the field and contribute to or use its development for novel applications in life sciences, industrial sensing, AR/VR, energy, automotive, etc.

References

1. P. Absil et al., Silicon photonics integrated circuits: a manufacturing platform for high density, low power optical I/O's. Opt. Express **23**(7), 9369–9378 (2015)
2. A. Subramanian et al., Low-loss singlemode PECVD silicon nitride photonic wire waveguides for 532–900 nm wavelength window fabricated within a CMOS pilot line. IEEE Photon. J. **5**(6), 2202809 (2013)
3. T. Shoji et al., Low loss mode size converter from 0.3 μm square Si wire waveguides to singlemode fibres. Electron. Lett. **38**(25), 1669–1670 (2002)
4. C. Alonso-Ramos et al., Efficient fiber-to-chip grating coupler for micrometric SOI rib waveguides. Opt. Express **18**(14), 15189–15200 (2010)
5. D. Vermeulen et al., Reflectionless grating couplers for Silicon-on-Insulator photonic integrated circuits. Opt. Express **20**(20), 22278–22283 (2012)
6. M. Antelius, K.B. Gylfason, H. Sohlström, An apodized SOI waveguide-to-fiber surface grating coupler for single lithography silicon photonics. Opt. Express **19**(4), 3592–3598 (2011)
7. Y. Wang et al., Focusing sub-wavelength grating couplers with low back reflections for rapid prototyping of silicon photonic circuits. Opt. Express **22**(17), 20652–20662 (2014)
8. J.H. Song, X. Rottenberg, Low-back-reflection grating couplers using asymmetric grating trenches. IEEE Photon. Technol. Lett. **29**(4), 389–392 (2017)
9. J.H. Song et al., Grating coupler design for reduced back-reflections. IEEE Photon. Technol. Lett. **30**(2), 217–220 (2018)
10. C.R. Doerr et al., Wide bandwidth silicon nitride grating coupler. IEEE Photon. Technol. Lett. **22**(19), 1461–1463 (2010)
11. J.H. Song et al., Grating devices on a silicon nitride technology platform for visible light applications. OSA Continuum **2**, 1155–1165 (2019)
12. ANSYS HFSS. [Online]. https://www.ansys.com/products/electronics/ansys-hfss. Accessed 17 Feb 2019
13. R. Adar et al., Phase coherence of optical waveguides. OSA (1993)
14. D. Bristow et al., Two-photon absorption and Kerr coefficients of silicon for 850–2200nm. Appl. Phys. Lett. **90**, 191104 (2007)
15. N. Hosseini et al., Stress-optic modulator in TriPleX platform using a piezoelectric lead zirconate titanate (PZT) thin film. Opt. Express **23**, 14018–14026 (2015)
16. A. Koen et al., Nanophotonic Pockels modulators on a silicon nitride platform. Nat. Commun. **9**, 3444 (2018)

Index

W. D. van Driel et al. (eds.), *Sensor Systems Simulations*,
https://doi.org/10.1007/978-3-030-16577-2

T

MIX
Papier aus verantwortungsvollen Quellen
Paper from responsible sources
FSC® C105338

If you have any concerns about our products,
you can contact us on
ProductSafety@springernature.com

In case Publisher is established outside the EU,
the EU authorized representative is:
Springer Nature Customer Service Center GmbH
Europaplatz 3, 69115 Heidelberg, Germany

Printed by Libri Plureos GmbH
in Hamburg, Germany